Modern Recording Techniques

Modern Recording Techniques is the bestselling, authoritative guide to sound and music recording. Whether you're just starting out or are looking to improve your skills, this book provides an in-depth guide to the art and technologies of music production and is a must-have reference for all audio bookshelves.

Using its familiar and accessible writing style, this new edition has been fully updated, presenting the latest production technologies and including detailed coverage of digital audio workstations (DAWs), networked audio, musical instrument digital interface (MIDI), signal processing and much more. *Modern Recording Techniques* is supported by a host of video tutorials, which provide additional listening and visual examples, making this text essential reading for students, instructors and professionals.

This updated tenth edition includes:

- Newly expanded "Art and Technology" chapters, providing more tips, tricks and insights for getting the best out of your recording, mixing, monitoring and mastering
- An expanded MIDI chapter to include MIDI 2.0
- More in-depth coverage of digital audio and the digital audio workstation
- Greater coverage of immersive audio, including Dolby Atmos Production

David Miles Huber is a 4× Grammy®-nominated producer and musician in the electronic dance and surround-sound genres, whose music has sold over the million mark. His latest music and collaborations can be heard at davidmileshuber.com and davidmileshuber.bandcamp.com.

Emiliano Caballero is a 2× Latin Grammy®-nominated mix engineer, recording engineer and music producer based in Los Angeles, California (www.emilianocaballero.com). He produces his own music under the name "Zelmar" (www.zelmarmusic.com) and is also one of the co-founders of the plugin company Leapwing Audio.

AUDIO ENGINEERING SOCIETY PRESENTS...

www.aes.org

Editorial Board

Chair: Francis Rumsey, Logophon Ltd.
Hyun Kook Lee, University of Huddersfield
Natanya Ford, University of West England
Kyle Snyder, University of Michigan

The MIDI Manual, 4th Edition
A Practical Guide to MIDI within Modern Music Production
David Miles Huber

Digital Audio Forensics Fundamentals
From Capture to Courtroom
James Zjalic

Drum Sound and Drum Tuning
Bridging Science and Creativity
Rob Toulson

Sound and Recording, 8th Edition
Applications and Theory
Francis Rumsey with Tim McCormick

Performing Electronic Music Live
Kirsten Hermes

Working with the Web Audio API
Joshua Reiss

Modern Recording Techniques, 10th Edition
A Practical Guide to Modern Music Production
David Miles Huber and Emiliano Caballero

For more information about this series, please visit: www.routledge.com/Audio
-Engineering-Society-Presents/book-series/AES

Modern Recording Techniques

A Practical Guide to Modern Music Production

Tenth Edition

David Miles Huber
Emiliano Caballero
Robert E. Runstein

Routledge
Taylor & Francis Group

NEW YORK AND LONDON

Designed cover image: East West Studios, Hollywood, California, photo by Brendan Dekora

Tenth edition published 2024
by Routledge
605 Third Avenue, New York, NY 10158

and by Routledge
4 Park Square, Milton Park, Abingdon, Oxon, OX14 4RN

Routledge is an imprint of the Taylor & Francis Group, an informa business

First edition published by H.W. Sams 1974
Ninth edition published by Routledge 2018

Library of Congress Cataloging-in-Publication Data
Names: Huber, David Miles, author. | Caballero, Emiliano, author. | Runstein, Robert E., author.
Title: Modern recording techniques : a practical guide to modern music production / David Miles Huber, Emiliano Caballero, Robert E. Runstein.
Description: Tenth edition. | New York, NY : Routledge, 2024. | Includes index.
Subjects: LCSH: Sound--Recording and reproducing. | Magnetic recorders and recording. | Digital audiotape recorders and recording.
Classification: LCC TK7881.4 .H783 2024 (print) | LCC TK7881.4 (ebook) | DDC 621.389/3--dc23/eng/20230321
LC record available at https://lccn.loc.gov/2023011911
LC ebook record available at https://lccn.loc.gov/2023011912

ISBN: 978-1-032-19716-6 (hbk)
ISBN: 978-1-032-19715-9 (pbk)
ISBN: 978-1-003-26053-0 (ebk)

DOI: 10.4324/9781003260530

Typeset in Giovanni
by Deanta Global Publishing Services, Chennai, India

Printed in Great Britain by Bell & Bain Ltd, Glasgow

Access the Support Material: youtube.com/modernrecordingtechniques

Contents

ACKNOWLEDGMENTS XXV

CHAPTER 1   Introduction..1

The Professional Studio Environment... 4

The Professional Recording Studio..5

The Control Room...7

The Project Studio.. 9

Making the Project Studio Pay for Itself.. 11

The Portable Studio..12

The iRevolution 13

The Retro Revolution 14

The Changing Faces of the Music Studio Business...........................15

Live/On-Location Recording: A Different Animal 17

Audio for Video and Film..18

Audio for Games 19

The DJ...19

The Times, They've Already Changed: Multimedia and the Web 20

Power to the People!... 20

Whatever Works for You...21

The People Who Make It All Happen 22

The Artist..22

Studio Musicians and Arrangers 22

The Producer 23

The Engineer..24

Assistant Engineer 24

Maintenance Engineer..25

Mastering Engineer 25

Studio Management...25

Music Law 26

Women and Minorities in the Industry 26

Behind the Scenes..27

Career Development 28

Self-Motivation 28

Networking: "Showing Up Is Huge"...29

So, What Are Some Good Ways to Get Started?..29

A Word on Professionalism 32

In Conclusion...**32**

CHAPTER 2   Sound and Hearing **35**

The Basics of Sound **35**

Waveform Characteristics..36

Amplitude 37

Frequency..38

Frequency Response 39

Velocity 40

Wavelength 40

Reflection of Sound..41

Diffraction of Sound..42

Phase...42

Phase Shift...43

Harmonic Content 45

Envelope...48

Loudness Levels: The Decibel...**49**

Logarithmic Basics 50

The Decibel...51

Sound-Pressure Level...51

Voltage 52

Power...52

The "Simple" Heart of the Matter 53

The Ear **54**

Threshold of Hearing 54

Threshold of Feeling 55

Threshold of Pain..55

Taking Care of Your Hearing 55

Psychoacoustics..**56**

Auditory Perception..56

Beats...58

Combination Tones 58

Masking..58

Perception of Direction 59

Perception of Space..61

Direct Sound 61

Early Reflections...61

Reverberation 62

CHAPTER 3   Studio Acoustics and Design **65**

Studio Types **66**

The Professional Recording Studio...66

The Audio-for-Visual and Media Production Environment 67
The Audio-for-Gaming Production Environment 67
The Project Studio 68

Primary Factors Governing Studio and Control Room Acoustics 69
Acoustic Isolation 70
  Walls 71
  Floors 74
  Risers 75
  Ceilings 76
  Windows and Doors 77
  ISO-Rooms and ISO-Booths 78
  Acoustic Partitions 79
Noise Isolation Within the Control Room 80

Symmetry in Control-Room Design 81

Frequency Balance 83
Reflections 83
Absorption 86
  High-Frequency Absorption 88
  Low-Frequency Absorption 89
  Flexible Surfaces 90

The Practical Side of Acoustics 91
Symmetry 91
25/25/50 92
Speaker Placement 93

Room Reflections and Acoustic Reverberation 93
Acoustic Echo Chambers 95

CHAPTER 4   Microphones: Design and Application 97

Microphone Design 98
The Dynamic Microphone 98
The Ribbon Microphone 99
  Further Developments in Ribbon Technology 100
The Condenser Microphone 101
  Powering a Condenser Mic 102
  External Power Supply 102
  Phantom Power 103
The Electret-Condenser Microphone 104

Microphone Characteristics 104
Directional Response 104
Frequency Response 108
Transient Response 109
Output Characteristics 110
Sensitivity Rating 110
Equivalent Noise Rating 110
Overload Characteristics 110

Microphone Impedance 111

Balanced/Unbalanced Lines **112**

Soldering, Baby! 113

Microphone Preamps **114**

Modeled Condenser Mic Systems **115**

Microphone Techniques **115**

Other Microphone Pickup Issues 117

Low-Frequency Rumble *117*

Proximity Effect *117*

Popping *118*

Off-Axis Pickup *118*

Pickup Characteristics as a Function of Working Distance **119**

Close Microphone Placement 120

Leakage *122*

Recording Direct 125

Distant Microphone Placement 126

Room Microphone Placement 129

Room Pickup in the Studio *130*

The Boundary Effect 131

"Reamping It" in the Mix *132*

Accent Microphone Placement *133*

Stereo and Immersive Mic Techniques 134

Spaced Pair *134*

X/Y *136*

M/S *136*

Decca Tree *138*

Surround Miking Techniques *138*

Ambient/Room Surround Mics *139*

Immersive Decca Tree *139*

Ambisonic Pickup *140*

Microphone Placement Techniques **141**

Brass Instruments 142

Trumpet *142*

Trombone *143*

Tuba *143*

French Horn *144*

Guitar 144

Acoustic Guitar *144*

Miking Near the Sound Hole *145*

Room and Surround Guitar Miking *145*

Nylon or Spanish Guitar 145

The Electric Guitar 146

Miking the Guitar Amp *146*

Recording Direct *147*

The Electric Bass Guitar 147

Keyboard Instruments 148
   Grand Piano..*148*
      Separation 149
   Upright Piano *150*
Electronic Keyboard Instruments 150
Percussion 151
   Drum Set *151*
   Miking the Drum Set *153*
   Kick Drum *154*
   Snare Drum *155*
   Overheads *155*
   Rack Toms *156*
   Floor Tom *157*
   Hi-Hat *157*
Tuned Percussion Instruments 157
   Congas and Hand Drums *157*
   Xylophone, Vibraphone and Marimba *158*
Stringed Instruments 158
   Violin and Viola *158*
   Cello *159*
   Double Bass..*159*
Voice 159
   Mic Tools for the Voice *160*
Woodwind Instruments 161
   Clarinet *161*
   Flute *162*
   Saxophone *162*
   Harmonica...*163*

Microphone Selection **163**
Shure SM57 163
Telefunken M81 164
AKG D112 164
Royer Labs R-121 165
Beyerdynamic M-160 165
AEA A440 166
Audio-Technica AT5045 166
AKG C214 167
Neumann TLM102 168
Warm Audio WA-47 168
Telefunken U47, C12 and ELA M251E 169

CHAPTER 5   The Analog Tape Recorder **171**
To Commit or Not to Commit It to Tape? **172**
The Medium of Magnetic Recording **173**
The Professional Analog ATR...174

The Tape Transport 174

The Magnetic Tape Head 176

Equalization 178

Bias Current 178

Monitoring Modes 179

Tape, Tape Speed and Head Configurations 180

Print-Through 181

Analog Tape Noise 182

Cleanliness 183

Degaussing 183

Editing Magnetic Tape 184

Backup and Archive Strategies **185**

Backing up Your Analog Project 185

Tape Restoration 186

Archive Strategies 186

Tape Availability 187

Tape Emulation Plug-Ins **187**

CHAPTER 6   Digital Audio Technology **189**

The Language of Digital **189**

Digital Basics 190

Sampling 191

Quantization 192

The Devil's in the Details 193

The Nyquist Theorem 193

Oversampling 194

Signal-to-Error Ratio 195

Dither 195

Fixed- versus Floating-Point Processing 197

The Digital Recording/Reproduction Process 197

The Recording Process **197**

The Playback Process **198**

Sound File Basics **199**

Sound File Bit Depths 200

Sound File Sample Rates 201

Professional Sound File Formats 202

Regarding Digital Audio Levels 203

Digital Audio Transmission **203**

AES/EBU 204

S/PDIF 205

SCMS 206

MADI 206

ADAT Lightpipe 207

TDIF 208

AES 67 208

Signal Distribution **209**
   What Is Jitter? 209
   Wordclock 210

CHAPTER 7   The Digital Audio Workstation **213**
Integration Now – Integration Forever! **214**
DAW Hardware **216**
   The Desktop Computer 219
   The Laptop Computer 219
   System Interconnectivity 220
      USB *221*
      Thunderbolt *222*
      FireWire *223*
      Audio over Ethernet *223*
   The Audio Interface 224
   Audio Driver Protocols 225
   Latency 226
   Need Additional I/O? 227
   DAW Controllers 228
   Hardware Controllers 228
   Instrument Controllers 229
   Touch Controllers 229
   Large-Scale Controllers 230
Sound File Formats **230**
   Sound File Sample and Bit Rates 231
   Sound File Interchange and Compatibility Between DAWs 231
DAW Software **232**
   Sound Recording and Editing 233
   Fixing Sound with a Sonic Scalpel 238
      Comping *240*
   MIDI Sequencing and Scoring 240
   Support for Video and Picture Sync 241
   Real-Time, On-Screen Mixing 241
      DSP Effects *242*
      DSP Plug-Ins *243*
      Accelerator Processing Systems *244*
      Fun With Effects *244*
      Equalization *244*
      Dynamics *245*
      Delay *245*
      Pitch and Time Change *246*
      ReWire *248*
   Mixdown and Effects Automation 249
   Exporting a Final Mixdown to File 250
Power to the Processor ... Uhhh, People! **250**

Get a Computer That's Powerful Enough 251

Make Sure You Have Enough Fast Memory 251

Keep Your Production Media Separate 252

Update Your Drivers ... With Caution! 252

Read Your Manuals 252

Going (at Least) Dual Monitor 252

Keeping Your Computer Quiet 253

Backup, Archive and Networking Strategies 254

    Computer Networking 255

Session Documentation 258

    Documenting Within the DAW 258

    Make Documentation Directories 259

Accessories and Accessorizing 259

Protect Your Investment 260

Protect Your Hardware 260

Protect Your Body 260

CHAPTER 8   Groove Tools and Techniques **263**

The Basics **264**

Pitch-Shift Algorithms 265

    Warping 266

Beat Slicing 266

Audio to MIDI 267

Groove Hardware **268**

Groove Software **269**

Looping Your DAW 269

Loop-based Audio Software 270

    ReWire 273

    Groove and Loop-Based Plug-Ins 274

Drum and Drum Loop Plug-Ins 274

Pulling Loops into a DAW Session **275**

iOS Groove Apps **276**

Groove Controllers **276**

DJ Software 277

Obtaining Loop Files from the Great Digital Wellspring **277**

CHAPTER 9   MIDI and Electronic Music Technology **279**

The Power of MIDI **280**

MIDI Production Environments 280

What Is MIDI? **283**

What MIDI Isn't **284**

System Interconnections **285**

The MIDI Cable 285

    MIDI Phantom Power 287

Wireless MIDI *287*

MIDI Jacks 287

MIDI Echo *288*

Typical Configurations 288

The Daisy Chain *289*

The Multiport Network *290*

MIDI 1.0 and 2.0 .... **291**

Exploring the MIDI 1.0 Spec **292**

The MIDI 1.0 Message 292

MIDI Channels 293

MIDI Modes 295

Channel Voice Messages 296

Explanation of Controller ID Parameters 299

System Messages 299

Exploring the MIDI 2.0 Spec **306**

Introduction to MIDI 2.0 306

The Three Bs *306*

Bidirectional Communication 307

Backwards Compatibility 307

Both Protocols 307

Higher Resolution for Velocity and Control Messages *308*

Tighter Timing *308*

Sixteen Channels Become 256 .... *309*

Built-in Support for "Per-Note" Events *309*

MIDI 2.0 Meets VST 3 .... *310*

MIDI Capability Inquiry (MIDI-CI) *310*

The Three Ps *311*

Profile Configuration 311

Property Exchange 312

Protocol Negotiation 312

The Universal MIDI Packet *313*

Message Types 313

Groups *313*

Jitter Reduction Timestamps 314

MIDI 1.0 Protocol Inside the Universal MIDI Packet *314*

MIDI 2.0 Protocol Messages 314

Expanded Resolution and Expanded Capabilities *315*

MIDI 2.0 Program Change Message 315

The Future of MIDI 2.0 315

MIDI and the Computer **316**

Connecting to the Peripheral World 316

The MIDI Interface *316*

Electronic Instruments **318**

Inside the Toys 318

Instrument and Systems Plug-Ins 319

Keyboards *320*
    The Synth 320
    Samplers 321
    Sample Libraries and DIY Sampling 323
The MIDI Keyboard Controller 324
The Drum Machine 325
MIDI Drum Controllers 327
Drum Replacement 328

Sequencing **329**
Integrated Hardware Sequencers 329
Software Sequencers 329
Basic Introduction to Sequencing 330
    Recording *331*
    Setting a Session Tempo *332*
    Changing Tempo *332*
    Click Track *332*
    Multitrack MIDI Recording *333*
    Punching In and Out *333*
    Step Time Entry *334*
    Drum Pattern Entry *334*
MIDI to Audio 335
Audio to MIDI 336
Saving Your MIDI Files 337
Again, We Can't Stress This Enough 337
Documentation 337
Editing 338
    Practical Editing Techniques *338*
        Transposition 339
        Quantization 339
        Humanizing 339
        Slipping in Time 340
        Editing Controller Values 340
        Playback 341
        Mixing a Sequence 342

Music Printing Programs **342**

CHAPTER 10 The iOS in Music Production **345**
Audio Inside the iOS **346**
Core Audio on the iOS 346
AudioBus 347
Audio Units for the iOS 347
Connecting the iOS to the Outside World **348**
Audio Connectivity 348
MIDI Connectivity 349
Recording Using iOS **349**

Handheld Recording Using iOS...349
Mixing With IOS...350
iDAWs...350
Taking Control of Your DAW Using the iOS............................351
The iOS on stage..**353**
iOS and the DJ..353
iOS as a Musical Instrument **354**
The Ability to Accessorize **355**

CHAPTER 11   Multimedia and the Web **357**
The Multimedia Environment..**358**
The Computer..358
Television and the Home Theater 358
Delivery Media...359
Networking...359
The Web 359
The Cloud...360
Physical Media **361**
The CD...361
The DVD...361
Blu-ray 363
The Flash Card and Memory USB Stick....................................363
Media Delivery Formats..364
Uncompressed Sound File Formats *364*
PCM Audio File Formats..*365*
Direct Streaming Digital (DSD) Audio....................................*365*
Compressed Codec SoundFile Formats *367*
Perceptual Coding *367*
MP3 368
MP4 369
WMA..369
AAC..370
FLAC..370
Tagged Metadata..370
MIDI..371
Standard MIDI Files *372*
General MIDI...*373*
Multimedia in the "Need for Speed" Era **373**
Streaming Audio over the Internet 376
On a Final Note **376**

CHAPTER 12   Synchronization **379**
Timecode **380**
Timecode Word..382

Sync Information Data *383*
Timecode Frame Standards *383*
3:2 Pulldown Rate.. *384*
Timecode Within Digital Media Production...385
Broadcast Wave File Format *385*
MIDI Timecode 386
MIDI Timecode Messages *386*
SMPTE/MTC Conversion... *387*
Timecode Production in the Analog Audio and Video Worlds **388**
LTC Refresh and Jam Sync 388
Synchronization Using SMPTE Timecode..390
SMPTE Offset Times...390
Distribution of SMPTE Signals 391
Timecode Levels *391*
Real-World Applications Using Timecode and MIDI Timecode **392**
Master/Slave Relationship 392
Video's Need for a Stable Timing Reference...393
Video Workstation or Recorder 394
Digital Audio Workstations...394
Routing Timecode to and From Your Computer 395
Analog Audio Recorders 396
A Simple Caveat..**396**
Keeping out of Trouble 396

CHAPTER 13   Amplifiers **399**
Amplification **399**
The Operational Amplifier...402
Preamplifiers...402
Equalizers 403
Summing Amplifiers 403
Distribution Amplifiers 404
Power Amplifiers...**404**
Power Amplifier Types..405
Class A Amplifier *405*
Class B Amplifier *406*
Class AB Amplifier *406*
Class D Amplifier.. *406*

CHAPTER 14   Power- and Ground-Related Issues **407**
Grounding Considerations..**407**
Power Conditioning...**409**
Multiple-Phase Power...410
Balanced Power 411
Hum, Radio Frequency (RF) and Electromagnetic Induction (EMI)........**411**

CHAPTER 15   Signal Processing **413**

The Wonderful World of Analog, Digital or Whatever **413**

The Whatever 413

Analog...414

Analog Recall... *414*

Digital 415

Plug-Ins... **415**

Plug-In Control and Automation 416

Signal Paths in Effects Processing **417**

Insert Routing 417

External Control Over an Insert Effect's Signal Path *418*

Send Routing 419

Vive la Difference 420

Sidechain Processing..420

Parallel Processing..421

Effects Processors **422**

Hardware and Plug-In Effects in Action 423

Equalization *423*

Peaking Filters..*424*

Shelving Filters *425*

High-Pass and Low-Pass Filters *425*

Equalizer Types *426*

Applying Equalization.. *427*

EQ in Action! *428*

Sound-Shaping Effects Devices and Plug-Ins................................431

Dynamic Range *431*

Dynamic Range Processors *432*

Compression...*433*

Multiband Compression *439*

Limiting...*439*

Expansion *441*

The Noise Gate *442*

Noise Reduction...443

Digital Noise Reduction.. *444*

Adaptive Filtering.. *444*

Fast Fourier Transform *446*

Spectral Analysis Noise Reduction *447*

Snap, Crackle and Pop.. *447*

Time-Based Effects **448**

Delay 448

Delay in Action: Less Than 15 ms... *449*

Delay in Action: 15 to 35 ms *450*

Delay in Action: More Than 35 ms *451*

Reverb 451

Reverb Types *452*

Psychoacoustic Enhancement...453
   Pitch Shifting *454*
   Time and Pitch Changes...*454*
Automatic Pitch Correction..455
Multiple-Effects Devices..456
Dynamic Effects Automation and Editing...457

CHAPTER 16 The Art and Technology of Monitoring 459

Active Versus Passive Listening **459**
   Tools for Listening 461

Subjectivity in the Audio World...**462**

Room, Speakers and Other Important Considerations...........................**462**
   A "Trusted" Space..462
   Monitor Speaker Types 465
      Far-Field Monitoring *465*
      Near-Field Monitoring..*466*
      Small Speakers...*467*
      Headphones...*468*
         Headphone and Monitor Environment Simulation Plug-ins...................*468*
      Earbuds...*469*
      In-Ear Monitoring...*469*
      Your Car *470*
      Alternate Environments *470*
   Spectral Reference..470
   Speaker Design 471
      Active Powered Versus Passive Speaker Design.............................*472*
      Speaker Polarity *473*

Monitoring **473**
   Balancing Speaker Levels..474

Monitor Volume...**475**

Monitor Level Control **476**
   Monitoring Configurations in the Studio 477
      1.0 Mono *477*
      2.0 Stereo..*478*
      2 + 1 (Stereo + Sub) *478*

CHAPTER 17 The Art and Technology of Recording 481

The Fundamental Basics...**481**
   The "Good Rule" 481
   The Transducer...482
   A Good Attitude 484
   Active Versus Passive Listening 484

The Recording Process **485**
   Preparation..485

The Power of Preparation 486
What's a Producer, and Do You Really Need One? 488
Going Into the Studio 489
Setting Up 491
A Word About Isolation 492
Recording 493
A Deeper Understanding of a Console's or DAW's Recording Path 495
Gain Level Optimization 496
Channel Input (Preamp) 497
Insert Point for a Hardware Console/Mixer 499
Virtual DAW Insert Point 499
Auxiliary Send Section 499
Virtual DAW Send 500
Equalization 501
Dynamics Section 503
Monitor Section 503
In-Line Monitoring 504
Direct Insert Monitoring 505
Separate Monitor Section 506
Channel Fader 506
Output Section 508
Channel Assignment 508
Grouping 509
Main Output Mix Bus 511
Main Monitor Level Section 512
Patch Bay 512
Metering 514
The Finer Points of Metering 515
The VU Meter 517
The Average/Peak Meter 518
Digital Console and DAW Mixer/Controller Technology 518
The Virtual Input Strip 519
The DAW Software Mixer Surface 520
OK, Back to the Task of Recording 521
Additional Thoughts 524
Fixing It in the Mix 526
A Word on Levels 527
Session Documentation 528
Monitoring the Mix 528
Latency 532
Working in a Combined Recording/Mix Space 532
Overdubbing 532
Options in Overdubbing 534
To Punch or Not to Punch 535
Reamping It! 536

CHAPTER 18   The Art and Technology of Mixing.................. **537**

The Art of Mixing.. **537**
Ear Training 538
Active Listening... *538*
Preparation... 539
Preparing for the Mix *539*

The Technology of Mixing **541**
Understanding the Underlying Concept of "the Mixing Surface"...... 543
The Mixer/Console Signal Path 544
Channel Input *544*
Insert Points *545*
Send Points *545*
Automation *545*
Write Mode 546
Read Mode 547
Drawn (Rubber Band) Automation.................................... 547
Grouping... *548*
Panning *548*
Channel Fader *549*
Gain Levels.. *549*
Channel Name *549*
Gain Structure, Baby!..549

Mixing and Balancing Basics **550**
Main Output Mix Bus 553

Some Final Words...**554**
Getting Too Close to the Mix.. 555
Quality Control (QC)... 556
Wherever You May Be, There You Are 556

CHAPTER 19   The Art and Technology of Mastering **557**

What Is Mastering? **558**
Help a Project to Sound "Right"... 559
Help Match the Levels and Overall Character Within a Project 559
Help With Overall Levels...560
Help With the Concept of Song Ordering 560
Help With the Concept of Song Timing 561

To Master or Not to Master – Was That the Question? **562**

Mastering the Details of a Project **563**
"Pre"paration.. 563
Providing Stems..564
Providing a Reference Track 564
To Be There, or Not to Be There 564

Common Tools of the Trade...**565**
Sound File Volume..565

Sound File Resolution 565

Dither 565

Relative Volumes 566

EQ 567

Dynamics 567

Compression in Mastering 568

Limiting in Mastering 569

Multiband Dynamic Processing in Mastering............................. 569

Loudness Units Full Scale 570

Mid/Side Processing 570

To Master or Not to Master Yourself – That's the Next Question! 572

The Zen of Self Self-Mastering? 573

Desert Island Mixes 573

Two-Step or One-Step (Integrated) Mastering Option 574

Understanding the Signal Chain 574

Mastering Plug-Ins 575

Mastering for the Internet 576

On a Final Note 576

CHAPTER 20   Immersive Audio 579

Immersive Audio: Past to the Present 580

Stereo Comes to Television 581

Theaters Hit Home 582

Today's Immersive Audio Experience 582

Immersive Layouts... 583

The LFE... 584

Bass Management in a Surround System 585

Speaker Placement and Setup.. 586

5.0 Surround Minus an LFE... 588

7.1 Speaker Placement.. 588

Adding Height Speakers.. 588

Immersive Soundbars... 588

Surround Channel Layouts and Assignments............................ 589

The Immersive Audio Interface 589

Speaker Level Calibration 591

Speaker Time Calibration 592

Monitoring in 5.1, 5.1.4, 7.1.4 and Beyond.............................. 592

The Basic Formats for Understanding Immersive Audio...................... 594

Binaural Immersive Audio 594

Ambisonic Immersive Audio... 594

Immersive Headphone Monitoring 595

Dolby Atmos.. 596

The Basics 598

ADM Authoring Tool.. 598

Renderer Tool 598
Atmos Levels 599

Mixing in Surround **599**

Reissuing Back Catalog Material **600**

CHAPTER 21 **Media Distribution and Manufacturing** **601**

Product Distribution **603**
Build Your Own Website 603
Uploading to Stardom 604
Streaming 605
Download Sites 606
Internet Radio 607

Money for Nothin' and the Chicks **607**
The Moral Question 608
Royalties and Other Business Issues 609
Do You Need a Music Lawyer? 610
Do You Need a Label? 611
Ownership of the Masters 611
Registering Your Work 611
Form SR.. *611*
Form PA.. *611*
Collecting the $$$ 612
Cryptocurrency 612

Product Manufacture **613**
The CD 614
DVD and Blu-ray Burning 617
Optical Disc Handling and Care 618
Vinyl 619
Disc Cutting... *619*
Disc-Cutting Lathe.. *620*
Cutting Head *620*
Pitch Control.. *621*
Vinyl Disc Plating and Pressing... *621*

Further Reading **622**

CHAPTER 22 **It's All About the Journey**...................................**623**

Wherever You May Be ... There You Are! **623**

Showing Up Is Huge!.. **624**

Being Yourself! **624**

The Various "Doors" of Life **625**

Out of This Experience, DMH Learned a Few Things **625**

Creating Your Own Reality... **625**

Personal and Professional Stumbling Blocks **626**

Reinventing Yourself... 627

Perception as It Deals With the Art of Hearing......................... 627

When Is Too Much Gear Simply Too Much? 628

The Bank Works for You, You Don't Work for the Bank...........628

It Does Takes Time..629

Happy Trails 629

You, Your Art and Your Well-Being 630

INDEX 633

Acknowledgments

David Miles Huber is a 4× Grammy-nominated producer and musician in the electronic dance and surround-sound genres, whose music has sold over the million mark. His dance performance style is energized and balanced out by lush beats and live acoustic instruments that combine to create a "Zen-Meets-Tech Experience". His latest music and collaborations can be heard at davidmileshuber.com and davidmileshuber.bandcamp.com.

I'd like to thank my husband, Daniel Butler, for putting up with the general rantin', ravin' and all-round craziness that goes into writing a never-ending book project. I'd also like to express my thanks to all of my friends and family in the United States and Europe, as well as to my music collaborators, who help me reach new heights. I'd like to give special thanks to my amazing buddy and collaborator Emiliano Caballero Fraccaroli (LA); the folks at Easy Street Records in West Seattle; Galaxy Studios (Belgium); Dominik Trampf, nhow Hotel (Berlin); Klaus, Drew, Michael, Yvonne and Uli (Berlin); Andre (Koln) and all the U67 boys around the world from Al's last Mix With the Masters class; The Recording Academy (the Grammy folks in LA/Pacific NW); Greg and the folks at Steinberg North America (LA); Native Instruments (Berlin/LA); Ableton (Berlin); Universal Audio and the folks at Plugin Alliance.

Emiliano Caballero is a 2× Latin Grammy®-nominated mix engineer, recording engineer and music producer based in Los Angeles, California. He produces his own music under the name "Zelmar" (www.zelmarmusic.com) and is also one of the co-founders of the plugin company Leapwing Audio.

I would like to thank my mentors: Dave Huber, Rafa Sardina, Al Schmitt (always in our hearts), George Massenburg, Arturo Vaquero, the studios and their staff where I got to learn my craft: Galaxy Studios, Electric Lady Studios, East West Studios, Capitol Studios … the friends that I made all over the world while living abroad, my family, young Emi and everyone who has been a part of my life journey, leading up to this day. I am grateful for the life I am living, doing what I most love and sharing my time with wonderful human beings.

A sincere thanks to you all from both of us for your kindness and support!

David Miles Huber
Emiliano Zelmar Caballero Fraccaroli
www.davidmileshuber.com
www.emilianocaballero.com

CHAPTER 1

Introduction

The world of modern music and sound production is multifaceted. It's an exciting world of creative individuals: musicians, engineers, producers, managers, manufacturers and businesspeople who are experts in such fields as music, acoustics, electronics, sales, production, broadcast media, multimedia, marketing, graphics, law and the day-to-day workings of the business of music. The combined efforts of these talented people work together to create a single end product: music that can be marketed to the masses. The process of turning a creative spark into a final product takes commitment, talent, a creative production team, a marketing strategy and, often, money. Throughout the history of recorded sound, the process of capturing music and transforming it into a marketable product has always been driven by changes in the art of music, production technology and cultural tastes.

In the past, the process of turning one's own music into a final product required the use of a commercial recording studio, which was (and still is) equipped with specialized equipment and a professionally skilled staff. With the introduction of the large-scale integrated (LSI) circuit, mass production and mass marketing (three of the most powerful forces in the Information Age), another option has arrived on the scene: the radical idea that musicians, engineers and/or producers can produce music in their own facility or home … on their own time. Along with this concept comes the realization that almost anyone can afford, construct and learn to master their own personal audio production facility. In short, we're living in the midst of a techno-artistic revolution that puts more power, artistic control and knowledge directly into the hands of artists and creative individuals from all walks of life … a fact that ensures that the industry will forever be a part of the creative life-force of *change*.

Those who are new to the world of modern digital audio and multitrack production, musical instrument digital interface (MIDI), mixing, remixing and the studio production environment should be aware that years of dedicated practice are often required to develop the skills that are needed to successfully master the art and application of these technologies. In short, it takes time to master the craft. A person new to the recording or project studio environment (Figures 1.1

DOI: 10.4324/9781003260530-1

and 1.2) might easily be overwhelmed by the amount and variety of equipment that's involved in the process; however, as you become familiar with the tools, toys and techniques of the recording process, a definite order to the studio's makeup will soon begin to emerge – with each piece of equipment and personal approach to production being designed to play a role in the overall scheme of making music and quality audio.

The goal of this book is to serve as a guide and reference tool to help you become familiar with the recording and production process. When used in conjunction with mentors, lots of hands-on experience, schooling, further reading, Web searching, soul searching and simple common sense, we hope this book will help introduce you to the equipment and day-to-day practices of the studio. Although it's taken the modern music studio over a hundred years to evolve to its current level of technological sophistication, we have moved into an important evolutionary phase in the business of music and its production: the digital age. Truly, this is an amazing time in production history, when we can choose between a vast array of powerful tools for fully realizing our creative and human potential in a cost-effective way. As always, patience and a nose-to-the-grindstone attitude are needed in order to learn how to use them effectively, but today's technology can free you up for the really important stuff: making music and audio productions. In my opinion, these are definitely the good ol' days!

FIGURE 1.1
The historic (but newly renovated) Capitol Records Recording Studios, Los Angeles, CA. (a) Studio A control room (courtesy of PMC Ltd., www.pmcspeakers.com). (b) Studio A (courtesy of Capitol Records, www.capitolrecords.com).

(a)

(b)

FIGURE 1.2
A couple of the many, many possible examples of a project studio (courtesy of Ableton AG, www.ableton .com).

TRY THIS: DIGGIN' DEEP INTO THE WEB

This book, by its very nature, is an overview of recording technology and production. It's a very in-depth look into the field, but there's absolutely no way that it can fully devote itself to all of the topics. However, we're lucky enough to have the Web at our disposal to help us dig deeper into a particular subject that we might not fully understand, or simply want to know more about. Giga-tons of sites can be found that are dedicated to even the most off-beat people, places, toys and things … and search engines can even help you find obscure information on how to fix a self-sealing stem-bolt on a 1905 sonic-driven nutcracker. As such, I strongly urge you to use the Web as an additional guide. For example, if there's a subject that you just don't get, look it up on www.wikipedia.org or simply Google it.

Of course, there's a wealth of info that can be found by searching the innumerable www.youtube.com

videos that relate to any number of hardware systems, software toys and production techniques. Further information relating to this book and the recording industry at large can also be found at www.modrec.com, as well as at youtube.com/modernr ecordingtechniques. Digging deeper into the Web will certainly provide you with a different viewpoint or another type of explanation, and having that "AH HA!" light bulb go off (as well as the "hokey pokey") is definitely what it's all about.

David Miles Huber (www.davidmileshuber.com)
Emiliano Caballero (www.emilianocaballero.com)

THE PROFESSIONAL STUDIO ENVIRONMENT

The commercial music studio is made up of one or more acoustic spaces that are specially designed and tuned for the purpose of capturing the best possible sound onto a recorded medium. In addition, these facilities are often structurally isolated in order to keep outside sounds from entering the room and being recorded (as well as to keep inside sounds from leaking out and disturbing the surrounding neighbors). In effect, the most important characteristics that go into the making and everyday workings of such a facility include:

- A professional staff
- Professional equipment
- Professional, yet comfortable, working environment
- Optimized acoustic and recording environment
- Optimized control room mixing environment

In this age of pro studios, project studios, digital audio workstations, groove tools and personal choices, it's easy to understand how the "different strokes for different folks" adage equally applies to recording, as the artistic and technological process can be approached in a number of different ways. The cost-effective environment of the project studio has also brought music and audio production to a much wider audience, thus making the process much more personal. If we momentarily set aside the monumental process of creating music in its various styles and forms, the process of creating, producing and distributing a music project will generally occur in seven distinct steps:

- Preparation
- Recording
- Overdubbing
- Mixdown
- Mastering
- Product manufacturing
- Marketing and sales

The focus of this book, in its basic form, is to familiarize you with the tools, toys and techniques that go into the recording of audio … as well as to introduce you to the seven steps that go into the making of audio for use in the various music and visual media that make up the artistry and business of audio. We urge you to take the time to read through each chapter, do additional dives into the various YouTube videos and Web resources on the many topics, talk among friends and share your insights … and, of course, get yourself into the studio, basement, garage or corner of your bedroom and experiment with the tools of your trade. In short, take the time to learn your butts off, while hopefully having lots of fun along the way.

The Professional Recording Studio

Professional recording studio spaces vary in size, shape and acoustic design (Figures 1.3 and 1.4) and usually reflect the personal taste of the owner or are designed to accommodate the music styles and production needs of clients, as shown by the following examples:

- A studio that records a wide variety of music (ranging from classical to rock) might have a large main room with smaller, isolated rooms off to the side for unusually loud or soft instruments, vocals, etc.
- A studio designed for orchestral film scoring might be larger than other studio types. Such a studio will often have high ceilings to accommodate the large sound buildups that are often generated by a large number of studio musicians.

FIGURE 1.3
Examples of a professional recording studio. (a) Berklee 160 Studio, Boston. (b) Trilogy Studios, San Francisco (courtesy of Walters-Storyk Design Group, www.wsdg.com).

(a)

(b)

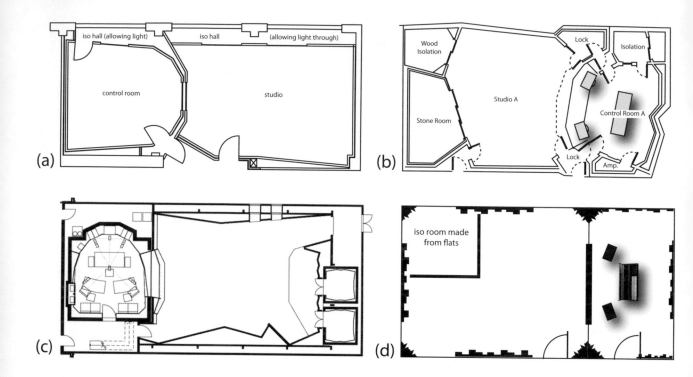

FIGURE 1.4
Basic studio floor plans. (a) KMR Audio Germany, Berlin (courtesy of KMR Audio, www.kmraudio.de; studio design by Fritz Fey, www. studioplan.de). (b) Paisley Park's Studio A, Chanhassen, MN (courtesy of Paisley Park Studios). (c) Wisseloord Main Hall, Hilversum, Netherlands (courtesy of Wisseloord, www.wisseloord.nl, acoustics and design by Jochen Veith, jv-acoustics, www. jv-acoustics.de). (d) Simple, basic studio layout.

- A studio used to produce audio for video, film dialogue, vocals and mixdown might consist of only a single, small recording space located off the control room for overdub purposes.

In fact, there is no secret formula for determining the perfect studio design. Each studio design (Figure 1.4) has its own sonic character, layout, feel and decor that are based on the personal tastes of its owners, the designer (if one was involved) and the going studio rates (based on the studio's return on investment and the supporting market conditions).

During the 1970s, studios were generally small. Because of the new development of (and over-reliance on) artificial reverb and delay devices, they tended to be acoustically "dead" in that the absorptive materials tended to suck the life right out of the room. The basic concept was to eliminate as much of the original acoustic environment as possible and replace it with artificial ambience.

Fortunately, as tastes began to change and music-makers grew tired of relying entirely upon artificial ambience, rooms (both large and smaller) began to revert back to the idea of basing their acoustics upon a combination of absorption and natural acoustic reflections. This use of balanced acoustics has revived the art of capturing the room's original acoustic ambience along with the actual sound pickup. In fact, through improved studio design techniques, we have learned how to achieve the benefits of both earlier and modern-day recording eras by building a room that provides a reasonable-to-maximum amount of isolation within the room (thereby reducing unwanted leakage from an instrument to other mics in the room) while encouraging higher-frequency reflections that can help give life and ambience to the overall sound. This natural balance of absorption and reflection is used to "liven up" the sound of an instrument or ensemble when they are recorded at a distance, a technique that has become popular when recording live rock drums, string sections, electric guitars, choirs, etc. Using close mic techniques, it's also possible to use one or more iso-booths or smaller rooms as a tool, should greater isolation be needed.

In short, it is this combination of the use of acoustic treatment, proper mic techniques and a personal insight into the instruments and artists within a room (combined with a sense of experimentation, experience and personal preferences) that can bring out the best in a recording facility.

In certain situations, a studio might not have a large recording space at all but simply a small or mid-sized iso-room for recording overdubs (this is often the case in facilities that are used in audio-for-visual post-production and/or music remixing). Project studios, mix rooms and newer "concept" studios might not have a separate recording space at all, opting to create an environment whereby the artists can record directly within the mixing/production space itself.

The Control Room

A recording studio's *control room* (Figures 1.5–1.7) serves a number of purposes in the recording process. Ideally, the control room is acoustically isolated from the sounds that are produced in the studio as well as from the surrounding, outer areas. It is optimized to act as a critical listening environment that uses carefully placed and balanced monitor speakers. This room also houses the majority of the studio's recording, control and effects-related equipment. At the heart of the control room is the recording console and/or digital audio workstation (DAW).

FIGURE 1.5
There are a lot more project control rooms than pro ones (courtesy of www.emiliano-caballero.com).

FIGURE 1.6
Morten Lindberg's 2L mix room (courtesy of 2L, www. 2L.no).

The *recording console* (also referred to as the *board* or *desk*, as seen in Figure 1.8a) can be thought of as an artist's palette for the artists, producer and/or recording engineer. The console allows the engineer to combine, control and distribute the input and output signals of most, if not all, of the devices found in the control room. The console's basic function is to allow any combination of mixing (variable control over relative amplitude and signal blending between channels), spatial positioning (left/right or surround-sound control over front, center, rear and sub), routing (the ability to send any input signal from a source to a destination) and switching for the multitude of audio input/output signals that are commonly encountered in an audio production facility … not to mention the fact that a console will also need to work in conjunction with a recording device. A DAW, as seen in Figure 1.8b), is a multi-channel media monster (audio, MIDI and video) that can work in conjunction with an outboard recording console or mixer, or it can work entirely on its own. DAWs are increasingly common fixtures within most control rooms, allowing us to work in an "in-the-box" stand-alone fashion.

FIGURE 1.7
Synchron Stage control room A, Vienna (courtesy of Walters-Storyk Design Group, www.wsdg.com).

The analog tape machine (24, 16, 8, 4 and 2 tracks, as seen in Figure 1.8c) is another way to capture sounds in the studio, using a way of working that's quite different (both functionally and sonically) than its digital counterpart.

(a)

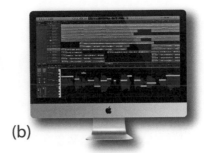
(b)

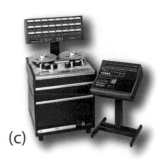
(c)

FIGURE 1.8
The heart(s) of the recording studio. (a) The recording console. (b) The digital audio workstation (DAW). (c) The analog tape recorder.

Tape machines might be located toward the rear of a control room, while a DAW might be located at the functional center of the workspace (if the DAW serves as the room's main recording/mixing device) or at the side of the console. Because of the added noise and heat generated by recorders, computers, power supplies, amplifiers and other devices, it's becoming more common for equipment to be housed in an isolated machine room that has a window and door adjoining the control room for easy access and visibility. In either case, DAW controller surfaces (which are used for computer-based remote control and mixing functions) and auto-locator devices (which are used for locating tape and media position cue points) are often situated in the control room near the engineer for easy access to all recording, mixing and transport functions. Effects devices (used to electronically alter and/or augment the character of a sound) and other signal processors are also placed nearby for easy accessibility (frequently being designed into an effects island or bay that's often located directly behind the console).

As with recording studio designs, every control room will usually have its own unique sound, feel, comfort factor and studio booking rate. Commercial control rooms often vary in design and amenities – from a room that's basic in form and function to one that is lavishly outfitted with the best toys. Again, the style and layout are a matter of personal choice; however, as you'll see throughout this book, there are numerous guidelines that can help you make the most of a recording space. It's really important to keep in mind that although the layout and equipment will always be important, it's the people (the staff, musicians and you) who will almost always play the most prominent role in capturing the feel of a performance and the heart of your clients.

THE PROJECT STUDIO

With the advent of affordable, high-quality digital audio workstations, plug-ins, controllers and speakers, it's a foregone conclusion that the vast majority of music and audio recording/production systems are being built and designed for personal use. The rise of the *project studio* (Figures 1.9–1.11) has brought about monumental changes in the business of music and professional audio, in

FIGURE 1.9
Happiness is recording your own band in the basement (courtesy of Yamaha Corporation of America; www.yamaha.com).

a way that has affected and altered almost every facet of the audio production community.

One of the greatest benefits of a project or portable production system centers on the idea that an artist can choose from a wide range of tools and toys to get the particular sounds that he or she likes on their own schedule, in their own way and without hiring out a pro studio. This technology is often extremely powerful, as the components combine to create a vast palette of sounds and handle a wide range of task-specific functions. Such systems often include a DAW computer for recording, MIDI sequencing, mixing and just about anything that relates to modern audio production, soft-/hardware electronic instruments, soft-/hardware effects devices, as well as speaker and/or headphone monitoring.

FIGURE 1.10
DMH's project studio when he's in Berlin circa 2020 (courtesy of DMH, www.davidmileshuber.com).

Systems like these are constantly being installed in the homes of almost all working and aspiring musicians, audio enthusiasts and DJs. Their sizes range from a corner in an artist's bedroom to a larger system that has been installed in a dedicated project studio. All of these system types can be designed to handle a wide range of tasks and have the important advantage of letting the artist produce his

FIGURE 1.11
The rustic look also works (courtesy of Steinberg Media Technologies GmbH, a division of Yamaha Corporation, www.steinberg.net).

or her music in a comfortable, cost-effective, at-home environment whenever the creative mood hits. Such production luxuries, which would have literally cost a fortune 30 years ago, are now within the reach of almost all working and aspiring musicians. This revolution has been carried out under the motto "You don't have to have a million-dollar studio to make good music". Truly, the modern-day project and portable studio systems offer such a degree of cost-effective power and audio fidelity that they can often match the production quality of a professional recording facility – all you need to supply is knowledge, care, dedication, patience and artistry.

Making the Project Studio Pay for Itself

Beyond the obvious advantage of being able to record when, where and how you want to in your own project studio, there are several additional benefits to working in a personal environment. Here are ways that a project studio can help subsidize itself, at any number of levels:

- Setting your own schedule and saving money while you're at it! An obvious advantage of a project studio revolves around the idea that you can create your own music on your own schedule. Part of the expense of using a professional studio comes from having to be practiced and ready to roll on a specific date or range of dates. Having your own project studio frees you up to lay down practice tracks and/or record when the mood hits, without having to worry about punching the studio's time clock.

- For those who are in the business of music, media production or the related arts business, the equipment, building and utility payments can be written off as a tax-deductible expense. Do some research and talk with a tax advisor; there are definitely advantages to writing off both personal and business studio deductions from your income tax.

- An individual artist or group might consider pre-producing a project in their own studio, allowing the time and expense billings to be a business tax deduction.
- The same artists might consider recording part or all of their production at their own project studio. The money saved (and deducted) could be later spent on a producer, better mixdown facility, professional freelance engineer, legal issues (such as copyright and contracts) … and let's not forget marketing.
- The "signed artist/superstar approach" refers to the mega-artist who, instead of blowing their advance royalties on lavish parties in a professional studio (a sure way never to see any money from your hard work), will spend the bucks on building their own professional-grade project studio. After the project has been recorded, the artist will still have a tax-deductible facility that can be operated as a business enterprise. Then, when the next project comes along, the artist will still have a personal facility where they can record, while the saved advance bucks on the new project can be put in the bank.
- The name of the game for all of the above approaches is all about being wise and financially responsible.

THE PORTABLE STUDIO

FIGURE 1.12
A portable studio can be set up just about anywhere (courtesy of Universal Audio, www.uaudio.com © 2023 Universal Audio, Inc. All rights reserved. Used with permission).

Of course, as laptops have grown in power, it has become a simple matter to load them with your favorite DAW software and audio interface, grab your favorite mics and headphones, put the entire system in your backpack and hit the road running. These systems (Figure 1.12) have actually gotten so powerful that they equal and can sometimes surpass large, tower-based studio systems, allowing you to record, edit, mix and produce on the go or in the studio without any compromises whatsoever. In the bedroom, on the beach or at a remote seaside island under battery or solar power – there are literally no limits to what these ever-growing production systems can do.

To take these ever-shrinking analogies to the *n*th degree, newer handheld recording systems can actually fit in your pocket, allowing you to sample and record sounds with professional results, using either their internal high-quality mics or, in some cases, external professional mics under phantom power. Truly, it is a small world after all (Figure 1.13)!

FIGURE 1.13
A portable handheld recorder can go with you anywhere to sample or capture the moment.

The iRevolution

In recent years, another studio revolution has taken place – the iOS revolution. In fact, some of the greatest changes in audio production today are coming about as a direct result of the introduction of devices like the iPad and other iOS devices. Of course, the strengths of these systems are that they are extremely portable, wireless and offer an ever-increasing amount of processing power. A huge by-product of all this is overall cost-effectiveness. As an example, years ago, a remote controller device for a DAW would set you back $1,200 or so ... now, it's a simple matter of getting out your pad and downloading a controller app from the "store", and you'll have as many (or more) functions as its "wired" hardware counterpart at a ridiculously low cost, or even for free (Figure 1.14).

As new apps (applications) come onto the market on a daily basis, the iOS revolution continues to make its mark on all forms of media production. Within the fields of music and audio production, a pad can be used for such applications as:

- Audio recording and mixing
- DAW and live mixing
- Electronic instruments
- Systems controllers
- Compositional tools

Obviously, as these devices become more powerful, they can be used for any number of on-the-go purposes that previously required a larger computer or laptop ... it is truly a technological revolution in the making.

The Retro Revolution

A revolution that has been making itself increasingly felt over the last decade at all levels of studio production is the desire for all things retro (Figures 1.15 and 1.16). Perhaps it's a need to revert back to our steampunk past, or just a nostalgia for simpler days, but retro is definitely in. So, what is retro? It's a desire to put older devices and techniques back into practice in our current productions, or to have new things that are designed in the style and function of yesteryear in our studios. Either way, it's often very cool to make use of these new, older toys in order to give our productions a fresh sound.

As with any art form, there is only one rule: there are none! Trends and what is "cool" come and go, but if you look into your heart and follow what's true to you, that sense of expression and feeling will be appreciated and felt by others.

A perfect example of this is the 2022 Grammy winner for Producer of the Year, Jack Antonoff. As he had started out with indie and underground bands and sounds, people in mainstream music didn't pay attention to his music until his unique way of imagining music and organic sounds started to become popular.

To us, the mainstream acceptance of all things retro goes above and beyond the reverence and interest in just older equipment – it has finally begun to further connect us with our use and interest in past techniques. This includes:

- The use of distance techniques when placing microphones
- The use of analog tape machines (or their modeled plug-in counterparts) to add an indefinable punch to our mixes
- The willingness to use older equipment to add a different sonic character to our sound
- The willingness to place a set of speakers in the studio and mic them so as to add "room sound" to a mix
- Placing a speaker/mic combination in the bathroom down the hall to get "that sound"
- Recording the guitar track in a huge gym down the street with your laptop to get a larger-than-life acoustic sound

FIGURE 1.16
London Bridge Studio Neve 8048 console (courtesy of London Bridge Studio, Seattle, WA, www.london-bridgestudio.com).

In short, I think retro is helping us to accept that all things don't always have to be new in order to be exciting and relevant. It can be a piece of equipment, it can be a mic technique, it can be expensive or it can cost nothing to use tools you already have at your disposal. All you need is a sense of adventure and a willingness to experiment.

THE CHANGING FACES OF THE MUSIC STUDIO BUSINESS

As we've noted, the role of the professional recording studio has begun to change as a result of upsurges in project studios, audio for video and/or film, multimedia and the Internet. These market forces have made it necessary for certain facilities to rethink their operational business strategies. Sometimes, this means that a studio will not be able to adapt to the changing times; however, for those who are able to react and diversify in the new digital age, new possibilities can be met with success, as is illustrated by the following examples:

- Personal production and home project studios have greatly reduced the need for an artist or producer to have constant and costly access to a professional facility. As a result, many pro studios now cater to artists and project studio owners who might have an occasional need for a larger space or better-equipped recording facility (e.g., for recording big drum sounds, string overdubs or an orchestral session). In addition, after an important project has been completed in a private studio, a professional facility might be needed to mix the production down into its final form. Most business-savvy studios are only too happy to capitalize on these new and constantly changing market opportunities.

- Upsurges in the need for audio for video, game and film post-production have created new markets that allow professional recording studios to provide services to the local, national and international broadcast and visual production communities. Creative studios often enter into lasting relationships with audio-for-visual and broadcast production markets, so as to thrive in the tough business of music, when music production alone might not provide enough income to keep a studio afloat.

- Studios are also taking advantage of Internet audio distribution techniques by offering Web development, distribution and other audio-for-Web services as an added incentive to their clients.

- A number of studios are also jumping directly into the business of music by offering advisory, business, networking and management services to artists and bands, sometimes signing the artists and funding tours in exchange for a piece of the business pie.

- A studio with several rooms might offer one of the rooms to an established engineer/mixer, offering it up as "Joe's Mix Room" in exchange for a roster of clients that comes with the territory of having "Joe" on board.

These and other aggressive marketing strategies (many of which may be unique to a particular area) are being widely adopted by commercial music and recording facilities to meet the changing market demands of new and changing media. No longer can a studio afford to place all of its eggs in one media basket. Tapping into changes in market forces and meeting them with new solutions are important for making it (or simply keeping afloat) in the business of modern music production and distribution. Make no mistake about it, starting, staffing and maintaining a production facility, as well as getting the clients' music heard, is serious work that requires dedication, stamina, innovation, guts and a definite dose of craziness.

OK, now for the hard part. Let's take a moment to say that all-important word again: business. With the onset of more creative changes, opportunities and options, the only thing that stays constant is change, right? For example, there has been a steady onslaught of technological advances in audio production (such as new software, portable recording and controller options that come with new generations of computer, laptop and pad technologies). However, beyond that, in reality, many (if not most) of the technical aspects of music and audio production have stayed the same. What *have* drastically changed are the business aspects of the industry – most notably in how music is distributed, marketed and consumed by the buying public.

We certainly don't have to remind you about how the traditional label distribution models have all changed in the wake of the download and streaming era. Over the last decade, the game has been continuously evolving in a way that keeps even the most seasoned industry professionals on their toes … is this necessarily a bad thing? I'm not convinced it is. Innovation and ingenuity have always been part of the creative process … it's what keeps things new and exciting. However, innovation doesn't always come easily and is often at a price. Recording studios are constantly being challenged to ride the wave of innovation – some make it, some are simply unable to adapt to the new world of the personal studio, the Internet and changing business models. In short, the reality is that the professional studio business is a tough one that requires that the studio make itself and its expertise relevant and marketable in the changing world of music and media production.

LIVE/ON-LOCATION RECORDING: A DIFFERENT ANIMAL

Unlike the traditional multitrack recording environment, where overdubs are often used to build up a song over time, *live/on-location recordings* are created on the spot, in real time, often during a single on-stage or in-the-studio performance event, sometimes with little or no studio post-production other than mixdown. A live recording might be very simple, possibly being recorded using only a few mics that are mixed directly to two or more tracks. Or, a more elaborate gig might call for a full-fledged multitrack setup, requiring the use of a temporary control room or fully equipped mobile recording van or truck (Figure 1.17). A more involved setup will obviously require a great deal of preparation and expertise, including a knowledge of sound reinforcement combined with the live recording techniques that are necessary to capture instruments in a manner that provides enough isolation between the tracks so as to yield control over the individual instruments during the mixdown phase.

FIGURE 1.17
Studio Metronome's live recording audio truck, Brookline, NH (courtesy of Metronome Media Group, www.studiometronome.com, Photo by Bennett Chandler).

Although the equipment and system setups will be familiar to any studio engineer, live recording differs from its more controlled studio counterpart in that it happens in a world where the motto of the day is "you only get one chance". When you're recording an event where the artist is spilling his or her guts to hundreds or tens of thousands of fans, it's critical for everything to run smoothly. Live recording usually requires a unique degree of preparedness, redundancy, system setup skills, patience and, above all, experience. It's all about capturing the sound and feel of the moment, the first time – and for a brave few, that's a very exciting thing.

AUDIO FOR VIDEO AND FILM

In this day and age, it's simply impossible to overlook the importance that quality audio plays in the production of film and video. With the introduction of complex surround playback formats, high-budget music scores and special effects production, audio-for-film has long been an established and specialized industry that has a dramatic effect on the movie-goer's experience. Most definitely, audio-for film is an art form that has touched and helped shape our world culture (Figures 1.18 and 1.19).

Prior to the advent of the DVD, Blu-ray and home theater surround sound, broadcast audio and the home theater experience were almost an afterthought in a TV tube's eye. However, with the introduction of these new technologies, audio has matured to being a highly respected part of video and visual media production. With the common use of immersive sound in the creation of movie soundtracks, along with the emerging popularity of Dolby Atmos in home and computer entertainment systems, the public has come to expect high levels of audio quality from their entertainment experience.

FIGURE 1.18
Skywalker Sound main control room (courtesy of Skywalker Sound, a division of Lucasfilm Ltd., www .skysound.com).

In modern-day production, MIDI, hard-disk recording, timecode and synchronization, automated mixdown and advanced effects have become everyday components of the audio-for-visual environment, requiring that professionals be highly specialized and skilled in order to meet the demanding schedules and production complexities.

FIGURE 1.19
Auditorium, Galaxy Studios,
Mol, Belgium (courtesy of
Galaxy Studios, www.galaxy.
be).

AUDIO FOR GAMES

Most of the robot-zappin', daredevil-flyin', Hufflepuff-boppin' addicts who are reading this book are very aware that one of the largest and most lucrative areas of media audio production is the field of scoring, designing and producing audio for computer games – Zaaaaaappppppppp! Like most subcategories within audio production, this field of expertise has its own set of technical and scheduling rigors and requirements that center as much around spreadsheets and databases as they do around audio equipment. With the tens of thousands of voice and music cues that are commonly required to make a game into a fully interactive experience, an entirely different skillset is often required of a gaming sound technician.

In addition, a rather interesting connection between orchestral recording and game audio has been steadily on the upsurge. Just as film makes use of heavy orchestral scoring for dramatic effect, game audio has also begun to score using big-budget and big-name orchestras to create a bigger-than-life storyline.

THE DJ

The days when DJs would bring their records to a gig and spin are pretty much gone (except for those who still prefer the retro, hands-on way of working). Now, on-stage systems are commonly made up of laptops, controllers, visual gear and other toys that have just as much in common with a project studio as they do with the theater or performance stage. In short, it's not just a pair of turntables anymore. From a performance perspective, the modern-day DJ can range from being someone who plays other people's music to one who creates and composes their own productions and then combines these sounds with the works of others.

A new development within the DJ community revolves around the use and distribution of stems (individual groupings of instruments that can combine together to make up a recording). Much in the same way that instruments, vocals, etc. can be grouped together and isolated in a session, it's possible for "stems" to be recorded to isolated tracks in a way that allows them to be individually remixed and/or isolated in ways that can create a whole new outlook on the track. Additionally, the use of stems in a DJ setting allows these various sub-grouped stems from various songs to be combined, mixed and mutilated in ways that can create a whole new song or composition. The rule here is: be inventive and have fun!

THE TIMES, THEY'VE ALREADY CHANGED: MULTIMEDIA AND THE WEB

With the integration of text, graphics, MIDI, digital audio and digitized video into almost every facet of the personal computer and mobile environment, the field of *multimedia* audio has become a fast-growing, established industry that represents an important and lucrative source of income for both creative individuals and production facilities alike. Of course, the use of audio-for-the-Web can take any number of forms:

- A record label might decide to offer a new release in a streaming-only format, requiring that the project be mastered in a way that best suits the medium.
- An online language dictionary might require that all of the translations be recorded so they can be easily pronounced and heard in high quality.
- An online download site might offer any number of music loops that can be downloaded and made into a personal remix that can be shared over the Web.
- A music sample library might make use of a major recording facility to record instruments for a sampler plug-in.
- The list is absolutely endless!

For decades, the industry has been crying foul over the breakup of the traditional record industry as we know it. In the early days of the Web, a new kid on the block came onto the scene ... the MP3. This "ripping" and playback codec made it possible for entire song libraries to be compressed (data-wise), uploaded, downloaded and streamed with relative ease. Such a simple beastie then progressed into a social animal that would itself bring an entire industry to its virtual knees. With media sharing came the eventual revolution of the social network – allowing people to connect with each other in totally new ways never before thought possible. With the sharing of information came the sharing and new networks for distributing music and visual media, allowing DMH to be currently listening to a melodic Ukrainian rapper on his phone (called a "Handy" in Deutschland) while walking on the Oberbaum Brücke in Berlin. It's a brave, new world after all, and we're all part of that big change!

POWER TO THE PEOPLE!

On a more personal and human front, with all of these amazing tools that are at our disposal, it totally makes sense that artists, producers and aspiring recording professionals will create art using the toys, tools and techniques that are affordable and understandable. However, technology isn't enough to create great art – a personal sense of drive, passion and ingenuity is also required. One more ingredient is also necessary to finish off this artistic "mix", namely, a personal and never-ending search for knowledge to improve your craft. This all-important ingredient can be gained by:

- Reading about the equipment choices that are available to you on the Web or in the ever-dwindling number of trade magazines that are available
- Visiting and talking with others of like (and dissimilar) minds about their equipment, techniques and personal working styles (conventions, industry organizations and social networks can be a really effective learning and networking tool)
- Enrolling in a recording course that best fits your needs, working style and budget
- Researching the type of equipment and room layout that best fits your needs and budget before you make any purchases and, if possible, getting your hands on equipment before you make any final purchases (i.e., checking them out at your favorite music store)
- Experience and time – always the best teacher

The more you take the time to familiarize yourself with the options and possibilities that are available to you, the less likely you are to be unhappy about how you've spent your hard-earned bucks after the fact. It is also important to point out that *having* the right equipment for the job isn't enough – it's also important to *take the time* to learn how to use your tools to their fullest potential. Whenever possible, read the manual and get your feet wet by taking the various settings, functions and options for a test spin long before you're under the time and emotional constraints of being in a session.

Whatever Works for You

As you begin to research the various types of recording and supporting systems that can be put to use in a project studio, you'll find that a wide variety of options are available. There are indeed hundreds, if not thousands, of choices for recording media, hardware types, software systems, speakers, effects devices – the list goes on. This should instinctually tell us that no one tool is right for the job. As with everything in art (even the business of an art), there are many personal choices that can be combined into a working system that's right for you. Whether you

- Work with a DAW or tape-based system
- Choose to use analog or digital effects equipment (or both)
- Are a Mac or PC kind of person (pretty much a nonissue these days)
- Use this type of software or that,

it all comes down to the bottom line of how does it sound? Does the music move you? How does it move the audience? How can it be sold? In truth, no prospective buyer will turn down a song because it wasn't recorded on such-and-such a machine, at such-and-such sample rate, using speakers made by so-and-so – it's the feel, baby. It's the emotion in the art that always seals the deal in the end.

THE PEOPLE WHO MAKE IT ALL HAPPEN

"One of the most satisfying things about being in the pro-fessional audio (and music) industry is the sense that you are part of a community."

Frank Wells, editor, Pro Sound News

When you get right down to the important stuff, the recording field is built around pools of talented individuals and service industries that work together toward a common goal: performing, producing, selling and enjoying music. As such, it's the people in the recording industry who make the business of music happen. Recording studios and other businesses in the industry aren't only known for the equipment that they have, but are more often judged by the quality, knowledge, vision and combined personalities of their staff. The following sections describe but a few of the ways in which a person can be involved in this multifaceted industry. In reality, the types and descriptions of a job in this techno-artistic industry are limited only by the imagination. New ways of expressing a passion for music production and sales are being created every day, and if you see a new opportunity, the best way to make it happen is to roll up your sleeves and "just do it".

The Artist

The strength of a recorded performance begins and ends with the artist. All of the technology in the world is of little use without the existence of the central ingredients of human creativity, emotion and individual technique. Just as the overall sonic quality of a recording is no better than its weakest link, it's the performer's job to see that music's main ingredient, its inner soul, is laid out for all to experience and hear. After all is said and done, a carefully planned and well-produced recording project is simply a gilded framework for the music's original drive, intention and emotion.

Studio Musicians and Arrangers

A project will often require additional musicians to add extra spice and depth to the artist's recorded performance. For example:

- An entire group of selected studio musicians might be called on to provide the best possible musical support for a high-profile artist or vocalist.

- A project might require musical ensembles (such as a choir, string section or background vocals) for a particular part or to give a piece a fuller sound.
- If a large ensemble is required, it might be necessary to call in a professional music contractor to coordinate all of the musicians and make the financial arrangements. The project might also require a music arranger, who can notate and possibly conduct the various musical parts.
- A member of a group might not be available or be up to the overall musical standards that are required by a project. In such situations, it's not uncommon for a replacement studio musician to be called in to fit the bill.

In situations like these, a project that's been recorded in a private studio might benefit from the expertise of a professional studio that has a larger recording room, an analog multitrack for that certain sound and/or an engineer who knows how to better deal with a complicated production scenario.

The Producer

Beyond the scheduling and budgetary aspects of coordinating a recording project, it's the job of a producer to help the artist and record company create the best possible recorded performance and final product that reflects the artist's vision. A producer can be hired for a project to fulfill a number of specific duties or might be given full, creative rein to help with any and all parts of the creative and business side of the process to get the project out to the buying public. More likely, however, a producer will act collaboratively with an artist or group to guide them through the recording process to get the best possible final product. This type of producer might:

- Help the artist (and/or record label) create the best possible recorded performance and final product that reflects the artist's vision. This will often include a large dose of musical input, creative insight and mastery of the recording process
- Assist in the selection of songs
- Help to focus the artistic goals and performance in a way that best conveys the music to the targeted audience
- Help to translate that performance into a final, salable product (with the technical and artistic help of an engineer and mastering engineer)

It's interesting to note that because engineers spend much of their working time with musicians and industry professionals with the intention of making their clients sound good, it's not uncommon for an engineer to take on the role of producer or co-producer (by default or by mutual agreement). Conversely, as producers and artists alike become increasingly knowledgeable about recording technology, it's increasingly common to find them on the other side of the glass, sitting behind the controls of a console.

Additionally, a producer might also be chosen for his or her ability to understand the process of selling a final recorded project from a business perspective to a label, to a film licensing entity or to the buying public. This type of producer can help the artist gain insights into the world of business, business law, budgeting and sales, always an important ingredient in the process.

Of course, in certain circumstances, a project producer might be chosen for his or her reputation alone and/or for giving a certain cachet to a project that can help put a personal "brand" on the project, thereby adding to the project's stature and hopefully helping to grab the public's attention.

One final thing is for certain: the artist and/or label should take time to study what type of outside producer is needed (if any) and then agree upon his or her creative and financial role in the project *before* entering into the creative process.

The Engineer

The role of an engineer can best be described as an interpreter between technology and art. He or she must be able to express the artist's music and the producer's concepts and intent through the medium of recording technology. In this world, both the music and the recording process itself are totally subjective and artistic in nature and rely on the tastes, experiences and feelings of those involved. During a recording session, one or more engineers can be used on a project to:

- Conceptualize the best technological approach for capturing a performance or music experience
- Translate the needs and desires of the artists and producer into a technological approach that best captures the music
- Document the process for other engineers or future production use
- Place the musicians in the desired studio positions
- Choose and place the microphones or pickup connections
- Set levels and balances on the recording console or DAW mixing interface
- Capture the performance (onto hard disk or tape) in the best possible way
- Overdub additional musical parts into the session that might be needed at a later time
- Mix the project into a final master recording in any number of media formats (mono, stereo and immersive)
- Help in meeting the needs for archiving and/or storing the project
- Last, but not least, be helpful, understanding and supportive in a way that can put those who are in a stressful situation at ease

In short, engineers use their talent and artful knowledge of recording media technology to convey the best possible finished sound for the intended media, the client and the buying public.

Assistant Engineer

Many studios often train future engineers (or build up a low-wage staff) by allowing them to work as assistants or interns who can offer help to staff and visiting freelance engineers. The assistant engineer might do microphone and headphone setups, run DAW or tape machine operations, carry out system patching, help with session documentation, do session breakdowns and (in certain cases) perform rough mixes and balance settings for the engineer on the console. With the proliferation of freelance engineers (engineers who are not employed by the studio but are retained by the artist, producer or record company to work on a particular project), the role of the assistant engineer has become even more important. It's often his or her role to guide freelance engineers through the technical aspects and quirks that are peculiar to the studio, and to generally babysit the technical and physical aspects of the place.

Traditionally, being an assistant has been a no- or low-wage job that can expose a "newbie" to a wide range of experiences and situations. With hard work and luck, many assistants have worked their way into the hot seat whenever an engineer quits or is unexpectedly ill. As in life, there are no guarantees in this position – you just never know what surprises are waiting around the next corner for those who rise to the occasion.

Maintenance Engineer

The maintenance engineer's job is to see that the equipment in the studio is maintained in top condition and regularly aligned and repaired when necessary. Of course, with the proliferation of project studios, cheaper mass-produced equipment, shrinking project budgets and smaller staff, most studios will not have a maintenance engineer on staff. Larger organizations (those with more than one studio) might employ a full-time staff maintenance engineer, whereas outside freelance maintenance engineers and technical service companies are often called in to service smaller commercial studios in both major and non-major markets.

Mastering Engineer

Often, a final master recording will need to be tweaked in terms of level, equalization (EQ) and dynamics so as to present the final "master" recording in the best possible sonic and marketable light. If the project calls for it, this job will fall to a mastering engineer, whose job it is to listen to and process the recording in a specialized, fine-tuned monitoring environment. Of course, mastering is a techno-artistic field in its own right. Beauty is definitely in the ear of the beholding client, and one mastering engineer might easily have a completely different approach to the sound and overall feel to a project than the next dude or dudette. However, make no mistake about it – the mastering of a project can have a profound impact on the final sound of a project, and the task of finding the right mastering engineer for the job should never be taken lightly. Further info on the subject can be found in Chapter 20.

Studio Management

Running a business in the field of music and audio production requires the special talents of businesspeople who are knowledgeable about the inner workings of promotion, the music studio, the music business and, above all, the people. It requires constant attention to quirky details that would probably be totally foreign to someone outside "the biz". Studio management tasks include:

- *Management:* The studio manager (who might or might not be the owner) is responsible for managerial and marketing decisions for all of the inner workings of the facility and its business.
- *Bookings:* This staff person keeps track of most of the details relating to studio booking, usage and billing.
- *Competent administration staff:* These folks keep everyone happy and running as smoothly as possible.

Note, however, that some or all of these functions often vary from studio to studio. These and other equally important staff members are necessary in order to successfully operate a commercial production facility on a day-to-day basis.

Music Law

It's never good for an artist, band or production facility to underestimate the importance of a music lawyer. When entering into important business relationships, it's always a good idea to have a professional ally who can help you, your band or your company navigate the potentially treacherous waters of a poorly or vaguely written contract. Such a professional can serve a wide range of purposes, ranging from the primary duties of looking after their clients' interests and ensuring that they don't sign their careers away by entering into a life of indentured, nonprofit servitude, all the way to introducing an artist to the best possible music label or distribution network.

Music lawyers, like many in this business, can be involved in the working of a business or career in many ways; hence, various fee scales are used. For example, a new artist might meet up with a friend who knows about a bright, young, freshly graduated music lawyer who has just passed the bar exam. By developing a relationship early on, there are any number of potential opportunities for building trust and making special deals that are beneficial to both parties, etc. On the other hand, a more established lawyer could help solicit and shop a song, band or artist more effectively within a major music, TV or film market. As with most facets of the biz, answers to these questions are often situational and require intuition, careful reference checking and the building of trust over time. Again, it's important to remember that a good music lawyer can be extremely important (at the right moment) and is the unsung hero of many successful careers.

Women and Minorities in the Industry

Ever since its inception, males have dominated the recording industry. I remember many sessions in which the only women on the scene were female artists, secretaries or studio groupies in short dresses. Fortunately, over the years,

women have begun to play a much more prominent role, both in front of and behind the glass, and in every facet of studio production and the business of music (Figure 1.20). Fortunately, in recent decades, most of the resistance to including new and fresh blood based on gender, race or sexual orientation into the business has greatly reduced. In the end, the most important thing that you can do to make it in "the biz" is to be sincere, work hard, play nice and simply be yourself.

FIGURE 1.20
Women's Audio Mission, an organization formed to assist women in the industry (courtesy of the Women's Audio Mission, www.womensaudiomission.org).

Behind the Scenes

In addition to the positions listed earlier, there are scores of other professionals who serve as a backbone for keeping the business of music alive and functioning. Without the many different facets that contribute to the making of the music business, the biz would be very, very different. A small sampling of the additional professional fields that help make it happen includes:

No matter who you are, where you're from or what your race, gender, sexual or planetary orientation is, remember this universal truth: If your heart's in it and you're willing to work hard enough, you'll make it (whatever you perceive "it" to be). Don't let them tell you (or tell yourself) otherwise.

- Artist management
- Artist booking agents
- A&R (artist and repertoire)
- Equipment design
- Equipment manufacturing
- Music and print publishing
- Distribution
- Web development
- Graphic arts and layout
- Audio company marketing
- Studio management

- Live sound
- Live sound tour management
- Acoustics
- Audio instruction
- Club management
- Sound system installation for nightclubs, airports, homes, etc.
- ... and a whole lot more!

This incomplete listing serves as a reminder that the business of making music is full of diverse possibilities and extends far beyond the notion that in order to make it in the biz, you'll have to sell your soul or be someone you're not. In short, there are many paths that can be taken in this techno-artistic business. Once you've found the one that best suits your own personal style, you can then begin the lifelong task of gaining knowledge and experience and pulling together an interactive network with those who are currently working in the field.

It's also important to realize that finding the career niche that's right for you might not happen overnight. You might try your hand at one aspect of production, only to find that your passion lies totally in another field. When and if this happens, don't beat yourself up. Finding the right career path that best fits your life might not be easy ... but it's super important. As the saying goes, "Wherever you may be, there you are!" Finding the path that's best for you is a lifelong ongoing quest; the general idea is to work hard, learn and enjoy the ride.

CAREER DEVELOPMENT

It's a sure bet that those who are interested in getting into the business of audio will quickly find out that it can be a tough nut to crack. For every person who makes it, a large number won't. In short, there are a lot of people who are waiting in line to get into what is perceived by many to be a glamorous biz. So, how do you get to the front of the line? Well, folks, here are the primary keys:

- Self-Motivation
- Networking

Self-Motivation

The business of art (the techno-art of recording and music being no exception) is one that's generally reserved for self-starters and self-motivated people. Even if you get a degree from XYZ college or recording school, there's absolutely no guarantee that your dream studio will be knocking at your door with an offer in hand (in fact, they most certainly won't). It takes a large dose of perseverance, talent and personality to make it.

This may sound strange, but one of the best ways to get into the biz is to simply jump in and start. In fact, you might try this little trick ... find a stick (or one of those Scottish swords, if you happen to have one hanging around) and get down on one knee, then "knight" yourself on the shoulder with the figurative "sword" and say: "I am now a _____ !" (Fill in the blank with whatever you want to be – engineer, artist, producer … whatever) and then say "arise Sir or Dame _____ … you are now a _____!" Simply become it … right there on the spot! Now, make up a business card, start a business and begin contacting artists to work with (or make the first step toward becoming the creative person you want to be). All you have to do is BELIEVE IN YOURSELF, work hard and follow the Golden Rule.

In fact, there are many ways to get to the top of your own personal mountain. For example, you could get a diploma from a school of education or from the school of hard knocks (it usually ends up being from a bit of both), but the goals and the paths are up to you.

Networking: "Showing Up Is Huge"

The other half of the success equation rests with your ability to network well with other people. As the venerable expression says, "It's not [only] WHAT you know, it's WHO you know". Maybe you have an uncle or a friend in the business, or a friend who has an uncle – you just never know where help or that initial break might come from next. This idea of getting to know someone who knows someone else is what makes the business and production world go around. So, don't be afraid to put your best face forward and start meeting people. If you want to work at XYZ Studios, hang out without being in the way. You never know, the engineer might need some help or might know someone who can help get you into the proverbial door. The longer you stick with it, the more people you'll meet … and eventually, you'll have a bigger and stronger network than you ever thought could be possible.

ANCIENT PROVERB

Being "in the right place at the right time" means being in the wrong (or right) place at the wrong time a thousand times! In short, "Showing up is HUGE"!

A friend of mine recently added a phrase to this proverb: "but you have to be ready!" If you're at the right place, at the right time and you're not ready to step up to the plate, then it's all been for naught.

So, What Are Some Good Ways to Get Started?

- Join an industry association such as The Recording Academy (Grammys), Grammy U (for students), Audio Engineering Society (AES), etc.

- Attend conventions and industry business functions (both nationally and in your area).
- Visit a favorite studio and get to know them (and make it easy for them to get to know you).
- Online social networking.

Of course, I've been assuming that you want to get into the production side of the recording business. But for those who just want to learn the tools and toys of recording technology from an artist's standpoint, these networking tools apply even more to those who want and need to get their names out to the music-consuming public. For business professionals, networking is essential – for the artist, it's the driving force of your life.

So, when do you start this grand adventure? When do you start building your career? The obvious answer is *RIGHT NOW*. If you're in school, you have already started the process. If you're just hanging out with like-minded biz folks and/ or joined a local or national organization, that, too, is an equally strong start. Whatever you do, don't wait until you graduate or until some magic date in the future, because putting it off will just put you that much further behind.

In addition to all this, make yourself visible. Try not to be afraid when sending out a link to your site (which should include your resume, music/mix examples, etc.) when asking for a job or any type of position. The worst thing they can do is say "No." You might also keep in mind that "No" could actually mean "No, not right now". You might actually ask if this is the case. If so, they might take your persistence into account before saying "No" two or three times. By picking a market and particular area, blanketing that area with resume/press kits and knocking on doors, you just never know what might happen. I know it's not easy, but if you fail, simply pick yourself up (again), reevaluate your strategies, and start pounding the streets (again). Just remember the self-motivation rule: "failing at something isn't a bad thing – not trying is!"

Here are a few additional networking and job placement tips to get you started:

- Make a Facebook or personal Web page (WordPress is an easy, free and powerful way to get started).
- Send out lots of resumes, or better yet, make an online bio/resume link on your page.
- Choose a mentor who you can rely on and talk to (sometimes they fall out of the sky, sometimes you have to develop the relationship over time).
- Pick the areas you want to live in (if that's a key factor).
- Pick the companies in that area and target them.
- Contact studios or companies in this area that might be looking for interns.
- Visit these places, just to hang out and see what they are like.
- Use your school counselors for intern placement.
- Always remember to follow up at least once, usually more.

To summarize … by now it should be painfully obvious that getting into music production, audio production and all things recording takes hard work, perseverance, blood, sweat, tears and laughter. For every person who builds a personal career in audio production, a large number won't make it. There are a lot of people waiting in line to get into what is perceived by many to be a glamorous biz. So, how do you get to the front of the line? Well, folks, just as the best way to get to Carnegie Hall is to practice – here are some key skills that are practically requirements:

- A ton of self-motivation
- Good networking and communication skills
- A good, healthy attitude
- An ever-present willingness to learn
- The realization that "showing up is always huge!"

The business of art (the techno-arts of recording and music production being no exception) is one that's generally reserved for self-starters. Even if you get a degree from XYZ College or recording school, there's absolutely no guarantee that anyone will be knocking on your door asking you to work for them. More often than not, it takes a large dose of perseverance, talent, personality and luck to make it.

There are many ways to get to the top of your own personal mountain. You could get a diploma from a school of education or from the school of hard knocks (it usually ends up being from both), but the goals and the paths are up to you. As a mentor of mine always said, "Failure isn't a bad thing – not trying is!"

Another huge part of the success equation lies in your ability to network with other people. As the venerable expression says, "It's not [only] what you know – it's who you know". Maybe you have an uncle or a friend in the business, or a friend who has an uncle – you just never know where help might come from next. This idea of getting to know someone, who knows someone else, who knows someone else, is what makes the business world go around. Don't be afraid to put your best face forward and start meeting people. If you want to play gigs around your region (or beyond), get to know a promoter or venue manager and hang out without being too much in the way. You never know – the music maven down the street might know someone who can help get your feet in the proverbial door. The longer you stick with it, the more people you'll meet, thereby making a bigger and stronger network than you thought would be possible.

As a close buddy of mine always says, "Showing up is huge!" It's the wise person who realizes that being in the right place at the right time means being at the wrong place hundreds and hundreds of times. You just never know when lightning is going to strike – just try to be prepared and stand under the right tree when it does.

Here are some more practical and immediate tips for musicians and producers:

- Build a personal and/or band website: making a great social network presence and/or creating your own personal site helps to keep the world informed of your gigs, projects, bio and general goings-on.

- Build a relationship with a music lawyer: many music lawyers are open to building relations that can be kicked into gear at a future time. Take the time to find a solicitor who is right for you. Does he or she understand your personal music style? If you don't have the bucks, is this person willing to work with you and your budget as your career grows?
- The same questions might be asked of a potential manager. This symbiotic relationship should be built with care, honesty and safeguards (which is just one of the many reasons you want to know a music lawyer).
- Copyright your music: always protect your music by registering it with the Library of Congress. It's easy and inexpensive and can give you peace of mind about knowing that the artistic property that you're sending out into the world is protected. Go to www.copyright.gov for more information (www.copyright.gov/forms). Additional organizations also exist that can help you get paid.
- On a personal note as a musician, I've come to realize that making music is about the journey – not necessarily the goal of being a star, or being the big man/woman on campus. It's about building friendships, collaborations, having good and bad times at gigs – and, of course, it's all about making music.

A Word on Professionalism

Before we close this beginning chapter, there's one more subject that we'd like to touch on – perhaps the most important one of all: professional demeanor. Without a doubt, the life and job of a typical engineer, producer or musician isn't always an easy one. It often involves long hours and extended concentration with people who, more often than not, are new acquaintances. In short, it can be a high-pressure job. On the flip side, it's one that's often full of new experiences, with demands that change on almost a daily basis, and often connects you with exciting people who feel passionately about their art and chosen profession.

It's been my observation (and that of many I've known) that the best qualities that can be exhibited by anyone in "the biz" are:

- Having an innate willingness to experiment
- Being open to new ideas (flexibility)
- Having a good sense of humor
- Having an even temperament (this often translates as "patience and understanding")
- Being open to communicating with others
- Being able to convey and understand the basic nuances of people from all walks of life and with many different temperaments

The best advice we can possibly give is to be open, be patient and above all, BE YOURSELF. Also, be extra patient with yourself. If you don't know something …

ask. If you make a mistake (trust me, you will; we all do), admit it and don't be hard on yourself. It's all part of the process of learning and gaining experience.

This last piece of advice might not be as popular as the others, but it might come in handy someday: it's important to be open to the fact that there are many, many aspects to music and sound production, and you may find that your career calling might be better served in another branch of the biz (other than the one that you've been studying or striving for). That's totally OK! Change is an important part of any creative process – that and taxes are the only constants you can count on!

IN CONCLUSION

Obviously, these tips are just part of an ever-changing list. The process of producing, recording and mixing in any type of studio environment is an ongoing, lifelong pursuit. Just when you think you've gotten it down, the technology or the nature of the project changes under your feet – hopefully, you'll be the better for it and will be open to learning a new process or piece of gear or software tech.

Far more than just the technology, the process of coming up with your own production style and applying these tools, toys and techniques in your own way is what makes us artists – whether you're in front of the proverbial glass or behind it. Over time, your own list of studio tips and tricks will grow. Take the time to write them down and pass them on to others, and be open to the advice of your friends and colleagues. Use the trade mags, conventions and the Web to lead you to new ideas. This way, you're opening yourself up to new insights on using the tools of your profession and to finding new ways of doing stuff. Learning is an ongoing process – try to have lots of fun along the way!

CHAPTER 2

Sound and Hearing

When we make a recording, in effect, we're actually capturing and storing sound into a memory media so that an original event or generated signal can be re-created at a later date. If we start with the idea that *sound* is actually a concept that corresponds to the brain's perception and interpretation of a physical auditory stimulus, the study of sound can be divided into four areas:

- The basics of sound
- The characteristics of the ear
- How the ear is stimulated by sound
- The psychoacoustics of hearing

THE BASICS OF SOUND

Sound arrives at the ear in the form of periodic variations in atmospheric pressure called *sound-pressure waves*. This is the same atmospheric pressure that's measured by the weather service, although the changes in pressure heard by the ear are simply too small in magnitude and fluctuate too rapidly to be observed on a barometer. An analogy of how sound waves travel in air can be demonstrated by bursting a balloon. Before we stick it with a pin, the molecular motion of the room's atmosphere is at a normal resting pressure. The pressure inside the blown-up balloon is much higher, though, and the molecules are compressed much more tightly together (Figure 2.1a), like people packed into a crowded subway car. When the balloon is popped – KAPOW! (Figure 2.1b) – the tightly compressed area under high pressure begins to exert an outward force on its molecular neighbors in an effort to move toward areas of lower pressure. When the neighboring set of molecules have been compressed, they will then exert an outward force on the next set of lower-pressured neighbors in a continuing ongoing outward motion (Figure 2.1c), which travels until the pressure stabilizes and the molecules have used up all their energy in the form of heat.

Likewise, as a vibrating mass (such as a guitar string, a person's vocal chords or a loudspeaker) moves outward from its normal resting state, it squeezes air

DOI: 10.4324/9781003260530-2

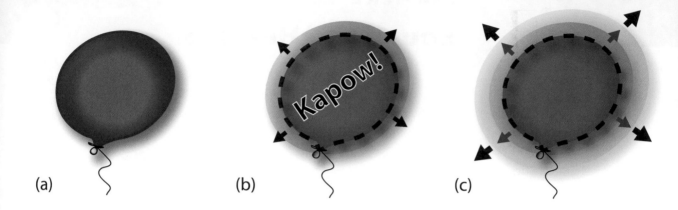

(a)   (b)   (c)

FIGURE 2.1

Wave movement in air as it moves away from its point of origin. (a) An intact balloon contains pressurized air. (b) When the balloon is popped, the compressed molecules exert a force on its outer neighbors in an effort to move to areas of lower pressure. (c) This exerted force continues outwards to the next set of molecules in an effort to move to areas of lower pressure.

molecules into a compressed area, away from the sound source. This causes the area being acted on to have a greater than normal atmospheric pressure, a process called *compression* (Figure 2.2a). As the vibrating mass moves inward from its maximum movement, an area with a lower than normal atmospheric pressure will be created, in a process called *rarefaction* (Figure 2.2b). As the vibrating body cycles through its inward and outward motions, areas of higher and lower compression states are then cyclically generated. These areas of high and low pressure will cause the oscillating wave to move outward from the sound source in the same way that the compressed wave moved outward from the burst balloon. It's interesting (and important) to note that the molecules themselves don't move through air at the velocity of sound – only the sound wave itself moves through the atmosphere in the form of high-pressure compression waves that continue to push against areas of lower pressure (in an outward direction). This outward pressure motion is known as *wave propagation*, which is the building block of sound.

Waveform Characteristics

A *waveform* is the graphic representation of a sound-pressure level or voltage level as it moves through a medium over time. In short, a waveform lets us see and explain the actual phenomenon of wave propagation in our physical environment and will generally have the following fundamental characteristics:

- Amplitude
- Frequency
- Velocity
- Wavelength
- Phase
- Harmonic content
- Envelope

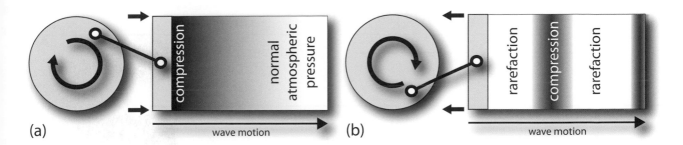

(a)   wave motion   (b)   wave motion

FIGURE 2.2
Effects of a vibrating mass on air molecules and their propagation. (a) Compression – air molecules are forced together to form a compression wave. (b) Rarefaction – as the vibrating mass moves inward, an area of lower atmospheric pressure is created.

These characteristics allow one waveform to be distinguished from another. The most fundamental of these are amplitude and frequency (Figure 2.3). The following sections describe each of these characteristics. It's important to note that although several math formulas have been included, it is by no means important that you memorize or worry about them. It's far more important that you grasp the basic principles of acoustics rather than fret over the underlying math.

AMPLITUDE

The distance above or below the centerline of a waveform (such as a pure sine wave) represents the *amplitude* level of that signal. The greater the distance or displacement from that centerline, the more intense the pressure variation, electrical signal level or physical displacement will be within a medium. Waveform amplitudes can be measured in several ways (Figure 2.4). For example, the measurement of the maximum signal level of a wave, either positive or negative, is called its *peak amplitude value* (or peak level). The combined measurement of the positive and negative peak signal levels is called the *peak-to-peak value*. The *root-mean-square (rms) value* was developed to determine a meaningful average level of a waveform over time (one that more closely approximates the level that's actually perceived by our ears and gives a better real-world measurement of overall signal amplitudes). The rms value of a sine wave can be calculated by squaring the amplitudes at points along the waveform and then taking the

FIGURE 2.3
Amplitude and frequency ranges of human hearing.

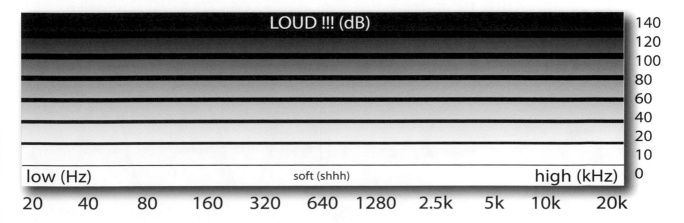

LOUD !!! (dB)   140 120 100 80 60 40 20 10

low (Hz)   soft (shhh)   high (kHz)   0

20   40   80   160   320   640   1280   2.5k   5k   10k   20k

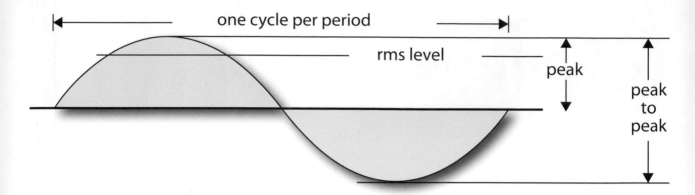

one cycle per period

rms level

peak

peak to peak

FIGURE 2.4

Graph of a sine wave showing the common ways to measure amplitude.

mathematical average of the combined results. The math isn't as important as the basic concept that the rms value of a perfect sine wave is equal to 0.707 times its instantaneous peak amplitude level. Because the square of a positive or negative value is always positive, the rms value will always be positive. The following simple equations show the relationship between a waveform's peak and rms values:

$$\text{rms voltage} = 0.707 \times \text{peak voltage}$$

$$\text{peak voltage} = 1.414 \times \text{rms voltage}$$

FREQUENCY

The rate at which an acoustic generator, electrical signal or vibrating mass repeats within a cycle of positive and negative amplitude is known as the *frequency* of that signal. As the rate of repeated vibration increases within a given time period, the frequency (and thus the perceived pitch) will likewise increase, and vice versa. One completed excursion of a wave (which is plotted over the 360° axis of a circle) is known as a *cycle* (Figure 2.5a). The number of cycles that occur within a second (which determines the frequency) is measured in hertz (Hz). The diagram in Figure 2.5b shows the value of a waveform as starting at zero (0°). At time $t = 0$, this value increases to a positive maximum value and then decreases negatively as it moves back towards its original zero point, where the process begins all over again in a repetitive fashion. A cycle can begin at any

FIGURE 2.5

Waveform motion over time. (a) Cycle divided into the 360° of a circle. (b) Amplitude over time.

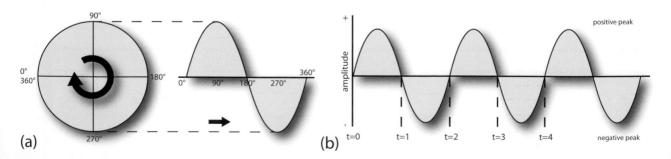

(a)

(b)

angular degree point on the waveform; however, to be complete, it must pass through a single 360° rotation and end at the same point as its starting value. For example, the waveform that starts at $t = 0$ and ends at $t = 2$ constitutes a cycle, as does the waveform that begins at $t = 1$ and ends at $t = 3$.

FREQUENCY RESPONSE

One of the ways that we can "look" at how an audio device might sound is by charting its output level over a frequency range using a visual rating called a *frequency response curve*. This curve is used to graphically represent how a device will respond to the audio spectrum (the 20- to 20,000-Hz range of human hearing) and thus, how it will affect a signal's overall sound. This is done by sending a reference tone that is at equal level at all frequencies to the device under test; any changes in level over frequency at the device's output will give us an accurate measure of any volume changes over frequency that the device will exhibit.

As an example, Figure 2.6 shows the frequency response of several unidentified devices. In these and all cases, the *x*-axis graphically represents frequency, while the *y*-axis represents the device's measured output signal. By feeding the input of an electrical device with a constant-amplitude reference signal that sweeps over the entire frequency spectrum, the results can then be charted on an amplitude versus frequency graph that can be easily read at a glance. If the measured signal

FIGURE 2.6
Frequency-response curves: (a) Curve showing a bass boost. (b) Curve showing a boost at the upper end. (c) Curve showing a dip in the midrange.

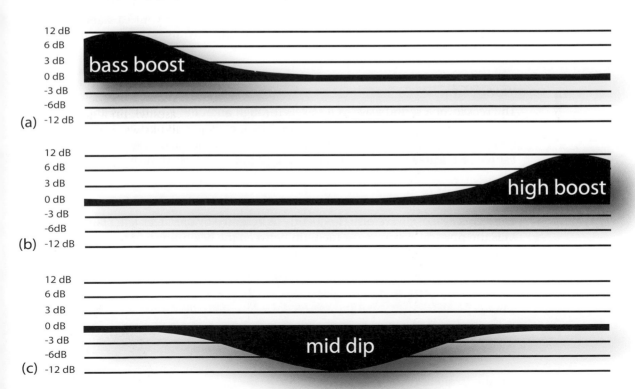

is the same level at all frequencies, the curve will be drawn as a flat, straight line from left to right (known as a *flat frequency response curve*). This indicates that the device passes all frequencies equally (with no frequency being emphasized or de-emphasized). If the output lowers or increases at certain frequencies, these changes will easily show up as dips or peaks in the chart.

Just as an electrical device (such as an amplifier, digital audio circuit, etc.) can be graphed over its input versus output range, an electro-acoustic device (such as a microphone, speaker, phono cartridge, etc.) can also be charted. This can be done in several ways: for example, a high-quality calibration speaker can feed equal-level test tones in a non-reverberant chamber to a microphone; any changes in level at the mic's output can be charted as its response curve. Conversely, a high-quality calibration microphone can be placed in a non-reverberant chamber to test the output of a speaker that is fed with equal-level tones over the spectrum, which will give us a reference versus test device output frequency level that can be charted on the graph.

Before continuing on, it should be pointed out that care should be taken when looking at graphs and all of the other types of specs that can be used to determine the "quality" of a device. It's important to keep in mind that sound is an art form, and that some of the best-sounding and most sought-after devices will often not stand up against modern-day specs (specifications). Indeed, the numbers on the spec sheet will not and cannot tell the whole story. Devices that "look" perfect on paper may sound dreadful, and "dreadful-looking" devices just might sound amazing. The moral of the story is to use your ears and the opinions of others that you trust over (or in addition to) the numbers that are printed on the spec sheet.

VELOCITY

The *velocity* of a sound wave as it travels through air at 68°F (20°C) is approximately 1,130 feet per second (ft/sec) or 344 meters per second (m/sec). This speed is temperature dependent and increases at a rate of 1.1 ft/sec for each Fahrenheit degree increase in temperature (2 ft/sec per Celsius degree).

WAVELENGTH

The *wavelength* of a waveform (frequently represented by the Greek letter lambda, λ) is the physical distance in a medium between the beginning and the end of a cycle. The physical length of a wave can be calculated using:

$$\lambda = V/f$$

where

λ is the wavelength in the medium

V is the velocity in the medium

f is the frequency (in hertz)

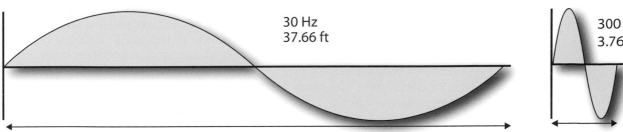

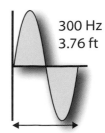

30 Hz
37.66 ft

300 Hz
3.76 ft

The time it takes to complete 1 cycle is called the *period* of the wave. To illustrate, a 30-Hz sound wave completes 30 cycles each second or 1 cycle every 1/30th of a second. The period of the wave is expressed using the symbol T:

$$T = 1/f$$

where T is the number of seconds per cycle.

Assuming that sound propagates at the rate of 1,130 ft/sec, all you need to do is divide this figure by the desired frequency. For example, the simple math for calculating the wavelength of a 30-Hz waveform would be $1,130/30 = 37.6$ ft long, whereas a waveform having a frequency of 300 Hz would be $1,130/300 = 3.7$ ft long (Figure 2.7). Likewise, a 1,000-Hz waveform would work out as being $1,130/1,000 = 1.13$ ft long, and a 10,000-Hz waveform would be $1,130/10,000 = 0.11$ ft long. From these calculations, you can see that whenever the frequency is increased, the wavelength decreases.

FIGURE 2.7
Wavelengths decrease in length as frequency increases (and vice versa).

REFLECTION OF SOUND

Much like a light wave, sound reflects off a surface boundary at an angle that's equal to (and in the opposite direction of) its initial angle of incidence. This basic property is one of the cornerstones of the complex study of acoustics. For example, Figure 2.8a shows how a sound wave reflects off a solid smooth surface in a simple and straightforward manner (at an equal and opposite angle). Figure 2.8b shows how a convex surface will splay the sound outward from its surface, radiating the sound outward in a wide dispersion pattern.

FIGURE 2.8
Incident sound waves striking surfaces with varying shapes: (a) Single-planed, solid, smooth surface. (b) Convex surface. (c) Concave surface. (d) 90° corner reflection.

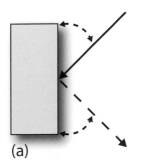

(a)

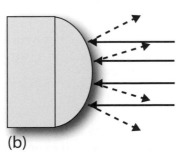

(b)

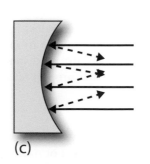

(c)

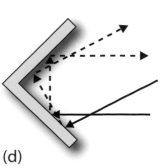

(d)

In Figure 2.8c, a concave surface is used to focus a sound inward toward a single point, while a 90° corner (as shown in Figure 2.8d) reflects patterns back at angles that are equal to their original incident direction. This holds true both for the 90° corners of a wall and for intersections where the wall and floor meet. These corner reflections help to provide insights into how volume levels often build up in the corners of a room (particularly at bass frequencies at wall-to-floor corner intersections).

TRY THIS: REFLECTION OF SOUND

1. Get out your cell phone, handy, mobile or whatever you call it in your country and begin playing some music while holding it in the air. How does it sound? … tinny and not that loud, right?
2. Now, place the cell phone against a flat wall. Did the sound change? Did it get fuller, louder and deeper (especially in the low end)?
3. Now, place it on a flat counter, against the wall. How did that affect the sound? Did the bass just get even deeper?
4. Now, place it on the floor, in the corner of the room. It just got louder, and the bass boosted even further, didn't it?
5. This is a function of the "boundary effect", whereby the reflections coming from the wall and corners combine together to make the overall acoustic signal louder … and often, more pronounced in the low end.

DIFFRACTION OF SOUND

Sound has the inherent ability to diffract around or through a physical acoustic barrier. In other words, sound can bend around an object in a manner that reconstructs the signal back into its original form in both frequency and amplitude. For example, in Figure 2.9a, we can see how a small obstacle will scarcely impede a larger acoustic waveform. Figure 2.9b shows how a larger obstacle can obstruct a larger portion of the waveform; however, past the obstruction, the signal bends around the area in the barrier's wake and begins to reconstruct itself. Figure 2.9c shows how the signal is able to radiate through an opening in a large barrier. Although the signal is greatly impeded (relative to the size of the opening), it nevertheless begins to reconstruct itself in wavelength and relative amplitude and begins to radiate outward as though it were a new point of origin. Finally, Figure 2.9d shows how a large opening in a barrier lets much of the waveform pass through relatively unimpeded.

PHASE

Because we know that a cycle can begin at any point on a waveform, it follows that whenever two or more waveforms are involved in producing a sound, their

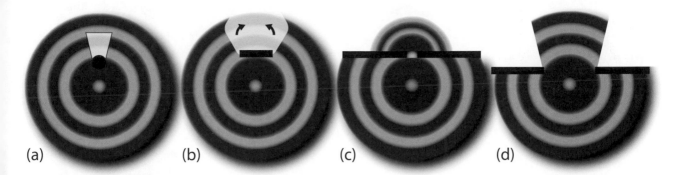

(a)   (b)   (c)   (d)

relative amplitudes can (and almost always will) be different at any one point in time. For simplicity's sake, let's limit our example to two pure tone waveforms (sine waves) that have equal amplitudes and frequency, but start their cyclic periods at different times. Such waveforms are said to be *out of phase* with respect to each other. Variations in *phase*, which are measured in degrees (°), can be described as a time delay between two or more waveforms. These delays are often said to have differences in relative phase degree angles (over the full rotation of a cycle, e.g., 90°, 180°, 270° or any angle between 0° and 360°). The sine wave (so named because its amplitude follows a trigonometric sine function) is usually considered to begin at 0° with an amplitude of zero; the waveform then increases to a positive maximum at 90°, decreases back to a zero amplitude at 180°, increases to a negative maximum value at 270°, and finally, returns back to its original level at 360°, simply to begin all over again.

We'd like you to take a bit of time to re-read the above sentence. It's the basic building block of acoustics and recorded sound. OK, let's carry on ….

Whenever two waveforms having the same frequency, shape and peak amplitude are completely in phase (meaning that they have no relative time difference), the newly combined waveform will have the same frequency, phase and shape, but will be double in amplitude (Figure 2.10a). If the same two waves are combined completely out of phase (having a phase difference of 180°), they will cancel each other out when added, resulting in a final relative value of zero amplitude (Figure 2.10b). If the second wave is only partially out of phase (by a degree other than 180°), the levels will be added at points where the combined amplitudes are positive and will be reduced in level where the combined results subtract from each other (Figure 2.10c).

Whenever two or more waveforms arrive at a single acoustic location or are electrically conducted through a cable out of phase, their relative signal levels will be added together to create a combined amplitude level at that specific point in time.

FIGURE 2.9
The effects of obstacles on sound radiation and diffraction. (a) A small obstacle will scarcely impede a longer wavelength signal. (b) A larger obstacle will obstruct the signal to a greater extent; the waveform will also reconstruct itself in the barrier's wake. (c) A small opening in a barrier will greatly impede a signal; the waveform will emanate from the opening and reconstruct itself as a new source point. (d) A larger opening allows sound to pass unimpeded, allowing it to quickly diffract back into its original shape.

Phase Shift

Phase shift is a term that describes one waveform's lead or lag time with respect to another. Basically, it results from a time delay between two (or more) waveforms

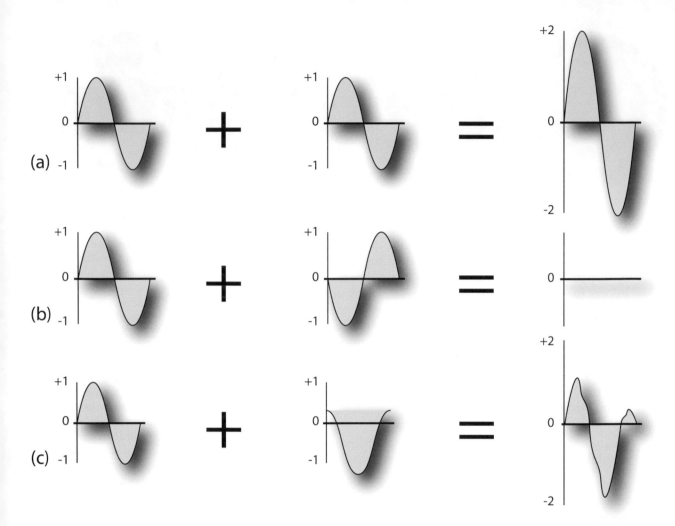

FIGURE 2.10
Combining sine waves of various phase relation-ships. (a) The amplitudes of in-phase waves increase in level when mixed together. (b) Waves of equal amplitude cancel completely (no output) when mixed 180° out of phase. (c) When partial phase angles are mixed, the signals will increase and decrease when acoustically or electrically combined together.

(with differences in acoustic distance being one form of this type of delay). For example, a 500-Hz wave completes one cycle every 0.002 sec. If you start with two in-phase, 500-Hz waves and delay one of them by 0.001 sec (half the wave's period), the delayed wave will lag the other by one-half a cycle, or 180° (resulting in a combined-waveform cancellation). Another example might include a single source that's being picked up by two microphones that have been placed at different distances, thereby creating a corresponding time delay when the signals are mixed together. Such a delay can also occur when a single microphone picks up direct sounds as well as those that are reflected off of a nearby boundary (such as a floor or wall). These signals will be in phase at frequencies where the path-length difference is equal to the signal's wavelength, and out of phase at those frequencies where the multiples fall at or near the half-wavelength distance (or any other possible summing combination). In all the above situations, these boosts and cancellations combine to alter the signal's overall frequency response at the pickup. For this and other reasons, you might want to keep acoustic leakage between microphones and reflections from nearby boundaries to a minimum whenever possible. It's important to remember, however, that there are no hard and fast rules. Some of the best engineers in the world make use of controlled (and sometimes uncontrolled) leakage to add "life" to the sound of a recording.

TRY THIS: PHASE

1. Go to youtube.com/modernrecordingtechniques and search for "phase".
2. Part one of the video details how two 1,000-Hz tones, when combined in phase, will sum together to create a signal that is 3 dB louder in level.
3. Part two shows how two 1,000-Hz tones, when combined 180° out of phase, will sum together to create a signal that completely cancels out in level.

or ...

1. Alternately, you could generate and save your own 1,000-Hz tone.

2. Load the file onto track 1 of the digital audio workstation (DAW) of your choice, making sure to place the file at the beginning of the track, with the signal panned center.
3. Load the same file again into track 2, and listen to both tracks combined. The result should be a summed signal that is 3 dB louder.
4. Flip the phase on track 2, and listen to the results. The combined tracks should cancel, resulting in no output.
5. Offsetting track 3 (relative to track 1) should produce varying degrees of cancellation.
6. Feel free to mix any phase combination to a file and view the resulting waveform. Interesting, huh?

HARMONIC CONTENT

Up to this point, our discussion has centered on the sine wave, which is composed of a single frequency that produces a pure sound at a specific pitch. Fortunately, musical instruments rarely produce pure sine waves. If they did, all of the instruments would sound the same, and music would be pretty boring. The factor that helps us differentiate between instrumental "voicings" is the presence of frequencies (called *partials*) that exist in addition to the fundamental pitch that's being played. Partials that are higher than the fundamental frequency are called *upper partials* or *overtones*. Overtone frequencies that are whole-number multiples of the fundamental frequency are called *harmonics*. For example, the frequency that corresponds to concert A is 440 Hz (Figure 2.11a). An 880-Hz wave is a harmonic of the 440-Hz fundamental because it is twice the frequency (Figure 2.11b). In this case, the 440-Hz fundamental is technically the first harmonic because it is one times the fundamental frequency, and the 880-Hz wave is called the second harmonic because it is two times the fundamental. The third harmonic would be

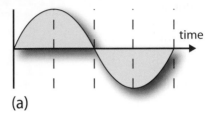
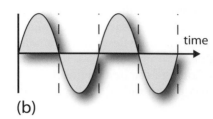
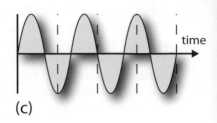

(a)   (b)   (c)

FIGURE 2.11

An illustration of harmonics: (a) 440 Hz – first harmonic "fundamental waveform". (b) 880 Hz – second harmonic. (c) 1,320 Hz – third harmonic.

three times 440 Hz, or 1, 320Hz (Figure 2.11c). Instruments, such as bells, xylophones and other percussion instruments, will often contain overtone partials that aren't harmonically related to the fundamental at all.

The ear perceives frequencies that are whole, doubled multiples of the fundamental as being related in a special way (a phenomenon known as the *musical octave*). As concert A is 440 Hz (A4), the ear hears 880 Hz (A5) as being the next highest frequency that sounds most like concert A. The next related octave above that will be 1,760 Hz (A6). Therefore, 880 Hz is said to be one octave above 440 Hz, and 1,760Hz is said to be two octaves above, etc. Because these frequencies are even multiples of the fundamental, they're known as *even harmonics*. Not surprisingly, frequencies that are odd multiples of the fundamental are called *odd harmonics*. In general, even harmonics are perceived as creating a sound that is pleasing to the ear, while odd harmonics will often create a dissonant, harsher tone.

TRY THIS: HARMONICS

1. Go to youtube.com/modernrecordingtechniques and search for "harmonics".
2. Part one shows the playback of the general tuning first harmonic (A440 Hz).
3. Part two plays back the first (440 Hz) and second harmonic (880 Hz). Do they sound related in nature?
4. Part three plays back the first (440 Hz) and third harmonic (1,320 Hz). Do they sound more dissonant?
5. Part four plays back the first (440 Hz) and fourth harmonic (1,760 Hz). How does that sound?
6. Part five plays back the first (440 Hz) and third harmonic (1,320 Hz) and the fifth harmonic (2,200 Hz). How does that sound?
7. Part six plays back the first (440 Hz) and fourth harmonic (1,760 Hz) and the sixth harmonic (2,640 Hz). How does that sound?

Because musical instruments produce sound waves that contain harmonics with various amplitude and phase relationships, the resulting waveforms bear little resemblance to the shape of a single-frequency sine wave. Therefore, musical waveforms can be divided into two categories: simple and complex. Square,

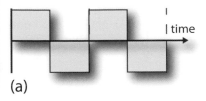 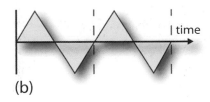 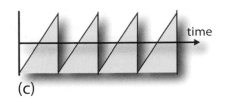

(a) (b) (c)

FIGURE 2.12
Simple waveforms: (a) Square
waves. (b) Triangle waves. (c)
Sawtooth waves.

triangle and sawtooth waves are examples of *simple waves* that contain a consistent harmonic structure (Figure 2.12). They are said to be simple because they're continuous and repetitive in nature. One cycle of a square wave looks exactly like the next, and they are symmetrical about the zero line.

Complex waves, on the other hand, represent practically all other sounds that are produced in music and nature. They almost never repeat and often are not symmetrical about the zero line. An example of a complex waveform (Figure 2.13) is one that's created by any naturally occurring sound (such as music or speech). Although complex waves are rarely repetitive in nature, "all sounds" can be mathematically broken down into a series of ever-changing combinations of individual sine waves (or re-synthesized from sine waves through a complex process known as Fourier analysis).

Regardless of the shape or complexity of the waveform that reaches the eardrum, the inner ear is able to perceive these component signals and then transmit the stimulus to the brain. This can be illustrated by passing a square wave through a bandpass filter that's set to pass only a narrow band of frequencies at any one time. Doing this would show that the square wave is composed of a fundamental frequency plus a number of harmonics that are made up of odd-number multiple frequencies (whose amplitudes decrease as the frequency increases). In Figure 2.14, we see how individual sine-wave harmonics can be combined together to form a square wave.

If we were to analyze the harmonic content of sound waves that are produced by a violin and compare them with the content of the waves that are produced by a viola (with both playing concert A440 Hz), we might come up with results similar to those shown in Figure 2.15. Notice that the violin's harmonics differ in both degree and intensity from those of the viola. These harmonics and their relative intensities are extremely important, as they determine an instrument's characteristic sound (which is called the instrument's *timbre*). If we changed an instrument's harmonic balance, the sonic character of that instrument would also be changed. For example, if the violin's upper harmonics were reduced, the violin would sound more like a viola.

FIGURE 2.13
Example of a complex
waveform.

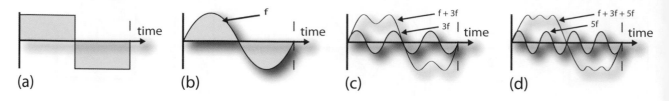

(a)   (b)   (c)   (d)

FIGURE 2.14

Breaking a square wave down into its odd-harmonic components: (a) Square wave with frequency f. (b) Sine wave with frequency f. (c) Sum of a sine wave with frequency f and a lower-amplitude sine wave of frequency 3f. (d) Sum of a sine wave of frequency f and lower-amplitude sine waves of 3f and 5f, which is beginning to resemble a square wave.

Because the relative harmonic balance is so important to an instrument's sound, the frequency response of a microphone, amplifier, speaker and all other elements in the signal path can have an effect on the timbral (tonal) balance of a sound. If the frequency response isn't flat, the timbre of the sound will be changed. For example, if the high frequencies are amplified less than the low and middle frequencies, then the sound will be duller than it should be. For this reason, a choice of mic, mic placement or equalizer can be used as a tool to vary the timbre of an instrument, thereby changing its subjective sound.

In addition to the variations in harmonic balance that can exist between instruments and their families, it is common for this harmonic balance to vary with respect to the direction that a sound wave radiates from an instrument. Figure 2.16 shows the principal radiation patterns as they emanate from a cello (as seen from both side and top views).

ENVELOPE

Timbre isn't the only characteristic that helps us differentiate between instruments. Each one produces a sonic amplitude *envelope* that works in combination with timbre to determine its unique and subjective sound. The envelope of an acoustic or electronically generated waveform can be described as characteristic variations in amplitude level that occur in time over the duration of a played note. This envelope (ADSR) is composed of four sections:

- *Attack* (A) refers to the time taken for a sound to build up to its full volume when a note is initially sounded.
- *Decay* (D) refers to how quickly the sound levels off to a sustain level after the initial attack peak.
- *Sustain* (S) refers to the duration of the ongoing sound that's generated following the initial attack decay.

FIGURE 2.15

Harmonic structure of concert A440: (a) Played on a viola. (b) Played on a violin.

(a)

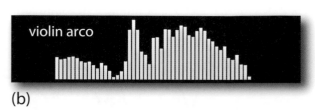

(b)

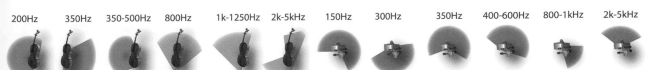

200Hz   350Hz   350-500Hz   800Hz   1k-1250Hz   2k-5kHz   150Hz   300Hz   350Hz   400-600Hz   800-1kHz   2k-5kHz

FIGURE 2.16
Radiation patterns of a cello over frequency as viewed from the side (top) and from overhead (bottom).

- *Release* (R) relates to how quickly the sound will decay once the note is released.

Figure 2.17a illustrates the envelope of a trombone note. The attack, decay times and internal dynamics produce a smooth, sustaining sound. A cymbal crash (Figure 2.17b) combines a high-level, fast attack with a longer sustain and decay that creates a smooth, lingering shimmer. Figure 2.17c illustrates the envelope of a snare drum. Notice that the initial attack is much louder than the internal dynamics, while the final decay trails off very quickly, resulting in a sharp, percussive sound.

It's important to note that the concept of an envelope often relies on peak waveform values, while the human perception of loudness is proportional to the average wave intensity over a period of time (rms value). Therefore, high-amplitude portions of the envelope won't make an instrument sound loud unless the amplitude is maintained for a sustained period. Short high-amplitude sections (transients) tend to contribute to a sound's overall character rather than to its loudness. By using a compressor or limiter, an instrument's character can often be modified by changing the dynamics of its envelope without changing its overall timbre.

LOUDNESS LEVELS: THE DECIBEL

The human ear operates over an energy range of approximately $10^{13}:1$ (10,000,000,000,000:1), which is an extremely wide range. Since it's difficult for us to conceptualize number ranges that are this large, a logarithmic scale has been adopted to compress the measurements into figures that are more manageable. The unit used for measuring sound-pressure level (SPL), signal level and relative changes in signal level is the *decibel* (*dB*), a term that literally means 1/10th of a Bell (an older telegraph and telephone transmission loss measurement unit that was named after Alexander Graham Bell, inventor of the

FIGURE 2.17
Various musical waveform envelopes: (a) Trombone. (b) Cymbal crash. (c) Snare drum, where A = attack, D = decay, S = sustain and R = release.

A D S R

(a)

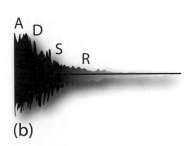

A D S R

(b)

A D S R

(c)

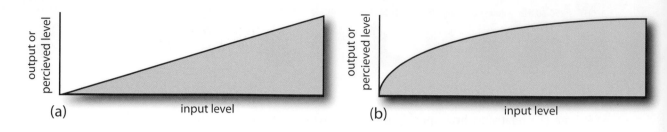

FIGURE 2.18
Linear and logarithmic curves: (a) Linear. (b) Logarithmic.

telephone). In order to develop an understanding of the decibel, we first need to examine logarithms and the logarithmic scale (Figure 2.18). The *logarithm* (*log*) is a mathematical function that reduces large numeric values into smaller, more manageable numbers. Because logarithmic numbers increase exponentially in a way that's similar to how we perceive the doubling of loudness levels (e.g., 2, 4, 8, 16, 32, 64, 128, 256 …), it expresses our perceived sense of volume more precisely than a linear curve can.

Before we delve into a deeper study of this important concept and how it deals with our perceptual senses, let's take a moment to understand the basic concepts and building block ideas behind the log scale, so as to get a better understanding of what examples such as "+3 dB at 10,000 Hz" really mean. I know that this can be a daunting subject to grasp, but be patient with yourself. Over time, the concept of the decibel will become as much a part of your working vocabulary as ounces, gallons and miles per hour.

Logarithmic Basics

In audio, we use logarithmic values to express the differences in intensities between two levels (often, but not always, comparing a newly measured level with a standard reference level). Because the standard numeric differences between these two levels can be really, really big, a simpler system makes use of representative values that are mathematical exponents of 10. To begin, finding the log of a number such as 17,386 without a calculator is not only difficult, it's unnecessary! All that's really important to help you along are three simple guidelines:

- The log of the number 2 is 0.3.
- When a number is an integral power of 10 (e.g., 100, 1000, 10,000), the log can be found simply by adding up the zeros in that number.
- Numbers that are greater than 1 will have a positive log value, while those less than 1 will have a negative log value.

Again, the first one is an easy fact to remember: the log of 2 is 0.3 – this will make sense shortly. The second one is even easier: the logs of numbers such as 100, 1,000 or 10,000,000,000,000,000 can be arrived at by simply counting up the zeros. The last guideline relates to the fact that if the measured value is lower than the reference value, the resulting log value will be negative. For example:

$$\log 2 = 0.3$$
$$\log 1/2 = \log 0.5 = -0.3$$
$$\log 10,000,000,000,000 = 13$$
$$\log 1,000 = 3$$
$$\log 100 = 2$$
$$\log 10 = 1$$
$$\log 0.1 = -1$$
$$\log 0.01 = -2$$
$$\log 0.001 = -3$$

All other numbers can be arrived at by using a scientific calculator (most computers and cell phones have one built in); however, it's unlikely that you'll ever need to know any log values beyond understanding the basic concepts that are listed here.

The Decibel

Now that we've gotten past the absolute bare basics, I'd like to break with tradition again and attempt an explanation of the decibel in a way that's less complex and relates more to our day-to-day needs in the sound biz. First off, the decibel is a logarithmic value that "expresses differences in intensities between two levels". From this, we can infer that these levels are expressed by several units of measure, the most common being SPL, voltage (V) and power (wattage, or W). Now, let's look at the basic math behind these three measurements.

Sound-Pressure Level

Sound-pressure level is the acoustic pressure that's built up within a defined atmospheric area (usually a square centimeter, or cm^2). Quite simply, the higher the SPL, the louder the perceived sound (Figure 2.19). In this instance, our measured reference (SPL_{ref}) is the threshold of hearing, which is defined as being the softest sound that an average person can hear. Most conversations will have an SPL of about 70 dB, while average home stereos are played at volumes ranging between 80 and 90 dB^{SPL}. Sounds that are so loud as to be painful have SPLs of

FIGURE 2.19
Chart of sound-pressure levels, typical A-Weighted Sound Levels dB re: 20μN/m² (courtesy of General Radio Company).

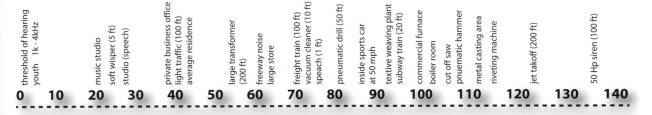

1/2 d (6dB higher in level)

d

2d (6dB lower in level)

FIGURE 2.20

Doubling the distance of a pickup will lower the perceived signal level by 6 dB^{SPL}, while halving the distance will increase it by 6 dB^{SPL}.

about 130 to 140 dB (10,000,000,000,000 or more times louder than the 0 dB reference). We can arrive at an SPL rating by using the formula:

$$dB^{SPL} = 20\log SPL/SPL_{ref}$$

where SPL is the measured sound pressure (in dyne/cm^2)

SPL_{ref} is a reference sound pressure (the threshold limit of human hearing, 0.02 millipascals = 2 ten-billionths of our atmospheric pressure)

From this, we feel that the major concept that needs to be understood is the idea that SPL figures change with the square of the distance (hence, the 20 log part of the equation). This means that whenever a source/pickup distance is doubled, the SPL level will be reduced by 6 dB (20 log 1/2 = 20 × –0.3 = –6 dB^{SPL}); as the distance is halved, it will increase by 6 dB (20 log 2/1 = 20 × 0.3 = 6 dB^{SPL}), as shown in Figure 2.20.

Voltage

Voltage can be thought of as the pressure behind electrons within a wire. As with acoustic energy, comparing one voltage level with another level (or reference level) can be expressed as dBv using the equation:

$$dBv = 20\log V/V_{ref}$$

where V is the measured voltage, and V_{ref} is a reference voltage (0.775 volts).

Power

Power is usually a measure of wattage or current and can be thought of as the flow of electrons through a wire over time. Power is generally associated with audio signals that are carried throughout an audio production system. Unlike SPL and voltage, the equation for signal level (which is often expressed in dBm) is:

$$dBm = 10\log P/P_{ref}$$

where P is the measured wattage, and P_{ref} is referenced to 1 milliwatt (0.001 watt).

The "Simple" Heart of the Matter

We're going to go out on a limb and state that when dealing with decibels, it's far more common for working professionals to deal with the concept of power. The dBm equation better expresses the spirit of the decibel term when dealing with the markings and measurements on an audio device or the numeric values in a computer dialog box. This is due to the fact that power is the unit of measure that's most often expressed when dealing with a setting on a fader or a dialog box; therefore, it's our personal opinion that the average working stiff only needs to grasp the following basic concepts, as shown in Figure 2.21:

- A 1-dB change is noticeable to most ears (but not by much).
- Turning something up by 3 dB will double the signal's level, but it will only be perceived as being 1¼ times as loud (definitely noticeable, but not as much of an increase in gain as you might think).
- Turning something down by 3 dB will halve the signal's level (likewise, halving the signal level won't decrease the perceived loudness as much as you might think).
- Turning something up by 10 dB will increase the signal's level tenfold but will be perceived as being only twice as loud.
- Since the log of an exponent of 10 can be easily figured by simply counting the zeros (e.g., the log of 1,000 is 3), increasing the signal's level tenfold will turn something up by 10 dB, 100-fold will yield a 20-dB increase, and 1,000 fold will yield a 30-dB increase, etc.

Everyone pretty much knows that it's unlikely that anyone will ever ask, "Would you please turn that up a hundred times?" It just won't happen! However, when a pro asks his or her assistant to turn the gain up by 20 dB, that assistant will often instinctively know what 20 dB is – and what it should sound like. There's a lot to know about the intricacies of the decibel, especially with regard to the differences as to how dBm, dBv and dB^{SPL} are calculated; however, we're saying

FIGURE 2.21
Various gain and perceived loudness changes on a fader.

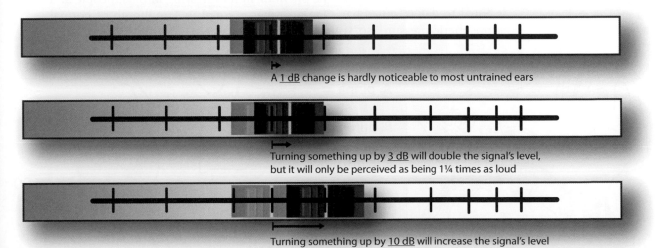

A 1 dB change is hardly noticeable to most untrained ears

Turning something up by 3 dB will double the signal's level, but it will only be perceived as being 1¼ times as loud

Turning something up by 10 dB will increase the signal's level 10-fold, but will be perceived as being only twice as loud

that the math really isn't nearly as important as the ongoing process of grasping an instinctive "feel" for the decibel and how it relates to relative levels within audio production.

THE EAR

A sound source produces acoustic waves by alternately compressing and rarefying the air molecules between it and the listener, causing fluctuations that fall above and below normal atmospheric pressure. The human ear (Figure 2.22) is an extremely sensitive transducer that responds to these pressure variations by way of a series of related processes that occur within the auditory organs – our ears. When these variations arrive at the listener, sound-pressure waves are collected in the aural canal by way of the outer ear's pinna. These are then directed to the eardrum, a stretched drum-like membrane, where the sound waves are changed into mechanical vibrations, which are transferred to the inner ear by way of three bones known as the hammer, anvil and stirrup. These bones act both as an amplifier (by significantly increasing the vibrations that are transmitted from the eardrum) and as a limiting protection device (by reducing the level of loud, transient sounds, such as thunder or fireworks explosions). The vibrations are then applied to the inner ear (cochlea – a tubular, snail-like organ that contains two fluid-filled chambers). Within these chambers are tiny hair receptors that are lined up in a row along the length of the cochlea. These hairs respond to certain frequencies depending on their placement along the organ, which results in the neural stimulation that gives us the sensation of hearing. Permanent hearing loss generally occurs when these hair/nerve combinations are damaged or as they deteriorate with age.

FIGURE 2.22
The ear. (a) Outer, middle and inner ear. (b) Showing the various frequency ranges that affect the hairs in the cochlea.

Threshold of Hearing

A convenient SPL reference is the *threshold of hearing*, which is the minimum sound pressure that produces the phenomenon of hearing in most people and is equal to 0.0002 microbar. One microbar is equal to 1 millionth of normal atmospheric pressure, so you can see that the ear is an amazingly sensitive

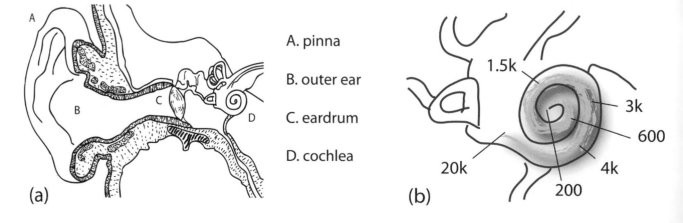

A. pinna

B. outer ear

C. eardrum

D. cochlea

(a) (b)

instrument. In fact, if the ear were any more sensitive, the thermal motion of molecules in the air would be audible! When referencing SPLs to 0.0002 micro-bar, this threshold level usually is denoted as 0 dB^{SPL}, which is defined as the level that an average person can hear at a specific frequency only 50% of the time.

Threshold of Feeling

An SPL that causes discomfort in a listener 50% of the time is called the *threshold of feeling*. It occurs at a level of about 118 dB^{SPL} between the frequencies of 200 Hz and 10 kHz.

Threshold of Pain

The SPL that causes pain in a listener 50% of the time is called the *threshold of pain* and corresponds to an SPL of 140 dB in the frequency range between 200 Hz and 10 kHz.

TAKING CARE OF YOUR HEARING

During the 1970s and early 1980s, recording studio monitoring levels were often turned up so high as to be truly painful. In the mid-1990s, a small band of powerful producers and record executives banded together to get the industry to successfully reduce these average volumes down to tolerable levels (85 to 95 dB) – a practice that generally continues to this day. Live sound venues and acts often continue the practice of raising house and stage volumes to chest-thumping levels (and beyond). Although these levels are exciting, long-term exposure can lead to temporary or even permanent hearing loss. Here are the different types of hearing loss:

- *Acoustic trauma*: This happens when the ear is exposed to a sudden, loud noise in excess of 140 dB. Such a shock could lead to permanent hearing loss.
- *Temporary threshold shift*: The ear can experience temporary hearing loss when exposed to long-term, loud noise.
- *Permanent threshold shift*: Extended exposure to loud noises in a specific or broad hearing range can lead to permanent hearing loss in those frequencies. In short, the ear becomes less sensitive to sounds in the damaged frequency range, leading to a reduction in perceived volume and intelligibility … What?

A simple fact to remember: once your hearing is gone (or permanently impaired), the damage is done – for good!

Here are a few hearing conservation tips (courtesy of the House Ear Institute, www.hei.org) that can help reduce hearing loss due to long-term exposure to sounds over 115 dB:

- Avoid hazardous sound environments; if they're not avoidable, wear hearing protection devices, such as foam earplugs, custom-molded earplugs or in-ear monitors.
- Monitor sound-pressure levels at or around 85 dB. The general rule to follow is that if you're in an environment where you must raise your voice to be heard, then you're monitoring too loudly and should limit your exposure times.
- Take 15-minute "quiet breaks" every few hours (especially if you're being exposed to levels above 85 dB).
- Musicians and other live entertainment professionals should avoid practicing at concert-hall levels whenever possible.
- Have your hearing checked periodically by a licensed audiologist.

PSYCHOACOUSTICS

The area of *psychoacoustics* deals with how and why the brain interprets a particular sound stimulus in a certain way. Although a great deal of study has been devoted to this subject, the primary device in psychoacoustics is the all-elusive brain, which is still largely unknown to present-day science.

Auditory Perception

From the outset, it's important to realize that the ear is a nonlinear device (meaning that what's received in your ears isn't always what you'll hear). It's also important to note that the ear's frequency response (its perception of timbre) changes with the loudness of the perceived signal. The "loudness" compensation switch found on many hi-fi preamplifiers is an attempt to compensate for this decrease in the ear's sensitivity to low- and high-frequency sounds at low listening levels.

The *Fletcher–Munson equal-loudness contour curves* (Figure 2.23) indicate the ear's average sensitivity to different frequencies at various levels. These indicate the sound-pressure levels that are required for our ears to hear frequencies along the curve as being equal in level to a 1,-000Hz reference level (measured in phons). Thus, to equal the loudness of a 1-kHz tone at 110 dB^{SPL} (a level typically created by a trumpet-type car horn at a distance of 1 m), a 40-Hz tone has to be about 6 dB louder, whereas a 10-kHz tone must be 4 dB louder in order to be perceived as being equally loud. At 50 dB^{SPL} (the noise level present in the average private business office), the level of a 40-Hz tone must be 30 dB louder and a 10-kHz tone 13 dB louder than a 1-kHz tone to be perceived as having the same volume. Thus, if a piece of music is mixed to sound great at a level of 85 to 95 dB, its bass and treble balance will actually be boosted when turned

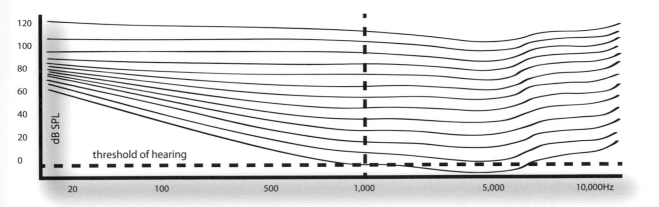

FIGURE 2.23
The Fletcher–Munson curve shows an equal loudness contour for pure tones as perceived by humans having an average hearing acuity. These "perceived" loudness levels (measured in phons) are charted relative to sound-pressure levels at 1,000 Hz.

up (often a good thing). If the same piece were mixed at 110 dB^{SPL}, it would sound great at that level but would be both bass and treble shy when played back at lower levels because no compensation for the ear's response was added to the mix. Over the years, it has generally been found that changes in apparent frequency balance are less apparent when monitoring at levels of 85 dB^{SPL}. In fact, many of the world's best Grammy award–winning mixers consistently mix their projects at levels that are even lower than that – the idea being that most systems (TVs, laptops, etc.) will be playing the music at these lower levels. If the mixes get turned up, the bass and high end will be exaggerated and will sound just that much better!

In addition to this, whenever it is subjected to sound waves that are above a certain loudness level, the ear can produce harmonic distortion that doesn't exist in the original signal. For example, the ear can cause a loud 1-kHz sine wave to be perceived as being a combination of 1-, 2-, 3-kHz waves, and so on. Although the ear might hear the overtone structure of a violin (if the listening level is loud enough), it might also perceive additional harmonics (thus changing the timbre of the instrument). This is one of several factors implying that sound monitored at very loud levels could sound quite different when played back at lower levels.

The loudness of a tone can also affect our ear's perception of pitch. For example, if the intensity of a 100-Hz tone is increased from 40 to 100 dB^{SPL}, the ear will hear a pitch decrease of about 10%. At 500 Hz, the pitch will change about 2% for the same increase in sound-pressure level. This is one reason why musicians find it difficult to tune their instruments when listening through loud headphones.

As a result of these nonlinearities in the ear's response, tones will often interact with each other rather than being perceived as being separate. Three types of interaction effects can occur:

- Beats
- Combination tones
- Masking

Beats

Two tones that differ only slightly in frequency and have approximately the same amplitude will produce an effect known as *beats*. This effect sounds like repetitive volume surges that are equal in frequency to the difference between these two tones. The phenomenon is often used as an aid for tuning instruments, because the beats slow down as the two notes approach the same pitch and finally stop when the pitches match. In reality, beats are a result of the ear's inability to separate closely pitched notes. This results in a third frequency that's created from the phase sum and difference values between the two notes.

TRY THIS: BEATS

1. Go to youtube.com/modernrecordingtechniques and search for "beats".
2. Part one is a playback of a 440-Hz tone.
3. Part two is a playback of a 440-Hz and 445-Hz tone. Can you hear the 5-Hz beat tone? (445 Hz – 440 Hz = 5 Hz)
4. Part three is a playback of a 445-Hz and 450-Hz tone. Can you hear the 5-Hz beat tone? (450 Hz – 445 Hz = 5 Hz)
5. Part four is a playback of a 440-Hz and 450-Hz tone. Can you hear the 10-Hz beat tone? (450 Hz – 440 Hz = 10 Hz)

Combination Tones

Combination tones result when two loud tones differ by more than 50 Hz. In this case, the ear perceives an additional set of tones that are equal to both the sum and the difference between the two original tones as well as being equal to the sum and difference between their harmonics. The simple formulas for computing the fundamental tones are:

$$\text{Sum tone} = f1 + f2$$

$$\text{Difference tone} = f1 - f2$$

Difference tones can be easily heard when they are below the frequency of both tones' fundamentals. For example, the combination of 2,000 Hz and 2,500 Hz produces a difference tone of 500 Hz.

Masking

Masking is the phenomenon by which loud signals prevent the ear from hearing softer sounds. The greatest masking effect occurs when the frequency of the sound and the frequency of the masking noise are close to each other. For example, a 4-kHz tone will mask a softer 3.5-kHz tone but has little effect on the audibility of a quiet 1,000-Hz tone. Masking can also be caused by harmonics of the masking tone (e.g., a 1-kHz tone with a strong 2-kHz harmonic might mask a 1,900-Hz tone). This phenomenon is one of the main reasons why stereo placement and equalization are so important to the mixdown process. An instrument that sounds fine by itself can be completely hidden or changed in character by louder instruments that have a similar timbre. Volume, equalization, mic choice or mic placement might have to be altered to make the instruments sound different enough to overcome any masking effect.

TRY THIS: MASKING

1. Go to youtube.com/modernrecordingtechniques and search for "masking".
2. Part one is a playback of a 1-kHz tone.
3. Part two is a playback of a 1-kHz and 4-kHz tone. Can you hear both of the tones clearly?

4. Part three is a playback of a 3,800-Hz tone.
5. Part four is a playback of a 3,800-Hz and 4-kHz tone. Can you hear both of the tones clearly?

Perception of Direction

Although one ear can't discern the direction of a sound's origin, two ears can. This capability of two ears to localize a sound source within an acoustic space is called *spatial* or *binaural localization*. This effect is the result of three acoustic cues that are received by the ears:

- Interaural intensity differences
- Interaural arrival-time differences
- The effects of the pinnae (outer ears)

Middle- to higher-frequency sounds originating from the right side (for example) will reach the right ear at a higher intensity level than the left ear, causing an *interaural intensity difference*. This volume difference occurs because the head casts an acoustic block or shadow, allowing only reflected sounds from surrounding surfaces to reach the opposite ear (Figure 2.24a). Because the reflected sound travels farther and loses energy at each reflection, the sound perceived by the left ear will naturally be greatly reduced, resulting in a signal that will be correctly perceived as coming from the right.

This acoustic blockage is relatively insignificant at lower frequencies, however, where wavelengths are large compared with the head's size, allowing the wave to easily bend around its acoustic shadow. For this reason, a different method

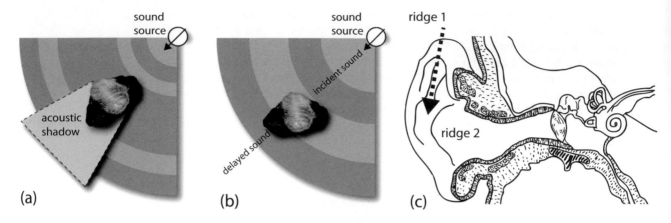

(a)

(b)

(c)

FIGURE 2.24

Localization of direction: (a) The head casts an acoustic shadow that helps with localization at middle to upper frequencies. (b) Interaural arrival-time differences allow us to discriminate direction at lower frequencies. (c) The pinna and its reflective ridges help in determining vertical location information.

of localization (known as *interaural arrival-time difference*s) is employed at lower frequencies (Figure 2.24b). In both Figure 2.24a and b, small time differences occur because the acoustic path length to the left ear is slightly longer than the path to the right ear. The sound pressure therefore arrives at the left ear at a later time than the right. This method of localization (in combination with interaural intensity differences) helps to give us lateral localization cues over the entire frequency spectrum.

Intensity and delay cues allow us to perceive the direction of a sound's origin, but not whether the sound originates from the front, behind or below. The pinna (Figure 2.24c), however, makes use of two ridges that reflect sound into the ear. These ridges introduce minute time delays between the direct sound (which reaches the entrance of the ear canal) and the sound that's reflected from the ridges (which varies according to source location). It's interesting to note that beyond 130° from the front of our face, ridge #1 is able to reflect and delay sounds by 0 and 80 microseconds (μsec), making rear localization possible. Ridge #2 has been reported to produce delays of between 100 and 330 μsec that help us to locate sources in the vertical plane. The delayed reflections from both ridges are then combined with the direct sound to produce frequency-response colorations that are compared within the brain to determine source location. Small movements of the head can also help provide additional position information.

If there are no differences between what the left and right ears hear, the brain assumes that the source is the same distance from each ear. This phenomenon allows us to position sound not only in the left and right loudspeakers but also monophonically between them. If the same signal is fed to both loudspeakers, the brain perceives the sound identically in both ears and deduces that the source must be originating from directly in the center. By changing the proportion that's sent to each speaker, the engineer changes the relative interaural intensity differences and thus creates the illusion of physical positioning between the speakers. This placement technique is known as *panning* (Figure 2.25).

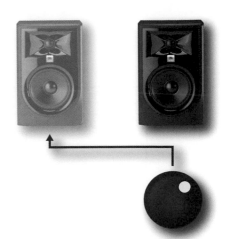

FIGURE 2.25
Pan pot settings and their
relative spatial positions.

Perception of Space

In addition to perceiving the direction of sound, the ear and brain combine together to help us perceive the size and physical characteristics of the acoustic space in which a sound occurs. When a sound is generated, a percentage reaches the listener directly without encountering any obstacles. A larger portion, however, travels out to the many surfaces that exist within an acoustic room or enclosure. If these surfaces are reflective, the sound is bounced back into the room and toward the listener. If the surfaces are absorptive, less energy will be reflected back to the listener. Three types of reflections are commonly generated within an enclosed space (Figure 2.26):

- Direct sound
- Early reflections
- Reverberation

DIRECT SOUND

In air, sound travels at a constant speed of about 1,130 feet per second, so a wave that travels from the source to the listener will follow the shortest path and arrive at the listener's ear first. This is called the *direct sound*. Direct sounds determine our initial perception of a sound source's location and size and convey the true timbre of the source.

EARLY REFLECTIONS

After the initial direct sound, the ear will begin to perceive sound reflections that bounce off surrounding, large-boundary surfaces within the room. These waves (which are called *early reflections*) must travel further than the direct sound to reach the listener and therefore arrive after the direct sound and from a multitude of directions. Early reflections give us clues as to the reflectivity, size and

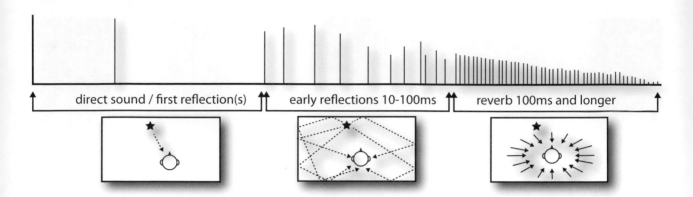

direct sound / first reflection(s)   early reflections 10-100ms   reverb 100ms and longer

FIGURE 2.26
The three distinct sound field types that are generated within an enclosed space.

general nature of an acoustic space. These sounds generally arrive at the ears less than 50 m/sec after the brain perceives the direct sound and are the result of reflections off of the largest, most prominent boundaries within a room. The time elapsed between hearing the direct sound and the beginning of the early reflections helps to provide information about the size of the performance room. Basically, the farther the boundaries are from the source and the listener, the longer the delay before it's reflected back to the listener, thus being perceived as a larger room.

Another aspect that occurs with early reflections is called *temporal fusion*. These early reflections arriving at the listener within 30 m/sec of the direct sound are not only audibly suppressed but are also fused with the direct sound. In effect, the ear can't distinguish these closely occurring reflections and considers them to be part of the direct sound. The 30-m/sec time limit for temporal fusion isn't absolute; rather, it depends on the sound's envelope. Fusion breaks down at 4 m/sec for transient clicks, whereas it can extend beyond 80 m/sec for slowly evolving sounds (such as a sustained organ note or legato violin passage). Although these early reflections are suppressed and fused with the direct sound, they still modify our perception of the sound, making it both louder and fuller.

Reverberation

Whenever room reflections continue to bounce off room boundaries, a randomly decaying set of sounds can often be heard after the source stops in the form of *reverberation* (Figure 2.27). A highly reflective surface absorbs less of the wave energy at each reflection and allows the sound to persist longer after the initial sound stops. Sounds reaching the listener 50 m/sec later in time are perceived as a random and continuous stream of reflections that arrive from all directions. These densely spaced reflections gradually decrease in amplitude and add a sense of warmth and body to a sound. Because it has undergone multiple reflections, the timbre of the reverberation is often quite different from the direct sound (with the most notable difference being a roll-off of high frequencies and a slight bass emphasis).

The time it takes for a reverberant sound to decrease to 60 dB below its original level is called its *decay time* or *reverb time* and is determined by the room's absorption characteristics. The brain is able to perceive the reverb time and timbre of the reverberation and uses this information to form an opinion on the hardness or softness of the surrounding surfaces. The loudness of the perceived direct sound increases rapidly as the listener moves closer to the source, while the reverberation levels will often remain the same, because the diffusion is roughly constant throughout the room. This ratio of the direct sound's loudness to the reflected sound's level helps listeners judge their distance from the sound source.

Whenever artificial reverb and delay units are used, the engineer can generate the necessary delay cues to roughly convince the brain that a sound was recorded in a huge, stone-walled cathedral when in fact, it was recorded in a small, absorptive room. To do this, the engineer programs the device to mix the original un-reverberated signal with the necessary early delays and random reflections. Adjusting the number and amount of delays on an effects processor gives the engineer control over all of the necessary parameters to determine the perceived room size, while decay time and frequency balance can help to determine the room's perceived surfaces. By changing the proportional mix of direct to processed sound, the engineer/producer can place the sound source at either the front or the rear of the artificially created soundscape.

FIGURE 2.27
Recording Hall 2 (the "smaller" hall) at Funkhaus, Berlin has a reverb time that can reach up to 4 seconds (courtesy of Funkhaus NelapeStrasse, Berlin, www.funkhaus-berlin.net).

CHAPTER 3

Studio Acoustics and Design

The *Audio Cyclopedia* defines *acoustics* as "a science dealing with the production, effects and transmission of sound waves; the transmission of sound waves through various mediums, including reflection, refraction, diffraction, absorption and interference; the characteristics of auditoriums, theaters and studios, as well as their design". We can see from this description that the proper acoustic design of music recording, project and audio-for-visual or broadcast studios is often no simple matter. A wide range of complex variables and interrelationships often come into play in the creation of a successful acoustic and monitoring design. When designing or redesigning an acoustic space, the following basic requirements should be considered:

- *Acoustic* isolation: this prevents external noises from transmitting into the studio environment through the air, ground or building structure. It can also prevent feuds that can arise when excessive volume levels leak out into the surrounding neighborhood.
- *Frequency balance*: the frequency components of a room shouldn't adversely affect the acoustic balance of instruments and/or speakers. Simply stated, the acoustic environment shouldn't alter the sound quality of the original or recorded performance.
- *Acoustic separation*: the acoustic environment should not interfere with intelligibility and should offer the highest possible degree of acoustic separation within the room (often a requirement for ensuring that sounds from one instrument aren't unduly picked up by another instrument's microphone).
- *Reverberation*: the control of sonic reflections within a space is an important factor for maximizing the intelligibility of music and speech. No matter how short the early reflections and reverb times are, they will add an important psychoacoustic sense of "space" in the sense that they can give our brain subconscious cues as to a room's size, number of reflective boundaries, distance between the source and listener, and so forth.
- *Cost factors*: not the least of all design and construction factors is cost. Multi-million-dollar facilities often employ studio designers and construction

DOI: 10.4324/9781003260530-3

teams to create a plush decor that's been acoustically tuned to fit the needs of both the owners and their clients. Owners of project studios and budget-minded production facilities, however, can all take full advantage of the same basic acoustic principles and construction techniques and apply them in cost-effective ways.

This chapter will discuss many of the basic acoustic principles and construction techniques that should be considered in the design of a music or sound production facility. I'd like to emphasize that any or all of these acoustical topics can be applied to any type of audio production facility and aren't only limited to professional music studio designs. For example, owners of modest project and bedroom studios should know the importance of designing a control room that's symmetrical and hopefully looks, feels and sounds good. It doesn't cost anything to know that if one speaker is in a corner and the other is on a wall, the perceived center image and frequency balance will be screwy. As with many techno-artistic endeavors, studio acoustics and design are a mixture of fundamental physics (in this case, mostly basic, dimensional mathematics) with an equally large dose of common sense and dumb luck. More often than not, acoustics is an artistic science that melds physics with the art of intuition and experience.

STUDIO TYPES

Although the acoustical fundamentals are the same for most studio design types, differences will often follow the form, function and budgets required by the tasks at hand. Some of the more common studio types include:

- Professional music studios
- Audio-for-visual production environments
- Audio-for-gaming production environments
- Project studios

The Professional Recording Studio

The *professional recording studio* (Figure 3.1) is first and foremost a commercial business, so its design, decor and acoustical construction requirements are often much more demanding than those of a privately owned project studio. In some cases, an acoustical designer and experienced construction team are placed in charge of the overall building phase of a professional facility. In others, the studio's budget is just too tight to hire such professionals, which places the studio owners and staff squarely in charge of designing and constructing the entire facility themselves. Whether you happen to have the luxury of building a new facility from the ground up or are renovating a studio within an existing shell, you could easily benefit from a professional studio designer's experience and skills. Such expert advice sometimes proves to be cost effective in the long run, because errors in design judgment can lead to cost overruns, lost business due to unexpected delays, or the unfortunate state of living with mistakes that could have been avoided.

(a)

(b)

FIGURE 3.1
Professional studio examples. (a) BiCoastal Music, Ossining, NY (courtesy of Russ Berger Design Group, www.rbdg.com). (b) London Bridge Studios big tracking room, Seattle (courtesy of London Bridge Studios, www.londonbridgestudio.com, Photo photo by Christopher Nelson).

The Audio-for-Visual and Media Production Environment

An audio-for-visual production facility is used for video, film, gaming and media post-production (often simply called "post") and includes such facets as music recording for film or other media (scoring), score mixdown, automatic dialog replacement (ADR – the replacement of on- and off-screen dialog in visual media) and Foley (the replacement and creation of on- and off-screen sound effects). As with the music studio, audio-for-visual production facilities can range from being high-end facilities that can accommodate the posting needs of network video or feature film productions (Figure 3.2) to a simple, budget-minded project studio that's equipped with video and a digital audio workstation. As with the music studio, audio-for-visual construction and design techniques often span a wide range of styles and scope in order to fit the budget needs at hand.

The Audio-for-Gaming Production Environment

With the ever-increasing popularity of having the gaming experience in the home, budgets and the need for improved audio in newer game releases, production facilities have sprung up that deal exclusively with the recording and post-production aspects of game audio. These can range from high-end facilities that resemble the high-end music and scoring studio to production houses that deal with the day-to-day creation and programming of the hundreds of thousands of audio clips that go into making a modern game. The production needs

FIGURE 3.2
PostWorks audio production and post facility, New York (courtesy of Avid Technology, Inc., www.avid.com).

of audio-for-gaming are usually different from those of music production; the required skills and need for attention to technical detail demand a high level of skill and dedication over a period of multiple months.

The Project Studio

It goes without saying that the vast majority of audio production studios fall into the project studio category. This basic definition of such a facility is open to interpretation. It's usually intended as a personal production resource for recording music, audio-for-visual production, multimedia production, voiceovers … you name it. Project studios can range from being fully commercial in nature to smaller setups that are both personal and private (Figure 3.3). All of these possible studio types have been designed with the idea of giving artists the flexibility of making their art in a personal, off-the-clock environment that's both cost and time effective. Of course, the design and construction considerations for creating a privately owned project studio will often differ from the design considerations for a professional music facility in two fundamental ways:

FIGURE 3.3
Gettin' it all going in the project studio. (a) Courtesy courtesy of Yamaha Corporation of America, www.yamaha.com (b) Courtesy courtesy of Universal Audio, www.uaudio.com © 2022 Universal Audio, Inc. All rights reserved. Used with permission.

- Building constraints
- Cost

(a)

(b)

FIGURE 3.4
DAWs and controllers at La-Rocc-A-Fella Center, North Hollywood (courtesy of Avid Technology, Inc., www.avid.com).

Generally, a project studio's room (or series of rooms) is built into an artist's home or a rented space where the construction and dimensional details are already defined. This fact (combined with inherent cost considerations) often leads the owner/artist to employ cost-effective techniques for sonically treating any deficiencies that occur within the room. Even if the room has little or no treatment, keep in mind that a basic knowledge of acoustical physics and room design can be a valuable and cost-effective tool as your experience, production needs and business requirements grow.

Modern-day digital audio workstations (DAWs) have squarely placed the Mac and PC in the middle of almost every pro and home project studio (Figure 3.4). In fact, in many cases, the DAW "IS" the project studio. With the advent of self-powered speaker monitors, cost-effective microphones and hardware DAW controllers, it's become a relatively simple matter to design a powerful production system into almost any existing space.

PRIMARY FACTORS GOVERNING STUDIO AND CONTROL ROOM ACOUSTICS

Regardless of which type of studio facility is being designed, built and used, a number of primary concerns should be addressed in order to achieve the best possible acoustic results. In this section, we'll take a close look at such important and relevant aspects of acoustics as:

- Acoustic isolation
- Symmetry in control room and monitoring design
- Frequency balance
- Absorption
- Reflection
- Reverberation

Although several mathematical formulas have been included in the following sections, it's by no means necessary that you memorize or worry about them. By far, I feel that it's more important that you grasp the basic principles of acoustics rather than worry about the underlying math. Remember: more often than not, acoustics is an artistic science that blends math with the art of intuition and experience.

Acoustic Isolation

Because most commercial and project studio environments make use of an acoustic space to record sound, it's often wise and necessary to employ effective isolation techniques into their design in order to keep external noises to a minimum. Whether that noise is transmitted through the medium of air (e.g., from nearby auto, train or jet traffic) or through solids (e.g., from air-conditioner rumbling, underground subways or nearby businesses), special construction techniques will often be required to dampen these extraneous sounds (Figure 3.5).

If you happen to have the luxury of building a studio facility from the ground up, a great deal of thought should be put into selecting the studio's location. If a location has considerable neighborhood noise, you might have to resort to extensive (and expensive) construction techniques that can "float" the rooms (a process that effectively isolates and decouples the inner rooms from the building's outer foundations). If there's absolutely no choice of studio location, and the studio happens to be located next to a recycling factory, just under the airport's main landing path or over the subway's uptown line, you'll simply have to give in to destiny and build acoustical barriers to these outside interferences.

FIGURE 3.5

Various isolation, absorption and reflective acoustical treatments for the construction of a recording/monitoring environment (courtesy of Auralex Acoustics, www .auralex.com; Courtesy of Gik Acoustics LLC, www .gikacoustics.com).

The reduction in the sound-pressure level (SPL) of a sound source as it passes through an acoustic barrier of a certain physical mass (Figure 3.6) is termed the *transmission loss* (TL) of a signal. This attenuation can be expressed (in dB) as:

$$TL = 14.5 \log M + 23$$

where *TL* is the transmission loss in decibels, and *M* is the surface density (or combined surface densities) of a barrier in pounds per square foot (lb/ft^2).

(a)

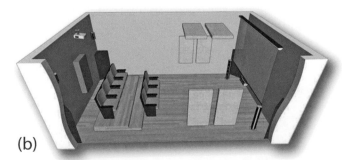
(b)

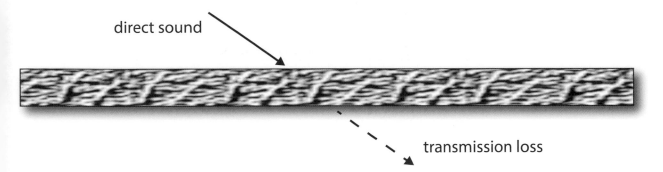

direct sound

transmission loss

FIGURE 3.6
Transmission loss refers to the reduction of a sound signal (in dB) as it passes through an acoustic barrier.

Because transmission loss is frequency dependent, the following equation can be used to calculate transmission loss at various frequencies with some degree of accuracy:

$$TL = 14.5 \log Mf - 16$$

where f is the frequency (in hertz).

Both common sense and the preceding two equations tell us that heavier acoustic barriers will yield a higher transmission loss. For example, Table 3.1 tells us that a 12-inch-thick wall of dense concrete (yielding a surface density of 150 lb/ft^2) offers a much greater resistance to the transmission of sound than can a 4-inch cavity filled with sand (which yields a surface density of 32.3 lb/ft^2). From the second equation (TL = 14.5 log Mf − 16), we can also draw the conclusion that for a given acoustic barrier, transmission losses will increase as the frequency rises. This can be easily illustrated by closing the door of a car that has its sound system turned up, or by shutting a single door to a music studio's control room. In both instances, the high frequencies will be greatly reduced in level, while the bass frequencies will be impeded to a much lesser extent. From this, the goal would seem to be to build a studio wall, floor, ceiling, window or door out of the thickest and most dense material that's available; however, expense and physical space often play roles in determining just how much of a barrier can be built to achieve the desired isolation. As such, a balance must usually be struck when using both space- and cost-effective building materials.

WALLS

When building a studio wall or reinforcing an existing structure, the primary goal is to reduce leakage (increase the transmission loss) through a wall as much as possible over the audible frequency range. This is generally done by:

- Building a wall structure that's as massive as is practically possible (in terms of both cubic and square foot density)
- Eliminating joints that can easily transmit sound through the barrier
- Dampening structures so that they are well supported by reinforcement structures and are free of resonances

Table 3.1	Surface Densities of Common Building Materials	
Material	**Thickness (inches)**	**Surface Density (lb/ft²)**
Brick	4	40.0
	8	80.0
Concrete (lightweight)	4	33.0
	12	100.0
Concrete (dense)	4	50.0
	12	150.0
Glass	—	3.8
	—	7.5
	—	11.3
Gypsum wallboard	—	2.1
	—	2.6
Lead	1/16	3.6
Particleboard	—	1.7
Plywood	—	2.3
Sand	1	8.1
	4	32.3
Steel	—	10.0
Wood	1	2.4

The following guidelines can be helpful in the construction of framed walls that have high transmission losses:

- If at all possible, the inner and outer wallboards should not be directly attached to the same wall studs. The best way to avoid this is to alternately stagger the studs along the floor and ceiling frame (i.e., framing 2 × 4 studs onto a 2 × 6 frame) so that the front/back facing walls aren't in physical contact with each other (Figure 3.7).
- Each wall layer should have a different density to reduce the likelihood of increased transmission due to resonant frequencies that might be sympathetic to both sides. For example, one wall might be constructed of two 5/8-inch gypsum wallboards, while the other wall might be underplayed with soft fiberboard that's also surfaced with two 3/4-inch gypsum wallboards.
- If you're going to attach gypsum wallboards to a single wall face, you can increase transmission loss by mounting the additional layers (not the first layer) with adhesive caulking rather than by using screws or nails.
- Spacing the studs 24 inches on center instead of using the traditional 16-inch spacing yields a slight increase in transmission loss.

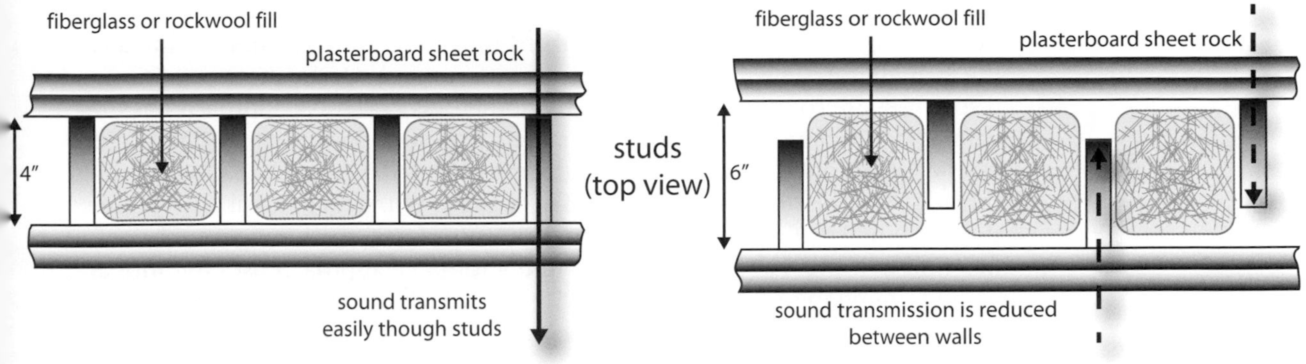

fiberglass or rockwool fill

plasterboard sheet rock

4"

studs
(top view)

6"

fiberglass or rockwool fill

plasterboard sheet rock

sound transmits
easily though studs

sound transmission is reduced
between walls

FIGURE 3.7
Double, staggered stud
construction greatly reduces
leakage by decoupling the
two wall surfaces from each
other: (a) Top view showing
walls that are directly tied to
wall studs (allowing sound
to easily pass through). (b)
Top view showing walls with
offset, non-touching studs
(so that sound doesn't easily
pass from wall to wall).

- To reduce leakage that might make it through the cracks, apply a bead of non-hardening caulk sealant to the inner gypsum wallboard layer at the wall-to-floor, wall-to-ceiling and corner junctions.

Generally, the same amount of isolation is required between the studio and the control room as is required between the studio's interior and exterior environments. The proper building of this wall is important so that an accurate tonal balance can be heard over the control-room monitors without promoting leakage between the rooms or producing resonances within the wall that would audibly color the signal. Optionally, a specially designed cavity, called a *soffit*, can be designed into the front-facing wall of the control room to house the larger studio monitors. This superstructure allows the main, far-field studio monitors to be mounted directly into the wall to reduce reflections and resonances in the monitoring environment.

It's important for a soffit to be constructed to high standards, using a multiple-wall or high-mass design that maximizes the density with acoustically tight construction techniques in order to reduce leakage between the two rooms. Cutting corners by using substandard (and even standard) construction techniques in the building of a studio soffit can lead to unfortunate side effects, such as wall resonances, rattles, and increased leakage. Typical wall construction materials include:

- *Concrete*: this is the best and most solid material, but it is often expensive, and it's not always possible to pour cement into an existing design.
- *Bricks* (*hollow-form or solid*): this excellent material is often easier to place into an existing room than concrete.
- *Gypsum plasterboard*: building multiple layers of plasterboard onto a double-walled stud frame is often the most cost- and design-efficient approach for reducing resonances and maximizing transmission loss. It's often a good idea to reduce these resonances by filling the wall cavities with Rockwool or fiberglass, while bracing the internal structure to add an extra degree of stiffness.

Studio monitors can be designed into the soffit in a number of ways. In one expensive approach, the far-field speakers' inner enclosure cavities are literally the walls of the control room's front wall concrete pour. Under these conditions, resonances are completely eliminated. Another less expensive approach has the studio monitors resting on poured concrete pedestals; in this situation, inserts can be cast into the pedestals that can accept threaded rebar rods (known as all-thread). By filing the rods to a chamfer (a sharp point), it's possible to adjust the position, slant and height of the monitors for final positioning into the soffit's wall framing. The most common and affordable approach uses traditional wood framing in order to design a cavity into which the speaker enclosures can be designed and positioned. Extra bracing, plasterboard and heavy construction should be used to reduce resonances.

FLOORS

For many recording facilities, the isolation of floor-borne noises from room and building exteriors is an important consideration. For example, a building that's located on a busy street and whose concrete floor is tied to the building's ground foundation might experience severe low-frequency rumble from nearby traffic. Alternatively, a second-floor facility might experience undue leakage from a noisy downstairs neighbor or, more likely, might interfere with a quieter neighbor's business. In each of these situations, increasing the isolation to reduce floor-borne leakage and/or transmission is essential. One of the most common ways to isolate floor-related noise is to construct a "floating" floor that's structurally decoupled from its subfloor foundation.

Common construction methods for floating a professional facility's floor use either neoprene "hockey puck" isolation mounts, U-Boat floor floaters (Figure 3.8a) or a continuous underlay, such as a rubberized floor mat. In these cases, the underlay is spread over the existing floor foundation and then covered with an overlaid plywood floor structure. In more extreme situations, this

FIGURE 3.8
Floor isolation treatments. (a) U-Boat™ floor beam float channels can be placed under a standard 2 × 4 floor frame to increase isolation. Floor floaters should be placed every 16 inches under a 2 × 4 floor joist (courtesy of Auralex Acoustics, www. auralex.com). (b) Basic guidelines for building a concrete floating floor using neoprene mounts.

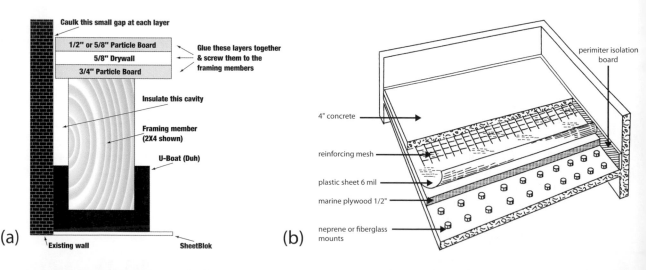

(a)

(b)

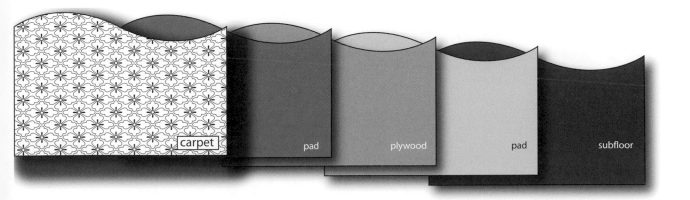

carpet   pad   plywood   pad   subfloor

superstructure could be covered with reinforcing wire mesh and finally topped with a 4-inch layer of concrete (Figure 3.8b). In either case, the isolated floor is then ready for carpeting, wood finishing, painting or any other desired surface.

An even more cost- and space-effective way to decouple a floor involves layering the original floor with a rubberized or carpet foam pad. A 1/2- or 5/8-inch layer of tongue-and-groove plywood or oriented strand board (OSB) is then laid on top of the pad. These should not be nailed to the subfloor; instead, they can be stabilized by glue or by locking the pieces together with thin metal braces. Another foam pad can then be laid over this structure and topped with carpeting or any other desired finishing material (**Figure 3.9**).

It is important that the floating superstructure be isolated from both the under-flooring and the outer wall. Failing to isolate these structures allows sounds to be transmitted through the walls to the subfloor, and vice versa (often defeating the whole purpose of floating the floor). These wall perimeter isolation gaps can be sealed with pliable decoupling materials such as widths of soft mineral fiberboard, neoprene, silicone or other pliable materials.

FIGURE 3.9
An alternative, cost-effective way to float an existing floor is by layering relatively inexpensive materials.

RISERS

As we saw from the equation TL = 14.5 log Mf – 16, low-frequency sound travels through barriers much more easily than does high-frequency sound. It stands to reason that strong, low-frequency energy is transmitted more easily than high-frequency energy between studio rooms, from the studio to the control room, or to outside locations. In general, the drum set is most likely to be the biggest leakage offender. By decoupling much of a drum set's low-frequency energy from a studio floor, many of the low-frequency leakage problems can be reduced. In most cases, the problem can be fixed by using a drum riser. Drum risers are available commercially (**Figure 3.10a**), or they can be easily constructed. In order to reduce unwanted resonances, drum risers should be constructed using 2 × 6-inch or 2 × 8-inch beams for both the frame and the supporting joists (spaced at 16 or 12 inches on center, as shown in **Figure 3.10b**). Sturdy 1/2- or 5/8-inch tongue-and-groove plywood panels should be glued to the supporting frames

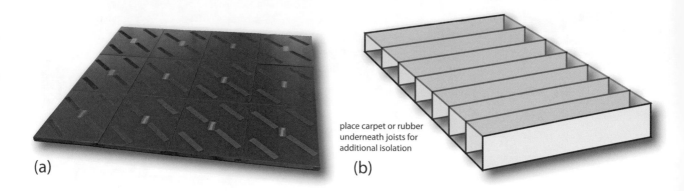

place carpet or rubber underneath joists for additional isolation

(a)   (b)

FIGURE 3.10
Drum/isolation risers. (a) HoverDeck™ 88 isolation riser (courtesy of Auralex Acoustics, www.auralex. com). (b) General construction details for a homemade drum riser.

with carpenter's glue (or similar wood glue) and then nailed or screwed down (using heavy-duty galvanized fasteners). When the frame has dried, rubber coaster float channels or (at the very least) strips of carpeting should be attached to the bottom of the frame, and the riser will be ready for action.

CEILINGS

Foot traffic and other noises from above a sound studio or production room are another common source of external leakage. Ceiling noise can be isolated in a number of ways. If foot traffic is your problem, and you're fortunate enough to own the floors above you, you can reduce this noise by simply carpeting the overhead hallway or by floating the upper floor. If you don't have that luxury, one approach to isolating ceiling-borne sounds is to hang a false structure from the existing ceiling or from the overhead joists (as is often done when a new room is being constructed). This technique can be fairly cost effective when spring or "Z" suspension channels are used (Figure 3.11). Z channels are often screwed to the ceiling joists to provide a flexible, yet strong, support to which a hanging wallboard ceiling can be attached. If necessary, fiberglass or other sound-deadening materials can be placed into the cavities between the overhead structures.

FIGURE 3.11
Ceiling isolator systems. (a) RSICSI-1 (Resilient Sound Isolation Clips) (courtesy of PAC International, Inc; www. pac-intl.com) (b) Z channels can be used to hang a floating ceiling from an existing overhead structure.

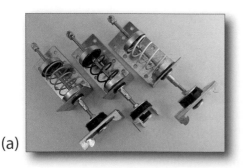

(a)

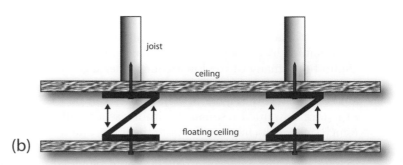

joist

ceiling

floating ceiling

(b)

WINDOWS AND DOORS

Access to and from a studio or production room area (in the form of windows and doors) can also be a potential source of sound leakage. For this reason, strict attention needs to be given to window and door design and construction. Visibility in a studio is extremely important within a music production environment. For example, when multiple rooms are involved, good visibility serves to promote effective communication between the producer or engineer and the studio musician (as well as among the musicians themselves). For this reason, windows have been an important factor in studio design since the beginning. The design and construction details for a window often vary with studio needs and budget requirements and can range from being deep, double-plate cavities that are built into double-wall constructions (Figure 3.12) to more modest prefab designs that are built into a single wall. Other, more expensive designs include floor-to-ceiling windows that create a virtual "glass wall" as well as those impressive ones that are designed into poured concrete soffit walls.

Access doors to and from the studio, control room and exterior areas should be constructed of solid wood or high-quality acoustical materials (Figure 3.13a), as solid doors generally offer higher TL values than their cheaper, hollow counterparts. No matter which door type is used, the appropriate seals, weather stripping and doorjambs should be used throughout so as to reduce leakage through

FIGURE 3.12

Detail for a practical window construction between the control room and studio. (a) Simplified drawing. (b) Detailed drawing (courtesy of Russ Berger Design Group, Inc., www.rbdg.com).

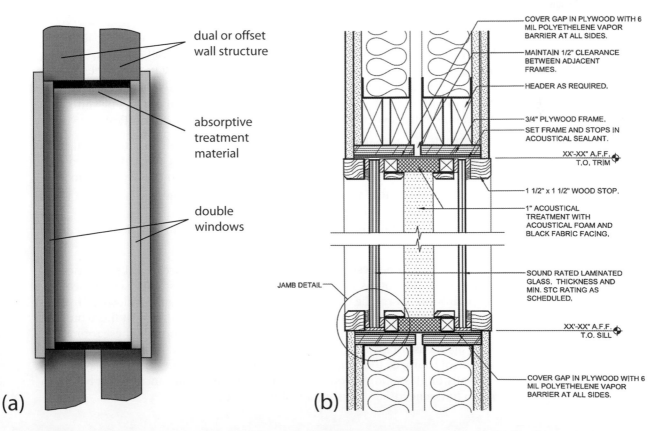

dual or offset wall structure

absorptive treatment material

double windows

COVER GAP IN PLYWOOD WITH 6 MIL POLYETHELENE VAPOR BARRIER AT ALL SIDES.

MAINTAIN 1/2" CLEARANCE BETWEEN ADJACENT FRAMES.

HEADER AS REQUIRED.

3/4" PLYWOOD FRAME. SET FRAME AND STOPS IN ACOUSTICAL SEALANT.

XX'-XX" A.F.F. T.O. TRIM

1 1/2" x 1 1/2" WOOD STOP.

1" ACOUSTICAL TREATMENT WITH ACOUSTICAL FOAM AND BLACK FABRIC FACING.

SOUND RATED LAMINATED GLASS. THICKNESS AND MIN. STC RATING AS SCHEDULED.

XX'-XX" A.F.F. T.O. SILL

COVER GAP IN PLYWOOD WITH 6 MIL POLYETHELENE VAPOR BARRIER AT ALL SIDES.

JAMB DETAIL

(a)

(b)

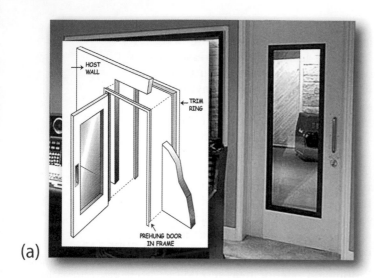

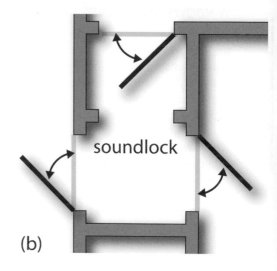

(a)   (b)

FIGURE 3.13
Door isolation systems. (a) A SoundSecure™ studio door (courtesy of ETSLindgren, www.etslindgren.com). (b) Example of a sound lock door system design.

the cracks. Whenever possible, double-door designs should be used to form an acoustical *sound lock* (Figure 3.13b). This construction technique dramatically reduces leakage because the air trapped between the two solid barriers offers high TL values.

ISO-ROOMS AND ISO-BOOTHS

Isolation rooms (iso-rooms) are acoustically isolated or sealed areas that are built into a music studio or just off a control room (Figure 3.14). These recording areas can be used to separate louder instruments from softer ones (and vice versa) in order to reduce leakage and to separate instrument types by volume to maintain control over the overall ensemble balance. For example:

- To eliminate leakage when recording scratch vocals (a guide vocal track that's laid down as a session reference), a vocalist might be placed in a small room while the rhythm ensemble is placed in the larger studio area.
- A piano or other instrument could be isolated from the larger area that's housing a full string ensemble.
- Vocals could be set up in the iso-room, while drums are being laid down in the main room. The possibilities are endless.

An iso-room can be designed to have any number of acoustical properties. By having multiple rooms and/or iso-room designs in a studio, several acoustical environments can be offered that range from being more reflective (live) to absorptive (dead), or a specific room can be designed to better fit the acoustical needs of a particular instrument (e.g., drums, piano or vocals). These rooms can be designed as totally separate areas that can be accessed from the main studio or control room, or they might be directly tied to the main studio by way of sliding walls or glass sliding doors. In short, their form and function can be put to use to fit the needs and personality of the session.

Isolation booths (*iso-booths*) provide the same type of isolation as an iso-room but are often much smaller. Often called *vocal booths*, these mini-studios are perfect for isolating vocals and single instruments from the larger studio. In fact, rooms that have been designed and built for the express purpose of mixing down a recording will often only have an iso-booth … and no other recording room. Using this space-saving option, vocals or single instruments can be easily overdubbed on site, and should more space be needed, a larger studio can be booked to fit the bill.

FIGURE 3.14
Studio incorporating multiple iso-rooms (courtesy of Blade Studios, www.bladestudios .com).

ACOUSTIC PARTITIONS

Movable acoustic *partitions* (also known as *flats* or *gobos*) are commonly used in studios to provide on-the-spot barriers to sound leakage. By partitioning a musician and/or instrument on one or more sides and then placing the mic inside the temporary enclosure, isolation can be greatly improved in a flexible way that can be easily changed as new situations arise. Acoustic partitions are currently available on the commercial market in various design styles and types for use in a wide range of studio applications (Figure 3.15). For those on a budget, or who have particular isolation needs, it's relatively simple to get out the workshop

FIGURE 3.15
Acoustic partition flat examples: (a) S5–2L "Sorber" baffle system (courtesy of ClearSonic Mfg., Inc., www .clearsonic.com). (b) Piano piano panel setup (courtesy of Auralex Acoustics, www. auralex.com).

(a)

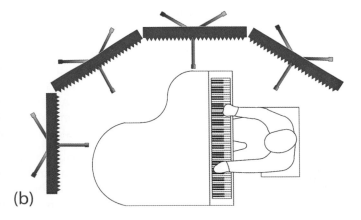

(b)

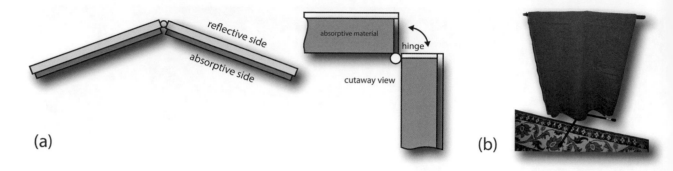

(a)   (b)

FIGURE 3.16
Examples of a homemade flat: (a) Homemade flat design. (b) The old "blanket and a mic boom" trick.

tools and make your own flats that are based around wood frames, fiberglass, Rockwool or other acoustically absorptive materials – and then decorate them with your favorite fabric coverings (Figure 3.16a).

If you can't get a flat when you need one, you can often improvise using common studio and household items. For example, a simple partition can be easily made on the spot by grabbing a mic/boom stand combination and retracting the boom halfway at a 90° angle to make a T-shape. Simply drape a blanket or heavy coat over the T-bar, and voilà - you've built a quick-'n'-dirty dividing flat (Figure 3.16b).

When using a partition, it's important to be aware that musicians need to be able to see each other, the conductor and the producer. Musicality and human connectivity almost always take precedence over technical issues.

Noise Isolation Within the Control Room

Isolation between rooms and the great outdoors isn't the only noise-related issue in the modern-day recording or project studio. The proliferation of computers, multitrack tape machines and cooling systems has created issues that present their own Grinch-like types of noise, Noise, NOISE! This usually manifests itself in the form of system fan noise, transport tape noise and computer-related sounds from central processing units (CPUs), case fans, hard drives and the like.

When it comes to isolating tape transport and system fan sounds, budget and size constraints permitting, it's often wise to build an iso-machine room or iso-closet that's been specifically designed and ventilated for containing such equipment. An equipment room that has easy-access doors that provide for current/future wiring needs can add a degree of peace-'n'-quiet and an overall professionalism that will make both you and your clients happy.

Within smaller studio or project studio spaces, such a room isn't always possible; however, with care and forethought, the whizzes and whirrs of the digital era can be turned into a nonissue that you'll be proud of. Here are a few examples of the most common problems and their solutions:

- Place the computer(s) in an isolated case, alcove or room (care needs to be taken to provide ventilation and to monitor the CPU/case temperatures so as not to harm your system).
- Connect the studio computers via a high-speed network to a remote server location.
- Replace fans with quieter ones. By doing some careful Web searching or by talking to your favorite computer salesperson, it's often possible to install CPU and case fans that are quieter than most off-the-shelf models.

SYMMETRY IN CONTROL-ROOM DESIGN

While many professional studios are built from the ground up using standard acoustic and architectural guidelines, most budget-minded production and project studios are often limited by their own unique sets of building, space and acoustic constraints. Even though the design of a budget, project or bedroom control room might not be acoustically perfect, if speakers are to be used in the monitoring environment, certain basic ground rules of acoustical physics must be followed in order to create a proper listening environment.

One of the most important acoustic design rules in a monitoring environment is the need for symmetrical reflections on all axes within the design of a control room or single-room project studio. In short, the center and acoustic *imaging* (ability to discriminate placement and balance in a stereo or surround field) is best when the listener, speakers, walls and other acoustical boundaries are symmetrically centered about the listener's position (often in an equilateral triangle). In a rectangular room, the best low-end response can be obtained by orienting the console and loudspeakers into the room's long dimension (Figure 3.17a). Of course, it's also possible to orient the mix position along the width of the room (Figure 3.17b). From a symmetrical standpoint, this is perfectly acceptable; however, it should again be pointed out that placing the mix position into the length of the room often allows better response and less potential for problems in the low end. Of course, these decisions are up to you and your particular layout situation. Should space or other room considerations come into play, centering the listener/monitoring position at a 45° angle within

FIGURE 3.17
Various acceptable symmetries in a monitoring environment. (a) Acoustic reflections must be symmetrical about the listener's position. In addition, orienting a control room along the long dimension can extend the room's low-end response. (b) Placing the mix position along the width of the room can also be an acceptable layout choice. (c) Placing the listening environment symmetrically in a corner is another example of how the left/right imagery can be improved over an off-center placement.

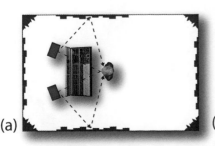
(a)

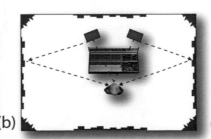
(b)

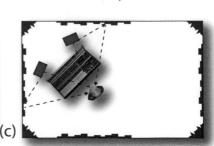
(c)

a symmetrical corner (Figure 3.17c) is another example of how the left/right imagery can be reasonably maintained.

With regard to setting up any production/monitoring environment, we'd like to first draw your attention to the need for symmetry in any critical monitoring environment. A symmetrical acoustic environment around the central mixing axis can work wonders toward creating a balanced left/right and surround image. Fortunately, this generally isn't a difficult goal to achieve. An acoustical and speaker placement environment that isn't balanced between the left-hand and right-hand sides will allow differing reflections, absorption coefficients and variations in frequency response. This can adversely affect the imaging and balance of your final mix. Further information on this important subject can be found later in this chapter; however, consider this your first heads-up on an important topic.

FIGURE 3.18

Center symmetry. (a) Placing the monitoring environment off center and in a corner will affect the audible center image, and placing one speaker in a 90° corner can cause an off-center bass buildup and adversely affect the mix's imagery. (b) Shifting the listener/monitoring position into the center will greatly improve the left/right imagery.

Should any primary boundaries of a control room (especially wall or ceiling boundaries near the mixing position) be asymmetrical from side to side, sounds heard by one ear will receive one combination of direct and reflected sounds, while the other ear will hear a different acoustic balance (Figure 3.18). This condition can drastically alter the sound's center image characteristics, so that when a sound is actually panned between the two monitor speakers, the sound will appear to be centered; however, when the sound is heard in another studio or standard listening environment, the imaging may be off center. To avoid this problem, care should be taken to ensure that both the side and ceiling boundaries are largely symmetrical with respect to each other and that all of the speaker-level balances are properly set.

While we're on the subject of the relationship between the room's acoustic layout and speaker placement, it's always wise to place near-field and all other speaker enclosures at points that are equidistant to the listener in the stereo and surround field. Whenever possible, speaker enclosures should be placed 1 to 2 feet away from the nearest wall and/or corner, which helps to avoid bass buildups that acoustically occur at boundary and corner locations. In addition

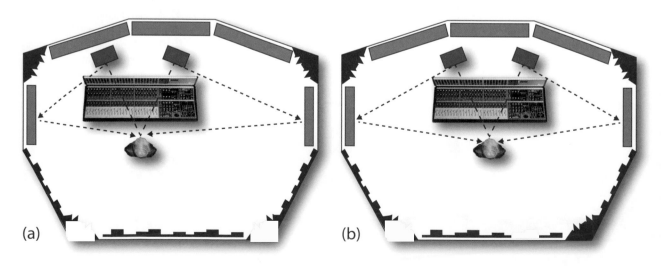

(a)   (b)

to strategic speaker placement, homemade or commercially available isolation pads can be used to reduce resonances that often occur whenever enclosures are placed directly onto a table or flat surface.

FREQUENCY BALANCE

Another important factor in room design is the need for maintaining the original *frequency balance* of an acoustic signal. In other words, the room should exhibit a relatively flat frequency response over the entire audio range without adding its own particular sound coloration. The most common way to control the tonal character of a room is to use materials and design techniques that govern the acoustical reflection and absorption factors.

Reflections

One of the most important characteristics of sound as it travels through air is its ability to reflect off a boundary's surface at an angle that's equal to (and opposite) its original angle of incidence (Figure 3.19). Just as light bounces off a mirrored surface or multiple reflections can appear within a mirrored room, sound reflects throughout room surfaces in ways that are often amazingly complex. Through careful control of these reflections, a room can be altered to improve its frequency response and sonic character.

In Chapter 2, we learned that sonic reflections can be controlled in ways that disperse the sound outward in a wide-angled pattern (through the use of a convex surface) or focus them on a specific point (through the use of a concave surface). Other surface shapes, on the other hand, can reflect sound back at various other angles. For example, a 90° corner will reflect sound back in the same direction as its incident source (a fact that accounts for the additive acoustic buildups at various frequencies at or near a wall-to-corner or corner-to-floor intersection).

The all-time winner of the "avoid this at all possible cost" award goes to constructions that include opposing parallel walls in their design. Such conditions give rise to a phenomenon known as *standing waves*. Standing waves (also known as room modes) occur when sound is reflected off parallel surfaces and travels back on its own path, thereby causing phase differences to interfere with a room's amplitude response (Figure 3.20a). Room modes are expressed as integer multiples of the length, width and depth of the room and indicate which multiple is being referred to for a particular reflection.

FIGURE 3.19
Sound reflects off a surface at an angle equal (and opposite) to its original angle of incidence, much as light will reflect off a mirror.

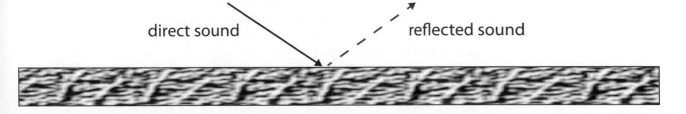

direct sound   reflected sound

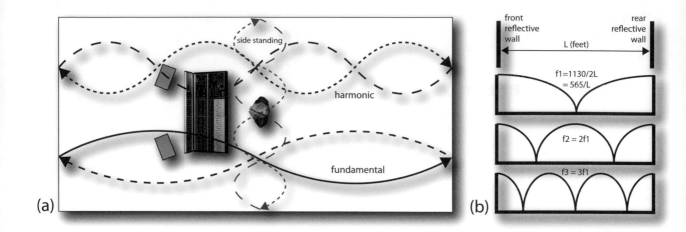

(a)   (b)

FIGURE 3.20

Standing waves within a room. (a) Reflective parallel surfaces can potentially cancel and reinforce frequencies within the audible spectrum, causing changes in its response. (b) The reflective, parallel walls create an undue number of standing waves, which occur at various frequency intervals ($f1$, $f2$, $f3$, $f4$, and so on).

Walking around a room with moderate to severe mode problems produces the sensation of increasing and/or decreasing volume levels at various frequencies throughout the area. These perceived volume changes are due to amplitude (phase) cancelations and reinforcements of the combined reflected waveforms at the listener's position. The distance between parallel surfaces and the signal's wavelength determines the nodal points that can potentially cause sharp peaks or dips at various points in the response curve (up to or beyond 19 dB) at the affected fundamental frequency (or frequencies) and upper harmonic intervals (Figure 3.20b). This condition exists not only for opposing parallel walls but also for all parallel surfaces (such as between the floor and ceiling or between two reflective flats). From this discussion, it's obvious that the most effective way to prevent standing waves is to construct walls, boundaries and ceilings that are nonparallel.

If the room in question is rectangular, or if further sound-wave dispersion is desired, diffusers can be attached to the wall and/or ceiling boundaries to help break up standing waves. Diffusers (Figure 3.21) are acoustical boundaries that reflect the sound wave back at various angles that are wider than the original incident angle (thereby breaking up the energy-destructive standing waves). In addition, the use of both nonparallel and diffusion wall construction can reduce extreme, recurring reflections and smooth out the reverberation characteristics of a room by building more complex acoustical pathways.

Flutter echo (also called *slap echo*) is a condition that occurs when parallel boundaries are spaced far enough apart that the listener is able to discern a number of discrete echoes. Flutter echo often produces a "boingy", hollow sound that greatly affects a room's sound character as well as its frequency response. A larger room (which might contain delayed echo paths of 50 m/sec or more) can have its echoes spaced far enough apart in time that the discrete reflections produce echoes that can actually interfere with the intelligibility of the direct sound. This will often result in a jumble of noise, and in these cases, a proper application of absorption and acoustic dispersion becomes critical.

 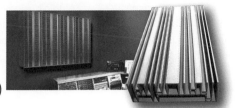 

(a)   (b)   (c)

FIGURE 3.21
Diffuser examples: (a) SpaceArray sound diffusers (courtesy of pArtScience, www.partscience.com). (b) Open-ended view of a Primacoustic™ Razorblade quadratic diffuser (courtesy of Primacoustic Studio Acoustics, www.primacoustic .com). (c) Art Diffusor sound diffusers Model E (courtesy of acousticsfirst, www.acous-ticsfirst.com).

When speaking of reflections within a studio control room, one long-held design concept relates to the concept of designing the room such that the rear of the room is largely reflective and diffuse in nature (acoustically "live"), while the front of the room is largely or partially absorptive (acoustically "dead"). This philosophy (known as LEDE; Figure 3.22) argues for the fact that the rear of the room should be largely reflective, providing for a balanced and diffuse environment that can help reinforce positive reflections which can add acoustic "life" to the mix experience. The front of the room would tend more toward the absorptive side in a way that reduces standing waves, flutter reflections and reflections from the rear of the speakers that would interfere with the overall response of the room.

FIGURE 3.22
LEDE control-room layout showing the live end toward the back of the room and the dead end (absorption) toward the front of the room.

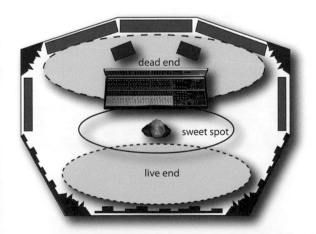 

It's important to realize that no two rooms will be acoustically the same or will necessarily offer the same design challenges. The one constant is that careful planning, solid design and ingenuity are the foundation of any good-sounding room. You should also keep in mind that numerous studio design and commercial acoustical product firms are available that offer assistance for both large and small projects. Getting professional advice can be a good thing.

Absorption

Another factor that often has a marked effect on an acoustic space involves the use of surface materials and designs that can absorb unwanted sounds (either across the entire audible band or at specific frequencies). The absorption of acoustic energy is, effectively, the inverse of reflection (Figure 3.23). Whenever sound strikes a material, the amount of acoustic energy that's absorbed relative to the amount that's reflected can be expressed as a simple ratio known as the material's *absorption coefficient*. For a given material, this can be represented as:

$$A = I_a/I_r$$

where I_a is the sound level (in dB) that is absorbed by the surface (often dissipated in the form of physical heat), and I_r is the sound level (in dB) that is reflected back from the surface.

The factor $(1 - a)$ is a value that represents the amount of reflected sound. This makes the coefficient a decimal percentage value between 0 and 1. If we say that a surface material has an absorption coefficient of 0.25, we're actually saying that the material absorbs 25% of the original acoustic energy and reflects 75% of the total sound energy at that frequency. A sample listing of these coefficients is provided in Table 3.2.

FIGURE 3.23
Absorption occurs when only a portion of the incident acoustic energy is reflected back from a material's surface.

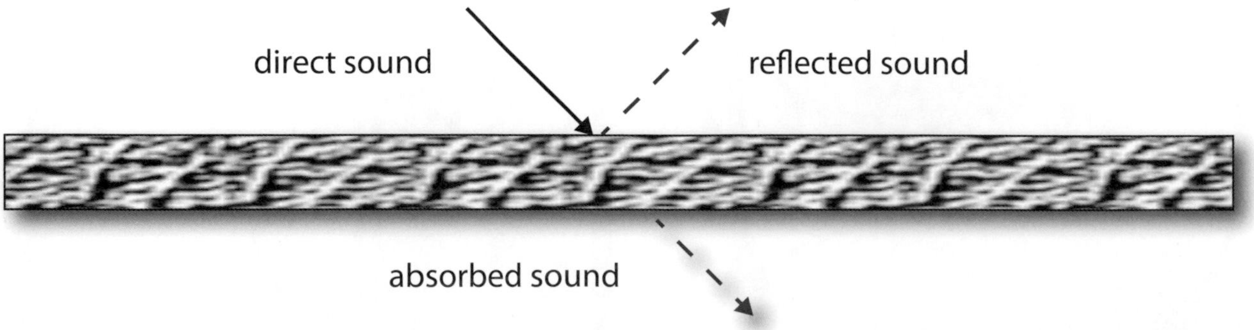

To determine the total amount of absorption that's obtained by the sum of all the absorbers within a total volume area, it's necessary to calculate the average absorption coefficient for all of the surfaces together. The *average absorption coefficient* (A_{ave}) of a room or area can be expressed as:

$$A_{ave} = s_1a_1 + s_2a_2 + \ldots s_na_n/S$$

direct sound   reflected sound

absorbed sound

| Table 3.2 | Absorption Coefficients for Various Materials. | | | | |

Material	Coefficients (Hz)				
	125	250	500	1000	2000
Brick, unglazed	0.03	0.03	0.03	0.04	0.05
Carpet (heavy, on concrete)	0.02	0.06	0.14	0.37	0.60
Carpet (with latex backing, on 40-oz hair-felt or foam rubber)	0.03	0.04	0.11	0.17	0.24
Concrete or terrazzo	0.01	0.01	0.015	0.02	0.02
Wood	0.15	0.11	0.10	0.07	0.06
Glass, large heavy plate	0.18	0.06	0.04	0.03	0.02
Glass, ordinary window	0.35	0.25	0.18	0.12	0.07
Gypsum board nailed to 2 × 4 studs on 16-inch centers	0.013	0.015	0.02	0.03	0.04
Plywood (3/8 8 inch)	0.28	0.22	0.17	0.09	0.10
Air (sabins/1000 1000 ft^3)	--	--	--	--	2.3
Audience seated in upholstered seats	0.08	0.27	0.39	0.34	0.48
Concrete block, coarse	0.36	0.44	0.31	0.29	0.39
Light velour (10 10 oz/yd^2 in contact with wall)	0.29	0.10	0.05	0.04	0.07
Plaster, gypsum, or lime (smooth finish on tile or brick)	0.44	0.54	0.60	0.62	0.58
Wooden pews	0.57	0.61	0.75	0.86	0.91
Chairs, metal or wooden, seats unoccupied	0.15	0.19	0.22	0.39	0.38

Note: These coefficients were obtained by measurements in the laboratories of the Acoustical Materials Association. Coefficients for other materials may be obtained from Bulletin XXII of the association.

where s_1, s_2, ..., s_n are the individual surface areas; a_1, a_2, ...a_n, are the individual absorption coefficients of the individual surface areas, and S is the total square surface area.

On the subject of absorption, one common misconception is that the use of large amounts of sound-deadening materials will reduce room reflections and therefore make a room sound "good". In fact, the overuse of absorption will often have the effect of reducing high frequencies, creating a skewed room response that is dull and bass-heavy, as well as reducing constructive room reflections, which are important to a properly designed room. In fact, with regard to the balance between reflection, diffusion and absorption, many designers agree that a balance of 25% absorption and 25% diffuse reflections is a good ratio that can help preserve the "life" of a room while reducing unwanted buildups.

HIGH-FREQUENCY ABSORPTION

The absorption of high frequencies is accomplished through the use of dense, porous materials, such as fiberglass, Rockwool, dense fabric and carpeting. These materials generally exhibit high absorption values at higher frequencies, which can be used to control room reflections in a frequency-dependent manner. Specially designed foam and acoustical treatments are also commercially available that can be attached easily to recording studio, production room or control room walls as a means of taming multiple room reflections and/or dampening high-frequency reflections.

In addition to buying commercial absorbers, it's very possible to put your handy shop tools to work by building your own cost-effective absorber panels (of any shape, depth and style). One straightforward way of making them is by using Rockwool as the basic ingredient for your homemade absorber:

1. Go to youtube.com/modernrecordingtechniques and search for "DIY Absorber" and/or search YouTube for any of the many videos that are available on the subject.
2. Measure the dimensions that you'll need to build your absorbers. Buy 1" × 4" fir boards that add up to your required dimensions (often, the hardware store will even cut them to suit your needs). You might also want to buy and measure your Rockwool bats at the same time (this might help you with determining your overall dimensions).
3. Lay the boards out on your workbench or protected table and drill pilot holes to the top and bottom frame edges; then, using a 2" sheetrock or other type of screw, screw the frames together.
4. Make your measurements for the amount of fabric that you'll want to stretch over the entire surface and around the edges, so that they stretch around the newly made box. The fabric can be of practically any type, but a nice, inexpensive fabric of your favorite color works well.
5. It always helps to iron the fabric before mounting it, just to get the wrinkles out.
6. Before attaching the fabric, you might want to see how the Rockwool fits into each frame. If all's ok, then begin carefully attaching the fabric to the frame with a heavy-duty staple gun, taking care that the fabric is tight, straight and looks good.
7. Once the fabric is attached and the Rockwool is inserted, it's ready to hang in your control room/studio wall.

When done right, these absorbers (Figure 3.24) can look professional and fit your specific needs at a fraction of their commercial equivalents, sometimes with better results.

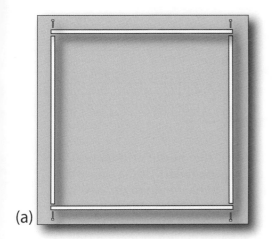

(b)

LOW-FREQUENCY ABSORPTION

It's important to note that materials which are absorptive in the high-frequency range may provide little resistance to the low-frequency end of the spectrum (and vice versa). In practice, however, this concept might not always point to the best or most cost-effective solutions. Care must always be taken when dealing with the low-end acoustic balance of a room. If a problem exists, our advice would be to take your time and approach the problem in a methodical and cost-effective fashion that would best suit your needs.

Various absorber types can be used to reduce low-frequency buildup at specific frequencies (and their multiples) within a room. These types of attenuation devices (known as *bass traps*) are available in a number of design types:

- Pressure-zone trap
- Quarter-wavelength trap
- Active trap
- Flexible materials

Pressure-zone trap: the pressure-zone bass trap absorber (Figure 3.25) works on the principle that sound pressure is doubled at large boundary points that are at 90° angles (such as walls and ceilings). By placing highly absorptive material at a boundary point (or points, in the case of a corner/ceiling intersection), the built-up pressure in the low-end spectrum can be partially absorbed. Of course, this type of absorber can be easily built (often using rockwool) or it can be commercially bought.

The *quarter-wavelength trap*: the quarter-wavelength bass trap (Figure 3.26) is an enclosure with a depth that's one-fourth the wavelength of the offending frequency's fundamental frequency and is often built into the rear facing wall, ceiling or floor structure and covered by a metal grating to allow foot traffic. The physics behind the absorption of a calculated frequency (and many of the

FIGURE 3.24
Homemade absorber panel:
(a) Showing fabric that's to be stretched over a wooden frame. (b) Once made, the Rockwool is placed inside the frame (which can be of any size or form), and it can be hung on the wall, lowered from the ceiling or placed in a corner.

(a)

(b)

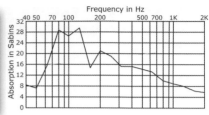

FIGURE 3.25

Pressure-zone trap. (a) A homemade version can be easily built up using rockwool, plywood and stretched fabric. (b) Realtrap MegaTraps commercial unit with absorption curve (courtesy of RealTraps, LLC, www. realtraps.com).

harmonics that fall above it) rests in the fact that the pressure component of a sound wave will be at its maximum at the rear boundary of the trap when the wave's velocity component is at a minimum. At the mouth of the bass trap (which is at a one-fourth wavelength distance from this rear boundary), the overall acoustic pressure will be at its lowest, while the velocity component (molecular movement) will be at its highest potential. Because the wave's motion (force) is greatest at the trap's opening, much of the signal can be absorbed by placing an absorptive material at that opening point. A low-density fiberglass lining can also be placed inside the trap to increase absorption (especially at harmonic intervals of the calculated fundamental).

Active trap: an active bass trap system makes use of a microphone, a low-frequency driver and a fast acting, band-limited amplifier to effectively create an inverse pressure wave that electronically "absorbs" low-end frequencies. Such a unit is actually capable of creating an effective area of absorption that is up to 40 times greater than its actual size.

FLEXIBLE SURFACES

FIGURE 3.26

A quarter-wavelength bass trap: (a) Physical concept design. (b) Sound is largely absorbed as heat, since the particle velocity (motion) is greatest at the trap's quarter-wavelength opening.

It's extremely important to realize that low frequencies are easily damped (reduced in level) by pliable materials. This occurs because a rooms' low-frequency energy will be absorbed (in the form of heat energy) as the material bends and flexes with the incident waveform.

This can be achieved on purpose with the use of homemade or commercially available absorbers (**Figure 3.27a**) … or it can unintentionally occur whenever

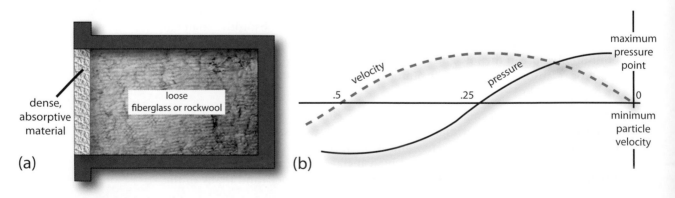

(a)

dense, absorptive material

loose fiberglass or rockwool

(b)

velocity

pressure

.5

.25

0

maximum pressure point

minimum particle velocity

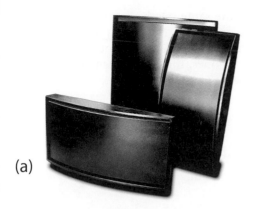

(a)

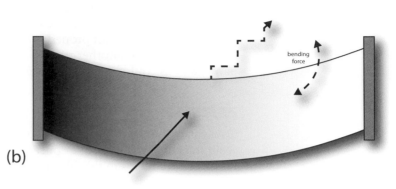

(b)

bending
force

surfaces within the production area flex. As an example, within DMH's personal studio, there is a large box that's built into the room that's necessary for the water system in the house. It doubles as a large bench at the back of the room that holds my bookshelves, acts as a big bench and actually looks really good. The box-like bench was really well constructed, but I neglected to securely fasten the front-facing board to the interior box frame (Figure 3.27b). As a result, the front face of the box was able to freely flex … and did a really good job of sucking a good part of the bass out of the room. After a while, I finally figured out the problem, secured the front face tightly to the frame with more screws, and the bass in the room likewise tightened up significantly. Moral of the story … make sure that all of your surfaces are securely locked down tightly to their frames.

FIGURE 3.27
Low-frequency absorption.
(a) Primacoustic™ Polyfuser, a combination diffuser and bass trap (courtesy of Primacoustic Studio Acoustics, www.primacoustic .com). (b) A large surface that can be "bent" by oncoming sound waves will often absorb low frequencies (either intentionally or not).

THE PRACTICAL SIDE OF ACOUSTICS

Here, both Emiliano and I would like to add our own personal 2 cents on this topic, which requires a good dose of practical physics, artistic insight, basic construction, intuition, art and above all … a good amount of common sense. Working to make your room sound as it should can range from being straightforward to frustratingly elusive. Changes to the room can also develop over time as you become more used to the overall layout, sound and character of your room. With this in mind, let's take some time to revisit a few ideas and concepts that can help keep your room on track.

Of course, it's important to keep in mind that these are our thoughts, opinions and experiences … There's absolutely no magic pill that works in all rooms, but these insights may help you to make your own informed opinions and then put them into practice. All the best!

Symmetry

It's definitely worth going back and re-reading the section on symmetry earlier in this chapter. If there's one thing that can get you off to a bad acoustical start, it would be ignoring the general idea that all sound within a production environment should arrive at the listener with the same relative level and frequency balance, as well as at the exact same relative time. Not doing so could result in a mix that is not properly centered. It might sound right in your room, but when heard in another room or over headphones, the relative center and pan positions could be off … not a good prospect.

25/25/50

Over the years, one of the primary guidelines that has been put into practice and seem to work for DMH is the concept of the 25/25/50 guideline. That's to say, a room that conforms to this idea would roughly contain:

- 25% absorption
- 25% diffusion
- 50% normal room

25% Absorption: this could take many forms, but well-placed, self-made or bought absorbers can be strategically placed around the room to reduce any large mid- to-high-frequency reflection nodes that might interfere with the room's overall response. Larger, corner absorption baffles could also be used, should the need absolutely arise.

25% Diffusion: this can likewise be done in any number of ways. You could buy pre-made structures, or you could simply make your own out of wood from the lumberyard or from your own scrap pile. For example, DMH constructed several hand-made wood diffusers (Figure 3.28a) that have been placed around the real part of the room, while wood-board original paintings have been placed at each side of the room at 15–20° angles to offset any side reflections that might be set up. Again, at the rear of the mix room, I chose to put in bookshelves (Figure 3.28b), which are used to hold all sorts of random items, in a way that makes for the perfect, utilitarian diffusion system.

50% Normal Room: beyond any problems with excessive reflections or obvious acoustic missteps, nothing beats a room that is comfortable and makes for a relaxing work environment. Go ahead, make it your own production haven

FIGURE 3.28
Homemade-made diffusers. (a) Hand-made wooden diffuser. (b) Placing bookshelves along the rear wall and putting "stuff" on them can provide both diffusion and a place for lots of general storage for things other than the non-diffuse surfaces of books.

(a)

(b)

that's conducive to being creative. One last piece of advice, though: a nice, good looking carpet can also go a long way towards reducing vertical standing waves and adding a relaxed, comfortable sense to the room.

On a personal note, over the years, DMH has become more and more attached to the LEDE (Live End Dead End) approach to control-room design. The general reason behind this approach is that it reduces reflections towards the front of the room (coming from behind the speakers, etc.) and emphasizes the good reflections that are hopefully being diffused from the rear of the room, being reflected back towards the listener to give a greater sense of life and openness to the overall sound experience. But, hey, that's just one opinion among many ... the only opinions that really count are yours and those of your customers/colleagues.

Speaker Placement

When it comes to speaker placement within the general soundfield area, the most important part of this equation is, of course, symmetry. Are the speakers placed with equal L/R distance to the listener? Is one speaker closer to a reflective or absorptive boundary that could cause a general L/R imbalance? Are your cables wired in phase with each other? These are just a few of the questions to ask yourself when setting up your system.

The next big question to ask relates to how close the speakers are to the rear and even the side walls. Are they so close to the back wall that a boundary bass-boost is set up? In this case, most active speaker systems will allow you to compensate in the low end for this problem; however, it's really important to realize that this form of compensation is generally a Band-Aid answer to a problem that's best fixed properly in the first place. Of course, we're talking about the moving of the speakers to a distance of a foot (⅓ meter) or greater. This will often eliminate the need for speaker EQ compensation and will allow the speaker's overall design to be heard in the best of all ways.

For further info on this subject, go to Chapter 19 (The Art and Technology of Mastering).

ROOM REFLECTIONS AND ACOUSTIC REVERBERATION

Another criterion for studio design is the need for a desirable room ambience and intelligibility, which is often contradictory to the need for good acoustic separation between instruments and their pickup. Each of these factors is governed by the careful control and tuning of the reverberation constants within the studio over the frequency spectrum.

Reverberation (*reverb*) is the persistence of a signal (in the form of reflected waves within an acoustic space) that continues after the original sound has ceased. The effect of these closely spaced and random multiple echoes gives us perceptible cues as to the size, density and nature of an acoustic space. Reverb also adds to the perceived warmth and spatial depth of recorded sound and plays an extremely important role in the perceived enhancement of music.

As was stated in the latter part of Chapter 2, the reverberated signal itself can be broken down into three components:

- Direct sound
- Early reflection
- Reverb

The direct signal is made up of the original, incident sound that travels from the source to the listener. Early reflections consist of the first few reflections that are projected to the listener off major boundaries within an acoustic space. These reflections generally give the listener subconscious cues as to the size of the room. (It should be noted that strong reflections off large, nearby surfaces can potentially have detrimental cancellation effects that can degrade a room's sound and frequency response at the listening position.) The last set of signal reflections makes up the actual reverberation characteristic. These signals are composed of random reflections that travel from boundary to boundary in a room and are so closely spaced that the brain can't discern the individual reflections. When combined, they are perceived as a single decaying signal.

Technically, reverb is considered to be the time that's required for a sound to die away to a millionth of its original intensity (resulting in a decrease over time of 60 dB), as shown by the following formula:

$$RT_{60} = V \times 0.049/AS$$

where RT is the reverberation time (in sec), V is the volume of the enclosure (in ft^3), A is the average absorption coefficient of the enclosure and S is the total surface area (in ft^2). As you can see from this equation, reverberation time is directly proportional to two major factors: the volume of the room and the absorption coefficients of the studio surfaces. A large environment with a relatively low absorption coefficient (such as a large cathedral) will have a relatively long RT_{60} decay time, whereas a small studio (which might incorporate a heavy amount of absorption) will have a very short RT_{60}.

The style of music and the room application will often determine the optimum RT_{60} for an acoustical environment. Reverb times can range from 0.25 sec in a smaller absorptive recording studio environment to 1.6 sec or more in a larger music or scoring studio. In certain designs, the RT_{60} of a room can be altered to fit the desired application by using movable panels or louvers or by placing carpets in a room. Other designs might separate a studio into sections that exhibit different reverb constants. One side of the studio (or separate iso-room) might be relatively non-reflective or dead, whereas another section or room could be much more acoustically live. The more reflective, live section is often used to bring certain instruments that rely heavily on room reflections and reverb, such as strings or an acoustic guitar, to "life". The recording of any number of instruments (including drums and percussion) can also greatly benefit from a well-designed acoustically live environment.

Isolation between different instruments and their pickups is extremely important in the studio environment. If leakage isn't controlled, the room's effectiveness becomes severely limited over a range of applications. The studio designs of the 1970s and 1980s brought about the rise of the "sound sucker" era in studio design. During this time, the absorption coefficient of many rooms was raised almost to an anechoic (no reverb) condition. With the advent of the music styles of the 1990s and a return to the respectability of live studio acoustics, modern studio and control-room designs have begun to increase in size and "liveness" (with a corresponding increase in the studio's RT_{60}). This has reintroduced the buying public to the thick, live-sounding music production of earlier decades, when studios were larger structures that were more attuned to capturing the overall acoustics of a recorded instrument or ensemble.

Acoustic Echo Chambers

Another physical studio design that was used extensively in the past (before the invention of artificial effects devices) for re-creating room reverberation is the *acoustic echo chamber*. A traditional echo chamber is an isolated room that has highly reflective surfaces into which speakers and microphones are placed.

The speakers are fed from an effects send, while the mic's reverberant pickup is fed back into the mix via an input strip of effects return. By using one or more directional mics that have been pointed away from the room's speakers, the direct sound pickup can be minimized. Movable partitions also can be used to vary the room's decay time. When properly designed, acoustic echo chambers have a very natural sound quality to them. The disadvantage is that they take up space and require isolation from external sounds; thus, size and cost often make it infeasible to build a new echo chamber, especially those that can match the caliber and quality of high-end digital reverb devices.

An echo chamber doesn't have to be an expensive, built-from-the-ground-up design. Actually, a temporary chamber can be made from a wide range of available acoustic spaces to pepper your next project with a bit of "acoustic spice". For example:

- An ambient-sounding chamber can be built by placing a Blumlein (crossed figure-8) pair or spaced stereo pair of mics in the main studio space and feeding a send to the studio playback monitors.
- A speaker/mic setup could be placed in an empty garage (as could a guitar amp/mic, for that matter).
- An empty stairwell often makes an excellent chamber.
- Any vocalist could tell you what'll happen if you place a singer or guitar speaker/mic setup in a nice bathroom with a tile shower.

From this, it's easy to see that ingenuity and experimentation are often the name of the makeshift echo/reverb game. In fact, there's nothing that says that the chamber has to be a real-time effect – for example, you could play back a song's effects track from a laptop DAW into a church's acoustic space and record the space to stereo tracks on the DAW, where they can be later placed into the mix. The limitless experimental options are totally up to you!

CHAPTER 4

Microphones

Design and Application

A *microphone* (often called a *mic*) is usually the first and most important device in a recording chain. A mic is a transducer that changes one form of energy (sound waves) into another corresponding form of energy (electrical signals). The quality of its pickup will often depend on external variables (such as placement, distance, instrument conditions and the acoustic environment). It also depends on design variables (such as the microphone's operating type, design characteristics and quality). These interrelated elements work together to affect the signal's overall sound quality.

In order to deal with the wide range of musical and acoustic situations that might come your way (not to mention your own personal taste), a large number of mic types, styles and designs can be pulled out of our "sonic toolbox" to get the job done in the best way. The truth is, microphone designs and types differ widely from one to the another. Some may have a certain sonic personality that allows them to work best with certain instruments and situations, while others may be suited for a broad range of applications. These "personalities" are then carefully chosen and placed by the engineer, producer and/or artist, using their intuition, experience and talent, to get the best possible sound from an acoustic source that best fits the application at hand.

The road to considering microphone choice and placement is best traveled by considering a few simple rules:

- Rule 1: *there are no rules, only guidelines*. Although guidelines can help you achieve a good pickup, don't hesitate to experiment in order to get a sound that best suits your needs or personal taste.
- Rule 2: *the overall sound of an audio signal is no better than the weakest link in the signal path*. If a mic or its placement doesn't sound as good as it could, make the changes to improve it BEFORE you commit it to disk, tape or whatever. More often than not, the concept of "fixing it later in the mix" will often put you in the unfortunate position of having to correct a situation after the fact rather than recording the best possible sound and/or performance during the initial session.

DOI: 10.4324/9781003260530-4

- Rule 3: *whenever possible, use the "Good Rule": good musician + good instrument + good performance + good acoustics + good mic + good placement = good sound.* This rule refers to the fact that a captured performance will only be as good as the performer, instrument, mic, mic placement and the entire signal chain that follows it. If any of these elements falls short of its potential, the track will suffer accordingly. However, if all of these links are the best that they can be, the recording will almost always be something that you can be proud of!

The "Good Rule"

Good musician + good instrument + good performance + good acoustics + good mic + good placement = good sound.

The miking of vocals and instruments (both in the studio and onstage) is definitely an art form. It's often a balancing act to get the most out of the Good Rule. Sometimes you'll have the best of all elements; at others, you'll have to work hard to make lemonade out of a situational lemon. The best rule of all is to use common sense and to trust your own instincts.

Before diving into the facts and placement techniques that deal with the finer points of microphone technology, we'd like to take a basic look at how microphones (and their operational characteristics) work. Why do we put this in the book? Well, from a personal standpoint, we feel that having a basic understanding of what happens "under the hood" will help you to have a better "mental" image of how a particular mic or mic technique will sound in a given situation. An operational understanding of how a mic works can combine with your own intuition, technical and sonic judgments to make the best artistic judgment at the time.

MICROPHONE DESIGN

A microphone is a device that converts acoustic energy into corresponding electrical voltages that can be amplified and recorded. In audio production, three transducer mic types can be used to accomplish this:

- Dynamic mic
- Ribbon mic
- Condenser mic

The Dynamic Microphone

In principle, the *dynamic mic* (Figure 4.1a and b) operates by using electromagnetic induction to generate an output signal. The simple *theory of electromagnetic induction* states that whenever an electrically conductive metal cuts across the flux lines of a magnetic field, a current of a specific magnitude and direction will be generated within that metal.

(a)

(b)

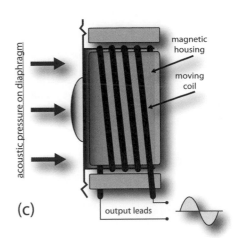

acoustic pressure on diaphragm

magnetic housing

moving coil

output leads

(c)

Dynamic mic designs (**Figure 4.1c**) generally consist of a stiff Mylar diaphragm of roughly 0.35-mil thickness. Attached to this diaphragm is a finely wrapped coil of wire (called a voice coil) that's precisely suspended within a strong magnetic field. Whenever an acoustic pressure wave hits the diaphragm's face, the attached voice coil is displaced and moves in proportion to the amplitude and frequency of the wave, causing the coil to cut across the lines of magnetic flux. According to the theory of electromagnetic induction, an analogous electrical signal (of a specific magnitude and direction) is then induced into the coil and across the output leads, thus producing an analog audio output signal.

The Ribbon Microphone

Like the dynamic microphone, the *ribbon mic* (**Figure 4.2a and b**) also works on the principle of electromagnetic induction. Older ribbon design types use a diaphragm of extremely thin aluminum ribbon (2 microns). Often, this diaphragm is corrugated along its length and is suspended within a strong field of magnetic flux (**Figure 4.2c**). Sound-pressure variations between the front and the back of the diaphragm cause it to move and cut across these flux lines, thereby inducing a current into the ribbon that's proportional to the amplitude and frequency of the acoustic waveform. Because the ribbon generates such a small output signal (when compared with the larger output that's generated by the multiple wire turns of a moving coil), its output signal is too low to drive a microphone input stage directly; thus, a step-up transformer (or amp in the case of an active ribbon mic) must be used to boost the output signal and impedance to an acceptable range.

Until recently, traditional ribbon technology could only be found on original, vintage mics (such as the older RCA and Cole ribbon mics). However, with the skyrocketing price of vintage mics and a resurgence in the popularity of the smooth, transient quality of the ribbon "sound", modern reproductions and entirely new ribbon mic designs have begun to spring up on the market.

FIGURE 4.1
The dynamic microphone: (a) The Shure 58 dynamic mic (courtesy of Shure Incorporated, www.shure.com, Images © 2017, Shure Incorporated – used with permission). (b) Telefunken M81 dynamic microphone (courtesy of Telefunken Elektroakoustik, www.telefunkenelektroakustik.com). (c) Inner workings of a dynamic microphone.

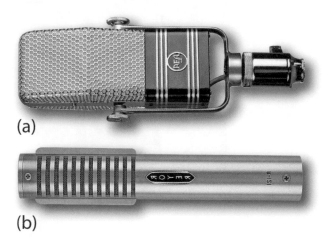

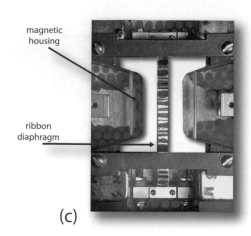

magnetic
housing

ribbon
diaphragm

(a)

(b)

(c)

FIGURE 4.2

The ribbon microphone:
(a) The AEA A440 ribbon
mic (courtesy of Audio
Engineering Associates,
www.ribbonmics.com). (b)
Royer Labs R-121 ribbon
microphone (courtesy of
Royer Labs, www.royerlabs.
com). (c) Cutaway detail of
a ribbon microphone (courtesy of Audio Engineering
Associates, www.ribbonmics
.com).

FURTHER DEVELOPMENTS IN RIBBON TECHNOLOGY

During the past several decades, certain microphone manufacturers have made
changes to original ribbon technologies by striving to miniaturize and improve
their basic operating characteristics. For example, the popular M160 (Figure 4.3)
and M260 ribbon mics from Beyerdynamic use a rare-earth magnet to produce
a capsule that's small enough to fit into a 2-inch grill ball (much smaller than a
traditional ribbon-style mic). The ribbon (which is corrugated along its length
to give it added strength and at each end to give it flexibility) is 3 microns thick,
about 0.08 inch wide, 0.85 inch long and weighs only 0.000011 ounce. A plastic
throat is fitted above the ribbon, which houses a pop-blast filter. Two additional
filters and the grill greatly reduce the ribbon's potential for blast and wind damage, a feature that has made these designs suitable for outdoor and handheld
use.

Other alterations to the traditional ribbon technology make use of phantom
power to supply power to an active, internal amplifier so as to boost the mic's
output to that of a dynamic or condenser mic without the need for a passive
transformer (an explanation of phantom power can be found in the next section
on condenser mics).

FIGURE 4.3

The Beyerdynamic M160
ribbon mic (courtesy of
Beyerdynamic, www.beyer-
dynamic.com).

(a)

(b)

The Condenser Microphone

Condenser mics (like those having capsules, which are shown in Figure 4.4) operate on an *electrostatic principle* rather than the electromagnetic principle used by a dynamic or ribbon mic. The capsule of a basic condenser mic consists of two plates: one very thin movable diaphragm and one fixed backplate. These two plates form a capacitor (or condenser, as it is still called in the UK and in many parts of the world). A capacitor is an electrical device that's capable of storing an electrical charge. The amount of charge that it can store is determined by its capacitance value and the voltage that's applied to it, according to the formula:

$$Q = CV$$

where Q is the charge (in coulombs), C is the capacitance (in farads) and V is the voltage (in volts).

At its most basic level, a condenser mic operates when a regulated DC power supply is applied between its diaphragm plates to create a capacitive charge. When sound acts upon the movable diaphragm, the varying distance between the plates will likewise create a change in the device's capacitance (Figure 4.5a). According to the preceding equation, if Q (the power supply charge) is constant

FIGURE 4.4
Inner details of a condenser mic: (a) AKG C214 condenser mic (courtesy of AKG Acoustics GmbH, www.akg.com). (b) Exposed diaphragm (courtesy of ADK, www.adkmic.com; photograph by K. Bujack).

FIGURE 4.5
Interactions as a result of diaphragm capacitance changes: (a) Output and potential relationships as a result of changing capacitance. (b) As a sound wave decreases the condenser spacing by d, the capacitance will increase, causing the voltage to proportionately fall (and vice versa).

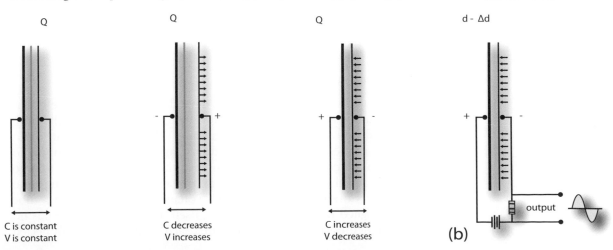

Q

Q

Q

d - Δd

C is constant
V is constant

(a)

C decreases
V increases

C increases
V decreases

output

(b)

and *C* (the diaphragm's capacitance) changes, then *V* (voltage across the diaphragm) will change in a proportional and inverse fashion, thereby giving us our output signal.

Since the charge (Q) is known to be constant, and the diaphragm's capacitance (C) changes with differences in sound pressure, the voltage (V) must change in inverse proportion.

Given that the capsule's voltage now changes in proportion to the sound waves that act upon it, boom! -- we now have a condenser mic that has an audio output signal!

The next trick is to tap into the circuit to capture these changes in output voltage. This is done by placing a high-value resistor across the diaphragm circuit. Since the voltage across the resistor will change in inverse proportion to the capacitance across the capsule plates, this signal will then become the feed for the mic's output (Figure 4.5b).

As the resulting signal has an extremely high impedance, it must be fed through a preamplifier in order to preserve the mic's frequency response characteristics. Since this amp must be placed at a point just following the resistor (often at a distance of 2 inches or less), it is almost always placed within the mic body in order to prevent hum, noise pickup and signal-level losses. This preamp (in addition to the need for a polarizing voltage source across the diaphragm leads) means that a powering voltage/current source must be included in the design.

POWERING A CONDENSER MIC

As you have read, all condenser microphones require a polarizing voltage as well as an amplifier that's required to step the impedance of the capsule down to a value that will work in everyday production applications. The three systems for dealing with these power requirements are:

- An external power supply
- Phantom power
- Electret self-charging system

EXTERNAL POWER SUPPLY

Older condenser microphones and modern-day tube reproductions are generally valued by studios and collectors alike for their sound, which results from even-harmonic distortion and "warm-sounding" characteristics that occur whenever vacuum tubes are used. These mics use an external power supply to provide power to the condenser diaphragm and internal amp circuits that go between the mic and the console/interface/mic preamp. The supply plugs into a 120/240-V power outlet to provide a high voltage/current supply to a tube (valve) that's designed into the mic's housing itself. Unfortunately, there is no standardization for these supplies, meaning that each supply must be connected to and stored with its associated mic (often leading to a jumble of cables on the studio floor, but that's the price we pay for our art).

PHANTOM POWER

Most modern professional condenser (and some ribbon) mics are designed to be powered directly from the console/interface/mic preamp through the use of a *phantom power supply*. Phantom power works by supplying a positive DC supply voltage of +48 V equally through both audio conductors (pins 2 and 3) of a balanced mic line to the condenser capsule and preamp (Figure 4.6). This voltage is equally distributed through identical value resistors so that no voltage differential exists between the two leads. The ground side of the circuit is then supplied to the capsule and preamp through the balanced cable grounding wire (pin 1).

Since the audio is only affected by potential differences between pins 2 and 3 (and not the ground signal on pin 1), the carefully matched +48 V powering potentials at these leads cancel out and are therefore not electrically "visible" to the input stage of a balanced mic preamp. Instead, only the balanced, alternating audio signal that's being simultaneously carried along the two audio leads (pins 2 and 3) will be seen by the audio circuitry.

The resistors (R) used for distributing power to the signal leads should be ¼-W resistors with a ±1% tolerance, and a value of 6.8 kΩ is often used within a 48-V system. In addition to precisely matching the supply voltages, these resistors also help to provide a degree of power isolation between other mic inputs on a console. If a signal lead were accidentally shorted to ground (which could happen if defective cables or unbalanced XLR cables were used), the power supply should still be able to deliver power to other mics in the system. If two or more inputs were accidentally shorted, however, the phantom voltage could then drop to levels that would be too low to be usable.

FIGURE 4.6
Schematic drawing of a phantom power wiring/cable system.

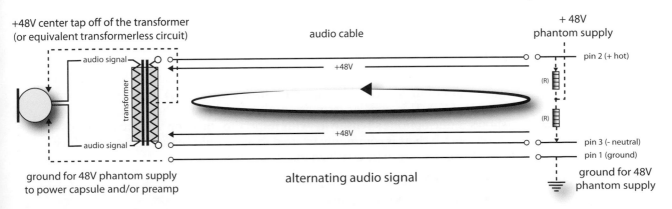

+48V center tap off of the transformer (or equivalent transformerless circuit)

audio cable

+ 48V phantom supply

audio signal

+48V

pin 2 (+ hot)

transformer

(R)

(R)

audio signal

+48V

pin 3 (- neutral)

pin 1 (ground)

ground for 48V phantom supply to power capsule and/or preamp

alternating audio signal

ground for 48V phantom supply

The Electret-Condenser Microphone

Electret-condenser mics work on the same operating principles as their externally polarized counterparts, with the exception that a static polarizing charge has been permanently set up between the mic's diaphragm and its backplate (using a process that works much like the static-cling that occurs when you take socks out of a dryer). Since the charge (*Q*) is permanently built into the capsule, no external source is required to power the diaphragm. However, as with a powered condenser mic, the capsule's output impedance is so high that a preamp will still be required to reduce it to a standard value. As a result, a battery, external powering source or standard phantom supply must be used to power the low-current amp.

TRY THIS: MIC TYPES

1. Go to youtube.com/modernrecordingtech niques, search for "Mic Types", and listen to the sound of each mic operating type.

2. If you'd like to DIY, then pull out several mics from each operating type and plug them in (if you don't have several types, maybe a studio, your school or a friend has a few you can borrow). Try each one on an instrument and/or vocal. Are the differences between operating types more noticeable than between models in the same family?

MICROPHONE CHARACTERISTICS

In order to handle the wide range of applications that are encountered in studio, project and on-location recording, microphones will often differ in their overall sonic, electrical and physical characteristics. The following section highlights many of these characteristics in order to help you choose the best mic for any given application.

Directional Response

The *directional response* of a mic refers to its sensitivity (output level) at various angles of incidence with respect to the front (on-axis) of the microphone (Figure 4.7a). This angular response can be graphically charted in a way that shows a microphone's sensitivity with respect to direction and frequency over 360°. Such a chart is commonly referred to as the mic's *polar pattern*. This directionality of a mic can be classified into two categories:

- Omnidirectional polar response
- Directional polar response

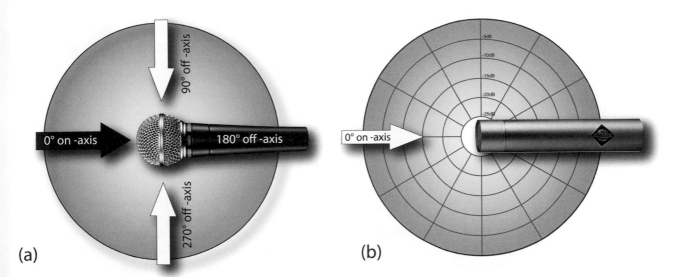

(a)   (b)

An *omnidirectional mic* (Figure 4.7b) is a pressure-operated device that's respon-sive to sounds which emanate from all directions. In other words, the diaphragm will react equally to all sound-pressure fluctuations at its surface, regardless of the source's location. Pickups that display *directional* properties are pressure-gradient devices, meaning that the pickup is responsive to relative differences in pressure between the front, back and sides of a diaphragm. For example, a purely pressure-gradient mic will exhibit a *bidirectional* polar pattern (commonly called a *figure-8 pattern*), as shown in Figure 4.8. Many of the older ribbon mics exhibit a bidirectional pattern, since their diaphragms are often exposed to sound waves from both the front and the rear axis. As such, they are equally sen-sitive to sounds that emanate from either direction (Figure 4.9a and b). Sounds from the rear will produce a signal that's 180° out of phase with an equivalent on-axis signal. Sound waves arriving 90° off-axis produce equal but opposite pressures at both the front and the rear of the ribbon (Figure 4.9c), resulting in a cancelation at the diaphragm and no output signal.

FIGURE 4.7
Directional axis of a micro-phone: (a) 0° is located at the front of the mic's capsule. (b) Graphic representation of a typical omnidirectional pickup pattern.

FIGURE 4.8
Graphic representation of a typical bidirectional pickup pattern.

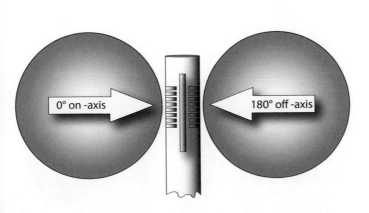

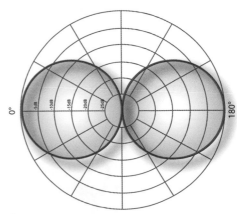

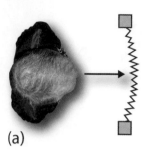

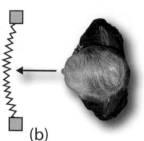

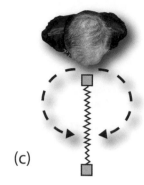

(a)

(b)

(c)

FIGURE 4.9
Sound sources on-axis and 90° off-axis at the ribbon's diaphragm. (a) The ribbon is sensitive to sounds at the front and (b) at the rear, while (c) sound waves from the sides (90° and 270°) off-axis are canceled.

Figure 4.10 graphically illustrates how the acoustical combination (as well as electrical and mathematical combination, for that matter) of a bidirectional (pressure-gradient) and omnidirectional (pressure) pickup can be combined to obtain various other directional pattern types. Actually, an infinite number of directional patterns can be obtained from this mixture, with the most widely known patterns being the *cardioid*, *supercardioid* and *hypercardioid* polar patterns (Figure 4.11).

Often, dynamic mics achieve a cardioid response (named after its heart-shaped polar chart, as shown in Figure 4.12) by incorporating a rear port into their design. This port serves as an acoustic labyrinth that creates an acoustic resistance (delay). In Figure 4.13a, a dynamic pickup having a cardioid polar response is shown receiving an on-axis (0°) sound signal. In effect, the diaphragm receives two signals: the incident signal, which arrives from the front, plus an acoustically delayed rear signal. In this instance, the on-axis signal exerts a positive pressure on the diaphragm and begins its travels 90° to a port located on the side of the pickup. At this port, the signal is delayed by another 90° (using an internal, acoustically resistive material or labyrinth). In the combined time it takes for the delayed signal to reach the rear of the diaphragm (180°), the on-axis signal moves on to the negative (180°) portion of its acoustic cycle and then begins

FIGURE 4.10
Directional combinations of various bidirectional and non-directional pickup patterns.

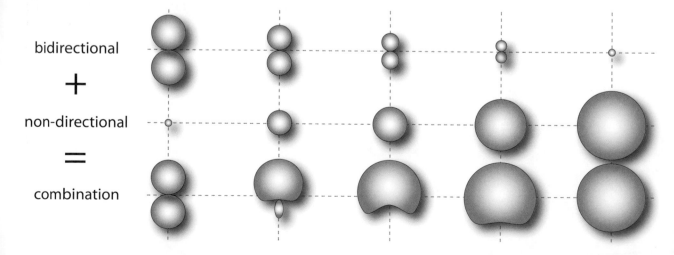

bidirectional

+

non-directional

=

combination

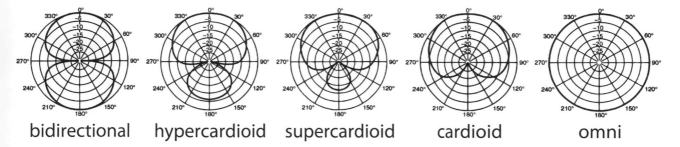

to exert a negative pressure on the diaphragm (pulling it outward). Since the delayed rear signal is also 180° out of phase at the rear of the diaphragm, it will begin to push it back outward, resulting in a reinforced positive output signal. In fact, this signal is actually increased in level due to the combined signals.

Conversely, when a sound arrives at the rear of the mic, it begins its trek around to the capsule's front. As the sound travels 90° to the side of the pickup, it is again delayed by another 90° before reaching the rear of the diaphragm. During this same delay period, the sound continues its journey around to the front of the mic. Since the acoustic pressures at the diaphragm's front and rear sides are equal and opposite in pressure, the diaphragm will be simultaneously pushed inward and outward with equal force, resulting in little or no movement, and therefore, will have little or no output signal (Figure 4.13b). The attenuation of such an off-axis signal, with respect to an equal on-axis signal, is known as its *front-to-back discrimination* and is rated in decibels.

By mounting two capsules back-to-back on a central backplate, a condenser mic can be electrically selected to switch from one polar pattern to another. For example, configuring these dual-capsule systems electrically in phase will create an omnidirectional pattern, while configuring them out of phase results in a bidirectional pattern. A number of intermediate patterns (such as cardioid and hypercardioid) can be created by electrically varying between these two polar states (in either continuous or stepped degrees), as was seen earlier in Figure 4.10.

FIGURE 4.11
Various polar patterns with output sensitivity plotted over the angle of incidence.

FIGURE 4.12
Graphic representation of a typical cardioid pickup pattern.

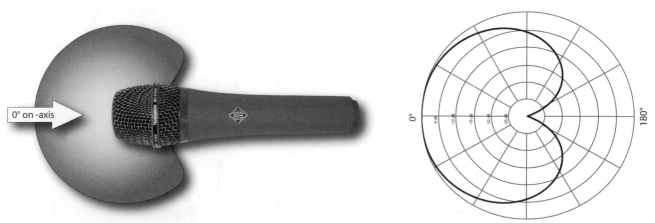

TRY THIS: POLAR PATTERNS

1. Go to youtube.com/modernrecordingtech niques, search for "Polar Patterns", and listen to the sound of each pattern.
2. If you'd like to DIY, then pull out several mics from each pattern type and plug them in (if you don't have several types, maybe a studio, your

school or a friend has a few you can borrow). You also might have a multi-pattern condenser around that you can use to switch between patterns. Just remember to mute the channel before you switch patterns to avoid the loud pops that often result.

Frequency Response

The on-axis *frequency-response curve* of a microphone is the measurement of its output over the audible frequency range when driven by a constant, on-axis input signal. This response curve (which is generally plotted as output level in dB over the 20- to 20,000-Hz frequency range) will often yield valuable information and can give clues as to how a microphone will react at specific frequencies.

A mic that's designed to respond equally to all frequencies is said to exhibit a flat frequency response (shown as the curve in Figure 4.14a). Others can be made to emphasize or de-emphasize the high-, mid- or low-end response of the audio spectrum (shown as the boost in the high-end curve in Figure 4.14b) so as to give it a particular sonic character. The solid frequency-response curves (as shown in both parts a and b) were measured on-axis and exhibit an acceptable response. However, the same mics might exhibit a "peaky" or erratic curve when measured off-axis (shown as the dotted curves). These signal colorations could

FIGURE 4.13

The directional properties of a cardioid microphone. (a) Signals arriving at the front (on-axis) of the diaphragm will produce a full output level. (b) Signals arriving from the rear of the diaphragm (90° – 90° = no output) will cancel each other out, resulting in a greatly reduced output.

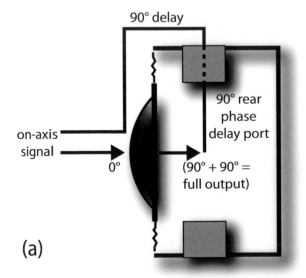

(a)

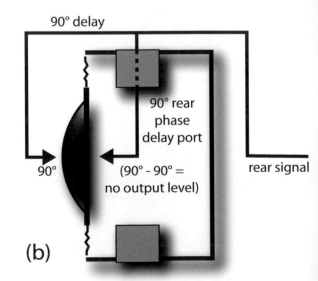

(b)

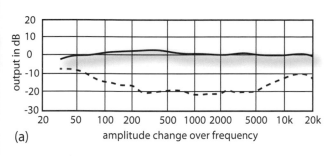

(a) amplitude change over frequency

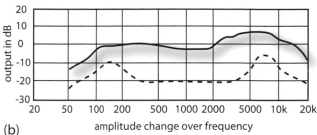

(b) amplitude change over frequency

affect their sound when operating in an area where off-axis sound (in the form of leakage) arrives at the pickup. Such colorations will often result in a tone quality change when the leaked signal is mixed in with other properly miked signals.

It's extremely important to keep in mind that there are many other variables that will determine how a mic will sound, some of which have no measurement standards. Some of the most coveted mics in the world have response characteristics that are far from flat. They might have a high-end peak, which helps to give them a "present sound", or they could have other personalities that make them perfect for a specific application. In short, looking at the specs can be helpful, but there is never a substitute for listening and making decisions based on what you hear with your own ears.

FIGURE 4.14

Frequency response curves: (a) Response curve of the AKG C460B/CK61 ULS. (b) Response curve of the AKG D321 (courtesy of AKG Acoustics GmbH., www.akg -acoustics.com).

Transient Response

A significant piece of data (which currently has no accepted standard of measure) is the *transient response* of a microphone (Figure 4.15). Transient response is the measure of how quickly a mic's diaphragm will react when it is hit by an acoustic wave front. This figure varies wildly among microphones and is a major reason for the difference in sound quality among the three pickup types. For example, the diaphragm of a dynamic mic can be quite large (up to 2.5 inches). With the additional weight of the coil of wire and its core, this combination can be very large in mass when compared with the power of the sound wave that drives it. Because of this, a dynamic mic can be very slow in reacting to a waveform; this often gives it a rugged, gutsy and less accurate sound. By comparison, the diaphragm of a ribbon mic is much lighter, so its diaphragm can react more quickly to a sound waveform, resulting in a clearer, more present sound. The

FIGURE 4.15

Transient response characteristics of a percussive woodblock using various microphone types: (a) Shure SM58 dynamic. (b) RCA 44BX ribbon. (c) AKG C3000 condenser.

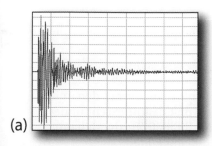

(a)

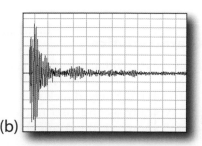

(b)

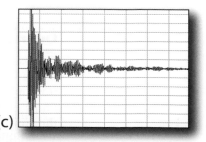

(c)

condenser pickup has an extremely light diaphragm, which varies in diameter from 1 inch to less than 0.25 inch and has a thickness of about 0.0015 inch. This means that the diaphragm offers very little mechanical resistance to a sound-pressure wave, allowing it to accurately track the wave over the entire frequency range – potentially giving it a present and clear sound.

Output Characteristics

A microphone's *output characteristics* refer to its measured sensitivity, equivalent noise, overload characteristics, impedance and other output responses.

Sensitivity Rating

A mic's *sensitivity rating* is the output level (in volts) that a microphone will produce, given a specific and standardized acoustic signal at its input (rated in dB sound-pressure levels [SPL]). This figure will then specify the amount of amplification that's required to raise the mic's signal to line level (often referenced to –10 dBv or +4 dBm) and allows us to judge the relative output levels between any two mics. A microphone with a higher sensitivity rating will produce a stronger output signal voltage than one with a lower sensitivity.

Equivalent Noise Rating

The *equivalent noise* rating of a microphone can be viewed as a device's electrical self-noise. It is expressed in dB^{SPL} or dBA (a weighted curve) as a signal that would be equivalent to the mic's self-noise voltage. As a general rule, the mic itself doesn't contribute much noise to a system when compared with the mixer's amplification stages, the recording system or media (whether analog or digital). However, with recent advances in mic preamp/mixer technologies and overall reductions in noise levels produced by digital systems, these noise ratings have become increasingly important. Interestingly enough, the internal noise of a dynamic or ribbon pickup is actually generated by the electrons that move within the coil or ribbon itself. Most of the noise that's produced by a condenser mic is generated by the built-in preamp. It almost goes without saying that certain microphone designs will have a higher degree of self-noise than others; thus, care should be taken in your microphone choices for critical applications (such as with classical recording or film production techniques).

Overload Characteristics

Just as a microphone is limited at low levels by its inherent self-noise, it's also limited at high SPLs by *overload distortion*. In terms of distortion, the dynamic microphone is an extremely rugged pickup, often capable of an overall dynamic range of 140 dB. Typically, a condenser microphone won't distort except under the most severe SPL; however, the condenser system differs from the dynamic in that at high acoustic levels, the capsule's output might be high enough to overload the mic's preamplifier. To prevent this, most condenser mics offer a switchable attenuation pad that immediately follows the capsule output and serves to reduce the signal level before the preamp's input, thereby reducing or eliminating overload distortion. When inserting such an attenuation pad into the circuit, keep in mind that the mic's signal-to-noise ratio will be degraded by the amount of applied attenuation; therefore, it's always wise to remove the inserted pad when using the microphone under normal level conditions.

Microphone Impedance

Microphones are designed to exhibit different *output impedances*. Output impedance is a rating that's used to help you match the output resistance of one device to the rated input resistance requirements of another device (so as to provide the best possible level and frequency response matching).

Impedance is measured in ohms (with its symbol being Ω or Z). The most commonly used microphone output impedances are 50, 150 and 250 Ω (low) and 20 to 50 kΩ (high). Each impedance range has its advantages. In the past, high-impedance mics were used because the input impedances of most tube-type amplifiers were high. A major disadvantage to using high-impedance mics is the likelihood that their cables will pick up electrostatic noise (like those caused by motors and fluorescent lights). To reduce such interference, a shielded cable is necessary, although this begins to act as a capacitor at lengths greater than 20 to 25 feet, which serves to reduce much of the high-frequency information that's picked up by the mic. For these reasons, high-impedance microphones are rarely used in the professional recording process.

Most modern-day systems, on the other hand, are commonly designed to accept a low-impedance microphone source. The lines of very-low-impedance mics (50 Ω) have the advantage of being fairly insensitive to electrostatic pickup. They are, however, sensitive to induced hum pickup from electromagnetic fields (such as those generated by AC power lines). This extraneous noise can be greatly reduced through the use of a twisted-pair cable, because the interference that's magnetically induced into the cable will flow in opposite directions along the cable's length and will cancel out at the console or mixer's balanced microphone input stage. Mic lines of 150 to 250 Ω are less susceptible to signal losses and can be used with cable lengths of up to several thousand feet. They're also less susceptible to electromagnetic pickup than the 50-Ω lines but are more susceptible to electrostatic pickup. As a result, most professional mics operate with an impedance of 200 Ω, using a shielded twisted-pair cable to reduce noise.

A number of high-end preamps that are now on the market offer a variable input impedance control, allowing the preamp to match its impedance to the mic's design characteristics. This can have an effect on the overall sound and operating characteristics of the pickup.

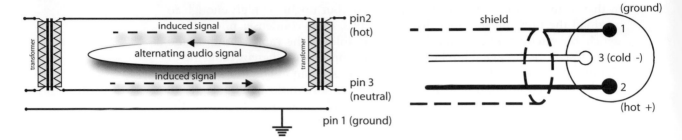

FIGURE 4.16
Wiring detail of a balanced microphone cable, whereby the induced signals travel down the wires in equal and opposite polarities that cancel at the transformer. The AC audio signals are left unaffected and are able to generate an output signal (courtesy of Loud Technologies Inc., www.mackie.com).

BALANCED/UNBALANCED LINES

In short, a *balanced line* uses three wires to properly carry audio. Two of the wires are used to independently carry the audio signal, while a third lead is used as a neutral ground wire. From a noise standpoint, whenever an electrostatic or electromagnetic signal is induced across the audio leads, it will be induced into both of the two audio leads at an equal level and polarity (Figure 4.16). Since the input of a balance device will only respond to the alternating voltage potentials between the two audio leads, the unwanted noise (which has the same polarity) will be canceled out.

Several connector types are used to route analog, balanced signals between devices:

- XLR connectors (Figure 4.17) are most commonly used to connect microphones to preamp, console and interface systems. They are also used to connect line-level connections between professional effects and audio systems devices. The two-conductor, balanced, XLR connector and cable specifies pin 2 as being positive (+ or hot) and pin 3 as being negative (– or neutral), with the cable ground being connected to pin 1.
- 1/4″ balanced connectors (Figure 4.18) are used to interface between both professional and semi-pro equipment at line level. These connectors are also sometimes used to provide a line-level balanced or unbalanced connection. The 1/4″ balanced connector and cable specifies the tip as being positive (+ or hot) and the middle sleeve as being negative (– or neutral), with the cable ground being connected to the connector's shield.
- TT (or bantam) connectors/cables allow balanced lines to be plugged into a patch bay for fast and easy access to the various line-level devices within a production facility.

FIGURE 4.17
Diagram for wiring a balanced cable to a balanced XLR connector.

If the hot and neutral pins of balanced mic cables are haphazardly pinned in a music or production studio, it's possible that any number of mics (and other

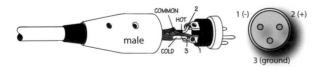

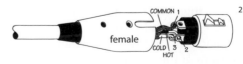

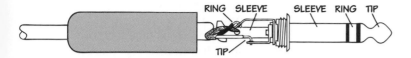

FIGURE 4.18
Physical and diagram wiring for a balanced (two-conductor) 1/4-inch phone connector.

equipment, for that matter) could be wired in opposite, out-of-phase polarities. For example, if a single instrument were picked up by two mics using two improperly phased cables, the instrument might totally or partially cancel when mixed to mono. For this reason, it's always wise to use a phase tester or volt-ohm meter to check the cable wiring throughout a pro or project studio complex.

High-impedance mics and most line-level instrument lines use unbalanced lines (Figure 4.19) to transmit signals from one device to another. In an unbalanced circuit, a single signal lead carries a positive current potential to a device, while a second, grounded shield (which is tied to the chassis ground) is used to complete the circuit's return path. When working at low signal levels (especially at mic levels), any noises, hums, buzzes or other types of interference that are induced into the signal path will be amplified along with the input signal.

Soldering, Baby!

Before we leave the section on mic cables and their types, there's a rather important request that both DMH and EC would like to make … and this is that you consider learning how to solder and make your own cables.

If you happen to have cables and connectors hanging around, it would be a huge bummer not to be able to get your trusty soldering iron and solder out and fix a bad connector or to make up a new connector that exactly fits your needs, instead of having to go online and wait for several days while a manufactured cable is being delivered.

So, here's the deal: simply go online and search for a beginner's kit that includes a simple hand soldering iron, some solder, and extra odds and ends to help with the process. If you want to go just a bit further, you might want to also order a

FIGURE 4.19
Unbalanced microphone circuit: (a) Diagram for wiring an unbalanced microphone (or line source) to a balanced XLR connector. (b) Diagram for wiring an unbalanced 1/4-inch phone connector. (c) Physical and diagram wiring for an unbalanced 1/4-inch phone connector. (d) Diagram for wiring an unbalanced phono (RCA) connector (courtesy of Loud Technologies Inc., www.mackie.com).

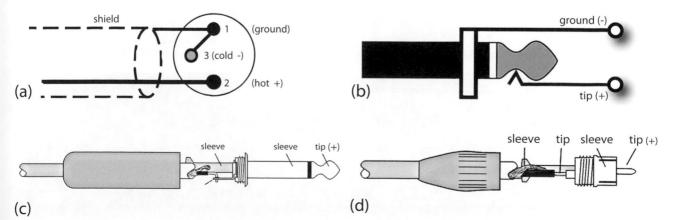

simple electronics kit that you can build yourself. Personally, DMH opted for a digital clock kit. It'll help you with learning how to build and solder simple circuit boards and in the end, you'll have a cool device that might be of use. The clock has been happily doing its thing in the bedroom for a few years now. Again, the goal is to learn while having fun.

MICROPHONE PREAMPS

FIGURE 4.20

Outboard microphone preamplifier examples. (a) PreSonus DigiMAX D8 8-channel microphone preamp with lightpipe (courtesy of Presonus Audio Electronics, Inc., www.presonus.com). (b) Rupert Neve Designs Portico 5024 4-channel mic preamp (courtesy of Rupert Neve Designs LLC, www.rupertneve.com). (c) Universal Audio 4-710d Four Channel Hybrid (Tube/Transistor) Preamplifier (courtesy of Universal Audio, www.uaudio.com © 2022 Universal Audio, Inc. All rights reserved. Used with permission).

Since the output signals of most microphones are at levels far too low to drive the line-level input stage of most recording systems, a mic preamplifier must be used to boost their signal to acceptable line levels (often by 30 to 70 dB). With the advent of improved technologies in analog and digital console design, hard-disk recorders, digital audio workstations (DAWs), signal processors and the like, low noise and distortion figures have become more important than ever. In this day and age, most of the mic "pre"s (pronounced "preeze") that are designed into an audio interface or production console are capable of providing a professional, high-quality sound. It's not uncommon, however, for an engineer, producer or artist to prefer a preamp which has a personal and "special sound" that can be used in a critical application to produce just the right tone for a particular application. In such a case, an outboard mic preamp might be chosen instead (**Figure 4.20**) for its characteristic sound, low noise or special distortion specs. These devices might make use of tube, FET and/or integrated circuit technology, and offer advanced features in addition to the basic variable input gain, phantom power and high-pass filter controls. As with most recording tools, the sound, color scheme, retro style, tube or transistor type, and budget level are up to the individual, the producer and the artist – it's completely a

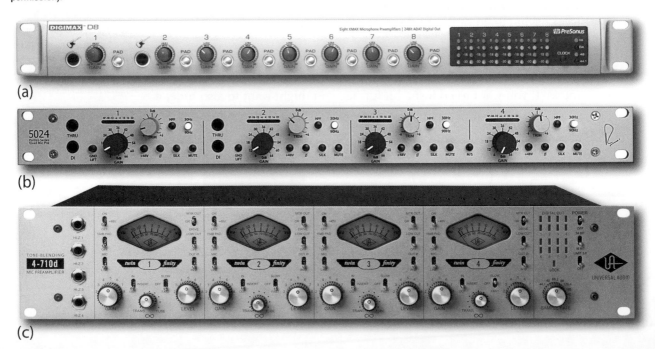

(a)

(b)

(c)

matter of personal style and taste, and this includes choosing to use the high-quality preamps that are built into your interface/mixer.

MODELED CONDENSER MIC SYSTEMS

Modeled condenser mic systems (Figure 4.21) make use of a single mic type that is designed to be used in conjunction with modeling software that can alter the frequency response and other characteristics to make the mic sound like other popular (and often revered) mic types.

Using a plug-in, instead of DSP processing built into the mic, has the major advantage of allowing the user to change all of the modeling settings, such as mic type and polar pattern. As these systems make use of dual capsules, it's possible to record the channel information of each capsule and then alter the overall model type, pickup response, polar response, off-axis response and other characteristics during mixdown after the performance has been recorded. It's even possible to record each capsule output to a separate track as a stereo pickup or to alter the various characteristics of each capsule such that a different mic emulation can be placed on either pickup side … even after the fact, during mixdown.

MICROPHONE TECHNIQUES

Most microphones have a distinctive sound character that's based on their specific type and design. A large number of types and models can be used for a variety of applications, and it's up to the engineer/artist to choose the right one for the job. In general, there are two particular paths that one can take when choosing the types and models of microphones for a studio's production toolbox. These can be placed into two categories:

- You can select a limited range of mics that are well suited for a wide range of applications
- You can acquire a larger collection of mics that are commonly perceived as being individually suited for a particular instrument or situation
- Or both

FIGURE 4.21
Townsend Labs Sphere L22™ microphone system for the UAD interface and DSP systems (courtesy of Universal Audio, www.uaudio. com © 2022 Universal Audio, Inc. All rights reserved. Used with permission).

The first approach is ideal for the project studio and those who are just starting out on a limited budget. This is also common practice among seasoned professionals who swear by a limited collection of their favorite mics that are chosen to cover a wide range of applications. These dynamic, ribbon and/or condenser mics can be used both in the project studio and in the professional studio to achieve the best possible sound on a budget.

The second approach (DMH often refers to it as the "Alan Sides" approach – go ahead, Google him) is better suited to the professional studio (and to personal mic collectors). This path is taken by those who have a need or desire to amass a large "dream collection" of carefully chosen mics that can be surgically used for particular applications. In the end, both approaches have their merits – indeed, it's usually wise to keep an open mind and choose a range of mic types that can best fit your needs, budget and personal style. However, you might consider buying a matched pair of mics, especially at first. This opens up your options for placing two mics in the room or on an instrument so you can record in stereo (overhead drums, stereo XY mics on a guitar, placing room mics, etc., come to mind just for starters).

Choosing the appropriate mic, however, is only half the story. The placement of a microphone will often play just as important a role in getting the right sound, and it is one of the engineer's most valued tools. Because mic placement is an art form, there is no right or wrong. Placement techniques that are currently considered "bad" might easily be accepted as being standard practice five years from now, and as new musical styles develop, new recording techniques will also tend to evolve, helping to breathe new life into music and production. The craft of recording should always be open to change and experimentation – two of the strongest factors that keep the music and the biz of music alive and fresh.

Here are several pieces of practical advice that can help get you started.

- If you're recording a musician (especially an experienced one), you might consider asking him or her how they've been recorded in the past. Do they have a favorite mic or technique that's often worked best for them? This tactic can help to put the artist at ease and give you insights into new studio miking techniques that can be a helpful production and educational tool.
- Think carefully before "printing" a recorded signal with effects directly to a track. The recording of an instrument with effects or dynamics can't be undone at a later time – so unless you're absolutely sure, it's often wise to add effects to the track later during mixdown, or as an option, you could print the signal to two tracks, one with and one without effects.
- Make use of the various aspects of a room's acoustics when recording an instrument. This could include the use of distance as a tool for changing the size and character of an instrument or group (which could be recorded to separate tracks for later blending within the mix).
- Although there are no rules for this type of creativity, you might keep in mind the instrument and traditional pickup style that's to be recorded so as to not go too far afield from the expected norm (unless you want to break the rules to lay a new path).

Other Microphone Pickup Issues

Before we move into the realm of microphone placement and pickup techniques, let's take a quick look at a few issues that are common to many pickups and placement situations. These are:

- Low-frequency rumble
- Proximity effect
- Popping
- Off-axis pickup

LOW-FREQUENCY RUMBLE

When using a mic in the studio or on location, *rumble* (low-frequency, high-level vibrations that occur in the 3- to 25-Hz region) can easily be transmitted from the floor of a studio, hall or unsupported floor, through the mic stand and directly to the mic. Sources such as passing trucks, air conditioners, subways or fans can be reduced or eliminated in a number of ways, such as:

- Using a shock mount to isolate the mic from the vibrating surface and floor stand
- Choosing a mic that displays a restricted low-frequency response
- Restricting the response of a wide-range mic by using a low-frequency roll-off filter

PROXIMITY EFFECT

Another low-frequency phenomenon that occurs in most directional mics is known as *proximity effect*. This common effect causes an increase in bass response whenever a directional mic is brought within 1 foot of the sound source. This bass boost (which is often most noticeable on vocals) proportionately increases as the distance decreases. To compensate for this effect (which is somewhat greater for bidirectional mics than for cardioids), a low-frequency roll-off filter switch (which is often located on the microphone body) can be used. If none exists, an external roll-off or equalizer can be used to reduce the low end. Finally, the directional mic can be swapped with an omni-direction one – yes, this actually works! Since an omni doesn't have a physical or electrical circuit for directionality, there won't be a level differential between the front and the back of the mic … meaning that the bass response won't be boosted. Any of these tools can be used to help restore the bass response to a flat and natural-sounding balance.

On a more positive note, this increase in bass response has long been appreciated by vocalists and DJs for its ability to give a full, "larger-than-life" quality to voices that are otherwise thin. In many cases, the use of a directional mic has become an important part of the engineer's, producer's and vocalist's toolbox.

TRY THIS: PROXIMITY EFFECT

1. Go to youtube.com/modernrecordingtech niques, search for "Proximity Effect", and listen to the sound of each mic operating type.
2. If you'd like to DIY, then pull out an omnidirectional, cardioid and bidirectional mic (or one that can be switched between these patterns).
3. Move in on each mic pattern type from distances of 3 feet to 6 inches (being careful of volume levels and problems that can occur from popping).
4. Does the bass response increase as the distance is decreased with the cardioid? With the bidirectional? What about the omni's low-end boost?

POPPING

Another annoying problem that's generally associated with directional mics that are exposed to wind or breath blasts is *popping*. Of course, we've all experienced the effect of talking into a mic that blasts a loud, low-frequency noise whenever we say or sing the letter "p" or "b". The best ways to reduce or eliminate popping are:

- To slip a pop filter over the head of a mic, which provides resistance to breath and wind blasts
- To place a blast/pop shield between the artist and the mic
- To place the mic slightly off-axis (off center) to the singer
- To replace the directional microphone with an omnidirectional mic when working at close distances

OFF-AXIS PICKUP

As we've previously seen, when sound arrives at a directional microphone from off-axis (from the sides, rear, etc.), there are degrees of reduction in level that occur due to acoustical or electrical delay phase cancelation at the diaphragm. These cancelations will vary in level depending upon the emanating direction of the sound source – meaning that a mic's off-axis frequency response may differ from its on-axis response.

This difference in response might not be a problem (or at least, something to watch out for) if it were not for acoustic leakage that arrives off-axis at a pickup from another source that's also miked (Figure 4.22). If this leakage is loud enough, the off-axis pickup will combine with the other instrument's pickup to change the sound's overall tone color – particularly if the original and leaked

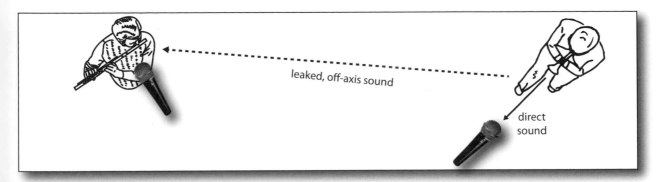

leaked, off-axis sound

direct
sound

FIGURE 4.22
Off-axis leakage can
combine with another direct
mic pickup to alter the tonal
character of a sound.

sounds are mixed to a single channel. None of this is meant to scare you; it's simply to point out that you should be aware of what can happen when an instrument or sound source is picked up by multiple mics.

One of the top engineers in the world (the late, great Al Schmitt) made extensive use of leakage to create a live, present, "you-are-there" feel to his recordings, often choosing to use omnidirectional mics (which actually have far less off-axis coloration due to the fact that no directional phase cancelation occurs). Of course, he used some of the best musicians in the world and recorded at the revered Capitol Studios, but that doesn't mean that you can't get amazing results by experimenting with the artist and any room using alternative mic techniques.

PICKUP CHARACTERISTICS AS A FUNCTION OF WORKING DISTANCE

In studio and soundstage recording, four fundamental styles of microphone placement are directly related to the working distance of a microphone from its sound source. These extremely important placement styles are as important as any tool in the toy box:

- Close miking
- Distant miking
- Accent miking
- Ambient miking

It's very interesting that with the resurgence of retro gear, the industry is finally coming around to the idea of changing the working distance as well as making use of a room's acoustics as an important part or "effect" within the recording process. As one recording artist put it:

> You can use the room to change the era of the sound. If you want a modern vocal sound, get right up on the mic. If you want a Motown vocal sound, step away from the mic three feet and sing harder.
>
> Ben Harper

On a personal note, both DMH and EC really can't stress enough the power of making use of placement, the room's acoustics and distance as a trilogy of tools for capturing the depth and feel of a performance.

Close Microphone Placement

When a *close microphone* placement is used, the mic is often positioned about 1 inch to 3 feet from a sound source. This commonly used technique generally yields two results:

- It creates a tight, present sound quality
- It effectively excludes the acoustic environment from being picked up

Because sound diminishes with the square of its distance from the sound source, a sound that originates 3 inches from the pickup will be much higher in level than one that originates 6 feet from the mic (Figure 4.23). Therefore, whenever close miking is used, only the desired on-axis sound will be recorded; extraneous, distant sounds (for all practical purposes) won't be picked up. In effect, the distant pickup will be masked by the closer sounds and/or will be reduced to such a relative level that it's well below the main pickup signal.

Probably the best analogy that can be given at this point is that of a microscope or an eyepiece. For this example, let's place a guitar player in the middle of a studio (feel free to "Try this" for yourself). If we look through our eyepiece from a distance of 6 feet (about 2 meters) at the guitar, we'll see the entire instrument and probably a good amount of the player. As we move further in, we'll see less and less of the player and even the guitar. As we move closer into the instrument, we'll start to see only certain parts of the instrument (the strings, the resonance hole, the player's fingers over the strings, etc.). Of course, as we move further in, only a microscopic part of the instrument will be seen.

FIGURE 4.23
Close miking reduces the effects of picking up the distant acoustic environment.

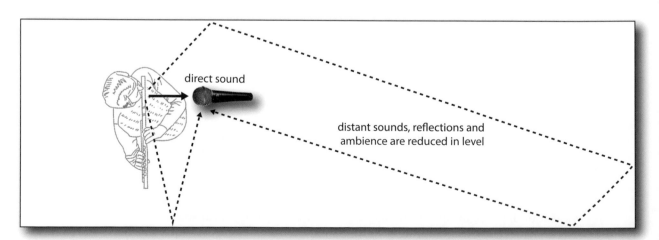

direct sound

distant sounds, reflections and ambience are reduced in level

Conversely, as we move out again past the 2-meter point, we'll start seeing more and more of the room, until the player, instrument and room all combine together to make up the overall pickup.

Of course, this visual analysis is directly analogous to how a mic pickup will "hear" the combined overall room/instrument sound – and will then sonically "zoom in" on parts of the instrument as the pickup is moved closer to the instrument and then closer still to specific parts of the instrument.

This analogy works for all types of acoustic spaces and instruments – we urge you to put a single mic and player out in the room and try zooming the pickup in and out for yourself.

TRY THIS: CLOSE MIKING

1. Go to youtube.com/modernrecordingtech niques, search for "Close Miking", and listen to the sound of each mic operating type.
2. If you'd like to DIY, then mic an acoustic instrument (such as a guitar or piano) at a distance of 1 to 3 inches.
3. Move (or have someone move) the mic over the instrument's body as it's being played, while

listening to variations in the sound. Does the sound change? What are your favorite and least favorite positions?
4. Now, pull the mic back by a foot or more – then to a distance of 6 feet (2 m) or more. How does the overall pickup change?

Because close mic techniques commonly involve distances of 1 to 6 inches (2.5 to 15 cm), the tonal balance (timbre) of an entire sound source often can't be picked up; rather, the mic might be so close to the source that only a small portion of the surface is actually picked up. This gives it a tonal balance that's very area specific (much like hearing the focused parts of an instrument through an acoustic microscope in our earlier explanation). At these close distances, moving a mic by only a few inches can easily change the pickup tonal balance. If this occurs, try using one or more of the following remedies:

- Move the microphone along the surface of the sound source until the desired balance is achieved
- Place the mic farther back from the sound source to allow a wider angle (thereby picking up more of the instrument's overall sound)
- Change the mic
- Equalize the signal until the desired balance is achieved

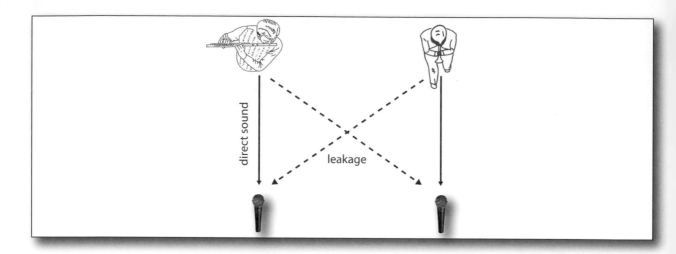

FIGURE 4.24
Leakage occurs due to the indirect pickup of a distant signal source.

FIGURE 4.25
The first console built for modern recording, at Bill Putnam's Universal Recorders in Chicago, circa 1950 (courtesy of Universal Audio, www.uaudio.com © 2022 Universal Audio, Inc. All rights reserved. Used with permission).

LEAKAGE

Whenever an instrument's mic is placed at enough of a distance to also pick up the sound of a nearby instrument, a condition known as leakage will occur (Figure 4.24). Whenever a signal is picked up by both its intended mic and a nearby mic (or mics), it's easy to see how the signals could be combined together within the mixdown process. Whenever this occurs, level and phase cancelations can sometimes make it more difficult to have control over the volume and tonal character of the involved instruments within a mix.

But as with everything in life, too much or too little might end up creating an imbalance in your recording. Think of leakage as a tool for you to create an atmosphere or to achieve a specific sound in your recording.

Some of the most interesting and original sounds came from the use of leakage as a tool. Back in the 50s or 60s, a recording desk didn't have many inputs (Figure 4.25), so musicians had to share the same mics and acoustic space; hence, both the engineers and the musicians had to learn how to balance on the fly. For example, listen to Dave Brubeck's "Take Five" or Frank Sinatra's "Live at the Sands", and you will understand what leakage that supports the music sounds like.

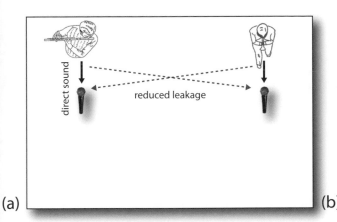
(a)

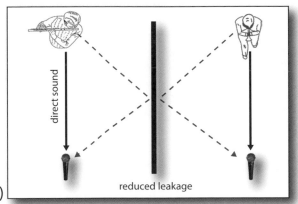
(b)

As a dear friend of ours, George Massenburg, says: a very sharp knife, in the hands of a master chef, is an instrument of precision and can create works of art … but in a less experienced set of hands, a finger just might get cut off!

So, in order to avoid the problems that can be associated with leakage, try the following:

- Place the mics closer to their respective instruments (Figure 4.26a).
- Use directional mics.
- Place an acoustic barrier (known as a flat, gobo or divider) between the instruments (Figure 4.26b). Alternatively, mics/instruments can be surrounded on several sides by sound baffles, and (if needed) a top can be draped over them.
- Spread the instruments farther apart.
- An especially loud (or quieter) instrument can be isolated by putting it in an unused iso-room or vocal or instrument booth. Electric instrument amps that are played at high volumes can also be recorded in an isolated room or area. The amp and mic can be covered with a blanket, iso-box or other sound-absorbing material so that there's a clear path between the amplifier and the mic.
- Separation can be achieved by plugging otherwise loud electronic instruments directly into the console via a direction injection (DI) box, thereby bypassing the miked amp.

Obviously, these isolation examples only hint at the number of possibilities that can occur during a session. For example, you might choose not to isolate the instruments and instead, place them in an acoustically "live" room. This approach will require that you carefully place the mics in order to control leakage; however, the result will often yield a live and present sound. As an engineer, producer and/or artist, the choices belong to you. Remember, the idea is to prepare your session, work out the kinks beforehand and simplify technology as much as possible in the studio, because Murphy's law is always alive and well in any production room.

FIGURE 4.26
Two methods for reducing leakage: (a) Place the microphones closer to their sources. (b) Use an acoustic barrier to reduce leakage.

Whenever individual instruments are being miked close (or semi-close), it's generally wise to follow the 3:1 *distance rule*.

3:1 Distance Rule

To reduce leakage and maintain phase integrity, this rule states that for every unit of distance between a mic and its source, a nearby mic (or mics) should be separated by at least three times that distance (Figure 4.27).

Some err on the side of caution and avoid leakage even further by following a 5:1 distance rule. As always, experience will be your best teacher. Although the close miking of a sound source offers several advantages, a mic should be placed only as close to the source as is necessary, not as close as possible. Unless care is taken and careful experimentation is done, miking too close can color the recorded tone quality of a source and/or yield an unnaturally dry pickup.

Again, it should be noted that a bit of "bleed" (a slang word for leakage) between mics just might be a good thing. With semi-distant and even multiple mics that are closely spaced, the pickup of a source by several pickups can add a sense of increased depth and sonic space. Having an overall distant set of mics in the studio can add a dose of natural ambience that can actually help to "glue" a mix together. The concept of minute phase cancelations and leakage in a mix isn't always something to be feared; it's simply important that you be aware of the effects that it can have on a mix … and use that knowledge to your advantage.

In addition to all these considerations, the placement of musicians and instruments will often vary from one studio and/or session to the next because of the room, the people involved, the number of instruments, isolation (or lack thereof) among instruments, and the degree of visual contact that's needed for creative communication. If additional isolation (beyond careful microphone

FIGURE 4.27

Example of the 3:1 microphone distance rule: "For every unit of distance between a mic and its source, a nearby mic (or multiple mics) should be separated by at least three times that distance".

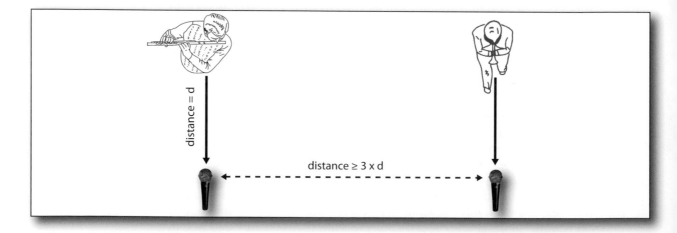

placement) is needed, flats and baffles can be placed between instruments in order to prevent loud sound sources from spilling over into other open mikes. Alternatively, the instrument or instruments could be placed into separate isolation (iso) rooms and/or booths, or they could be overdubbed separately at a later time.

During a session that involves several musicians or more, the setup should allow them to see and interact with each other as much as possible. It's extremely important that they be able to give and receive visual cues so they can better "feel the vibe". The instrument/mic placement, baffle arrangement and possibly, room acoustics (which can often be modified by placing absorbers in the room) will depend on the engineer's and artists' personal preferences as well as the type of sound the producer wants – so, it's always a good idea to consult with the producer and/or band before any studio setup is attempted.

Again, it should be pointed out that some of the world's top engineers/producers would also say that "leakage can be your friend" in certain circumstances. This hints at the idea that when the "good rule" is in full effect, and the room's acoustics match the style of the recording, pulling the carefully chosen mics away from the sound source to pick up more of the room (along with other instrument leakage) just might add life to the overall sound. One example of this that comes to mind is the leakage that was sometimes used to add "life" to the late Al Schmitt's jazz recordings, made at Capitol Records' famous studios in LA. Another rather novel approach to positive leakage was used by the late Bruce Swedien (Michael Jackson's then engineer) when he would set up mics in the studio for 12 strings, have 6 players sit in the first few rows (with all the mics open), then have the 6 move to the last rows (again, with all mics open) and then combine the two overdubs, allowing the leakage to add to the overall liveness of the sound. Should Michael J. be required to layer his voice in multiple layers, Bruce would have him move his position from the mic so that the time differences would make for a more life-like combined track. These are but a few of the many tricks that can be applied to add "life" to a recording and mix.

Recording Direct

Should there be a problem with miking an instrument, due to leakage, sonic preference or bad acoustics in a room, the signal of an electric or electronic instrument (guitar, keyboard, etc.) can alternatively be "directly injected" into a console, recorder or DAW without the use of a microphone. This option could produce a cleaner, more present sound by bypassing the distorted components of a head/amp combination, or the feed could be taken after the amp's feed. In the project or recording studio, the DI box (Figure 4.28) serves to interface an instrument with an analog output signal to a console or recorder in the following ways:

- It reduces an electric or electronic instrument's line-level output signal to mic level for direct insertion into the console's mic input jack.

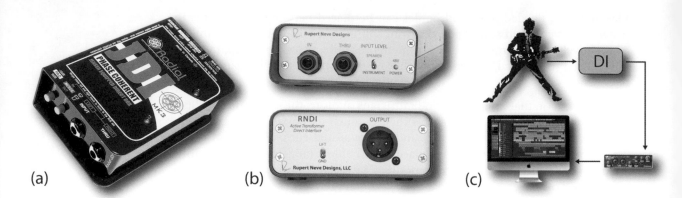

(a) (b) (c)

FIGURE 4.28
The DI box: (a) Radial JDI passive direct box (courtesy of Radial Engineering, www. radialeng.com). (b) Rupert Neve Designs RNDI Active Transformer Direct Interface (courtesy of Rupert Neve Designs, www.rupertneve. com). (c) DI box electric guitar/DI/interface connection.

- It changes an instrument's unbalanced, high-source impedance line to a balanced, low-source impedance signal that's needed by the console's input stage.
- It often can electrically isolate the audio signal paths between the instrument and mic/line preamp stages (thereby reducing the potential for ground-loop hum and buzzes).

Most commonly, the instrument's output is plugged directly into the DI box (where it's stepped down in level and impedance), and the box's output is then fed into the mic "pre"s of a console or DAW. If a "dirtier" sound is desired, certain boxes will allow high-level input signals to be taken directly from the amp's speaker output jack. It's also not uncommon for an engineer, producer and/or artist to combine the punchy, full sound of a mic with the present crispness of a direct sound. These signals can then be combined onto a single track or recorded to separate tracks (thereby giving more flexibility in the mixdown stage). The ambient image can be "opened up" even further by mixing a semi-distant or distant mic (or stereo pair) with the direct (and even the close-miked amp) signal. This ambient pickup can be mixed either into a stereo field or into an immersive field to fill out the sound.

When recording a guitar, the best tone and lowest hum pickup for a direct connection occurs when the instrument volume control is fully turned up. Because guitar tone controls often use a variable treble roll-off, leaving the tone controls at the full treble setting and using a combination of console equalization (EQ) and different guitar pickups to vary the tone will often yield the maximum amount of control over the sound. Note that if the treble is rolled off at the guitar, boosting the highs with EQ will often increase pickup noise.

Distant Microphone Placement

Within the study of modern recording and production techniques, one of the less understood aspects of miking is distant placement. This is a shame, as having an insight into distant microphone techniques unlocks the key to a wide range of tools and tricks that can be useful within the studio, soundstage and concert hall.

Let's begin our study into distance by taking a visual and experimental approach to the subject:

- If we were to look through a viewfinder that was placed very, very close to an instrument, we might only see an area of black.
- If we pull back by only 2 inches, we might see some wood but still not understand what we're looking at. We could start moving around and see that we're looking at an instrument, but we'll only be getting a small part of the picture.
- Pulling back to 4 inches, we'll finally see that we were looking into the sound hole of an acoustic guitar. Taking some time to look around, we might get more information as to what type, how many strings, etc.
- Pulling back to a foot, we can start seeing more of the instrument, and we might get some sense of its size and type. We might even get a glimpse of the artist.
- Moving to a distance of 4 feet, we can finally see the whole instrument and much of the artist. Maybe we can see part of the room that it's being played in.
- Finally, moving back to a distance of 12 or more feet, we can see much of the room, the artist and the instrument. In short, we can see "the big picture".

As you might have guessed, this visual analogy also applies to microphone placement and technique. Moving in close gives us a microscopic sense of the sound of an instrument, whereby we can move around and listen for just that certain, localized sound. Pulling back gives us a better overall sense of the sound of the instrument as a whole. Pulling back to a greater distance begins to introduce the general acoustics of the room, giving a sense of the instrument in its natural-sounding environment.

Distant miking techniques can be used to serve two functions:

- To combine an actual acoustic room environment with close or modern studio mic techniques to give a sense of acoustic "life" to a studio recording.
- To capture the recording environment as a whole. That's to say that the instrument or ensemble will be captured along with its natural acoustic environment to create a combined representation of what the listener would hear in that room.

Distant miking is quite often used in classical and other traditional styles of recording to pick up large instrumental ensembles (such as a symphony orchestra or choral ensemble). In this application, the pickup will largely rely on the acoustic environment to help achieve a natural, ambient sound. The mic or mics should be placed at a distance so as to strike an overall balance between the ensemble's direct sound and the room's acoustics. This approach will result in a balance that's determined by a number of factors, including the size of the sound source, its overall volume level, mic distance and placement … as well as the acoustic and reverberant characteristics of the room.

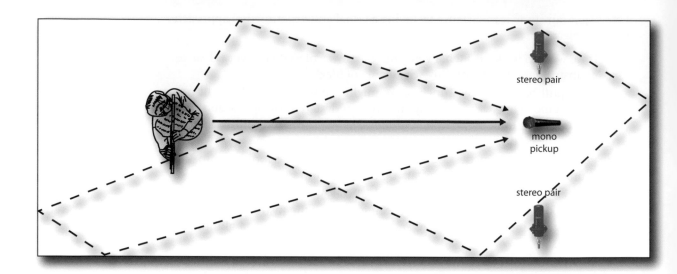

stereo pair

mono pickup

stereo pair

FIGURE 4.29
One of the many possible distant pickup miking examples.

With distant microphone placement (Figure 4.29), multiple mics can be positioned within the acoustic space at a distance that will pick up the proper balance between the instrument/ensemble and the room. Often, these pickups will be spaced in order to capture a good stereo image of the instrument/ensemble within the room. Larger ensembles or instruments (such as a pipe organ) might require a greater spacing in order to best capture the "width" of the event. In other cases, the mics might be placed in a coincident fashion (placed very close together in a specific placement pattern) so as to use directionality and level differences to capture the width, breadth and imagery of the overall acoustic sound of being in the room. These coincident stereo and surround techniques will be covered later in this chapter.

As for microphone placement when using distant techniques, there are several approaches that can yield natural and excellent results. Let's start off by taking a careful look at stereo miking techniques (which are also covered later in this chapter). An in-depth look at stereo mic techniques is not placed here in the distant section for the simple reason that stereo mic techniques can be invaluable (and to many, like DMH, indispensable) in all types of close, distant and any other type of placement settings. Once you've looked at stereo and immersive mic techniques, the next aspect that needs to be considered is the music style, production approach and application. For example:

- A classical recording might rely strictly on a very simple (but carefully chosen and laid out) mic choice and placement. This approach assumes that the recording hall is acoustically matched to your music style, whereby the musicians and the hall become one perfectly combined sonic experience.
- In a large recording studio room setting, a set of overall distant pickups might be combined with semi-close (accent) mics that can accentuate and make certain instruments more present. This approach is often taken during the recording of a film score, whereby a large room sound needs to be

balanced out with a closer mic technique that can be mixed in a controlled multitrack mixdown setting.

- In a music-for-games setting, careful control over the recorded sound might be required, allowing sounds to be carefully mixed to match the game play while also giving the option of having a distant larger-than-life music mix.

Room Microphone Placement

Miking a room (*room miking*) places the pickup at such a distance that the reverberant character of the room sound will be predominant and can then be mixed in with the main signal or be recorded to its own tracks later during mixdown. This distant mic technique can be used to enhance a recording in a number of ways, such as:

- In a studio recording, ambient room microphones can be used in the studio to add a sense of space or natural acoustics back into the sound
- In a live concert recording, ambient mics can be placed in a hall to restore the natural reverberation that's often lost with close miking techniques
- In a live concert recording, room mics can be placed over the audience to pick up their reaction and applause

TRY THIS: AMBIENT MIKING

1. Mic an instrument or its amp (such as an acoustic or electric guitar) at a traditional distance of 6 inches to 1 foot.
2. Place a stereo mic pair (in an X/Y and/or spaced configuration) in the room, placed at 6 to 12 feet (or further in a large room) away from the instrument.
3. Mix the two pickup types together. Does it "open" the sound up and give it more space? Does it muddy the sound up or breathe new life into it?
4. If you're lucky enough to be surround-capable, place the ambient tracks to the rear. Does it add an extra dimension of space? Does it alter the recording's definition?

Although overall room miking techniques are often used in classical recording, it's never a good idea to underestimate the power and effect that distant miking can have within a modern production. Recording instruments using standard close mic techniques definitely has its own advantages (reduced leakage, better control over problematic pickups, greater presence and immediacy); however, these sounds can occasionally end up being a bit too "on-your-face" – even when artificial delay, reverb and other effects are added. Adding actual room sound to the mix can add a degree of realism and depth that often just can't be duplicated using artificial processing.

ROOM PICKUP IN THE STUDIO

Placing a distant mic or (even better) a mic pair within a large room or studio can help bring instruments such as drums, guitars, pianos and other percussion to life. For example, when recording drums, you might consider placing a stereo mic pair at a distance within the room, placed fairly high. If the other instruments (such as bass, guitar, piano) are isolated away, this will give you a distant drum sound that can help make the instrument sound really big or even huge! Even if the instruments are in the same room, this distant pair might just be what's needed to give the sound that extra live cohesiveness, air or live "punch". Of course, it goes without saying that these distant room mics will need to be recorded onto their own separate tracks, thereby allowing decisions to be made later during mixdown.

Besides the standard, tried-and-true approaches to recording that are often applied to our own projects, DMH has come to rely upon several approaches to recording that have helped make the best and most open-sounding recordings that he possibly can. In addition to a deep appreciation for session preparation, here are DMH's personal techniques for approaching a recording:

- I often record an instrument in stereo (usually XY) – I have practically my entire life. This approach is done only by a handful of engineers. But the list of those who do is impressive. My reasoning is that stereo miking spreads the sound, giving it a natural "spread" in a mix. In short it adds width, depth and ambience to a sound, even when used in a close mic setting.

- Whenever the instrument is on its own in a room (of any size, including an iso-room), I will place a distant stereo (and sometimes quad) pair in the room. These tracks can be mixed in with the instrument during mixdown to give a larger, fuller, "you-are-there" sound. This pickup array can also be used in an immersive mix to enhance the overall size and depth of the pickup.

- If the instrument has a MIDI [musical instrument digital interface] out jack I'll always record that MIDI track within the DAW's session. At a later time, it is then possible or even easy to change the sound, fix a note or replay the track through a speaker that's placed in the studio and re-record it through a huge amp stack. Literally, the sky's the limit.

TRY THIS: COMBINING VARIOUS DISTANT/PICKUP TECHNIQUES

1. Have someone grab their favorite electric guitar/ amp setup and place it in the studio.
2. Place a DI on the guitar and record it direct.
3. Place a close mic at 3–6 inches from the amp speaker cabinet and send that to a track.

4. Place an ambient mic or stereo pair 6–8 feet from the speaker and send that to a track or tracks.
5. Place a stereo mic pair (XY or Blumlein) in the room at a good distance and height, then send them to a stereo track.
6. In mixdown, listen to the pickups individually and then combine them. How does the sound develop? How does the room sound affect the sound?
7. If it's available, insert an amp plug-in into the direct track and try various amp combinations.

Does it sound different or better than the actual amp pickups?
8. If it's available, insert a room simulation plug-in into the close amp track or tracks and blend that in instead of the actual room mics. How does that affect the "liveness" of the guitar?
9. Play around with the various combinations and have fun learning how distance can affect your overall sound.

The Boundary Effect

If a distant mic is used to pick up a portion of the room sound, placing it at a random height can result in a hollow sound due to phase cancelations that occur between the direct sound and delayed sounds that are reflected off the floor and other nearby surfaces (Figure 4.30). If these delayed reflections arrive at the mic at a time that's equal to one-half a wavelength (or at odd multiples thereof), the reflected signal will be 180° out of phase with the direct sound. These reflections could produce dips in the signal's pickup response that might adversely color the signal. Since the reflected sound is at a lower level than the direct sound (as a result of traveling farther and losing energy as it bounces off a surface), the cancelation will only be partially complete. Raising the mic will have the effect of reducing reflections (due to the increased distances that the reflected sound must travel), while conversely, moving the mic close to the floor will reduce the path length and raise the range in which the frequency cancelation occurs. In practice, a height of 1/8 inch or less will raise the cancelation above 10 kHz. One such microphone design type, known as a *boundary microphone* (Figure 4.31), places an electret-condenser or condenser diaphragm well within these low height restrictions. For this reason, this mic type might be a good choice for use as an overall distant pickup when the mics need to be out of sight (i.e., when placed on a floor, wall or large boundary).

FIGURE 4.30
Resulting frequency response from a microphone that receives a direct and delayed sound from a single source.

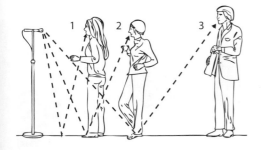

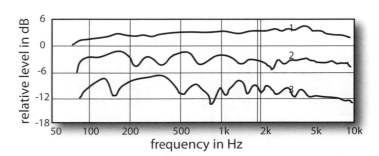

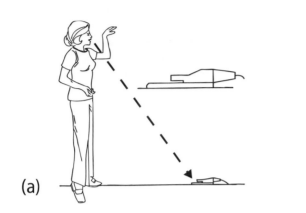 

(a) (b)

FIGURE 4.31
The boundary microphone system: (a) Mic placement. (b) The PZM-6D boundary microphone (courtesy of Crown International, Inc., www.crownaudio.com).

FIGURE 4.32
Figure showing how a direct recording can be "reamped" in a studio, allowing complete tonal, mic placement and acoustical control, after the fact! (courtesy of Radial Engineering, www.radialeng. com).

"REAMPING IT" IN THE MIX

Another way to alter the sound of a track that has already been recorded by injecting a new sense of acoustic space into an existing take is to *reamp* a track. The "reamp" process (originally conceived in 1993 by recording engineer John Cuniberti and now owned by Radial Engineering; www.radialeng.com) lets us record a guitar's signal directly to a track using a DI during the recording session and then play this cleanly recorded track back through a miked guitar amp/speaker, allowing it to be re-recorded to new tracks at another time (Figure 4.32).

The re-recording of an instrument that has already been recorded directly gives us total flexibility for changing the final, recorded amp and mic sound at a later time. For example, it's well known that it's far easier to add an effect to a "dry" track (one that's recorded without effects) during mixdown than to attempt to remove an effect after it's been printed to track. Whenever reamping is used at a later time, it's possible to audition any number of amps, using any number of effects and/or mic settings, until the desired sound has been found. This process allows the musician to concentrate solely on getting the best recorded performance without having to spend extra time getting the perfect guitar, amp, mic and room sound. Leakage problems in the studio are also reduced, as no mics are used in the direct pickup process.

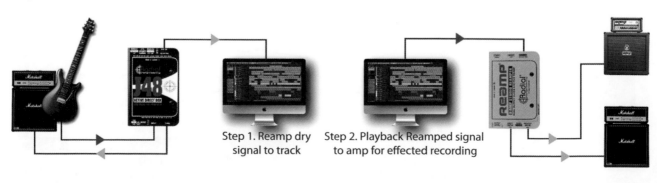

Step 1. Reamp dry signal to track   Step 2. Playback Reamped signal to amp for effected recording

Although the concept of recording an instrument directly and playing the track back through a miked amp at a later time is relatively new, the idea of using a room's sound to fill out the sound of a track or mix isn't. The reamp concept takes this idea a bit further by letting you go as wild as you like later in the production process. For example, you could re-record a single, close-miked guitar amp and then go back at a later time and layer a larger, distant stack on top of the original track. An electronic guitarist could take the process even further by recording his or her MIDI guitar directly and to a sequenced MIDI track. In this way, the reamp, synth patch and edit combinations could be virtually limitless.

TRY THIS: REAMP EXPERIMENT

1. Have someone grab their favorite electric guitar and record a track direct using a DI.
2. After the recording in step 1 is done, play the recorded track back through a guitar amp in the studio (feel free to turn it up).
3. Place a mic at a distance of 4–8 feet and record that track (you could also use the various placement settings that were used in the previous distant/pickup DIY).
4. When you combine the DI and reamped tracks together, did that inject new "life" into the track?
5. You might see what happens when you swap one amp/speaker combo for another and try the experiment again. The process of reamping in the studio opens up a lot of post-production possibilities – have fun!

ACCENT MICROPHONE PLACEMENT

When using distant mic techniques to record an ensemble, sometimes a softer instrument or one that is to perform an important solo can get lost or slightly buried within the overall, larger pickup. In such a situation, an *accent mic* might be needed (Figure 4.33). This happens by placing a mic at a closer distance to the solo or important instrument that needs a little pickup help. This mic is then carefully mixed in with the overall distant pickups to fill out the overall sound and bring out the problem instrument.

As you might expect, care needs to be taken when placing an accent mic, as the tonal and ambient qualities of a pickup will sound very different when it is placed close to an instrument versus when it is placed at a distance. For example, if a solo instrument within an orchestra needs an extra mic for added volume and presence, placing the mic too close would result in a pickup that sounds overly present, unnatural and out of context with the distant, overall orchestral pickup. To avoid this pitfall, a compromise in distance should be struck. The microphone should be placed at a reasonably close range to an instrument or

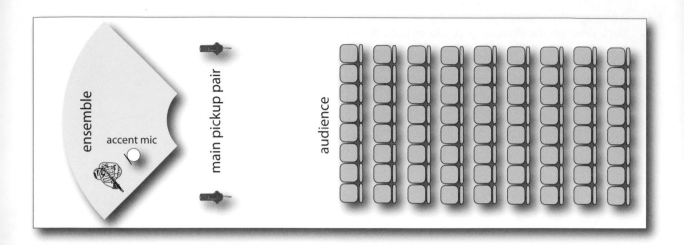

FIGURE 4.33
Accent microphone placed at proper compromise distance.

section within a larger ensemble, but not so close as to have an unnatural sound. The amount of accent signal that's introduced into the mix should sound natural relative to the overall pickup, and a good accent mic should only add presence to a solo passage and not stick out as a separate, identifiable pickup.

Stereo and Immersive Mic Techniques

For the purpose of this discussion, the term *stereo and immersive* (4.0, 5.1, 7.1 and 9.1 surround sound) *miking techniques* refers to the use of two or more microphones so as to obtain a coherent, multichannel sonic "image" of a recording. These techniques can be used in close, distant or room miking of single instruments, vocals, large or small ensembles, within on-location or studio applications – in fact, the only limitation is your imagination. The five fundamental multichannel miking techniques are:

- Spaced pair
- X/Y
- M/S
- Decca tree
- Multi-array (can be an arrangement of up to 11 mics)
- Ambisonic pickup

SPACED PAIR

Spaced microphones (Figure 4.34) can be placed in front of an instrument or ensemble (in a left/right fashion) to obtain an overall stereo image. This technique places the two mics (of the same type, manufacturer and model) anywhere from only a few feet to more than 30 feet apart (depending on the size of the instrument or ensemble) and uses time and amplitude cues in order to create a stereo image. The primary drawback to this technique is the strong potential for phase discrepancies between the two channels due to differences in a sound's

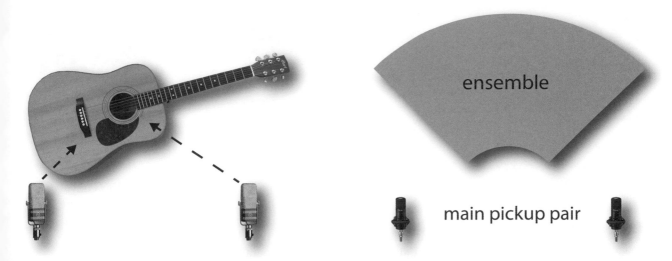

arrival time at one mic relative to the other. When mixed to mono, these phase discrepancies could result in variations in frequency response and even the partial cancelation of instruments and/or sound components in the pickup field.

Another pickup type that makes use of a spaced mic pair is the A/B stereo recording technique. This makes use of two mics that are used to pick up the same pickup source and are often spaced between 3 and 10 feet apart. This distance, which can make use of omni or cardioid mics, works by the time-of-arrival (phase) and level (amplitude) differences that are received by the mics. Thus, the location cues that are perceived by our ears and brain make use of the level and time-of-arrival differences between the mics to create a stereo spread when panned full left and right.

This stereo technique can be used for picking up the overall sound of a room (capturing a full ensemble) or drums (capturing the overall sound of a set), or it can be used to pick up larger instruments (such as a piano, as seen in Figure 4.35). In the latter case, two mics can be placed respectively over the distance of the piano to pick up the preferred upper and lower string registers.

FIGURE 4.34
Spaced stereo miking technique examples.

FIGURE 4.35
A/B stereo miking technique example for a grand piano.

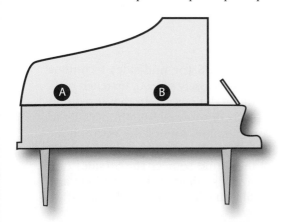

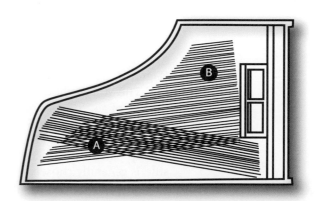

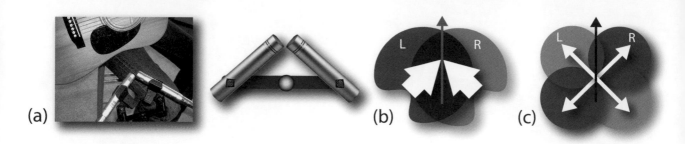

(a) (b) (c)

FIGURE 4.36

X/Y stereo miking patterns: (a) Technique using an X/Y crossed cardioid pair. (b) Crossed cardioid pattern. (c) Blumlein crossed figure-8 pattern.

X/Y

X/Y stereo miking is an intensity-dependent system that uses only the cue of amplitude to discriminate direction. With the X/Y coincident-pair technique (Figure 4.36), two directional microphones of the same type, manufacturer and model are placed with their grills as close together as possible (without touching) and facing at angles to each other (generally between 90° and 135°). The midpoint between the two mics is pointed toward the source, and the mic outputs are equally panned left and right. Even though the two mics are placed together, the stereo imaging is excellent – often better than that of a spaced pair. In addition, due to their proximity, no appreciable phase problems arise. Most commonly, X/Y pickups use mics that have a cardioid polar pattern, although the Blumlein technique is being increasingly used. This technique (which is named after the unheralded inventor of stereo audio and radar, Alan Dower Blumlein) uses two crossed bidirectional mics that are offset by 90° to each other. This simple technique often yields excellent ambient results for the pickup of the overall room ambience within a studio or concert hall, while also being a good choice for picking up sources that are placed "in the round". These ambient pickup styles can be mixed into a stereo or surround-sound production to provide a natural reverb and/or ambience.

Stereo microphones that contain two diaphragms in the same case housing are also available on the new and used market. These mics are either fixed (generally in a 90° pattern) or are designed so that the top diaphragm can be rotated by 180° (allowing the adjustment of various coincident X/Y, Blumlein and M/S angle configurations).

M/S

Another coincident-pair system, known as the M/S (or mid-side) technique (Figure 4.37a), is similar to X/Y in that it uses two closely spaced, matched pickups. The M/S method differs from the X/Y method, however, in that it requires the use of an external transformer, active matrix or software plug-in in order to work. In the classic M/S stereo miking configuration, one of the microphone capsules is designated the M (mid) position pickup and is generally a cardioid pickup pattern that faces forward, toward the sound source. The S (side) capsule is generally chosen as a figure-8 pattern that's oriented sideways (90° and 270°) to the on-axis pickup (i.e., with the null facing forward, into the cardioid's

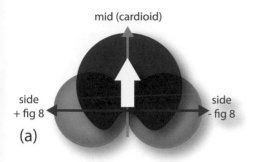

(a)   (b)

main axis). In this way, the mid capsule picks up the direct sound, while the side figure-8 capsule picks up ambient and reverberant sound. These outputs are then combined through a sum-and-difference decoder matrix either electrically (through a transformer matrix) or mathematically (through a digital M/S plug-in), which then resolves them into a conventional X/Y stereo signal: (M + S = left) and (M – S = right).

One advantage of this technique is its absolute monaural compatibility. When the left and right signals are combined, the sum of the output will be (M + S) + (M – S) = 2M. That's to say, the side (ambient) signal will be canceled, but the mid (direct) signal will be accentuated. Since it is widely accepted that a mono signal loses its intelligibility with added reverb, this tends to work to our advantage. Another amazing side benefit of using M/S is the fact that it lets us continuously vary the mix of mid (direct) to side (ambient) sound that's being picked up either during the recording (from the console location) or even at a later time during mixdown, after it's been recorded! These are both possible by simply mixing the ratio of mid to side that's being sent to a hardware or software matrix decoder (Figure 4.37b). In a mixdown scenario, all that's needed is to record the mid on one track and the side on another. (It's often best to use a digital recorder, because phase delays associated with the analog recording process can interfere with decoding.) During mixdown, routing the M/S tracks to the decoder matrix allows you to make important decisions regarding stereo width and depth at a later, more controlled date.

At this point, I'd like to offer up a personal fact on stereo miking: DMH shares a common working mic practice with both Bruce Swedien (Michael Jackson, etc.) and Allen Sides (Ella Fitzgerald, Joni Mitchell, Phil Collins, Frank Zappa, Ice-T, Sheryl Crow, Michael Jackson, Mary J. Blige and too many others to mention). We all record many of our tracks in stereo. Why? First off, think of recording a stereo synthesizer using only one channel. We all know that the effects, width and depth of that synth were programmed to come across best in stereo. The same thing actually occurs when recording acoustic instruments. When two mics are used, a larger acoustic width can be picked up when panned left/right. This often results in a more expansive, "live-sounding" instrument, voice or soundscape. In Allen Sides' words – "I still record everything in stereo. I don't use pan pots if I can avoid it. It makes things sound more exciting". Go ahead and try it for yourself and see what you think.

FIGURE 4.37

M/S stereo microphone technique: (a) M/S mic configuration. (b) K-Stereo Ambience Recovery with M/S controls highlighted (courtesy of Universal Audio, www.uaudio.com © 2022 Universal Audio, Inc. All rights reserved. Used with permission).

TRY THIS: STEREO MIC TEST

1. Place a single mic in front of an instrument at a traditional distance/placement.
2. Place an additional stereo mic pair (near the same position for an X/Y pickup or at a desired placement for a spaced configuration).
3. Listen to the mono and stereo pickups in an A/B fashion. How do the two compare?

DECCA TREE

Although not as commonly used as the preceding stereo techniques, the *Decca tree* is a time-tested, classical miking technique that uses both time and amplitude cues in order to create a coherent stereo image. Attributed originally to Decca engineers Roy Wallace and Arthur Haddy in 1954, the Decca tree (Figure 4.38) originally consisted of three omnidirectional mics (originally, Neumann M50 mics were used). In this arrangement, a left and right mic pair is placed 3 feet apart, and a third mic is placed 1.5 feet out in front and panned in the center of the stereo field. Still favored by many in orchestral situations as a main pickup pair, the Decca tree is most commonly placed on a tall boom, above and behind the conductor. According to lore, when Haddy first saw the array, he remarked, "It looks like a bloody Christmas tree!" The name stuck.

SURROUND MIKING TECHNIQUES

FIGURE 4.38
Decca tree microphone array. (a) Hardware mount (courtesy of Audio Engineering Associates). (b) In a studio setting (courtesy of Galaxy Studios, www.galaxy.be). (c) Arrangement in the front segment of a circle, with optional counterweight at the rear.

With the advent of 5.1, 7.1 and 5.1.4 immersive-sound production, it's certainly possible to make use of a surround console or DAW to use multiple-pickup surround mic techniques in order to capture the actual acoustic environment and then translate that into a surround mix. Just as the number of techniques and personal styles increases when miking in stereo compared with mono, the number of placement and technique choices will likewise increase when miking a source and room in immersive surround. Although guidelines have been and will continue to be set, both placement and mixing styles are definitely more

(a)

(b)

R

L

C

of an art than a science. For more info on surround, see Chapter 21, Immersive Audio.

AMBIENT/ROOM SURROUND MICS

As was said earlier in the room pickup section, placing a spaced or coincident pair of room mics in a studio can work wonders to add a sense of space to a stereo recording. Well, the same room mics can also come in really handy if a surround mix is made of the project. By simply panning the room mics to the rear of the soundscape, the overall ambience will come alive in the mix, giving an added sense of space to an ensemble group, drum set or instrument overdub. So, during a recording or an overdub, the addition of a pair of room mics can come in handy for both your stereo mixes and any immersive mixes that you might make in the future – double score!

IMMERSIVE DECCA TREE

The first approach (which doesn't actually fall under the Decca tree category) involves the use of four cardioid mics that are spaced at 90° angles, representing L, R, Ls, Rs, with the on-axis point being placed 45° between the L and R mics. This "quad" configuration can be made by mounting the mics offset by 90° from each other on three (or two crossed) stereo bars (Figure 4.39a).

One of the most logical techniques for capturing an ensemble or instrument in a basic 5.1 surround setting places five mics onto a modified Decca tree (Figure 4.39b). This ingenious and simple system adds two rear-facing mics to the existing three-mic Decca tree system. Another simpler approach is to place five cardioid mics in a circle, such that the center channel faces toward the source, thereby creating a simple setup that can be routed L, C, R, Ls and Rs.

One last approach uses an advanced cage-type configuration (Figure 4.39c), whereby the bottom-paced mics are laid out in the standard 5.1 fashion (as before). The top mics are then laid out in a quad configuration for picking up the height room information.

FIGURE 4.39
Surround mic placements. (a) A simple four-microphone (quad) approach to surround miking can be easily made by using three stereo bars. (b) Alternatively, five cardioid microphones can be arranged in a circular pattern (with the center microphone facing toward the source) to create a modified, mini-surround Decca tree. (c) Immersive 5.1, 7.1, 5.1.4 and higher cube mic layouts can also be experimented with when recording ensembles.

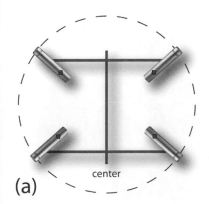

(a) center

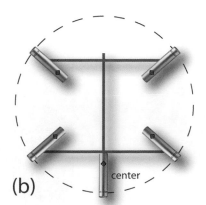

(b) center

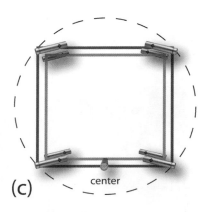

(c) center

Setup details for this last setting can range from a simple placement of the mics in an outward-facing fashion to far more complicated layouts involving various cube sizes that are carefully configured, with the mics being placed in various angles and directions. Further info on immersive Decca cube layouts can be found on the web.

AMBISONIC PICKUP

In short, an Ambisonic mic, encoding and reproduction system allows a 360° immersive soundfield to be picked up by an array of mics (four, eight or more, as seen in Figure 4.40a) that can then be combined and manipulated at the time of recording or during mixdown. In this latter scenario, complete control over direction, directionality and other ambient control factors is achieved by recording each capsule onto its own, separate track.

Developed by a small team of academics in Britain in the 1970s, this process is based upon an expansion of the general concepts of A. D. Blumbein's Mid-Side stereo technique. It basically makes use of phase- and time-related information to encode and decode ambient acoustic information using advanced mathematics. The Ambisonic encoding process can be done at various levels of accuracy. The most basic (first order) makes use of four mics and/or four channels of mix information to deliver a sense of direction to the listener, while the highest degree of perceptual processing is done using third-order (16-channel) processing.

It should be noted that a multi-mic array is not necessary in order to make use of Ambisonics; special Ambisonic-related panners are now commonly integrated into most DAWs, which allow a mono or stereo source to be panned into an immersive field (with height) and then mixed in an Ambisonic environment that's popularly listened to in binaural over headphones (Figure 4.40b). Further info on the Ambisonic process can be found in Chapter 21.

FIGURE 4.40
Rode NT-SF1 Ambisonic mic. (a) Capsule array. (b) Immersive decoding software plug-in (courtesy of Rode, www.rode.com).

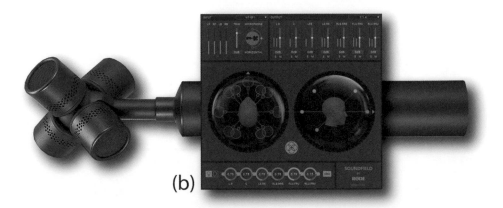

(a)

(b)

MICROPHONE PLACEMENT TECHNIQUES

The following sections are meant to be used as a general guide to mic placement for various acoustic and popular instruments (most often in close and semi-close applications). It's important to keep in mind that these are only guidelines. Several general application and characteristic notes are detailed in Table 4.1, and descriptions of several popular mics are outlined toward the end of this chapter in the "Microphone Selection" section to help give insights into placement and techniques that might work best in a particular application.

As a general rule, choosing the best mic for an instrument or vocal will ultimately depend on the sound you're searching for. For example, a dynamic mic will often yield a "rugged" or "punchy" character (which is often further accentuated by the proximity of bass boost that's generally associated with a directional mic). A ribbon mic will often yield a mellow sound that ranges from being open and clear to slightly "croony" depending on the type and distances involved.

Table 4.1	Microphone Selection Guidelines
Needed Application	**Required Microphone Choice and/or Characteristic**
Natural, smooth tone quality	Flat frequency response
Bright, present tone quality	Rising frequency response
Extended lows	Dynamic or condenser with extended low response
Extended highs (detailed sound)	Condenser
Increased "edge" or midrange detail	Dynamic
Extra ruggedness	Dynamic or modern ribbon/condenser
Boosted bass at close distances	Directional mic
Flat bass response up close	Omnidirectional mic
Less leakage, feedback, acoustics	Directional or omnidirectional mic at close distances
Enhanced pickup of room acoustics	Place mic or stereo pair at greater working distances
Reduced handling noise	Omni, vocal or directional microphone with shock mount
Reduced breath popping	Omni or directional mic with pop filter
Distortion-free pickup of loud sound	Dynamic or condenser with high maximum SPL rating
Noise-free pickup of quiet sound	Condenser with low self-noise and high sensitivity

Condenser mics are often characterized as having a clear, present and full-range sound that varies with mic design, grill options and capsule size. Before jumping into this section, we'd like to again take time to refer you back to the "Good Rule", at the beginning of this chapter, for anyone who wants to be a better engineer, producer and/or musician:

The "Good Rule"

Good musician + good instrument + good performance + good acoustics + good mic + good placement = good sound.

Starting with an experienced, rehearsed and ready musician who has a quality instrument that's well tuned and in shape is the best insurance for getting the best possible sound. Let's think about this for a moment. Say that we have a live rhythm session that involves drums, piano, bass guitar and scratch vocals. All of the players are the best around, except for the drummer, who is new to the studio process. Unfortunately, you've now signed on to teach the drummer the ropes of proper drum tuning, studio interaction and playing under pressure. It goes without saying that the session might go far less smoothly than it otherwise would, as you'll have to take the extra time to work with the player to get the best possible sound. Once you're rolling, it'll also be up to you or the producer to pull a professional performance out of someone who's new to the field – and the session will probably suffer for it.

Don't get us wrong, musicians have to start somewhere, but an experienced, capable musician who comes into the studio with a great instrument that's tuned and ready to go (and who might even clue you in on some sure-fire mic and placement techniques for the instrument) is simply a joy from a sound, performance, time and budget-saving standpoint. Simply put, if you and/or the project's producer have prepared enough to get all your "goods" lined up, the track will have a much better chance of being something that everyone can be proud of. Just as with the art of playing an instrument, preparation, careful mic choice, placement and "style" in the studio are a few of the fundamental calling cards of a good engineer. Experience simply comes with time and the willingness to experiment. Be patient, learn, listen and have fun, and you too will eventually rise to the professional occasion.

Brass Instruments

The following sections describe many of the sound characteristics and miking techniques that are encountered in the brass family of instruments.

TRUMPET

The fundamental frequency of a trumpet ranges from E3 to D6 (165 to 1,175 Hz) and contains overtones that stretch upward to 15 kHz. Below 500 Hz, the sounds emanating from the trumpet project uniformly in all directions; above 1,500 Hz, the projected sounds become much more directional; and above 5 kHz, the dispersion emanates at a tight 30° angle from in front of the bell. The formants of a trumpet (the relative harmonic and resonance frequencies that give an instrument its specific character) lie at around 1 to 1.5 kHz and at 2 to 3 kHz. Its tone can be radically changed by using a mute (a cup-shaped dome that fits directly over the bell), which serves to dampen frequencies above 2.5 kHz. A conical mute (a metal mute that fits inside the bell) tends to cut back on frequencies below 1.5 kHz while encouraging frequencies above 4 kHz. Because of the high sound-pressure levels that can be produced by a trumpet (up to 130 dB^{SPL}), it's best to place a mic slightly off the bell's center at a distance of 1 foot or more (Figure 4.41). When closer placements are needed, a –10 to –20-dB pad can help prevent input overload at the mic or console preamp input. Under such close working conditions, a windscreen can help protect the diaphragm from wind blasts.

TROMBONE

Trombones come in a number of sizes; however, the most commonly used "bone" is the tenor, which has a fundamental note range spanning from E2 to C5 (82 to 523 Hz) and produces a series of complex overtones that range from 5 kHz (when played medium loud) to 10 kHz (when overblown). The trombone's polar pattern is nearly as tight as the trumpet's: frequencies below 400 Hz are distributed evenly, whereas its dispersion angle increases to 45° from the bell at 2 kHz and above. The trombone most often appears in jazz and classical music. The "Mass in C Minor" by Mozart, for example, has parts for soprano, alto, tenor and bass trombones. This style obviously lends itself to the spacious blending that can be achieved by distant pickups within a large hall or studio. On the other hand, jazz music often calls for closer miking distances. At 2 to 12 inches, for example, the trombonist should play slightly to the side of the mic to reduce the chance of overload and wind blasts. In the miking of a trombone section, a single mic might be placed between two players, acoustically combining them onto a single channel and/or track.

TUBA

The bass and double-bass tubas are the lowest pitched of the brass/wind instruments. Although the bass tuba's range is actually a fifth higher than the double

FIGURE 4.41
Typical close and semi-distant microphone placement for a single trumpet.

bass, it's still possible to obtain a low fundamental of B (31 Hz). A tuba's overtone structure is limited; its top response ranges from 1.5 to 2 kHz. The lower frequencies (around 75 Hz) are evenly dispersed; however, as frequencies rise, their distribution angles reduce. Under normal conditions, this class of instruments isn't miked at close distances. A working range of 2 feet or more, slightly off-axis to the bell, will generally yield the best results.

FRENCH HORN

The fundamental tones of the French horn range from B1 to B5 (62 to 988 Hz). Its "oo" formant gives it a round, broad quality that can be found at about 340 Hz, with other frequencies falling between 750 Hz and 3.5 kHz. French horn players often place their hands inside the bell to mute the sound and promote a formant at about 3 kHz. A French horn player or section is traditionally placed at the rear of an ensemble, just in front of a rear, reflective stage wall. This wall serves to reflect the sound back toward the listener's position (which tends to create a fuller, more defined sound). An effective pickup of this instrument can be achieved by placing an omni- or bidirectional pickup between the rear, reflecting wall and the instrument bells, thereby receiving both the direct and reflected sound. Alternatively, the pickups can be placed in front of the players, thereby receiving only the sound that's being reflected from the rear wall.

Guitar

The following sections describe the various sound characteristics and techniques that are encountered when miking the guitar.

ACOUSTIC GUITAR

The popular steel-strung, acoustic guitar has a bright, rich set of overtones (especially when played with a pick). Mic placement and distance will often vary from instrument to instrument and may require experimentation to pick up the best tonal balance. A balanced pickup can often be achieved by placing the mic (or an X/Y stereo pair) at a point slightly off-axis and above or below the sound hole at a distance of between 6 inches and 1 foot (Figure 4.42).

In this situation, EC would probably choose a pencil or small diaphragm condenser microphone, whereas DMH would probably reach for a larger condenser … this goes to show that it's art and there's no right way to do any of this.

If you're using a stereo recording technique, XY for example, remember that the pickup can be sensitive to horizontal movements, and the center image can tilt to either the L or the R side (i.e., if your guitar player turns to the left or right while playing, there might be a shift in the stereo image). An alternative method is using a vertical XY (with the mics turned 90°), which is less sensitive to musician movements.

Condenser mics are often preferred for their smooth, extended frequency response and excellent transient response. The smaller-bodied classical guitar is normally strung with nylon or gut and is played with the fingertips, giving it a warmer, mellower sound than its steel-strung counterpart. To make sure that the instrument's full range is picked up, place the mic closer to the center of the bridge, at a distance of between 6 inches and 1 foot.

FIGURE 4.42
One of the many possible mic placements for the guitar (in either a mono or X/Y stereo configuration).

MIKING NEAR THE SOUND HOLE

The sound hole (located at the front face of a guitar) serves as a bass port, which resonates at the lower frequencies (around 80 to 100 Hz). Placing a mic too close to the front of this port might result in a boomy and unnatural sound; however, miking close to the sound hole is often popular on stage or around high acoustic levels because the guitar's output is highest at this position. To achieve a more natural pickup under these conditions, the microphone's output can be rolled off at the lower frequencies (5 to 10 dB at 100 Hz).

ROOM AND SURROUND GUITAR MIKING

An effective way to translate an acoustic guitar to the wide stage of surround (if a big, full sound is what you're after) is to record the guitar using X/Y or spaced techniques stereo (panned front L/R) and pan the guitar's electric pickup (or added contact pickup) to the center speaker. Extra ambient mics could also be used during an overdub to add room ambience to a stereo mix or to be panned to the rear in a surround mix.

Nylon or Spanish Guitar

When recording a nylon string guitar, it's good to keep in mind the sound of the instrument; it's a combination of the fingers on the right hand, the resonance of the guitar and also the workings of the left hand.

An AB setup using pencil condenser microphones for flamenco or classical guitar might be a good way to start: one of the microphones around the right shoulder area and the other one near the 12th fret. Another idea is to record an AB or XY in a vertical plane, across the 12th fret. This vertical pickup style has two advantages: if the player moves around, there is little shift in the stereo image, plus this area provides an overall balanced sound. If the AB were in the horizontal axis, one capsule might be closer to the body (resulting in a more boomy sound), whereas the other capsule might have a very different tone.

Remember: for a good stereo acoustic guitar balance, the capsules of the microphones shouldn't be placed too far from each other, otherwise the recorded sound might act more like a "double mono" pickup. As always, trust your ears!

The Electric Guitar

The fundamentals of the average 22-fret guitar extend from E2 to D6 (82 to 1174 Hz), with overtones that extend much higher. All of these frequencies might not be amplified, because the guitar chord tends to attenuate frequencies above 5 kHz (unless the guitar has a built-in low-impedance converter or low-impedance pickups). The frequency limitations of the average guitar loudspeaker often add to this effect because their upper limit is generally restricted to below 5 or 6 kHz.

MIKING THE GUITAR AMP

The most popular guitar amplifier used for recording is a small practice-type amp/speaker system. These high-quality amps often help the guitar's suffering high end by incorporating a sharp rise in the response range at 4 to 5 kHz, thus helping to give it a clean, open sound. High-volume, wall-of-sound speaker stacks are less commonly used in a session because they're harder to control in the studio and in a mix. By far the most popular mic type for picking up an electric guitar amp is the cardioid dynamic. A dynamic tends to give the sound a full-bodied character without picking up extraneous amplifier noises. Often, guitar mics will have a pronounced presence peak in the upper frequency range, giving the pickup an added clarity. For increased separation, a microphone can be placed at a working distance of 2 inches to 1 foot. When miking at a distance of less than 4 inches, mic/speaker placement becomes slightly more critical (Figure 4.43). For a brighter sound, the mic should face directly into the center of the speaker's cone. Placing it off the cone's center tends to produce a more mellow sound while reducing amplifier noise. During an overdub, or when the amp is placed in an iso-booth/room, the ambient image can be "opened up" even further by mixing a semi-distant mic (at 6' – 2 meters) with the DI signal (and/or even with the close-miked amp signal). An additional room stereo pair pickup can be mixed either into a stereo field or into the rear soundfield of an immersive mix to further fill out the sound.

Isolation cabinets have also come onto the market, which are literally sealed boxes that house a speaker or guitar amp/cabinet system as well as an internal mic mount. These systems are used to reduce leakage and to provide greater control over instrument levels within a recording studio or control room during a session.

RECORDING DIRECT

A DI box is often used to feed the output signal of an electric guitar directly into the mic input stage of a recording console or mixer. By routing the direct output signal to a track, a cleaner, more present sound can be recorded. This technique also reduces the leakage that results from having a guitar amp in the studio and even makes it possible for the guitar to be played in the control room or project studio. The output signal can be taken directly from the guitar or other type of instrument (Figure 4.44a), or it can be taken from the amp's high-level or direct output (Figure 4.44b) and routed to the track.

A combination of both the direct and miked signals will often result in a sound that adds the characteristic fullness of a miked amp with the extra "bite" that a DI tends to give. These may be combined onto a single track or whenever possible, assigned to separate tracks, allowing greater control during mixdown (Figure 4.44c).

FIGURE 4.43
Miking an electric guitar cabinet directly in front of or off center to the cone. A semi-distant mic can be used to pick up the entire cabinet, while a room mic setup can be used to capture ambience.

The Electric Bass Guitar

The fundamentals of an electric bass guitar range from about E1 to F4 (41.2 to 343.2 Hz). If it's played loudly or with a pick, the added harmonics can range upward to 4 kHz. Playing in the "slap" style or with a pick gives a brighter, harder attack, while a "fingered" style will produce a mellow tone. In modern music production, the bass guitar is often recorded direct for the cleanest possible

FIGURE 4.44
Direct recording of an electric guitar: (a) Direct recording. (b) The signal can be taken from the amp's line or mic-level output. (c) Combined direct and miked signal.

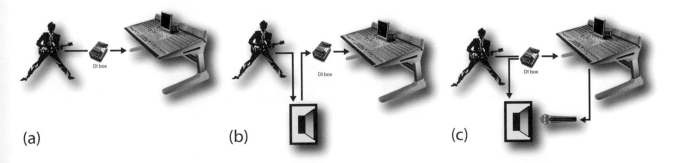

(a)   (b)   (c)

sound. As with the electric guitar, the electric bass can be either miked at the amplifier or picked up through a DI box. If the amp is miked, dynamic mics usually are chosen for their deep, rugged tones. The large-diaphragm dynamic designs tend to subdue the high-frequency transients. When combined with a boosted response at around 100 Hz, these large-diaphragm dynamics give a warm, mellow tone that adds power to the lower register. Equalizing a bass can sometimes increase its clarity, with the fundamental being affected from 125 to 400 Hz and the harmonic punch being from 1.5 to 2 kHz. A compressor is commonly used on electric and acoustic basses. It's a basic fact that the signal output from the instrument's notes often varies in level, causing some notes to stand out while others dip in volume. A compressor having a smooth input/output ratio of roughly 4:1, a fast attack (8 to 20 milliseconds) and a slower release time (1/4 to 1/2 second) can often smooth out these levels, giving the instrument a strong, present and smooth bass line.

Keyboard Instruments

The following sections describe the various sound characteristics and techniques that are encountered when miking keyboard instruments.

GRAND PIANO

The grand piano is an acoustically complex instrument that can be miked in a variety of ways depending on the style and preferences of the artist, producer and/or engineer. The overall sound emanates from the instrument's strings, soundboard and mechanical hammer system. Because of its large surface area, a minimum miking distance of 4 to 6 feet is needed for the tonal balance to fully develop and be picked up; however, leakage from other instruments often means that these distances aren't practical or possible. As a result, pianos are often miked at distances that favor such instrument parts as:

- Strings and soundboard, often yielding a bright and relatively natural tone
- Hammers, generally yielding a sharp, percussive tone
- Soundboard holes alone, often yielding a sharp, full-bodied sound

In modern music production, two basic grand piano styles can be found in the recording studio: the concert grand, which traditionally has a rich and full-bodied tone (often used for classical music and ranging in size up to 9 feet in length), and the studio grand, which is more suited for modern music production and has a sharper, more percussive edge to its tone (often being about 7 feet in length).

Figure 4.45 shows a number of miking positions that can be used in recording a grand piano. Although several mic positions are illustrated, it's important to keep in mind that these are only guidelines from which to begin. Your own personal sound can be achieved through mic choice and experimentation with mic placement.

 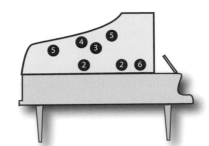 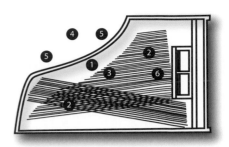

FIGURE 4.45
Possible miking combinations for the grand piano.

- Position 1: The mic is attached to the partially or entirely open lid of the piano. The most appropriate choice for this pickup is the boundary mic, which can be permanently attached or temporarily taped to the lid. This method uses the lid as a collective reflector and provides excellent pickup under restrictive conditions (such as on stage and during a live video shoot).
- Position 2: Two mics are placed in a spaced stereo configuration at a working distance of 6 inches to 1 inch. One mic is positioned over the low strings, and one is placed over the high strings.
- Position 3: A single mic or coincident stereo pair is placed just inside the piano between the soundboard and its fully or partially open lid.
- Position 4: A single mic or stereo coincident pair is placed outside the piano, facing into the open lid (this is most appropriate for solo or accent miking).
- Position 5: A spaced stereo pair is placed outside the lid, facing into the instrument.
- Position 6: A single mic or stereo coincident pair is placed just over the piano hammers at a working distance of 4 to 8 inches to give a driving pop or rock sound.

A condenser or extended-range dynamic mic is most often the preferred choice when miking an acoustic grand piano, as those types of mics tend to accurately represent the transient and complex nature of the instrument. Should excessive leakage be a problem, a close-miked cardioid (or cardioid variation) can be used; however, if leakage isn't a problem, backing away to a compromise distance (3 to 6 feet) can help capture the instrument's overall tonal balance.

Separation

Separation is often a problem associated with the grand piano whenever it is placed next to noisy neighbors. Separation, when miking a piano, can be achieved in the following ways:

- Place the piano inside a separate isolation room.
- Place a flat (acoustic separator) between the piano and its louder neighbor.

- Place the mics inside the piano and lower the lid onto its short stick. A heavy moving or other type of blanket can be placed over the lid to further reduce leakage.
- Overdub the instrument at a later time. In this situation, the lid can be removed or propped up by the long stick, allowing the mics to be placed at a more natural-sounding distance.

UPRIGHT PIANO

You would expect the techniques for this seemingly harmless piano type to be similar to those for its bigger brother. This is partially true. However, because this instrument was designed for home enjoyment and not performance, the mic techniques are often very different. Since it's often more difficult to achieve a respectable tone quality when using an upright, you might want to try the following methods:

- *Miking over the top*: place two mics in a spaced fashion just over and in front of the piano's open top, with one over the bass strings and one over the high strings. If isolation isn't a factor, remove or open the front face that covers the strings in order to reduce reflections and therefore, the instrument's characteristic "boxy" quality. Also, to reduce resonances, you might want to angle the piano out and away from any walls.
- *Miking the kickboard area*: for a more natural sound, remove the kickboard at the lower front part of the piano to expose the strings. Place a stereo spaced pair over the strings (one each at a working distance of about 8 inches over the bass and high strings). If only one mic is used, place it over the high-end strings. Be aware, though, that this placement can pick up excessive foot-pedal noise.
- *Miking the upper soundboard area*: to reduce excessive hammer attack, place a microphone pair at about 8 inches from the soundboard, above both the bass and high strings. In order to reduce muddiness, the soundboard should be facing into the room or should be moved away from nearby walls.

Electronic Keyboard Instruments

Signals from most electronic instruments (such as synthesizers, samplers and drum machines) are often taken directly from the device's line-level output(s) and inserted into a console, either through a DI box or directly into a channel's line-level input. Alternatively, the keyboard's output can be plugged directly into the recorder or interface line-level inputs. The approach to miking an electronic organ can be quite different from the techniques just mentioned. A good Hammond or other older organ can sound wonderfully "dirty" through miked loudspeakers. Such organs are often played through a Leslie cabinet (Figure 4.46a and b), which adds a unique, Doppler-based vibrato. Inside the cabinet is a set of rotating speaker baffles that spin on a horizontal axis and in

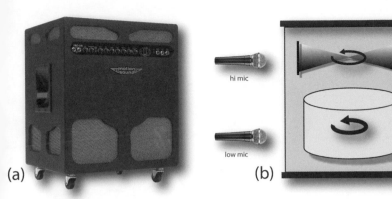

(a)   (b)   (c)

turn, produce a pitch-based vibrato as the speakers accelerate toward and away from the mics. The upper high-frequency speakers can be picked up by either one or two mics (each panned left and right), with the low-frequency driver being picked up by one mic. Motor and baffle noises can produce quite a bit of wind, possibly creating the need for a windscreen and/or experimentation with placement. Of course, there are a number of Leslie and rotary effect plug-ins on the market that can be inserted into a track at any time (and for any instrument) for effect (Figure 4.46c).

Percussion

The following sections describe the various sound characteristics and techniques that are encountered when miking drums and other percussion instruments.

DRUM SET

The standard drum kit (Figure 4.47) is often at the foundation of modern music, because it provides the "heartbeat" of a basic rhythm track; consequently, a proper drum sound is extremely important to the outcome of most music projects. Generally, the drum kit is composed of the kick drum, snare drum, high toms, low tom (one or more), hi-hat and a variety of cymbals. Since a full kit is a series of interrelated and closely spaced percussion instruments, it often takes real skill to translate the proper spatial and tonal balance into a project. The larger-than-life driving sound of the acoustic rock drum set that we've all become familiar with is the result of an expert balance among playing techniques, proper tuning and mic placement.

During the past decades, drums have undergone a substantial change with regard to playing technique, miking technique and choice of acoustic recording environment. In the 1960s and 1970s, the drum set was placed in a small isolation room called a drum booth. This booth acoustically isolated the instrument from the rest of the studio and had the effect of tightening the drum sound because of the limited space (and often dead acoustics). The drum booth also physically isolated the musician from the studio, which often caused the musician to feel

FIGURE 4.46
A Leslie speaker cabinet creates a unique vibrato effect by using a set of rotating speaker baffles that spin on a horizontal axis. (a) Modern portable rotary amp with built-in microphones and three XLR outputs (courtesy of Motion Sound, www.motion-sound.com). (b) Miking the rotating speakers of a Leslie cabinet. (c) Rotary Leslie effect plug-in (courtesy of Steinberg Media Technologies GmbH, a division of Yamaha Corporation, www.steinberg.net).

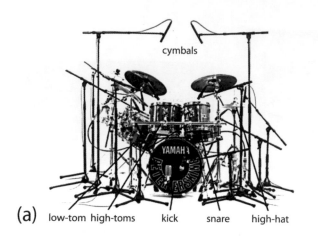

(a) low-tom high-toms kick snare high-hat

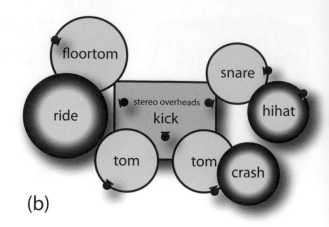

(b)

FIGURE 4.47

The drum set: (a) Peter Erskine's studio drum kit (courtesy of Beyerdynamic, www.beyerdynamic.com). (b) Traditional drum placement and mic positioning.

removed and less involved in the action. Today, many engineers and producers have moved the drum set out of smaller iso-rooms and back into larger open studio areas where the sound can fully develop and combine with the studio's own acoustics. In many cases, this effect can be exaggerated by placing a distant mic pair in the room (a technique that often produces a fuller, larger-than-life sound, especially in immersive).

Before a session begins, the drummer should tune each drum while the mics and baffles for the other instruments are being set up. Each drumhead should be adjusted for the desired pitch and for constant tension around the rim by hitting the head at various points around its edge and adjusting the lugs for the same pitch all around the head. Once the drums are tuned, the engineer should listen to each drum individually to make sure that there are no buzzes, rattles or resonant after-rings. Drums that sound great in live performance may not sound nearly as good when being close miked. In a live performance, the rattles and rings are covered up by the other instruments and are lost before the sound reaches the listener. Close miking, on the other hand, picks up the noises as well as the desired sound.

If tuning the drums doesn't bring the extraneous noises or rings under control, duct or masking tape can be used to dampen them. Pieces of cloth, dampening rings, paper towels or a wallet can also be taped to a head in various locations (which are determined by experimentation) to eliminate rings and buzzes. Although head damping has been used extensively in the past, present methods use this damping technique more discreetly and will often combine dampening with proper design and tuning styles (all of which are the artist's personal choice).

During a session, it's best to remove the damping mechanisms that are built into most drum sets, because they apply tension to only one spot on the head and unbalance its tension. These built-in dampeners often vibrate when the head is hit and are a chief source of rattles. Removing the front head and placing a blanket or other damping material inside the drum (so that it's pressing against the head) can often dampen the kick drum. Adjusting the amount of material can vary the sound from being a resonant boom to a thick, dull thud. Kick drums are usually (but not always) recorded with their front heads removed, while other drums are recorded with their bottom heads either on or off. Tuning the drums is more difficult if two heads are used, because the head tensions often interact; however, they will often produce a more resonant tone. After the drums have been tuned, the mikes can be put into position. It's important to keep the mics out of the drummer's way, or they might be hit by a stick or moved out of position during the performance.

MIKING THE DRUM SET

After the drum set has been optimized for the best sound, the mics can be placed into their pickup positions (Figure 4.48). Because each part of the drum set is so different in sound and function, it's often best to treat each grouping as an individual instrument. In its most basic form, the best place to start when miking a drum set is to start with the fundamental "groups." These include:

- Position 1: placing a mic on the kick
- Position 2: placing a mic on the snare drum
- Position 3: at an absolute minimum, the entire drum set can be adequately picked up using only four mics by adding two overhead spaced pickups
- Position 4: or a coincident pair can be placed over the set
- Position 5: one or two mics placed on the high toms
- Position 6: placing a mic on the low tom

A mic's frequency response, polar response, proximity effect and transient response should be taken into account when matching it to the various drum groups. In fact, mic placement is your best friend when it comes to getting the best balance between drum "instruments" (and each is best thought of as percussion instruments that work together to make an overall kit), while a mic's polar response is another friend when it comes to isolating a specific drum, etc., onto a track with maximum rejection and overall leakage (for gaining better control over the various drums within the overall drum mix). Dynamic range is another important consideration when miking drums. Since a drum set is capable of generating extremes of volume and power (as well as softer, more

FIGURE 4.48
Typical microphone placements for a drum set: (a) Side view showing a "bare bones" jazz kit. (b) Front view showing a basic rock setup. (c) Top view showing a basic rock setup.

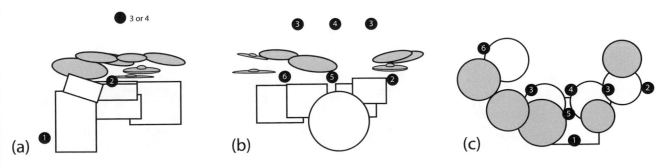

subtle sounds), the chosen mics must be able to withstand strong peaks without distorting and yet still be able to capture the more delicate nuances of a sound.

Since the drum set usually is one of the loudest sound sources in a studio setting, it's often wise to place it on a solidly supported riser. This reduces the amount of low-end "thud" that can otherwise leak through the floor into other parts of the studio. Depending on the studio layout, the following drum scenarios may occur:

- The drums could be placed in their own room, isolated from other instruments.
- To achieve a bigger sound, the drums could be placed in the large studio room while the other instruments are placed in smaller iso-rooms or are recorded directly.
- To reduce leakage, the drums could be placed in the studio, while being enclosed by 4-foot (or higher) divider flats.

KICK DRUM

FIGURE 4.49
Kick drum mic placements. (a) Placing a microphone inside the kick at a close distance results in a sharp, deep attack (moving the mic placement will affect the tone). (b) Placing the mic at a distance just outside the kick drumhead to bring out the low end and natural fullness. (c) Do both and record them to their own tracks for greater control and choices during mixdown.

The kick drum adds a low-energy drive or "punch" to a rhythm groove. This drum has the capability to produce low frequencies at high SPL, so it's necessary to use a mic that can both handle and faithfully reproduce these signals. Often, the best choice for the job is a large-diaphragm dynamic mic. Since proximity effect (bass boost) occurs when using a directional mic at close working distances, and because the drum's harmonics vary over its large surface area, even a minor change in placement can have a profound effect on the pickup's overall sound. Moving the mic closer to the head (Figure 4.49) can add a degree of warmth and fullness, while moving it farther back often emphasizes the high-frequency "click". Placing the mic closer to the beater emphasizes the hard "thud" sound, whereas an off-center pickup captures more of the drum's characteristic skin tone. A dull and loose kick sound can be tightened to produce a sharper, more defined transient sound by placing a blanket or other damping material inside the drum shell firmly against the beater head. Cutting back on the kick's equalization at 300 to 600 Hz can help reduce the dull "cardboard" sound, whereas boosting from 2.5 to 5 kHz adds a sharper attack, "click" or "snap". It's also often

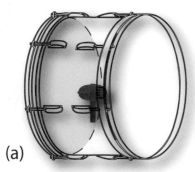

(a)

(b)

(c)

a good idea to have a can of WD-40® or other light oil handy in case squeaks from some of the moving parts (most often the kick pedal) get picked up by the mics.

SNARE DRUM

Commonly, a snare mic is aimed just inside the top rim of the snare drum at a distance of about 1 inch (Figure 4.50). The mic should be angled for the best possible separation from other drums and cymbals. Its rejection angle should be aimed at either the hi-hat or rack toms (depending on leakage difficulties). Usually, the mic's polar response is cardioid, although bidirectional and super-cardioid responses might offer a tighter pickup angle. With certain musical styles (such as jazz), you might want a crisp or "bright" snare sound. This can be achieved by placing an additional mic on the snare drum's bottom head and then combining the two mics into a single track. Because the bottom snare head is 180° out of phase with the top, it's almost always a wise idea to reverse the bottom mic's phase polarity. When playing in styles where the snare springs are turned off, it's also wise to keep your ears open for snare rattles and buzzes that can easily leak into the snare mic (as well as other mics). The continued ringing of an "open" snare note (or for any other drum type, for that matter) can be dampened in several ways. Dampening rings, which can be purchased at music stores, are used to reduce the ring and to deepen the instrument's tone. If there are no dampening rings around, the tone can be dampened by taping a billfold or similar-sized folded paper towel to the top/side of the drumhead, a few inches off its edge.

OVERHEADS

Overhead mics are generally used to pick up the high-frequency transients of cymbals with crisp, accurate detail while also providing an overall blend of the entire drum kit. Because of the transient nature of cymbals, a condenser

FIGURE 4.50
Snare drum mic placements. (a) With a mic placed over and just inside the top rim. (b) With the top mic placed, a lower mic can be placed and flipped 180° out of phase with the top mic so as to pick up the snare springs (in engaged).

(a)

(b)

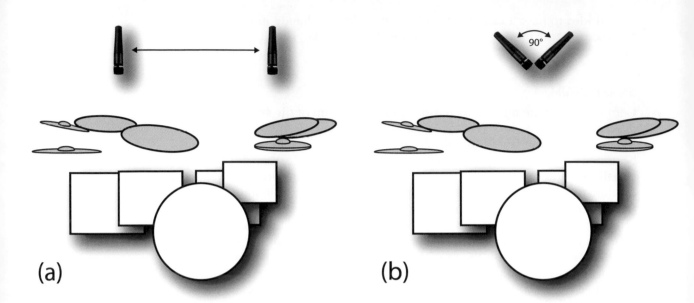

(a)　　　　　　　　　　　　　　　　(b)

FIGURE 4.51

Typical stereo overhead pickup positions: (a) Spaced pair technique. (b) X/Y coincident technique.

mic is often chosen for its accurate high-end response. Overhead mic placement can be very subjective and personal. One type of placement is the spaced pair, whereby two mics are suspended above the left and right sides of the kit. These mics are equally distributed about the L/R cymbal clusters so as to pick up their respective instrument components in a balanced fashion (Figure 4.51a). Another placement method is to suspend the mics closely together in a coincident fashion (Figure 4.51b). This often yields an excellent stereo overhead image with a minimum of the phase cancelations that might otherwise result when using spaced mics. Again, it's important to remember that there are no rules for getting a good sound. If only one overhead mic is available, place it at a central point over the drums. (Note: now is the time that my friend George Massenburg would want me to remind you to make sure that the overheads are equidistant to the kick, so as to avoid phase problems.) Lastly, if you're using a number of pickups to close mic individual components of a kit, there might be times when you won't need overheads at all (the leakage spillover just might be enough to do the trick).

RACK TOMS

The upper rack toms can be miked either individually (Figure 4.52a) or by placing a single mic between the two at a short distance (Figure 4.52b). When miked individually, a "dead" sound can be achieved by placing the mic close to the drum's top head (about 1 inch above and 1 to 2 inches in from the outer rim). A sound that's more "live" can be achieved by increasing the height above the head to about 3 to 6 inches. If isolation or feedback is a consideration, a hypercardioid pickup pattern can be chosen. Another way to reduce leakage and to get a deep, driving tone (with less attack) is to remove the tom's bottom head and place the mic inside, 1 to 6 inches away from the top head.

(a)

(b)

FLOOR TOM

Floor toms can be miked similarly to the rack toms. The mic can be placed 2 to 3 inches above the top and to the side of the head, or it can be placed inside 1 to 6 inches from the head. Again, a single mic can be placed above and between the two floor toms, or each can have its own mic pickup (which often yields a greater degree of control over panning and tonal color).

HI-HAT

The "hat" usually produces a strong, sibilant energy in the high-frequency range, whereas the snare's frequencies often are more concentrated in the midrange. Although moving the hat's mic won't change the overall sound as much as it would on a snare, you should still keep the following three points in mind:

- Placing the mic above the top cymbal will help pick up the nuances of sharp stick attacks.
- The open and closing motion of the hi-hat will often produce rushes of air; consequently, when miking the hat's edge, angle the mic slightly above or below the point where the cymbals meet.
- If only one mic is available (or desired), both the snare and the hi-hat can be simultaneously picked up by carefully placing the mic between the two, facing away from the rack toms as much as possible. Alternatively, a fig-ure-8 mic can be placed between the two with the null axis facing toward the cymbals and kick.

Tuned Percussion Instruments

The following sections describe the various sound characteristics and techniques that are encountered when miking tuned percussion instruments.

CONGAS AND HAND DRUMS

Congas, tumbas and bongos are single-headed, low-pitched drums that can be individually miked at very close distances of 1 to 3 inches above the head and

FIGURE 4.52
Miking toms. (a) Individual miking of a rack tom. (b) Single microphone placement for picking up two toms.

2 inches in from the rim, or the mics can be pulled back to a distance of 1 foot for a fuller, "live" tone. Alternatively, a single mic or X/Y stereo pair can be placed at a point about 1 foot above and between the drums (which are often played in pairs). Another class of single-headed, low-pitched drums (known as hand drums) isn't necessarily played in pairs but is often held in the lap or strapped across the player's front. Although these drums can be as percussive as congas, they're often deeper in tone and often require that the mic(s) be backed off in order to allow the sound to develop and/or fully interact with the room. In general, a good pickup can be achieved by placing a mic at a distance of 1 to 3 feet in front of the hand drum's head. Since a large part of the drum's sound (especially its low-end power) comes from its back hole, another mic can be placed at the lower port at a distance of 6 inches to 2 feet. Since the rear sound will be 180° out of phase from the front pickup, the mic's phase should be reversed whenever the two signals are combined.

XYLOPHONE, VIBRAPHONE AND MARIMBA

The most common way to mic a tuned percussion instrument is to place two high-quality condenser or extended-range dynamic pickups above the playing bars at a spaced distance that's appropriate to the instrument size (following the 3:1 general rule). A coincident stereo pair can help eliminate possible phase errors; however, a spaced pair will often yield a wider stereo image.

Stringed Instruments

Of all the instrumental families, stringed instruments are perhaps the most diverse. Ethnic music often uses instruments that range from being single stringed to those that use highly complex and developed systems to produce rich and subtle tones. Western listeners have grown accustomed to hearing the violin, viola, cello and double bass (both as solo instruments and in an ensemble setting). Whatever the type, stringed instruments vary in their design type and in construction to enhance or cut back on certain harmonic frequencies. These variations are what give a particular stringed instrument its own characteristic sound.

VIOLIN AND VIOLA

The frequency range of the violin runs from 196 Hz to above 10 kHz. For this reason, a good mic that displays a relatively flat frequency response should be used. The violin's fundamental range is from G3 to E6 (196 to 1,300 Hz), and it is particularly important to use a mic that's flat around the formant frequencies of 300 Hz, 1 kHz and 1,200 Hz. The fundamental range of the viola is tuned a fifth lower and contains fewer harmonic overtones. In most situations, the violin's or viola's mic should be placed within 45° of the instrument's front face. The distance will depend on the particular style of music and the room's acoustic condition. Miking at a greater distance will generally yield a mellow, well-rounded

(a)

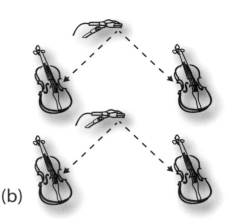

(b)

tone, whereas a closer position might yield a scratchy, more nasal quality – the choice will depend on the instrument's tone quality. The recommended miking distance for a solo instrument is between 3 and 8 feet, over and slightly in front of the player (Figure 4.53). Under studio conditions, a closer mic distance of between 2 and 3 feet is recommended. For a fiddle or jazz/rock playing style, the mic can be placed at a close working distance of 6 inches or less, as the increased overtones help the instrument to cut through an ensemble. Under PA (public address) applications, distant working conditions are likely to produce feedback (since less amplification is needed). In this situation, an electric pickup, contact or clip-type microphone can be attached to the instrument's body or tailpiece.

FIGURE 4.53
Miking a violin. (a) Example of a typical mic placement for the violin. (b) Example of miking for a small, four-piece string section.

CELLO

The fundamental range of the cello is from C2 to C5 (65 to 520 Hz), with overtones up to 8 kHz. If the player's line of sight is taken to be 0°, then the main direction of sound radiation lies between 10° and 45° to the right. A quality mic can be placed level with the instrument and directed toward the sound holes. The chosen microphone should have a flat response and be placed at a working distance of between 6 inches and 3 feet.

DOUBLE BASS

The double bass is one of the orchestra's lowest-pitched instruments. The fundamentals of the four-string type reach down to E1 (41 Hz) and up to around middle C (260 Hz). The overtone spectrum generally reaches upward to 7 kHz, with an overall angle of high-frequency dispersion being ±15° from the player's line of sight. Once again, a mic can be aimed at the f holes at a distance of between 6 inches and 1.5 feet.

Voice

From a shout to a whisper, the human voice is a talented and versatile sound source that displays a dynamic and timbrel range that's matched by few other instruments. The male bass voice can ideally extend from E2 to D4 (82 to 294 Hz) with sibilant harmonics extending to 12 kHz. The upper soprano voice can range upward to 1,050 Hz with harmonics that also climb to 12 kHz.

When choosing a mic and its proper placement, it's important to step back for a moment and remember that the most important "device" in the signal chain is the vocalist. Let's assume that the engineer/producer hasn't made the classic mistake of waiting until the last minute (when the project goes over budget and/or into overtime) to record the vocals. Good, now the vocalist can relax and concentrate on a memorable performance. Next step is to concentrate on the vocalist's "creature comforts". How are the lighting and temperature settings? Is the vocalist thirsty? Once done, you can go about the task of choosing your mic and its placement to best capture the performance.

The engineer/producer should be aware of the following traps that are often encountered when recording the human voice:

- *Excessive dynamic range*: this can be solved either by mic technique (physically moving away from the mic during louder passages) or by inserting a compressor into the signal path. Some vocalists have dynamics that range from whispers to normal volumes to practically screaming – all in a single passage. If you optimize your recording levels during a moderate-volume passage and the singer begins to belt out the lines, then the levels will become too "hot" and will distort. Conversely, if you set your recording levels for the loudest passage, the moderate volumes will be buried in the music. The solution to this dilemma is to place a compressor in the mic's signal path. The compressor automatically "rides" the signal's gain and reduces excessively loud passages to a level that the system can effectively handle. (See Chapter 12 for more information about compression and devices that alter dynamic range.) Don't forget that in the digital domain, it's often possible to apply compression later within the mixdown stage.

- *Sibilance*: this occurs when sounds such as f, s and sh are overly accentuated. This often is a result of tape saturation and distortion at high levels or slow tape speeds. Sibilance can be reduced by inserting a frequency-selective compressor (known as a de-esser) into the chain or through the use of moderate equalization.

- *Excessive bass boost due to proximity effect*: this bass buildup often occurs when a directional mic is used at close working ranges. It can be reduced or compensated for by increasing the working distance between the source and the mic, by using an omnidirectional mic (which doesn't display a proximity bass buildup), or through the use of equalization.

MIC TOOLS FOR THE VOICE

Some of the most common tools in miking are used for fixing problems that relate to picking up the human voice and to room isolation.

Explosive popping p and b sounds often result when turbulent air blasts from the mouth strike the mic diaphragm. This problem can be avoided or reduced by:

- Placing a pop filter over the mic
- Placing a mesh windscreen between the mic and the vocalist
- Taping a pencil in front of the mic capsule so as to break up the "plosive" air blasts
- Using an omnidirectional mic (which is less sensitive to popping but might cause leakage issues)

Reducing problems due to leakage and inadequate isolation can be handled in any number of ways, including:

- Choice of directional pattern (i.e., choosing a tighter cardioid or hypercardioid pattern can help reduce unwanted leakage)
- Isolating the singer with a flat or portable isolation cage
- Isolating the singer in a separate iso-booth
- Overdubbing the vocals at a later time, keeping in mind that carefully isolated "scratch" vocals can help glue the band together

Woodwind Instruments

The flute, clarinet, oboe, saxophone and bassoon combine to make up the woodwind class of instruments. Not all modern woodwinds are made of wood, nor do they produce sound in the same way. For example, a flute's sound is generated by blowing across a hole in a tube, whereas other woodwinds produce sound by causing a reed to vibrate the air within a tube.

Opening or covering finger holes along the sides of the instrument controls the pitch of a woodwind by changing the length of the tube and therefore, the length of the vibrating air column. It's a common misunderstanding that the natural sound of a woodwind instrument radiates entirely from its bell or mouthpiece. In reality, a large part of its sound often emanates from the finger holes that span the instrument's entire length.

CLARINET

The clarinet commonly comes in two pitches: the B-flat clarinet, with a lower limit of D3 (147 Hz), and the A clarinet, with a lower limit of C3 (130 Hz). The highest fundamental is around G6 (1,570 Hz), whereas notes an octave above middle C contain frequencies of up to 150 Hz when played softly. This spectrum can range upward to 12 kHz when played loudly. The sound of this reeded woodwind radiates almost exclusively from the finger holes at frequencies between 800 Hz and 3 kHz; however, as the pitch rises, more of the sound emanates from the bell. Often, the best mic placement occurs when the pickup is aimed toward the lower finger holes at a distance of 6 inches to 1 foot (Figure 4.54a).

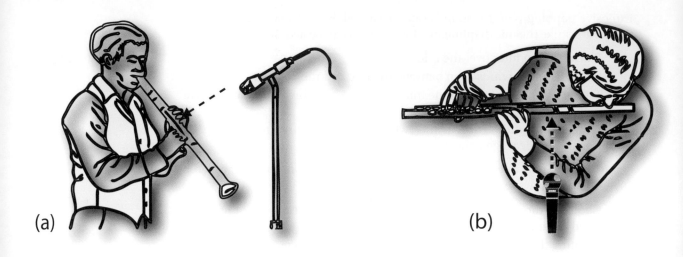

(a)

(b)

FIGURE 4.54

Typical woodwind placement. (a) Mic position for the clarinet. (b) Mic position for a flute.

FLUTE

The flute's fundamental range extends from B3 to about C7 (247 to 2,093 Hz). For medium–loud tones, the upper overtone limit ranges between 3 and 6 kHz. Commonly, the instrument's sound radiates along the player's line of sight for frequencies up to 3 kHz. Above this frequency, however, the radiated direction often moves outward 90° to the player's right. When miking a flute, placement depends on the type of music being played and the room's overall acoustics. When recording classical flute, the mic can be placed on-axis and slightly above the player at a distance of between 3 and 8 feet. When dealing with modern musical styles, the distance often ranges from 6 inches to 2 feet. In both circumstances, the microphone should be positioned at a point 1/3 to 1/2 the distance from the instrument's mouthpiece to its footpiece. In this way, the instrument's overall sound and tone quality can be picked up with equal intensity (Figure 4.54b). Placing the mic directly in front of the mouthpiece will increase the level (thereby reducing feedback and leakage); however, the full overall body sound won't be picked up, and breath noise will be accentuated. If mobility is important, an integrated contact pickup can be used, or a clip mic can be secured near the instrument's mouthpiece.

SAXOPHONE

Saxophones vary greatly in size and shape. The most popular sax for rock and jazz is the S-curved B-flat tenor sax, whose fundamentals span from B2 to F5 (123 to 698 Hz), and the E-flat alto, which spans from C3 to G5 (130 to 784 Hz). Also within this family are the straight-tubed soprano and sopranino, as well as the S-shaped baritone and bass saxophones. The harmonic content of these instruments ranges up to 8 kHz and can be extended by breath noises up to 13 kHz. As with other woodwinds, the mic should be placed roughly in the middle of the instrument at the desired distance and pointed slightly toward the bell (Figure 4.55). Keypad noises are considered to be a part of the instrument's

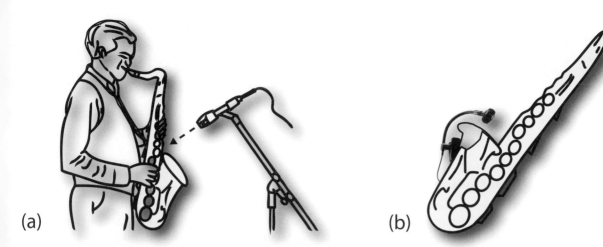

(a)

(b)

FIGURE 4.55
Typical microphone positions for the saxophone: (a) Standard placement. (b) Typical "clip-on" placement.

sound; however, even these can be reduced or eliminated by aiming the microphone closer to the bell's outer rim.

HARMONICA

Harmonicas come in all shapes, sizes and keys – and are divided into two basic types: the diatonic and the chromatic. Their pitch is determined purely by the length, width and thickness of the various vibrating metal reeds. The "harp" player's habit of forming his or her hands around the instrument is a way to mold the tone by forming a resonant cavity. The tone can be deepened and a special "wahing" effect can be produced by opening and closing a cavity that's formed by the palms; consequently, many harmonica players carry their preferred microphones with them rather than being stuck in front of an unfamiliar mic and stand.

MICROPHONE SELECTION

The following information is meant to provide insights into a limited number of professional mics that are used for music recording and professional sound applications. This list is by no means complete, as literally hundreds of mics are available, each with its own particular design, sonic character and application.

Shure SM57

The SM57 (Figure 4.56) is widely used by engineers, artists, touring sound companies, etc. for instrumental and remote recording applications. The SM57's midrange presence peak and good low-frequency response make it useful for use with vocals, snare drums, toms, kick drums, electric guitars and keyboards.

Specifications:

- *Transducer type*: moving-coil dynamic
- *Polar response*: cardioid

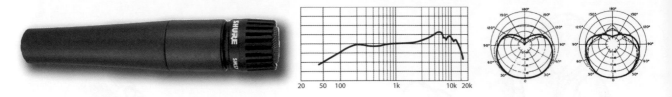

FIGURE 4.56
Shure SM57 dynamic microphone (courtesy of Shure Brothers, Inc., www.shure .com, images © 2023, Shure Incorporated – used with permission).

- *Frequency response*: 40 to 15,000 Hz
- *Equivalent noise rating*: –7.75 dB (0 dB = 1 V/microbar)

Telefunken M81

The Telefunken M81 (Figure 4.57) has been likened to a condenser microphone and has become a staple for vocal and snare drum applications, especially in the world of touring and live performance. As an alternative to the extended top-end capabilities of the Telefunken M80, the linear-response M81 is a tool that is a bit less specialized, giving the microphone more universal application ability.

Specifications:

- *Transducer type*: moving-coil dynamic
- *Polar response*: supercardioid
- *Capsule*: 25 mm
- *Frequency response*: 50 Hz/18 kHz
- *Sensitivity*: 1.54 mV/Pa
- *Maximum SPL*: 135 dB

AKG D112

Large-diaphragm cardioid dynamic mics, such as the AKG D112 (Figure 4.58), are often used for picking up kick drums, bass guitar cabinets and other low-frequency, high-output sources.

Specifications:

FIGURE 4.57
Telefunken M81 dynamic microphone (courtesy of Telefunken Elektroakustik, Inc., www.telefunkenelekt roakustik.com).

- *Transducer type*: moving-coil dynamic
- *Polar response*: cardioid
- *Frequency response*: 30 to 17,000 Hz
- *Sensitivity*: –54 dB ± 3 dB re. 1 V/microbar

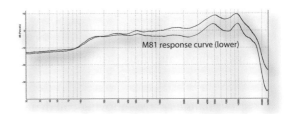

M81 response curve (lower)

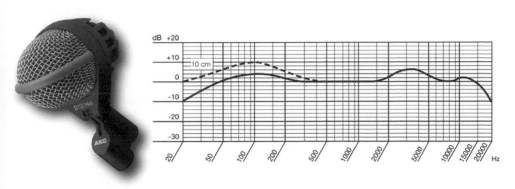

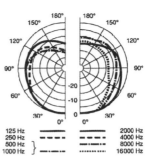

Royer Labs R-121

The R-121 is a ribbon mic with a figure-8 pattern (Figure 4.59). Its sensitivity is roughly equal to that of a good dynamic mic, and it exhibits a warm, realistic tone and flat frequency response. Made using advanced materials and cutting-edge construction techniques, its response is flat and well balanced; the low end is deep and full without getting boomy, mids are well defined and realistic, and the high-end response is sweet and natural sounding.

Specifications:

- *Acoustic operating principle*: electrodynamic pressure-gradient ribbon
- *Polar pattern*: figure-8
- *Generating element*: 2.5-micron aluminum ribbon
- *Frequency response*: 30 to 15,000 Hz ± 3 dB
- *Sensitivity*: −54 dBV re. 1 V/Pa ± 1 dB
- *Output impedance*: 300 Ω at 1 K (nominal); 200 Ω optional
- *Maximum SPL*: >135 dB

Beyerdynamic M-160

The Beyer M-160 ribbon microphone (Figure 4.60) is capable of handling high SPL without sustaining damage while providing the transparency that often is

FIGURE 4.58
AKG D112 dynamic microphone (courtesy of AKG Acoustics GmbH., www.akg.com).

FIGURE 4.59
Royer Labs R-121 ribbon microphone (courtesy of Royer Labs, www.royerlabs.com).

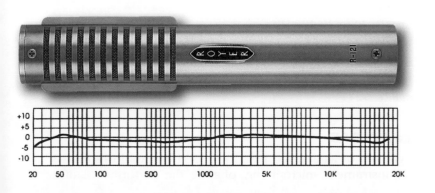

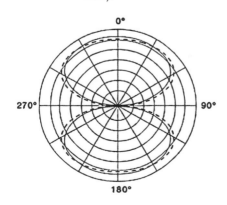

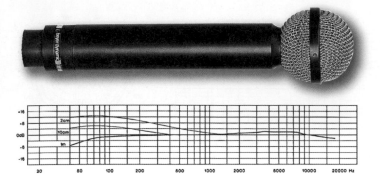

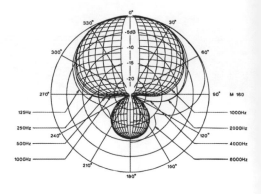

FIGURE 4.60

Beyerdynamic M-160 ribbon microphone (courtesy of Beyerdynamic, www.beyer-dynamic.com).

inherent in ribbon mics. Its hypercardioid response yields a wide-frequency response/low-feedback characteristic for both studio and stage.

Specifications:

- *Transducer type*: ribbon dynamic
- *Polar response*: hypercardioid
- *Frequency response*: 40–18,000 Hz
- *Sensitivity*: 52 dB (0 dB = 1 mW/Pa)
- *Equivalent noise rating*: −145 dB
- *Output impedance*: 200 Ω

AEA A440

The AEA model R44 and A440 ribbon microphones (the latter is shown in Figure 4.2a) carry forward the classic tradition of the venerable 1930s RCA-44 into the twenty-first century. The R44 series even uses "new old-stock" RCA rib-bon material and features a re-engineered output transformer that recaptures the sonic signature of the original RCA-44 and surpasses it in frequency response and dynamics.

Specifications:

- *Transducer type*: ribbon dynamic
- *Polar response*: bidirectional
- *Frequency response*: 20 to 20,000 Hz
- *Sensitivity*: 30 mV (−33.5 dBV)/Pa (1 Pa = 94 dB[SPL])
- *Equivalent noise rating*: signal-to-noise ratio: 88 dB (A) (94 dB[SPL] minus equivalent noise)
- *Output impedance*: 92 Ω

Audio-Technica AT5045

Available as a single mic or as a stereo pair, the AT5045 is Audio-Technica's premier condenser instrument microphone, offering the performance of a

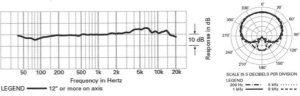

large-diaphragm, side-address condenser in a convenient, stick-type body. The AT5045 boasts Audio-Technica's largest single diaphragm. The instrument microphone's circuitry allows the AT5045 to achieve an unprecedented dynamic range (Figure 4.61).

- *Element*: fixed-charge back plate, permanently polarized condenser
- *Polar pattern*: cardioid
- *Frequency response*: 20–20,000 Hz
- *Open circuit sensitivity*: –35 dB (17.7 mV) re 1 V at 1 Pa
- *Maximum input sound level*: 149 dB^{SPL}, 1 kHz at 1% THD (total harmonic distortion)

FIGURE 4.61
Audio-Technica AT5045 condenser instrument microphone (courtesy of Audio-Technica U.S., www.audio-technica.com).

AKG C214

The AKG C214 (Figure 4.62) is the younger brother of the legendary C414 condenser. Engineered for highest linearity and neutral sound, this mic uses the same 1-inch capsule as the C414 in a single-diaphragm, cardioid-only design.

Specifications:

- *Transducer type*: condenser
- *Polar response*: omnidirectional, wide cardioid, cardioid, hypercardioid, figure-8 and four intermediate settings
- *Bass cut filter slope*: 12 dB/octave at 40 Hz and 80 Hz; 6 dB/octave at 16 Hz
- *Frequency response*: 20 to 20,000Hz
- *Sensitivity*: 20 mV/Pa

FIGURE 4.62
AKG C 214 condenser microphone (courtesy of AKG Acoustics GMBH, www.akg.com).

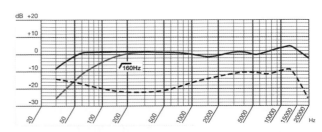

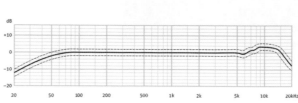

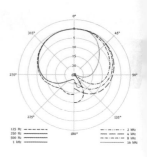

FIGURE 4.63

Neumann TLM 102 condenser microphone (courtesy of Georg Neumann GMBH, www.neumann.com).

Neumann TLM102

The TLM102 (Figure 4.63) has a newly developed large-diaphragm capsule (cardioid) with a maximum sound-pressure level of 144 dB, which permits the recording of percussion, drums, amps and other very loud sound sources. For vocals and speech, a slight boost above 6 kHz provides for excellent presence of the voice in the overall mix. Up to 6 kHz, the frequency response is extremely linear, ensuring minimal coloration and a clearly defined bass range. The capsule has an elastic suspension for the suppression of structure-borne noise.

Specifications:

- *Transducer type*: condenser
- *Polar response*: cardioid
- *Frequency response*: 20 to 20,000 Hz
- *Sensitivity*: 11 mV/Pa

Warm Audio WA-47

The WA-47 is an authentic reproduction of the most coveted tube microphone, hailed as the grandfather of professional recordings and timeless studio sound. Some of the most famous albums of all time feature vintage 47 microphones, with an artist roster that goes from Sinatra to Adele with many small acts like The Beatles in between. The WA-47 is a serious tool for professional recording. The WA-47 features a gold-sputtered, large, dual-diaphragm, single-backplate K47-style capsule, and the overall mic is meticulously designed with premium components to authentically reproduce that coveted sonic profile and feature set at an unprecedented price (Figure 4.64).

Specifications:

- *Frequency range*: 20 Hz – 20 kHz
- *Output impedance*: 200 Ohms
- *Rated load impedance*: ≥2 kOhms
- *S/N ratio*: 82 dBA

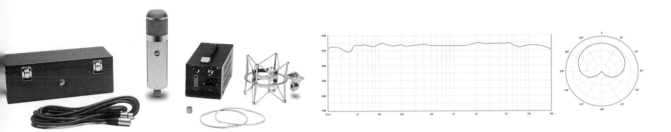

FIGURE 4.64
Warm Audio WA-47 authentic reproduction of one of the most coveted tube microphones (courtesy of Warm Audio LLC, www.warmaudio.com).

- *Equivalent noise*: 10 dBA (IEC651)
- *Gotham 5 Meter GAC-7, 7-pin tube microphone cable*
- *Total weight*: 9 lbs
- *Mic diameter*: 60 mm

Telefunken U47, C12 and ELA M251E

Telefunken (Figure 4.65) offers historic recreations of classic microphones (in addition to its own line of microphones) that are based around the distinctive tube mic sound, blending vintage style and sound with the reliability of a modern-day microphone.

FIGURE 4.65
Telefunken U47, C12 and ELA M251E classic condenser microphone recreations (courtesy of Telefunken Elektroakoustik, www.telefunkenelektroakustik.com).

CHAPTER 5

The Analog Tape Recorder

From its inception in Germany in the late 1920s (Figure 5.1) and its American introduction by Jack Mullin in 1945, the *analog tape recorder* (or *ATR*) had steadily increased in quality and universal acceptance, to the point that professional and personal studios had totally relied upon magnetic media for the storage of analog sound onto reels of tape. With the dawning of the project studio and computer-based digital audio workstations (DAWs), the use of two-channel and multitrack ATRs has steadily dwindled, to the point where no new analog tape machine models are currently being manufactured. In short, recording to analog tape has steadily become a high-cost, future-retro, specialty process for getting a "certain sound". That being said, the analog recording process is still highly regarded and even sought after by many studios for its characteristic sound and by others as a raised fist against the onslaught of the "evil digital empire". Without delving into the ongoing debate of the merits of analog versus digital, I think it's fair to say that each has its place and its own distinct type of sound and application in audio and music production. Although professional analog recorders are usually much more expensive than their digital counterparts, as a general rule, a properly aligned, professional analog deck will have a particular sound that's often described as being full, punchy, gutsy and "raw". In fact, the limitations of tape are often used as a form of "artistic expression" to get a certain sound.

Truth be known, for this and previous editions, there have been many who have asked that we delete this chapter. I know that many will never have the need for these amazing tools ... but we have kept it in for those of you who might have the need, might want the option or would simply like to know more about how an ATR works. In short, you never know when you might get a call from a supergroup's manager who wants to record to 24tk tape. The analog tape recorder isn't dead yet – and probably won't be for some time.

DOI: 10.4324/9781003260530-5

FIGURE 5.1
One of John T. (Jack) Mullin's WWII vintage German Magnetophones, which was confiscated during WWII and brought to the United States (courtesy of the late John T. Mullin).

TO COMMIT OR NOT TO COMMIT IT TO TAPE?

Before we delve into the inner workings of the analog tape recorder, let's take a moment to discuss ways in which the analog tape sound can be taken advantage of in the digital and project studio environment. Before you go out and buy your own deck, however, there are other cost-effective ways to get "that sound" on your own projects. For example:

- In recent times, a growing number of plug-ins have become available that can emulate (or approximate) the harmonic and overdriven sound of an analog tape track. Further info on such tape plug-ins can be found later in this chapter and in the book's chapter on signal processing.

- Rent a studio that has an analog multitrack for a few hours or days. You could record specific tracks to tape, transfer existing digital tracks to tape or dump an entire final mixdown to tape. For the cost of studio time and a reel of tape, you could inject your project with an entirely new type of sound.

- Rent an analog machine from a local studio equipment service. For a rental fee and basic cartage charges, you could reap the benefits of having an analog ATR for the duration of a project without any undue financial and maintenance overhead.

A few guidelines should also be kept in mind when recording and/or transferring tracks to or from a multitrack recorder:

- Obviously, high recording levels add to that sought-after "overdriven" analog sound; however, driving a track too hard (hot) can actually kill a track's definition or "air". The trick is often to find a center balance between the right amount of saturation, distortion and dynamic range.

- Noise reduction can be a good thing, but it can also diminish what is thought of as that "classic analog sound". Newer, wide–tape width recorders (such as ATR Services' ATR-102 1-inch, 2-track and the 108C 2-inch, 8-track recorder), as well as older 2-inch, 16-track recorders, can provide improved definition without the need for noise reduction.

THE MEDIUM OF MAGNETIC RECORDING

At a basic level, an analog audio tape recorder can be thought of as a sound recording device that has the capacity to store audio information onto a magnetizable tape-based medium and then play this information back at a later time. By definition, analog refers to something that's "analogous", similar to or comparable to something else. An ATR is able to transform an electrical input signal directly into a corresponding magnetic energy that can be stored onto tape in the form of magnetic remnants. Upon playback, this magnetic energy is then converted back into a corresponding electrical signal that can be amplified, mixed, processed and heard.

The recording medium itself is composed of several layers of material, each serving a specific function (**Figure 5.2**). The base material that makes up most of a tape's thickness is often composed of polyester or polyvinyl chloride (PVC), which is a durable polymer that's physically strong and can withstand a great deal of abuse before being damaged. Bonded to the PVC base is the all-important layer of magnetic oxide. The molecules of this oxide combine to create some of the smallest known permanent magnets, which are called *domains*. On an unmagnetized tape, the polarities of these domains are randomly oriented over the entire surface of the tape (**Figure 5.3a**). The resulting energy force of this random magnetization at the reproduce head is a general cancellation of the combined domain energies, resulting in no signal at the recorder's output (except for the tape noise that occurs due to the residual domain energy output … hisssssssss).

When a signal is recorded, the resulting magnetization from the record head polarizes the individual domains (at varying degrees in positive and negative angular directions) in such a way that their average magnetism produces a much larger combined magnetic flux (**Figure 5.3b**). When the tape is pulled across the playback head at the same, constant speed at which it was recorded, this alternating magnetic output is then converted back into an alternating signal that can then be amplified and further processed for reproduction.

FIGURE 5.2
Structural layers of magnetic tape.

1. tape coating
2. magnetic oxide
3. polyester base (pvc)
4. antistatic backing

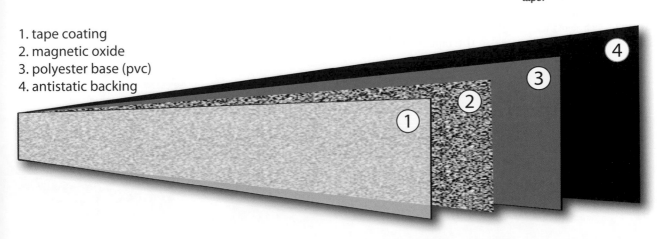

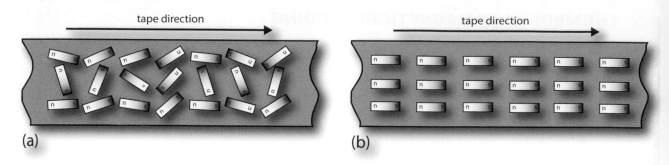

(a) (b)

FIGURE 5.3

Orientation of magnetic domains on unmagnetized and magnetized recording tape. (a) The random orientation of an unmagnetized tape results in no output. (b) Magnetized domains result in an average flux output at the magnetic head.

The Professional Analog ATR

Professional analog ATRs can be found in 2-, 4-, 8-, 16- and 24-track formats. Each configuration is generally best suited to a specific production and post-production task. For example, a 2-track ATR is generally used to record the final stereo mix of a project (Figure 5.4), whereas 8-, 16- and 24-track machines are obviously used for multitrack recording (Figure 5.5). Although no professional analog machines are currently being manufactured, a number of machines are being reconditioned or can be found on the used market in varying degrees of working condition.

THE TAPE TRANSPORT

The process of recording audio onto magnetic tape depends on the transport's capability to pass a precise length of tape over the record head at a specific and constant speed, with a uniform tension (Figure 5.6). During playback, this relationship is maintained by again moving the tape across the heads at the same speed, thereby preserving the program's original pitch, rhythm and duration.

FIGURE 5.4

Examples of various ATRs that can still be found. (a) Otari Mx-5050. (b) Ampex ATR-102 1-inch stereo. (c) Studer A-800 24-tk.

This constant speed and tension movement of the tape across a head's path is initiated by simply pressing the Play button. The drive can be disengaged at any time by pressing the Stop button, which applies a simultaneous breaking force to both the left and right reels. The Fast Forward and Rewind buttons cause the tape to rapidly shuttle in the respective directions in order to locate a specific point. Initiating either of these modes usually engages the tape lifters, which raise the tape away from the heads (definitely an ear-saving feature). Once the

(a) (b) (c)

FIGURE 5.5
Analog multitrack machines (right) coexisting with their digital tape counterparts (left) (courtesy of Galaxy Studios, www.galaxy.be).

play mode has been engaged, pressing the Record button allows audio to be recorded onto any selected track or tracks.

Beyond these basic controls, you might expect to run into several differences between transports (often depending on the machine's age). For example, older recorders might require that both the Record and Play buttons be simultaneously pressed in order to go into record mode;, while others may begin recording when the Record button is pressed while already in the Play mode.

On certain older professional transports (particularly those wonderful Ampex decks from the 1950s and 1960s), stopping a fast-moving tape by simply pressing the Stop button might stretch or destroy a master tape, because the inertia is simply too much for the electro-mechanical brake to deal with. In such a situation, a procedure known as "rocking" the tape is used to prevent tape damage. The deck can be rocked to its stop position by engaging the fast-wind mode in the direction opposite the current travel direction until the tape slows down to a reasonable speed … at which point it's safe to press the Stop button. Go ahead – thread a used, blank tape onto a machine and try it for yourself.

In recent decades (1980 and later), tape transport designs have made use of total transport logic (TTL), which places transport and monitor functions under microprocessor control. This has a number of distinct advantages. For example, with TTL, the recorder can sense the tape speed and direction and then

FIGURE 5.6
Relationship of time to the physical passage of recording tape.

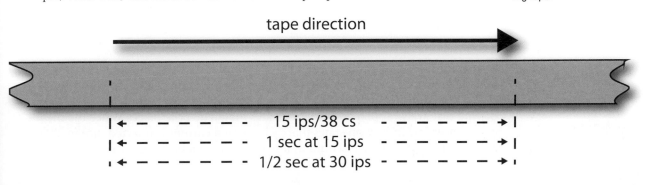

automatically rock the transport motors until the tape can safely be stopped, or it might slow the tape to a point where the deck can seamlessly slip into play, record or stop mode.

Most modern ATRs are equipped with a control that allows the tape to be shuttled at various wind speeds in either direction. This allows a specific cue point to be located by listening to the tape at varying play speeds, or this option can be used to gently and evenly wind the tape onto its reel at a slower speed for long-term storage. The Edit button (which can be found on certain professional machines) often has two operating modes: stop-edit and dump-edit. If the Edit button is pressed while the transport is in the stop mode, the left and right tape reel brakes are released, and the tape sensor is bypassed. This makes it possible for the tape to be manually rocked back and forth until the edit point is found. Often, if the Edit button is pressed while in the play mode, the take-up drive will be disengaged, and the tape sensor is bypassed. This allows unwanted sections of tape to be spooled off the machine (and into the trash can) while listening to the material as it's being discarded during playback.

A safety tape guide switch, which is incorporated into all professional transports, initiates the stop mode when it senses the absence of tape along its guide path; thus, the recorder stops automatically at the end of a reel or if the tape should accidentally break. This switch might be built into the tape-tension sensor arm, or it might exist in the form of a light beam that's interrupted when tape is present.

Most newer, professional ATRs are equipped with automatic tape counters that accurately read out time in hours, minutes, seconds and sometimes frames (00:00:00:00). Many of these recorders have digital readout displays that double as tape-speed indicators when in the "vari-speed" mode. This function incorporates a control that lets you vary the tape speed from fixed industry standards. On many tape transports, this control can be continuously varied over a ±20% range from the 7½, 15 or 30 ips (inches per second) standards.

THE MAGNETIC TAPE HEAD

Most professional analog recorders use three magnetic tape heads, each of which performs a specialized task:

- Record
- Reproduce
- Erase

The function of a *record head* (Figure 5.7a) is to electromagnetically transform analog electrical signals into corresponding magnetic fields that can be permanently stored onto magnetic tape. In short, the input current flows through coils of wire that are wrapped around the head's magnetic pole pieces. Since the theory of magnetic induction states that "whenever a current is injected into metal, a magnetic field is created within that metal", a magnetic force is caused to flow

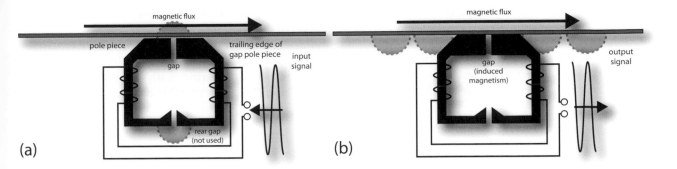

(a)   (b)

FIGURE 5.7
The record/playback heads.
(a) Record. (b) Play.

through the coil, into the pole pieces and across the head gap. Like electricity, magnetism flows more easily through some media than through others. The head gap between poles creates a break in the magnetic field, thereby creating a physical resistance to the magnetic "circuit". Since the gap is in physical contact with the moving magnetic tape, the tape's magnetic oxide offers a lower resistance path to the field than does the non-magnetic gap. Thus, the flux path travels from one pole piece, into the tape and to the other pole. Since the magnetic domains retain their polarity and magnetic intensity as the tape passes across the gap, the tape now has an analogous magnetic "memory" of the recorded event.

The reproduce or *playback head* (Figure 5.7b) operates in a way that's opposite to the record head. When a recorded tape track passes across the reproduce head gap, a magnetic flux is induced into the pole pieces. Since the theory of magnetic induction also states that "whenever a magnetic field cuts across metal, a current will be set up within that metal", an alternating current is caused to flow through the pickup coil windings, which can then be amplified and processed into a larger output signal.

It's important to note that the reproduce head's output is nonlinear because this signal is proportional to both the tape's average flux magnitude and the rate of change of this magnetic field. This means that the rate of change increases as a direct function of the recorded signal's frequency. Thus, the output level of a playback head is effectively doubled for each doubling in frequency, resulting in a 6-dB increase in output voltage for each increased octave. The tape speed and head-gap width work together to determine the reproduce head's upper-frequency limit, which in turn, determines the system's overall bandwidth. The wavelength of a signal that's recorded onto tape is equal to the speed at which tape travels past the reproduce head, divided by the frequency of the signal; therefore, the faster the tape speed, the higher the upper-frequency limit. Likewise, the smaller the head gap, the higher the upper-frequency limit.

The function of the *erase head* is to effectively reduce the average magnetization level of a recorded tape track to zero, thereby allowing the tape track to be re-used or recorded over. After a track is placed into the record mode, a high-frequency, high-intensity sine-wave signal is fed into the erase head (resulting in a tape that's being saturated in both the positive- and negative-polarity directions). This alternating saturation occurs at such a high speed that it serves to randomize any magnetic pattern that previously existed on the tape. As the tape moves away from the erase head, the intensity of the magnetic field decreases over time, leaving the domains in a random orientation, with a resulting average magnetization or output level that's as close to zero as tape noise will allow.

EQUALIZATION

Because the analog recording process isn't linear, equalization is needed to achieve a flat frequency-response curve when using magnetic tape. The 6-dB-per-octave boost that's inherent in the playback head's response curve requires that a complementary equalization cut of 6 dB per octave be applied within the playback circuit (see Figure 5.8).

BIAS CURRENT

In addition to the nonlinear changes that occur in playback level relative to frequency, another discrepancy in the recording process exists between the amount of magnetic energy that's applied to the record head and the amount of magnetism that's retained by the tape after the initial recording field has been removed. As Figure 5.9a shows, the magnetization curve of tape is linear between points A and B, as well as between points C and D. Signals greater than A and D have reached the saturation level and are subject to clipping distortion. Conversely, signals falling within the B to C range are too low in flux level to adequately magnetize the domains during the recording process (i.e., they're not strong enough to force the individual magnets to change orientation). For this reason, it's important that low-level signals be boosted in level so that they're pushed into the linear range. This boost is applied by mixing an AC bi*as current* (Figure 5.9b) with the audio signal. This bias current is applied by mixing the incoming audio signal with an ultrasonic sine-wave signal (often between 75 and 150 kHz). The

FIGURE 5.8

A flat frequency playback curve is the result of complementary equalization in the playback circuit.

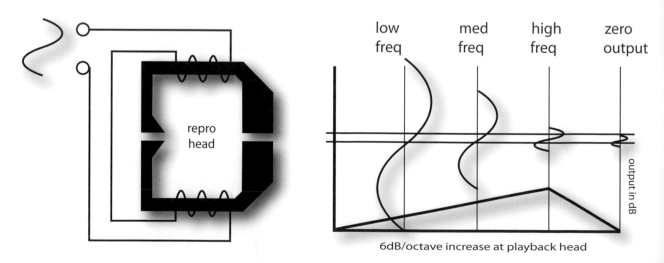

repro head

low freq   med freq   high freq   zero output

output in dB

6dB/octave increase at playback head

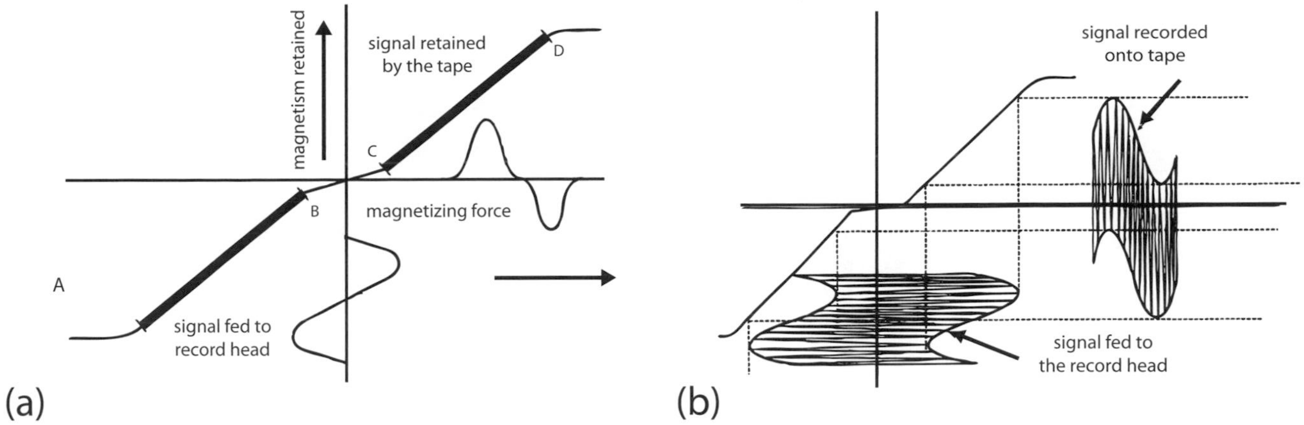

(a)

(b)

FIGURE 5.9
The effects of bias current
on recorded linearity: (a)
Magnetization curve showing
distortion at lower levels.
(b) After bias is applied, the
signal is boosted back into
the curve's linear regions.

combined signals are then amplitude modulated in such a way that the overall magnetic flux levels are given an extra "oomph", which effectively boosts the signal above the nonlinear zero-crossover range and into the linear (free of distortion) portion of the curve. In fact, if this bias signal weren't added, distortion levels would be so high as to render the analog recording process useless.

MONITORING MODES

The output signal of a professional ATR channel can be switched between three working modes:

- Input
- Reproduce
- Sync

In the *input* (source) *mode*, the signal at the selected channel output is derived from its input signal. Thus, with the ATR transport in any mode (including stop), it's possible to meter and monitor the signal that's present at a channel's selected input. In the reproduce mode, the output and metering signal is derived from the playback head. This mode can be useful in two ways: it allows previously recorded tapes to be played back, and it enables the monitoring of material off of the tape while in the record mode. The latter provides an immediate quality check of the ATR's entire record and reproduce process. The *sync mode* (originally known as selective synchronization, or sel-sync) is a required feature in analog multitrack ATRs because of the need to record new material on one or more tracks while simultaneously playing back tracks that have been previously recorded (during a process called *overdubbing*). Here's the deal – using the record head to lay down one or more tracks while listening to previously recorded tracks off the reproduce head would actually cause the newly recorded track(s) to be out of sync with the others on final playback (due to the physical distance between the two heads, as shown in Figure 5.10a). To prevent such a time lag, all of the reproduced tracks must be monitored off the record head at the same time that the new tracks are being laid down onto the same head. Since the record

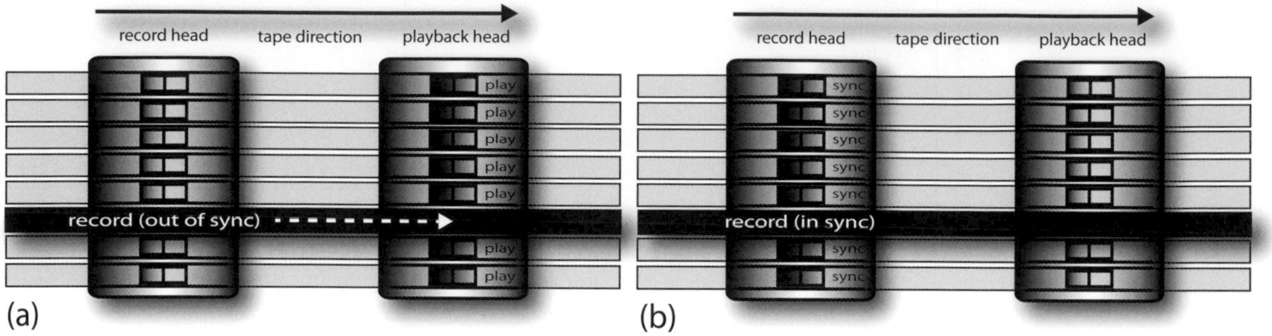

record head   tape direction   playback head

play
play
play
play
play

record (out of sync) ━━ ━━ ━━ ━━ ━━ ▶

play
play

(a)

record head   tape direction   playback head

sync
sync
sync
sync
sync

record (in sync)

sync
sync

(b)

FIGURE 5.10
The sync mode's function. (a) In the monitored playback mode, the recorded signal lags behind the recorded signal, thereby creating an out-of-sync condition. (b) In the sync mode, the record head acts as both record and playback head, bringing the signals into sync.

head is used for both recording and playback, there is no physical time lag and thus, no signal delay (Figure 5.10b).

Tape, Tape Speed and Head Configurations

Professional analog ATRs are currently available in a wide range of track- and tape-width configurations. The most common analog configurations are 2-track mastering machines that use tape widths of 1/4 inch, 1/2 inch and even 1 inch, as well as 16- and 24-track machines that use 2-inch tape. Figure 5.11 details many of the tape formats that can be currently found. Optimal tape-to-head performance characteristics for an analog ATR are determined by several parameters: track width, head-gap width and tape speed. In general, track widths are on the order of 0.080 inch for a 1/4-inch 2-track ATR; 0.070 inch for 1/2-inch 4-track, 1-inch 8-track, and 2-inch 16-track formats; or 0.037 inch for the 2-inch 24-track format. As you might expect, the greater the recorded track width, the greater the amount of magnetism that can be retained by the magnetic tape, resulting in a higher output signal and an improved signal-to-noise (S/N) ratio. The use of wider track widths also makes the recorded track less susceptible to signal-level dropouts.

The most common tape speeds used in audio production are 15 ips (38 cm/sec) and 30 ips (76 cm/sec). Although 15 ips will eat up less tape, 30 ips has gained wide acceptance in recent years for having its own characteristic sound (often having a tighter bottom end) as well as a higher output and lower noise figures

FIGURE 5.11
Analog track configurations for various tape widths.

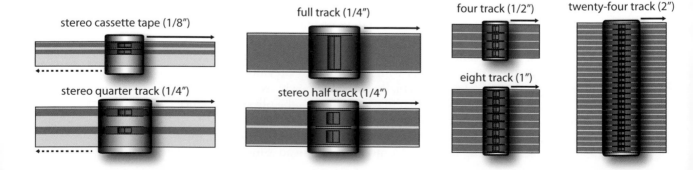

stereo cassette tape (1/8")

stereo quarter track (1/4")

full track (1/4")

stereo half track (1/4")

four track (1/2")

eight track (1")

twenty-four track (2")

(which in certain cases eliminates the need for noise reduction). On the other hand, 15 ips has a reputation for having a more "gutsy", rugged sound.

PRINT-THROUGH

A form of deterioration in a recording's quality, known as *print-through*, begins to occur almost immediately after a recording has been made. This effect is the result of the transfer of a recorded signal from one layer of tape to an adjacent outer track layer by means of magnetic induction, which gives rise to an audible false signal or echo on playback. The effects of print-through are greatest when recording levels are very high, and the effect decreases by about 2 dB for every 1-dB reduction in signal level. The extent of this condition also depends on such factors as length of storage, storage temperature and tape thickness (tapes with a thicker base material are less likely to have severe print-through problems).

Because of the effects of print-through, the standard method for professionally storing a recorded analog tape is in the *tails-out* position, using the following method:

1. Professional analog tape should *always* be stored tails-out (onto the right-hand take-up reel).
2. Upon playback, the tape should be rewound onto the left-most "supply reel".
3. During playback, feed the tape back onto the right-hand take-up reel, after which time it can again be removed for storage.
4. If the tape has been continuously wound and rewound during the session, it's often wise to rewind the tape and then smoothly play or slow-wind the tape onto the take-up reel, after which time it can be removed for storage.

So, why do we go through all this trouble? When a tape is improperly stored "heads-out", the signal will print through to the outer tape layer. Upon play-back, the signal will be heard as an unnatural pre-echo that can readily be heard. If a tape is properly stored using the tails-out method (Figure 5.12), the

FIGURE 5.12
Recorded analog tapes should always be stored in the tails-out position.

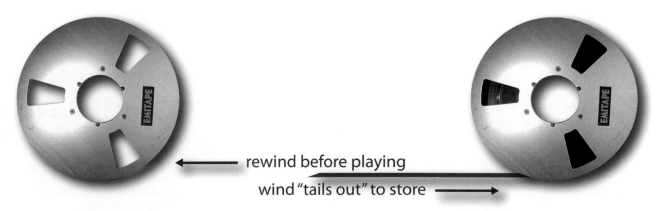

rewind before playing

wind "tails out" to store

print-through will bleed to the outer layers, causing the signal to bleed in such a way that the echo will follow the original signal. This will result in a decay that's subconsciously perceived by the listener as a natural after-echo instead of as a distracting pre-echo.

ANALOG TAPE NOISE

The roughly 60-dB S/N limitation that's imposed on a conventional analog ATR audio track is dictated by saturation (at the high–signal level end) and tape hiss at the low end (which is heard when the overall recorded level of the program is too low). Should an optimum level produce an unacceptable amount of noise, the engineer is faced with several options: record at a higher level (with the possibility of increased distortion) or change the signal's overall dynamic range by raising low-level signals above the noise (often with compression or limiting).

Analog tape noise might not be much of a problem when dealing with 1 or 2 tracks in an audio production, but the combined noise and other distortions that can occur when 8, 16, 24 or 48 tracks are combined can range from being bothersome to downright unacceptable. The following types of noise are often major contributors to the problem:

- Tape and amplifier noise
- Crosstalk between tracks
- Print-through
- Modulation noise

Modulation noise is a high-frequency component that causes sonic "fuzziness" by introducing sideband frequencies that can distort the signal. This noise-based distortion is due to the magnetic and mechanical properties of the analog recording process itself and actually increases as recorded levels rise. This noise is often higher in level than you might expect, and when combined with *asperity noise* (additional sideband frequencies that are also introduced by the analog record/playback process), these distortions definitely play a role in what is called the "analog sound".

TRY THIS: ANALOG TAPE MODULATION AND ASPERITY NOISE

1. Go to youtube.com/modernrecordingtechniques and search for "modulation noise"

or …

1. Feed a 0-VU, 1-kHz test tone to a track on a professional analog recorder.

2. Listen to the recorder's source (input) signal through the monitors at a moderate level.

3. Switch the recorder to monitor the tone from the track's playback (tape) head while in record. Does it sound different?

These analog-based noises can be reduced to acceptable levels by using different combinations of the following actions:

- Increasing the tape speed in order to record at higher flux levels (which might or might not help).
- Using an ATR with wider recorded tracks (i.e., allowing higher record/playback levels and reduced crosstalk specs).
- Using noise reduction hardware systems (such as Dolby-A and DBX NR). These are older systems from "back in the day" when analog was king. They are not easy to get hold of and may require quite a bit of setup and understanding before being inserted into the analog tape chain.

It should be noted that by using tape formulations that combine low noise and high output (resulting in an increased S/N ratio of 3 dB or more), noise levels can be reduced even further. However, when all's said and done, making most or all of these improvements might be simply too costly and impractical. In the end, it's important to realize that noise is simply an inherent part of the analog recording process, and the goal is to minimize its effects as much as possible … not necessarily to get rid of it.

CLEANLINESS

It's very important for the magnetic recording heads and all moving parts of an ATR transport deck to be kept free from dirt and oxide shed. Oxide shed occurs when friction causes small particles of magnetic oxide to flake off and accumulate on surface contacts. This accumulation is most critical at the surface of the magnetic recording heads, since even a minute separation between the magnetic tape and the heads can cause high-frequency *separation loss*. For example, a signal that's been recorded at 15 ips and has an oxide shed buildup of 1 mil (0.001 inch) on the playback head will cause a 15-kHz signal to drop by 55 dB below its standard operating level. Denatured (isopropyl) alcohol or an appropriate cleaning solution should be used to clean transport tape heads and guides (with the exception of the machine's pinch roller and other rubber-like surfaces) at regular intervals.

DEGAUSSING

Magnetic tape heads are made from a magnetically soft metal, which means that the alloy is easily magnetized, but once the coil's current is removed, the core shouldn't retain any of this magnetism. Small amounts of residual magnetism, however, will build up over time, which can actually partially erase high-frequency signals from a master tape. For this reason, all of the tape heads should be demagnetized after 10 hours of operation with a professional head demagnetizer. This handheld device works much like an erase head in that it saturates the magnetic head with a high-level alternating signal that randomizes residual magnetic flux. Once a head has been demagnetized (after 5 to 10 seconds), it's important to move the tool to a safe distance from the tape heads at a speed of less than 2 inches per second before turning it off, so as to avoid inducing a larger magnetic flux back into the head. Before an ATR is aligned, the magnetic tape heads should *always* be cleaned and demagnetized in order to obtain accurate readings and to protect your expensive reference alignment tapes.

Editing Magnetic Tape

Obviously, editing magnetic tape is *quite* different from editing digital audio on a DAW. It's a *very* hands-on thing (all puns intended). The process of editing tape is not entirely destructive, but it's close to it, and it requires a good ear, good eye-to-hand coordination and a sharp razor blade.

The process makes use of a tape edit block (Figure 5.13), special adhesive tape (that's specially formulated not to "bleed" its sticky adhesive onto tape heads, rollers and other parts in the tape's path) and a grease pencil. The process goes much like this:

1. First, locate the in and out edit points on the tape. It's up to the operator to determine a point just before where things go wrong and then to find the good part later in the tape, where the music is to be "picked up" in order to fix the mistake, etc.

2. Once the general areas where the edit is to be made are located on the tape, locate the first edit point. The machine can then be placed into "Edit Mode" (which disengages the reel brakes and tape lifters). This allows the tape to be manually "rocked" forwards and backwards to listen for the edit point. These audible cues are often a kick, snare or percussive sound; it can also be a sung word, literally anything. You just keep rocking the tape back and forth across the playback head until you've zeroed in on the exact point to be edited.

3. At this point, you take your grease pencil (black or white can be used) and gently but firmly mark the tape's backing at the exact point where the tape rests over the playback head.

4. Next, play or locate the tape to where the edit is to be picked up – manually locate the part to be kept in the edit and make your out edit mark on the tape.

FIGURE 5.13

A tape editing block is used to cut and splice together analog tape in the editing process.

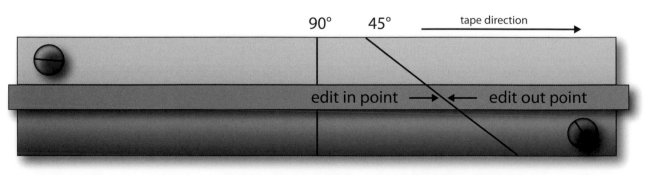

5. Now, you can wind back and go about the process of locating the first part of the edit, placing the exact "edit in point" on the edit block (almost always over the 90° angle) and using a fresh single-sided razor blade, cut across the tape with a smooth, clean movement.

6. Locate the "edit in point" on the tape, place it on the edit block at a point that butts up to your previous cut, and cut the tape.

7. Now, all you need to do is cut off about 1–1.5 inches of the special editing adhesive tape and bind the two pieces together, rewind the tape (remember to take the machine out of Edit Mode) to a point before your edit, and listen to your masterpiece.

8. Obviously, during this learning phase, it's important to take your time, save the unused tape just in case you miss your marks, and if possible, do a lot of practice tries with material that's not important (i.e., record some of your favorite music onto tape and have fun with it), and possibly have someone who's used to cutting tape standing around to help you get the hang of it (if you can find such a person).

BACKUP AND ARCHIVE STRATEGIES

In this day of hard drives and the cloud, we've all come to know the importance of backing up our data. With important music and media projects, it's equally important to create a backup copy of your analog projects in case of an unforeseen catastrophe or as added insurance that future generations can enjoy the fruits of your work.

Backing up Your Analog Project

The one basic truth that can be said about analog magnetic tape is that this medium has withstood the test of time. With care and reconditioning, tapes that have been recorded in the 1940s have been fully restored, allowing us to preserve and enjoy some of the best music of the day. On the other hand, digital data has two points that aren't exactly in its favor:

- Data that resides on hard drives isn't the most robust of media for storage over time. Even optical media (which can be rated to last over 100 years) haven't really been proven to last.
- Even if the data remains intact, with the ever-increasing advances in computer technology, who's to say that the media, drives, programs, session formats and file formats will be around in 10 years, let alone 50!

These warnings aren't slams against digital, just precautions against the march of technology versus the need for media preservation. For these reasons, media preservation is a top priority for such groups as the Recording Academy's Producers and Engineers Wing (P&E Wing), as well as for many major record labels – so much so that many stipulate in their contracts that multitrack sessions (no matter what the original medium) must be transferred and archived to 2-inch multitrack analog tape.

When transferring digital tracks to an analog machine, it's always wise to make sure that the recorder has been properly calibrated and that reference tones (1 kHz, 10 kHz, 16 kHz and 100 Hz) have been recorded at the beginning of the tape. When copying from analog to analog, both machines should be properly calibrated, and the calibration source for the newly recorded tones should be the master tape. If an SMPTE track is required, be sure to stripe the copy with a clean, jam-sync code. The backing up of analog tapes and/or digital data usually isn't a big problem – unless you've lost your original masters. In this situation, a proper safety master can be the difference between panic and peace.

Tape Restoration

Besides the fact that analog tape has a characteristic "sound", another factor that works in its favor is that fact that tape has been proven to be a robust medium that can withstand the effects of time … often for many decades. There can be a potential downside to this longevity, however. Tape (be it analog or digital in nature) consists of three basic components:

- Iron oxide, which contains the domain particles that retain its information in the form of magnetism
- A binder, which is a form of glue
- A carrier, which is the actual plastic tape itself

Over time, the glue that holds it all together can actually begin to break down and cause the oxide to shed away from the plastic tape itself. In fact, there are horror stories that tell of the oxide actually separating from the tape on several iconic master tapes when being copied to digital … literally self-destructing in the last attempts to save them to digital.

Fortunately, in many cases, this deterioration can be temporarily stabilized through a process called "baking". This loving name comes from the fact that the binder can actually be rejuvenated by placing the tape in a convection oven at 135 to 150 degrees Fahrenheit over a period of three to eight hours. As an alternative to using an oven, a dehydrator that's meant to dry out fruit, veggies or meat can also be used.

Of course, this involves a bit of art and magic and needs to be carried out with extreme care, so it's definitely best that you do as much research as possible before even attempting the baking of any magnetic tape media.

Archive Strategies

Just as it's important to back up your media, it's also important that both the original and backup media be treated and stored properly. Here are a few guidelines:

- As stated earlier, always store the tapes tails-out.
- Store the tapes to be stored onto the take-up storage reel at slow-wind or play speeds (to minimize tape edge wear).
- Store the boxes vertically. If they're stored horizontally, the outer tape edges could get bent and become damaged.
- Media storage facilities exist that can store your masters or backups for a fee. If this isn't an option, store them in an area that's cool and dry (e.g., no temperature extremes, in low humidity, no attics or basements).
- Store your masters and backups in separate locations. In case of a fire or other disaster, one would be lost, but not both (always a good idea with digital data as well). In recent times, a major label ignored this warning, and many precious masters (along with their backups) were lost in a fire.

Tape Availability

It's important to note that since analog tape is no longer a mainstream medium, it follows that there are no longer many manufacturing companies to make the actual tape stock. As of this writing, there are only a few manufacturers left in the world. It stands to follow logically that old-stock and used tape are still widely in circulation and can often be bought online.

TAPE EMULATION PLUG-INS

As DAW technology and methods for precisely modeling hardware devices in software have advanced, new types of plug-ins that can directly emulate the sound of analog tape machines have come onto the market. These devices are capable of closely mimicking the sonic character, tape noise, distortion and changes with virtual tape formulation and bias settings of any number of tape machines and model makes "virtually", allowing us to plug in the sound of our favorite deck onto any workstation track or bus (Figure 5.14).

FIGURE 5.14
Analog audio tape emulation plug-ins: (a) Magneto MkII (courtesy of Steinberg Media Technologies GmbH, a division of Yamaha Corporation, www.steinberg.net). (b) Slate Digital Virtual Tape Machines (courtesy of Slate Digital, www.slatedigital.com). (c) Ampex ATR 102 tape emulation plug-in for the Apollo family of interfaces and the UAD effects processing card (courtesy of Universal Audio, www.uaudio.com, ©2022 Universal Audio, Inc. All rights reserved. Used with permission).

(a)

(b)

(c)

CHAPTER 6

Digital Audio Technology

There's absolutely no doubt digital audio technology has changed the way that all forms of media are produced, gathered and distributed. As a driving force behind human creativity, expression and connectivity, the impact of digital media production is simply HUGE, and it is an integral part of both the medium and the message within modern-day communications.

As with most other media, these changes have been brought about by the integration of the personal computer and portable devices into the modern-day project studio environment. Newer generations of computers and related hardware peripherals have been integrated into both the pro and project studio to record, fold, spindle and mutilate audio with astonishing ease. This chapter is dedicated to helping you get familiar with the various digital system types and their relation to the modern-day music production studio.

THE LANGUAGE OF DIGITAL

Although digital audio is a varied and complex field of study, the basic theory behind the magic curtain isn't all that difficult to understand. At its most elementary level, it's simply a process by which numeric representations of analog signals (in the form of voltage levels) are encoded, processed, stored and reproduced over time through the use of a *binary number system*.

Just as English-speaking humans communicate by combining any of 26 letters together into groupings known as "words" and manipulate numbers using the decimal (base 10) system, the system of choice for a digital device is the binary (base 2) system. This numeric system provides a fast and efficient means for manipulating and storing digital data. By translating the alphabet, base 10 numbers or other form of information into a binary language form, a digital device (such as a computer or microprocessor) is used to perform calculations and tasks that would otherwise be cumbersome, less cost effective and/or downright impossible to perform in the analog domain.

DOI: 10.4324/9781003260530-6

D, O, G = (0100 0100)(0100 1111)(0100 0111) =

(alpha-bits) (digital words) (Moxie)

FIGURE 6.1
Digital and analog equivalents for an awesome four-legged, little animal, "Moxie".

To illustrate, let's take a look at how a human construct can be translated into a digital language (and back). If we type the letters D, O and G into a word processor, the computer will quickly go about the task of translating these keystrokes into a series of 8-bit digital words represented as [0100 0100], [0100 1111] and [0100 0111]. On their own, these digits don't mean much, but when these groups are put together, they form a word that represents a four-legged animal that's always glad to see you (Figure 6.1).

In a similar manner, a digital audio system works by sampling (measuring) the instantaneous voltage level of an analog signal at very precise intervals over time and then converting these samples into a series of encoded "words" that digitally represent the analogous voltage levels at each time interval. By successively measuring changes in an analog signal's voltage level (over time), this successive stream of representative digital words can then be stored in a form that accurately represents the original analog signal. Once stored, this data can be processed and reproduced in ways that continue to change the face of audio production.

It's interesting to note that binary data can be encoded as logical 1 "on" or 0 "off" states, using various methods to encode data as:

- Voltage or no voltage (circuitry)
- Magnetic flux or no flux (hard disk or tape)
- Reflection off a surface or no reflection (CD, DVD or other optical disc form)
- Electromagnetic waves (broadcast transmission)

From this, you'll hopefully begin to get the idea that human forms of communication (i.e., print, visual and audible media) can be translated into a digital form that can be easily understood and manipulated by a processor. Once the data has been recorded, stored and/or processed, the resulting binary data can be reconverted into a form that can be easily understood by us humans (such as a display readout, system control, text-to-speech, controller interaction, you name it). If you think this process of changing one form of energy into an analogous form (and then back again) sounds like the general definition of a transducer – you're right!

Digital Basics

The following sections provide a basic overview of the various stages that are involved in the encoding of analog signals into equivalent digital data and the subsequent converting of this data back into its original analog form.

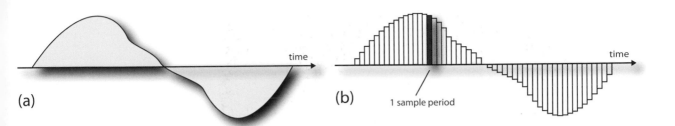

(a)

(b)   1 sample period

The encoding and decoding phases of the digitization process revolve around two processes:

- Sampling (the component of time)
- Quantization (the signal-level component)

FIGURE 6.2
Representations of an audio signal: (a) An analog signal is continuous in nature. (b) A digital signal makes use of periodic sampling to encode information.

In a nutshell, sampling is a process that affects the overall bandwidth (frequency range) that can be encoded within a sound file, while quantization refers to the volume level and resolution (overall quality and distortion characteristics) of an encoded signal compared with the original analog signal at its input.

SAMPLING

In the world of analog audio, signals are recorded, stored and reproduced as changes in voltage levels that vary over time in a continuous fashion (Figure 6.2a). The digital recording process, on the other hand, doesn't operate in this manner; rather, digital recording operates by taking periodic samples of an analog audio waveform over time (Figure 6.2b) and then calculating each of these snapshot samples into grouped binary words that digitally represent these voltage levels (as accurately as possible) as they change over time.

During this process, an incoming analog signal is sampled at discrete and precisely timed intervals (as determined by the sample rate). At each interval, this analog signal is momentarily "held" (frozen in time) while the converter goes about the process of determining what the voltage level actually is. This is done with a degree of accuracy that's defined by the quality of the converter's circuitry and the chosen bit depth. The converter then generates a binary-encoded word that's numerically equivalent to the analog voltage level at that point in time (Figure 6.3). Once this is done, the converter can store the representative word into a memory medium (disk, disc, random-access memory [RAM], tape, etc.), release its hold and then go about the task of determining the values of the next sampled voltage. The process is then continuously repeated throughout the recording process.

Within a digital audio system, the *sampling rate* is defined as the number of measurements (samples) that are periodically taken over the course of a second. Its reciprocal (sampling time) is the elapsed time that occurs between sampling periods. For example, a sample rate of 44.1 kHz corresponds to a sample time of 1/44,100 of a second.

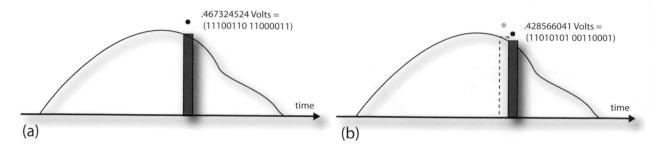

FIGURE 6.3

The sampling process. (a) The analog signal is momentarily "held" (frozen in time) while the converter goes about the process of determining the voltage level at that point in time and then converts that level into a binary-encoded word that's numerically equivalent to the original analog voltage level. (b) Once this digital information is processed and stored, the sample is released, and the next sample is held as the system again goes about the task of determining the level of the next sampled voltage — and so forth, and so forth, and so forth over the course of the recording.

This process can be likened to a photographer who takes a series of action sequence shots. As the number of pictures taken in a second increases, the accuracy of the captured event will likewise increase until the resolution is so great that you can't tell that the continuous and (hopefully) compelling movie is really a series of successive, discrete pictures. Since the process of sampling is tied directly to the component of time, the sampling rate of a system determines its overall bandwidth (**Figure 6.4**), meaning that a recording made at a higher sample rate will be capable of storing a wider range of frequencies (effectively increasing the signal's bandwidth at its upper limit).

QUANTIZATION

Quantization represents the amplitude component of the digital sampling process. It is used to translate the voltage levels of a continuous analog signal (at discrete sample points over time) into binary digits (bits) for the purpose of manipulating or storing audio data in the digital domain. By sampling the amplitude of an analog signal at precise intervals over time, the converter determines the exact voltage level of the signal (during a sample interval, when the voltage level is momentarily held), and then outputs this signal level as an equivalent set of binary numbers (as a grouped word of n-bits length), which represent the originally sampled voltage level (**Figure 6.5**). The resulting word is used to encode the original voltage level with as high a degree of accuracy as can be permitted by the word's bit length and the system's overall design.

Currently, the most common binary word lengths for audio are 16-bit (i.e., 11111101 10000011, having a theoretical dynamic range of 96.33 dB) and 24-bit resolutions (i.e., 11111101 10000011 10101110, having a theoretical dynamic range of 144.49 dB). In addition, computers and signal-processing devices are capable of performing calculations internally at the 32- and 64-bit resolution levels. This internal calculation headroom at the bit level helps to reduce errors whenever high track counts are summed together. In addition, this added internal headroom helps to reduce errors in performance whenever the multiple data streams are processed in real time. Since the internal bit depth is higher, these resolutions can be preserved (instead of being dropped by the system's hardware or software processing functions), with the final result being an n-bit data stream that's relatively free of errors. When implemented properly, this increase in summing and real-time processing accuracy can result in a system that's audibly superior in overall sound performance and reduced distortion.

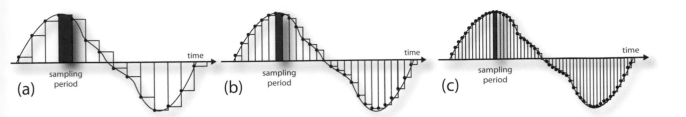

FIGURE 6.4
Discrete time sampling. (a) Whenever the sample rate is set too low, important data between sample periods will be lost. (b) As the rate is increased, more frequency-related data can be encoded. (c) Increasing the sampling frequency further can encode the recorded signal with an even higher bandwidth range, with increased sample accuracy.

THE DEVIL'S IN THE DETAILS

The details of the digital audio record/playback process can get quite complicated; however, the essential basics are:

- Sampling (in the truest sense of the word) analog voltage levels at precise intervals in time
- Conversion of these samples into a digital word value that most accurately represents these voltage levels
- Storing these numeric sample equivalents within a digital memory device

Upon playback, these digital words are then converted back into discrete voltages (again, at precise intervals in time), allowing the originally recorded signal voltages to be re-created, processed and played back.

Although the basic concept behind the sample-and-hold process is relatively straightforward, delving further into the process can quickly bog you down in the language of high-level math and physics. Luckily, there are a few additional details relating to digital audio that can be discussed at a basic level.

THE NYQUIST THEOREM

The Nyquist theorem is a basic rule that relates to the sampling process and states that:

FIGURE 6.5
The instantaneous amplitude of the incoming analog signal is broken down into a series of discrete voltage steps, which are then converted into equivalent binary-encoded words.

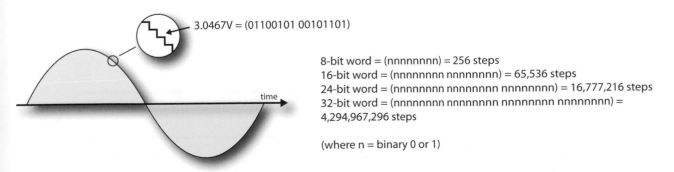

3.0467V = (01100101 00101101)

8-bit word = (nnnnnnnn) = 256 steps
16-bit word = (nnnnnnnn nnnnnnnn) = 65,536 steps
24-bit word = (nnnnnnnn nnnnnnnn nnnnnnnn) = 16,777,216 steps
32-bit word = (nnnnnnnn nnnnnnnn nnnnnnnn nnnnnnnn) = 4,294,967,296 steps

(where n = binary 0 or 1)

In order for the desired frequency bandwidth to be faithfully encoded in the digital domain, the selected sample rate must be at least twice as high as the highest frequency to be recorded (sample rate ≥2× highest frequency).

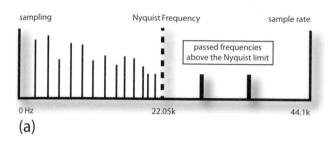

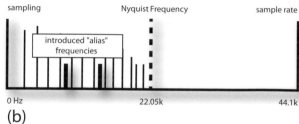

(a)   (b)

FIGURE 6.6

Frequencies that enter into the digitization process above the Nyquist half-sample frequency limit can introduce harmonic distortion. (a) Frequencies greater than 2× the sampling rate limit are passed into the sampling process. (b) This will result in "alias" frequencies that are introduced back into the audio band as distortion.

FIGURE 6.7

Anti-alias filtering. (a) An ideal filter would have an infinite attenuation at the 20-kHz Nyquist cutoff frequency. (b) Real-world filters require an additional frequency "guardband" in order to fully attenuate unwanted frequencies that fall above the half-bandwidth Nyquist limit.

In plain language, should frequencies that are greater than twice the sample rate be allowed into the sampling process, these frequencies would be higher than the sample rate can faithfully capture. When this happens, the successive samples won't be able to accurately capture these higher frequencies, but instead, will actually be recorded as false or "alias" frequencies that aren't actually there – and will be heard as harmonic distortion (Figure 6.6). As such, from a practical point of view, this would mean that an audio signal with a bandwidth of 20 kHz would require that the sampling rate be at least 40 kHz samples/sec.

In order to eliminate the effects of *aliasing*, a low-pass filter is placed into the circuit before the sampling process takes place. In theory, an ideal filter would pass all frequencies up to the Nyquist cutoff frequency and then prevent any frequencies above this point from passing. In the real world, however, such a "brick wall" filter doesn't really exist. For this reason, a slightly higher sample rate must be chosen in order to account for the cutoff slope that's required for the filter to be effective (Figure 6.7). As a result, an audio signal with a bandwidth of 20 kHz will actually be sampled at a standardized rate of 44.1 samples/sec, while a bandwidth of roughly 22 kHz would require the use of a sampling rate of at least 48 kHz samples/sec, etc.

OVERSAMPLING

This sampling-related process is commonly used in professional and consumer digital audio systems to improve the Nyquist filter's anti-aliasing characteristics.

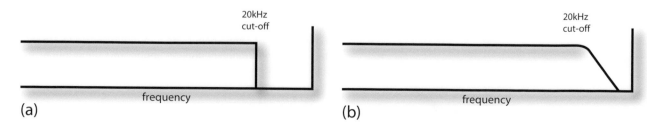

(a)   (b)

Oversampling increases the effective sample rate by factors ranging between 12 and 128 times the original rate. There are three main reasons for this process:

- Nyquist filters can be expensive and difficult to properly design. By increasing the effective sample bandwidth, a simpler and less expensive filter can be used.
- Oversampling generally results in a higher-quality analog-to-digital (A/D) and digital-to-analog (D/A) converter that sounds better.
- Since multiple samples are taken of a single sample-and-hold analog voltage, the average noise level will be lower.

Following the sample stage, the sampled data is digitally scaled back down to the target data rate and bandwidth for further processing and/or storage.

SIGNAL-TO-ERROR RATIO

The signal-to-error ratio is used to measure the quantization process. A digital system's *signal-to-error ratio* is closely akin (although not identical) to the analog concept of signal-to-noise (S/N) ratio. Whereas S/N ratio is used to indicate the overall dynamic range of an analog system, the signal-to-error ratio of a digital audio device indicates the degree of accuracy that's used to capture a sampled level and its step-related effects.

Although analog signals are continuous in nature, as we've read, the process of quantizing a signal into an equivalent digital word is not. Since the number of discrete steps that can be encoded within a digital word limits the accuracy of the quantization process, the representative digital word can only be an approximation (albeit an extremely close one) of the original analog signal level. Given a properly designed system, the signal-to-error ratio for a signal coded with n bits is:

$$\text{Signal-to-error ratio} = 6n + 1.8(\text{dB})$$

Therefore, the theoretical signal-to-error ratio for the most common bit rates will yield a dynamic range of:

8-bit word = 49.8 dB	24-bit word = 145.8 dB
16-bit word = 97.8 dB	32-bit word = 193.8 dB
20-bit word = 121.8 dB	64-bit word = 385.8 dB

DITHER

A process known as *dither* is commonly used during the recording or conversion process to increase the overall bit resolution (and therefore low-level noise and signal clarity) of a recorded signal when converting from a higher to a lower bit rate.

Technically, dither is the addition of very small amounts of randomly generated noise to an existing bit stream that allows the S/N and distortion figures to fall

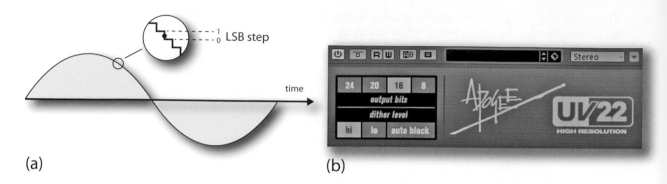

(a) **(b)**

FIGURE 6.8

Dither in action. (a) Values falling below the "least significant bit" level cannot be encoded without the use of dither. (b) Apogee UV22 dither plug-in for Steinberg Cubase and Nuendo (courtesy of Steinberg Media Technologies GmbH, a division of Yamaha Corporation, www.steinberg.net).

to levels that approach their theoretical limits. The process makes it possible for low-level signals to be encoded at less than the data's least significant bit level (less than a single quantization step, as shown in Figure 6.8a). You heard that right – by adding a small amount of random noise into the A/D path, we can actually:

- Improve the resolution of the conversion process below the least significant bit level
- Reduce harmonic distortion in a way that greatly improves the signal's performance

The concept of dither relies on the fact that noise is random. By adding small amounts of randomization into the quantization process, there is an increased probability that the D/A converter will be able to guess the least significant bit of a low-level signal more accurately. This is due to the fact that the noise shapes the detected sample in such a way that the sample-and-hold (S/H) circuitry can "guess" the original analog value with greater precision.

Dither is often applied to an application or process to reduce quantization errors that result in slight increases in noise and/or fuzziness that could otherwise creep into a bit stream. For example, when multiple tracks are mixed together within a digital audio workstation (DAW), it's not uncommon for digital data to be internally processed at 32- and 64-bit depths. In situations like this, dither is often used to smooth and round the data values so that the low-level (least significant bit) resolutions won't be lost when they are interpolated back to their original target bit depths.

Applications and DAW plug-ins can be used to apply dither to a sound file or master mix so as to reduce the effects of lost resolution due to the truncation of least significant bits (Figure 6.8b). For example, mastering engineers might experiment with applying dither to a high-resolution file before saving or exporting it as a 16-bit final master. In this way, noise will be reduced, and the sound file's overall clarity can be increased. When in doubt, it has been found that adding "triangular" dither to a file that is to be bit-reduced is the safest overall noise-shaping filter that can be used.

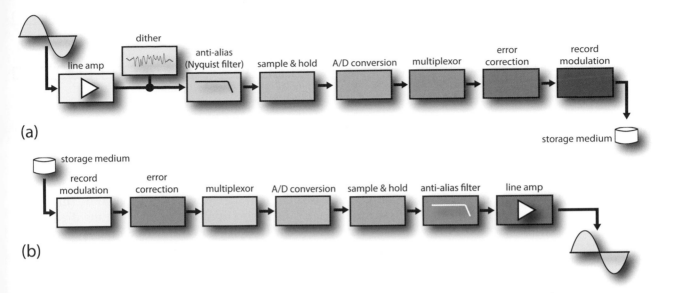

(a)

(b)

FIGURE 6.9
The digital audio chain: (a) recording. (b) Reproduction.

FIXED- VERSUS FLOATING-POINT PROCESSING

Many of the newer digital audio and DAW systems make use of floating-point arithmetic in order to process, mix and output digital audio. The advantage of the use of floating over fixed-point DSP calculations is that the former are able to use numeric "shorthand" in order to process a wider range of values at any point in time. In short, they are able to easily move or "float" the decimal point of a very large number in a way that can represent it as a much smaller value. By doing so, the processor is able to internally calculate much larger bit-depth values (i.e., 32- or 64-bit) with relative ease and increased data resolution.

The Digital Recording/Reproduction Process

The following sections provide a basic overview of the various stages within the process of encoding analog signals into equivalent digital data (Figure 6.9a) and then converting this data back into its original analog form (Figure 6.9b).

THE RECORDING PROCESS

In its most basic form, the *digital recording chain* includes a low-pass filter, a sample-and-hold circuit, an analog-to-digital converter, and circuitry for signal coding and error correction. At the input of a digital sampling system, the analog signal must be band limited with a low-pass filter so as to stop frequencies that are greater than half the sample rate frequency from passing into the A/D conversion circuitry. Such a stop-band (anti-aliasing) filter generally makes use of a sharp roll-off slope at its high-frequency cutoff point (oversampling might be used to simplify and improve this process).

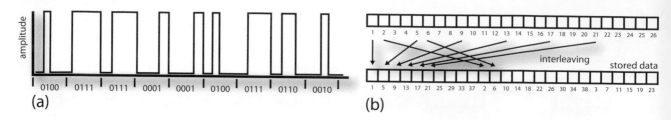

FIGURE 6.10
Data conditioning in the
digital recording process: (a)
Pulse-code modulation. (b)
Example of interleaved error
correction.

Following the low-pass filter, a *sample-and-hold* (S/H) circuit freezes and measures the analog voltage level that's present during the sample period. This period is determined by the sample rate (i.e., 1/44,100th of a second for a 44.1-K rate). At this point, computations (a series of computerized guessing games) are performed to translate the sampled voltage into an equivalent binary word. This step in the A/D conversion is one of the most critical components of the digitization process, because the sampled DC voltage level must be quickly and accurately quantized into an equivalent digital word (to the closest step level).

Once the sampled signal has been converted into its equivalent digital form, the data must be conditioned for further data processing and storage. This conditioning includes data coding, data modulation and error correction. In general, the binary digits of a digital bit-stream aren't directly stored onto a recording medium as raw data; rather, data coding is used to translate the data (along with synchronization and address information) into a form that allows the data to be most efficiently and accurately stored to a memory or storage medium. The most common form of digital audio data coding is *pulse-code modulation*, or PCM (Figure 6.10a).

The density of stored information within a PCM recording and playback system is extremely high, so much so that any imperfections (such as dust, fingerprints or scratches that might adhere to the surface of any magnetic or optical recording medium) would cause severe or irretrievable data errors. To keep these errors within acceptable limits, several forms of *error correction* are used (depending on the media type). One method uses redundant data in the form of parity bits and check codes in order to retrieve and/or reconstruct lost data. A second method uses error correction that involves interleaving techniques, whereby data is deliberately scattered across the digital bit stream according to a complex mathematical pattern. The latter has the effect of spreading the data over a larger surface of the recording medium, thereby making the recording medium less susceptible to dropouts (Figure 6.10b). In fact, it's a simple truth that without error correction, the quality of most digital audio media would be greatly reduced or (in the case of the CD and DVD) rendered almost useless.

THE PLAYBACK PROCESS

In many respects, the digital reproduction chain works in a manner that's complementary to the digital encoding process. Since most digital media encode data onto media in the form of highly saturated magnetic transition states or

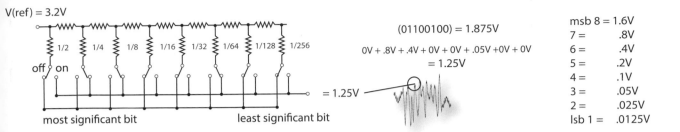

V(ref) = 3.2V

1/2 1/4 1/8 1/16 1/32 1/64 1/128 1/256

off on

most significant bit   least significant bit

= 1.25V

(01100100) = 1.875V

0V + .8V + .4V + 0V + 0V + .05V +0V + 0V
= 1.25V

= 1.25V

msb 8 = 1.6V
7 =   .8V
6 =   .4V
5 =   .2V
4 =   .1V
3 =   .05V
2 =   .025V
lsb 1 = .0125V

optical reflections, the recorded data must first be reconditioned in a way that restores the digital bit stream back into its original, modulated binary state (i.e., a transitional square wave). Once this is done, the encoded data can be de-inter-leaved (reassembled) back into its original form, where it can be converted back into PCM data, and the process of D/A conversion can take place.

Within the D/A conversion process, a stepped resistance network (sometimes called an R/2R network) is used to convert the representative words back into their analogous voltage levels for playback. During a complementary S/H period, each bit within the representative digital word is assigned to a resistance leg in the network (moving from the most significant to the least significant bit). Each "step" in the resistance leg is then designed to successively pass one-half the ref-erence voltage level as it moves down the ladder toward the LSB (Figure 6.11). The presence or absence of a logical "1" in each step is then used to turn on each successive voltage leg. As you might expect, when all the resistance legs are properly summed together, their voltages equal a precise level that matches (as closely as possible) the originally recorded voltage during the recorded sample period. Since these voltages are reproduced over time in precise intervals (as determined by the sample rate), you end up with successively changing voltages that make up the system's playback signal!

Following the conversion process, a final, complementary low-pass filter is inserted into the signal path. Again, following the principle of the Nyquist theo-rem, this filter is used to smooth out any of the step-level distortions that are introduced by the sampling process, resulting in a waveform that faithfully rep-resents the originally recorded analog waveform.

SOUND FILE BASICS

Within the digital audio community, there are, of course, standardized conven-tions that allow us to uphold general levels of professionalism and file inter-change between systems. There are definite ranges of personal choice as to what general file sample rate, bit depth and other specs might work best for you. Of course, there is a widespread movement that advocates the use of sound file specs that adhere to the general ideas of "high-definition audio". This ideology advocates the idea that only high–sample rate, high–bit depth audio should be used in the production of professional and/or high-quality audio. These are, of course, lofty goals – to capture, process and play back audio with as high a degree of audio quality and faithfulness to the music as is possible. I will leave the general debate over rates, bit depths and general specs in your capable hands, as tons has been written about this subject, and by now, you may have formed your own opinion about how you should work or about what will work best for you.

On a personal note, I will say this: most people who really know the ins and outs of digital audio conversion will tell you that determining the quality of a system is not always about "the numbers". That's to say it's not "always" about how high the sample rates are or how impressive the digital specs of a system will look on paper; it's far more often about the overall design quality of your system's converters, how well your system's internal system's jitter is kept to a minimum, etc.

I know that it's quite unpopular to say this, but it's a simple fact that a high-quality digital audio converter/DAW system running at 24/44.1 (for example) can sound as good as or better than a high-resolution 24/96 converter/DAW system that's less well designed or has higher than normal jitter within its design, or even one that's been improperly cabled or improperly linked to an external wordclock. In short, it's not always about the spec or sample rate numbers or about how the system's design looks on paper. In the end, it's the system's overall design quality that will determine how the conversion/DAW system will "sound". If you start there, the added idea of capturing your sound with sound file rates and settings that match your personal needs (or sonic requirements) will only put you that much further ahead of the game.

Sound File Bit Depths

The *bit depth* of a digitally recorded sound file directly relates to the number of quantization steps that are encoded into the bit stream. As a result, the bit rate (or bit depth) is directly correlated to the:

- Accuracy by which a sampled level (at one point in time) is encoded
- Signal-to-error figure … and thus, the overall dynamic range of the recorded signal

If the bit rate is too low to accurately encode the sample, the resolution will lead to quantization errors, which will result in distortion. Increasing the bit depth will often improve quantization errors in the record/playback process, generally leading to reduced distortion and an increase in the sound's overall transparency. The following are the most commonly used within the pro, project and general audio production community:

- *16 bits*: The long-time standard of consumer and professional audio production, 16 bits is the chosen bit depth of the CD-audio standard (offering a theoretical dynamic range of 97.8 dB). It is generally considered to be the minimum depth for high-quality professional audio production. Assuming that high-quality converters are used, this rate is capable of loss-less audio recording while conserving memory storage requirements.

- *20 bits*: Before the 24-bit depth came onto the scene, 20 bits was considered to be the standard for high–bit depth resolution. Although it's used less commonly now, it can still be found in high-definition audio recordings (offering a theoretical dynamic range of 121.8 dB).
- *24 bits*: Offering a theoretical dynamic range of 145.8 dB, this standard bit depth is often used in high-definition audio applications, often in conjunction with the 96k sample rate (i.e., 24/96).
- *32-bits and 32-bit float*: Compared with fixed-point files (16- or 24-bit), 32-bit float files store numbers in a floating-point format. This is fundamentally different from fixed point because numbers are stored with "scientific notation", using decimal points and exponents (for example, "1.4563 × 100" instead of "1,456,300"). This difference is significant because much larger and smaller numbers can be represented compared with a fixed-point representation. The formatting and encoding of the 32-bit word are not intuitive – it has been optimized for computers to be able to perform common math functions rather than for human readability. The first bit indicates a positive or negative value, the next 8 bits indicate the exponent, and the last 23 bits indicate the mantissa. More info is available by searching the web for the IEEE-754 floating-point format.

Further reading on sound file and compression codec specifics can be found in Chapter 11 (Multimedia).

Sound File Sample Rates

The *sample rate* of a recorded digital audio bit stream directly relates to the resolution at which a recorded sound will be digitally captured. Using the film analogy, if you capture more samples (frames) of a moving image as it moves through time, you'll end up with a more accurate representation of that recorded event. If the numbers of samples are too low, the resolution will be "lossy" and will distort the event. On the other hand, taking too many picture frames could result in a recorded bandwidth that's so high that the audience won't be able to discern any advantages that the extra information has to offer.

This analogy relates perfectly to audio, because the choice of sample rate will be determined by the bandwidth (number of overall frequencies that are to be captured) versus either the amount of storage and processing time that's needed to save the data to a memory storage medium or possibly, the time that will be required to up/download a file through a transmission and/or online data stream. The following are the most commonly used in the professional, project and audio production community:

- *32k*: This rate is often used by broadcasters to transmit/receive digital data via satellite. With its overall 15-kHz bandwidth and reduced data requirements, certain devices also use it in order to conserve memory. Although the pro community doesn't generally use this rate, it's surprising just how good a sound can be captured at 32k (given a high-quality converter).

- *44.1k*: The long-time standard of consumer and basic pro audio production, 44.1 is the chosen rate of the CD-audio standard. With its overall 20-kHz bandwidth, the 44.1k rate is generally considered to be the minimum sample rate for professional high-definition audio production. Assuming that high-quality converters are used, this rate is capable of recording lossless audio while conserving memory storage requirements.

- *48k*: This standard was adopted early on as a standard sample rate for professional audio applications (particularly when dealing with early hardware devices). It's also the adopted standard rate for use within professional video, DVD and broadcast production.

- *88.2k*: As a simple multiple of 44.1, this rate is often used within productions that are intended to be high-resolution products.

- *96k*: This rate has been adopted as the de-facto sample rate for high-resolution recordings and is a common rate for Blu-ray high-definition audio.

- *192k*: This high-resolution rate is increasingly used within pro audio production; however, the storage and media processing requirements are quite high (but not beyond the limits of modern-day computers).

Professional Sound File Formats

Although several formats exist for encoding and storing sound file data, only two have been universally adopted by the industry:

- Wave (.wav) format
- Audio Interchange File (AIFF) (.aif) format

These standardized formats make it easier for files to be exchanged between compatible media devices.

The most common file type is the Wave (or .wav) format. Developed for the Microsoft Windows operating system, this universal file type supports mono, stereo and multichannel files at a wide range of uncompressed resolutions and sample rates. Wave files contain PCM coded audio that follows the Resource Information File Format (RIFF) spec, which allows extra user information to be embedded and saved within the file itself.

The newly adopted Broadcast Wave format, which has been adopted by the Producers and Engineers wing (www.recordingacademy.com/producers-engineers-wing) as the preferred sound file format for DAW production and music archiving, allows metadata and timecode-related positioning information to be directly embedded within the sound file's data stream, making it easier for these wav files to be placed within music and media productions using precise timecode position placement.

Apple's Audio Interchange File (AIFF or .aif) format likewise supports mono, stereo and multichannel embedded sound files with 8-, 16- and 24-bit depths over a wide range of sample rates. Like Broadcast Wave files, AIFF files can also contain embedded text strings; however, unlike BWF, it's not capable of containing timecode-stamped information within the file itself.

Regarding Digital Audio Levels

Over the decades, the trend toward making recordings that are as loud as possible has totally pervaded the industry, to the point that it has been given the name of "The Loudness Wars". Not only is this practice used in mastering to make a song or project stand out in an on-the-air or in-the-device playlist; it has also been grandfathered in from the analog tradition of recording a track as "hot" as possible to get the best noise figures and "punch" out of a track. All of this is arguably well and good, except for the fact that whenever a track has been recorded at too high a level, you're not adding extra "punch" to the track at all – in fact, all you are adding is distortion, and a rather nasty-sounding one at that.

The dynamic range of a digital recording ranges from its theoretical floor signal level of (00000000) (00000000) (00000000), for a 24-bit recording, to its full-scale (dBFS) headroom ceiling of (11111111) (11111111) (11111111). It almost goes without saying that having average or peak levels that go above full scale can easily ruin a recording by adding clipping distortion. Since the overall dynamic range of a digital file can be 97.8 dB for a 16-bit file and 145.8 dB for a 24-bit file, it's generally a good idea to reduce your levels so that they peak at around –12 (or even at levels approaching –20dB). This will generally accurately capture the peaks without clipping and without adding any appreciable amount of noise into the mix.

One of the finer points that you should be aware of regarding overall recording levels (as well as levels that get summed together whenever all of the tracks in your mix get added together) is the math that's used by your DAW to deal with the general levels and gain structure within your session mix. Not all DAWs are created equal on this matter. Some DAWs make use of floating-point math (a digital shorthand that allows the software to deal with wide ranges in number calculations) to increase their overall internal gain structure, while other DAWs are not quite so forgiving, often leading to increased distortion when the tracks are summed within a mix. It's just a point for your information that you might want to be aware of within the mixing process when using your DAW.

Although digital audio level guidelines now exist within the broadcast community (to some extent), pro and project studio level guidelines are few and far between. In recent years, metering plug-ins and better DAW output metering has come onto the market, allowing you to better keep track of your overall level and metering needs (Figure 6.12).

DIGITAL AUDIO TRANSMISSION

In the digital age, it's become increasingly common for audio data to be distributed throughout a connected production system in the digital domain. In

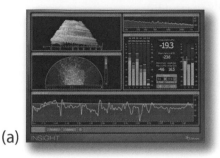

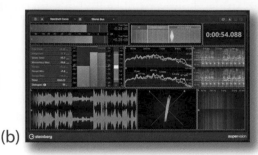

(a)   (b)

FIGURE 6.12

Several present-day DAW and digital systems metering options: (a) iZotope Insight Essential Metering Suite (courtesy of iZotope Inc., www.izotope.com). (b) Steinberg Cubase and Nuendo SuperVision and main output metering display (courtesy of Steinberg Media Technologies GmbH, a division of Yamaha Corporation, www.steinberg.net).

this way, digital audio can be transmitted in its original numeric form and (in theory) without any degradation throughout a connected path or system. When looking at the differences between the distribution of digital and analog audio, it should be kept in mind that unlike its analog counterpart, the transmitted bandwidth of digital audio data occurs in the megahertz range; therefore, it actually has more in common with video signals than with the lower-bandwidth range of analog audio. This means that care must be exercised to ensure that impedance values are more closely matched and that quick-fix solutions don't occur (for example, using a Y-cord to split a digital signal between two devices is a major no-no). Failure to follow these precautions could seriously degrade or deform the digital signal, causing increased jitter and unwanted distortions.

Due to these tight restrictions, several digital transmission standards have been adopted that allow digital audio data to be quickly and reliably transmitted between compliant devices. These include such protocols as:

- AES/EBU
- S/PDIF
- SCMS
- MADI
- ADAT lightpipe
- TDIF

AES/EBU

The AES/EBU (Audio Engineering Society and the European Broadcast Union) protocol has been adopted for the purpose of transmitting digital audio between professional digital audio devices. This standard (which is most often referred to as simply an AES digital connection) is used to convey two channels of interleaved digital audio through a single, three-pin XLR mic or line cable in a single direction. This balanced configuration connects pin 1 to the signal ground, while pins 2 and 3 are used to carry signal data. AES/EBU transmission data is low impedance in nature (typically 110 Ω) and has digital burst amplitudes that range between 3 and 10 V. These combined factors allow a maximum cable

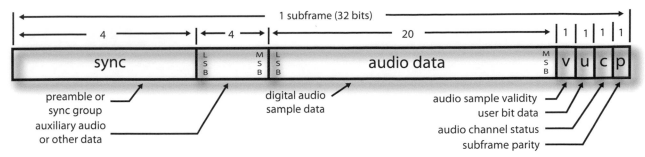

FIGURE 6.13
AES/EBU subframe format.

length of up to 328 feet (100 meters) at sample rates of less than 50 kHz without encountering undue signal degradation.

Digital audio channel data and subcode information is transmitted in blocks of 192 bits that are organized into 24 words (each being 8 bits long). Within the confines of these data blocks, two sub-frames are transmitted during each sample period that convey information and digital synchronization codes for both channels in an L-R-L-R fashion. Since the data is transmitted as a self-clocking bi-phase code (**Figure** 6.13), wire polarity can be ignored. In addition, whenever two devices are directly connected, the receiving device will usually derive its reference-timing clock from the digital source device.

In the late 1990s, the AES protocol was amended to include the "stereo 96k dual AES signal" protocol. This was created to address signal degradations that can occur when running longer cable runs at sample rates above 50 kHz. To address the problem, the dual AES standard allows stereo sample rates above 50 kHz (such as 24/96) to be transmitted over two synchronized AES cables (with one cable carrying the L information and the other carrying the R).

S/PDIF

The S/PDIF (Sony/Philips Digital Interface) protocol has been widely adopted for transmitting digital audio between consumer digital audio devices and their professional counterparts. Instead of using a balanced three-pin XLR cable, the popular S/PDIF standard has adopted the single-conductor, unbalanced phono (RCA) connector (**Figure** 6.14a), which conducts a nominal peak-to-peak voltage level of 0.5 V between connected devices, with an impedance of 75 Ω. In addition to using standard RCA cable connections, S/PDIF can also be transmitted between devices using Toslink optical connection lines (Figure 6.14b), which are commonly referred to as "lightpipe" connectors.

As with the AES/EBU protocol, S/PDIF channel data and subcode information are transmitted in blocks of 192 bits consisting of 12 words that are each 16 bits long. A portion of this information is reserved as a category code that provides the necessary setup information (sample rate, copy protection status and so on) to the copying device. Another portion is set aside for transmitting audio data that's used to relay track indexing information (such as start ID and program ID numbers), allowing this relevant information to be digitally transferred from

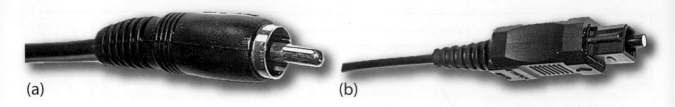

(a)   (b)

FIGURE 6.14

S/PDIF connectors: (a) RCA coax connection. (b) Toslink optical connection.

the master to the copy. It should be noted that the professional AES/EBU protocol isn't capable of digitally transmitting these codes during a copy transfer.

In addition to transmitting two channels in an interleaved L-R-L-R fashion, S/PDIF is able to communicate multichannel data between devices. Most commonly, this shows up as a direct 5.1 or 7.1 surround-sound link between a DVD player and an audio receiver/amplifier playback system (via either an RCA coax or an optical connection).

SCMS

Initially, certain digital recording devices (such as a DAT recorder) were intended to provide consumers with a way to make high-quality recordings for their own personal use. Soon after its inception, however, for better or for worse, the recording industry began to see this new medium as a potential source of lost royalties due to home copying and piracy practices. As a result, the RIAA (Recording Industry Association of America) and the former CBS Technology Center set out to create a "copy inhibitor". After certain failures and long industry deliberations, the result of these efforts was a process that has come to be known as the *Serial Copy Management System*, or SCMS (pronounced "scums"). This protocol was incorporated into many consumer digital devices in order to prohibit the unauthorized copying of digital audio at 44.1 kHz (of course, SCMS doesn't apply to the making of analog copies). Note also that with the demise of the DAT format (as both a recording and a mass music distribution medium), SCMS has fallen completely out of favor with both the consumer and pro audio communities.

So, what is SCMS? Technically, it's a digital protection flag that is encoded in byte 0 (bits 6 and 7) of the S/PDIF subcode area. This flag can have only one of three possible states:

- Status 00: no copy protection, allowing unlimited copying and subsequent dubbing
- Status 10: no more digital copies allowed
- Status 11: allows a single copy to be made of this product, but that copy cannot be copied

MADI

The MADI (Multichannel Audio Digital Interface) standard was jointly proposed as an AES standard by representatives of Neve, Sony and SSL as a straightforward,

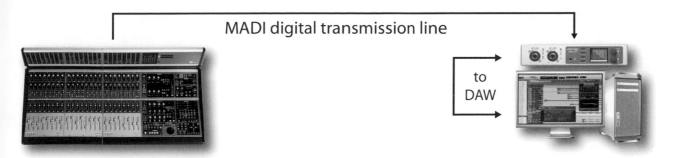

MADI digital transmission line

to
DAW

FIGURE 6.15
MADI offers a clutter-free
digital interface connection
between multitrack devices.

clutter-free digital interface connection between multitrack devices (such as a high-end workstation, digital tape recorder or mixing console, as shown in Figure 6.15). The transmission rate of 100 Mbit/sec provides for an overall bandwidth that's capable of handling the following number of channels in a single direction (one sender, one receiver):

- 56 channels at 32 kHz to 48 kHz (with ±12.5% pitch shift capabilities)
- 64 channels at 32 kHz to 48 kHz (no pitch shift)
- 28 channels at 64 kHz to 96 kHz (with ±12.5% pitch shift capabilities)

The linearly encoded digital audio is connected via a single 75-Ω, video-grade coaxial cable at distances of up to 120 feet (50 meters) or at greater distances whenever a fiber-optic cable is used.

In short, MADI makes use of a serial data transmission format that's compatible with the AES/EBU twin-channel protocol (whereby the data, Status, User and parity bit structures are preserved) and sequentially cycles through each channel (i.e., starting with Ch. 0 and ending with Ch. 55).

ADAT Lightpipe

A wide range of audio interface, mic preamps and older modular digital multitrack recorders currently use the *lightpipe* system for transmitting multichannel audio via a standardized optical cable link. These connections make use of standard Toslink connectors and cables to transmit up to eight channels of uncompressed digital audio at resolutions up to 24 bit at sample rates up to 48k over a sequential, optical bit stream. In what is called the S/MUX IV mode, a lightpipe connection can also be used to pass up to four channels of digital audio at the higher sample rates of 88.2 and 96k.

Although these connections are identical to those that are used to optically transmit S/PDIF stereo digital audio, the data streams are incompatible with each other. Lightpipe data isn't bidirectional, meaning that the connection can only travel from a single source to a destination. Thus, two cables will be needed to distribute data both to and from a device. In addition, synchronization data is imbedded within the digital data stream, meaning that no additional digital audio sync connections are necessary in order to lock device timing clocks; however, transport and timecode information is not transmitted and will require

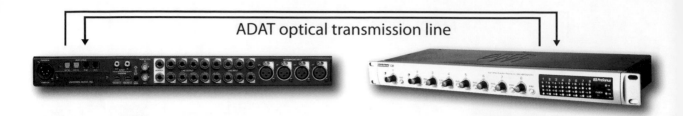

ADAT optical transmission line

FIGURE 6.16
ADAT optical I/O interconnection drawing between the audio interface and a multichannel preamp.

additional connections should transport control over an older ADAT or modular digital multitrack recorder be needed.

It's interesting to note that although the Alesis ADAT tape-based format has virtually disappeared into the sunset, the lightpipe protocol lives on as a preferred way of passing multichannel digital audio to and from multichannel mic preamps to a lightpipe-equipped digital audio interface (Figure 6.16). This allows a stream of up to 8 digital outs to be easily inserted into a suitable audio interface, with a single optical cable, thus increasing the number of inputs by 8 (or 16, should 2 ADAT I/O interconnections be available).

TDIF

The TDIF (Tascam Digital InterFace) is a proprietary Tascam format that uses a 25-pin D-sub cable to transmit and/or receive up to eight channels of digital audio between compatible devices. Unlike the lightpipe connection, TDIF is a bidirectional connection, meaning that only one cable is required to connect the eight ins and outs of one device to another. Although systems that support TDIF-1 cannot send and receive sync information (a separate wordclock connection is required for that), the more recent TDIF-2 protocol is capable of receiving and transmitting digital audio sync through the existing connection without any additional cabling.

AES 67

AES67 is a technical standard for audio over IP and audio over Ethernet (AoE) interoperability. The standard was developed by the Audio Engineering Society and first published in September 2013. It is a layer 3 protocol suite based on existing standards and is designed to allow interoperability between various IP-based audio networking systems such as RAVENNA, Livewire, Q-LAN and Dante.

AES67 promises interoperability between previously competing networked audio systems and long-term network interoperation between systems. It also provides interoperability with layer 2 technologies, like Audio Video Bridging (AVB). Since its publication, AES67 has been implemented independently by several manufacturers and adopted by many others.

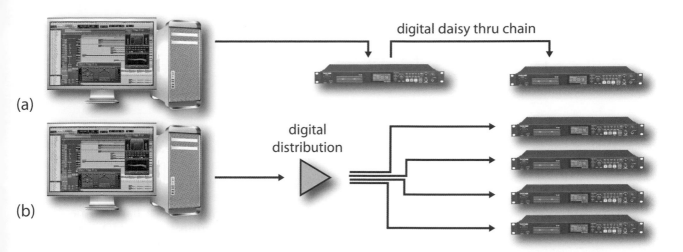

FIGURE 6.17
Digital audio distribution: (a) Several devices connected in a daisy-chain fashion. (b) Multiple devices connected using a distribution device.

SIGNAL DISTRIBUTION

If copies are to be made from a single, digital audio source, or if data is to be distributed throughout a connected network using AES/EBU, S/PDIF or MADI digital transmission cables, it's possible to daisy chain the data from one device to the next in a straightforward fashion (Figure 6.17a). This method works well if only a few devices are to be chained together. However, if several devices are connected together, time-base errors (known as *jitter*) might be introduced into the path, with the possible side effects being added noise, distortion and a slightly "blurred" signal image. One way to reduce the likelihood of such time-base errors is to use a digital audio distribution device that can route the data from a single digital audio source to a number of individual device destinations (Figure 6.17b).

What Is Jitter?

One of the more important and misunderstood characteristics surrounding the topic of digital audio transmission is the need for a stable timing element within the digital data stream itself. That is to say, a clean, high-quality digital audio signal relies upon a timing element (clock) that is highly stable, one that samples and processes the sample at exactly the right intervals (according to the chosen sample rate). If an unstable timing element is introduced into the circuit, signal degradation in the form of reduced audio quality (most likely heard as a blurred audio image, reduced resolution, increased noise and audible ticks and pops) can occur.

Let's take a closer look at how jitter affects digital and our audio production lives in practice. In a perfectly designed audio circuit that is perfectly interfaced to its digital neighbors, the clock would occur at exactly the right timing intervals. The clock would instruct the A/D converter to begin the guessing game

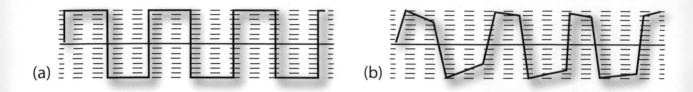

FIGURE 6.18

Example of time-base errors:
(a) A theoretically perfect
digital signal source. (b) The
same signal with jitter errors.

of finding the audio signal's voltage levels and then converting that level to an equivalent set of digital audio words – all at a precisely repeating, periodic rate (Figure 6.18a). The problem is that digital audio, contrary to popular opinion, is most often not perfect at all; indeed, undesired variation in the timing circuit of an A/D converter requires a great deal of skill and dedication to quality in order to reduce these timings to acceptably low levels. Systems that have not been carefully designed will often introduce these timing errors into the A/D conversion process (Figure 6.18b), thereby capturing digital audio that can range from having a "veiled" sound image to a sound that can be harsh and even dirty in nature. The unfortunate part is that whenever a circuit that has an erratic timing circuit is used to capture the sound, there is little or nothing that can be done to restore the sound to a quality state.

This concept of jitter becomes more complicated when you add the idea that digital connections to the outside world (i.e., cable quality, lengths, impedances, etc.) can have an effect upon the timing of the digital signal (which is often in the megahertz range). Care should be taken to use quality cabling (within reason) and to be aware of cable matching impedances.

I'm now going to break with tradition and offer up an observation that I've made over my career (and it happens to coincide with the opinions of many who are at the top levels of high-quality interface design) – I'm referring to the fact that the overall design quality of the converters within an interface will be the most important factor when it comes to capturing a killer-quality audio signal. I've noticed first-hand that an awesome-sounding converter or audio interface at lower sampling rates can sound better than an average interface running at higher resolution sample and even higher bit rates. It's not always about the specs, numbers and how high the file resolution is; it's primarily about the quality and care in design of the converters and overall A/D and D/A circuit design. Therefore, when you take an awesome interface and increase the sample/bit-rate resolution, everything jumps up to the highest possible level. Again, it's not always about the numbers; more often, it's about how the device "sounds".

Wordclock

One aspect of digital audio recording that never seems to get enough attention is the need for locking (synchronizing) the sample clocks within a connected digital audio system to a single timing reference. Left unchecked, it's possible that such gremlins as clicks, pops and jitter (oh my!) would make their way into the audio chain. Through the use of a single master timing reference known as *wordclock*, the overall sample-and-hold conversion timings during both the A/D and the D/A process for all digital audio channels and devices within the system will occur at exactly the same point in time.

How can you tell when the wordclock between multiple devices either isn't connected or is improperly set? If you hear ticks, clicks and pops over the monitors that sound like "pensive monkeys pounding away at a typewriter", you probably have a wordclock problem. When this happens, stop and deal with the issue, as those monkeys will almost certainly make it into your recordings.

To further illustrate the need for a master timing source, let's assume that we're in a room that has four or five clocks, and none of them display the same time! In situations like this, you'll never quite know what the time really is, as the clocks could be running at different speeds, or they could be running at the same speed but be set to different times. Trying to accurately keep track of the time would end up being a jumbled nightmare. On the other hand, when all of these clocks are locked to a single master clock, there will be only one master timing reference for the entire system.

In a manner similar to the distribution of timecode, there can only be one master wordclock reference within a connected digital distribution network (Figure 6.19). This reference source can be derived from a digital mixer, sound-card or any desired source that can transmit a stable wordclock. Often, this reference pulse is chained between the involved devices through the use of BNC and/or RCA connectors, using low-capacitance cables (often 75-Ω, video-grade coax cable is used, although this cable grade isn't always necessary with shorter cable runs).

It's interesting to note that wordclock isn't generally needed when making a digital copy directly from one device to another (via such protocols as AES, S/PDIF, MADI or TDIF2), because the timing information will be embedded within the data bit stream itself. Only when we begin to connect multiple devices that share and communicate digital data throughout a production network do we begin to see the immediate need for wordclock.

FIGURE 6.19
Example of wordclock distribution showing that there can be only one master clock within a digital production network.

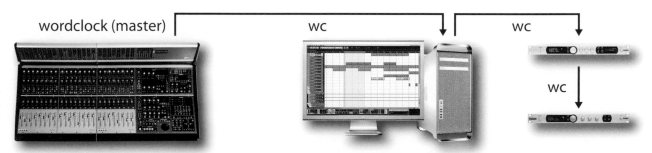

wordclock (master)   WC   WC   WC

It almost goes without saying that there will often be differences in connections and parameter setups from one system to the next. In addition to proper cabling, impedance and termination considerations throughout the network, specific hardware and software setups may be required in order to get all the device blocks to communicate properly. In order to better understand your particular system's setup (and to keep frustration to a minimum), it's always a good idea to keep all of your device's physical or pdf manuals close at hand.

Finally, external wordclock generators have come onto the market that provide an extremely stable timing reference for the digital clocking path. This can be viewed in several ways. It can be viewed as a tool to clean up and improve the sound of your A/D and D/A circuitry, or it could be viewed as an expensive device that wouldn't be necessary had the interface been properly designed in the first place. I will say this – it's been my experience that no two interfaces sound alike and that the choice of interface is an extremely important decision. It is a crucial transducer in the recording chain (along with your mics, your speakers and your skills). Try to make your choices wisely in a way that best matches your needs and budget.

CHAPTER 7

The Digital Audio Workstation

Over the history of modern recording, the style, form and function of digital audio production have changed to meet the challenges of faster processors, reduced size, larger drives, improved hardware systems and the ongoing push of marketing forces to sell, sell, sell! As a result, there is a wide range of system types that are designed for various purposes, budgets and production styles. As new technologies and programming techniques continue to turn out new hardware and software systems at a dizzying pace, many of the long-held production limitations have vanished as increased track counts, processing power and affordability have changed the way we see the art of production itself. In recent years, no single term implies these changes more than the *digital audio workstation* (DAW).

In recent years, the DAW has come to signify an integrated computer-based digital recording system that commonly offers a wide and ever-changing number of production features such as:

- Advanced multitrack recording, editing and mixdown capabilities
- Musical instrument digital interface (MIDI) sequencing, edit and score capabilities
- Integrated video and/or video sync capabilities
- Integration with peripheral hardware devices such as controllers, digital signal processing (DSP) acceleration systems, MIDI and audio interface devices
- Plug-in DSP support
- Support for plug-in virtual instruments
- Support for integrating timing, signal routing and control elements with other production software (ReWire)

Truth of the matter is, by offering an astounding amount of production power for the buck, these software-based programs (Figures 7.1 through 7.3) and their associated hardware devices have revolutionized the faces of professional,

DOI: 10.4324/9781003260530-7

FIGURE 7.1
Pro Tools hard-disk editing workstation for the Mac or PC (courtesy of Avid Technology, Inc., www.avid.com).

FIGURE 7.2
Cubase and Nuendo Media Production System for the Mac or PC (courtesy of Steinberg Media Technologies GmbH, a division of Yamaha Corporation, www.steinberg.net).

project and personal studios in a way that touches almost every life within the sound production communities.

INTEGRATION NOW – INTEGRATION FOREVER!

Throughout the history of music and audio production, we were raised with the idea that certain devices were only meant to perform a single task: A recorder records and plays back, a limiter limits and a mixer mixes. Fortunately, the age of the microprocessor has totally broken down these traditional lines in a way that has created a breed of digital chameleons that can change their functional colors as needed to match the task at hand. Along these same lines, the DAW isn't so much a device as a systems concept that can perform a wide range of audio production tasks with relative ease and speed. Some of the characteristics that can (or should be) offered by a DAW include:

- *Integration*: One of the biggest features of a workstation is its ability to provide centralized control over the digital audio recording, editing, processing and signal routing functions within the production system. It should also provide for direct communications with production-related hardware

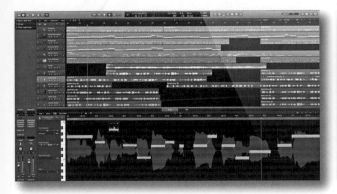

FIGURE 7.3
Logic DAW for the Mac (courtesy of Apple Inc., www. apple.com).

and software systems, as well as transport and sync control to/from external media devices.

- *Communication*: A DAW should be able to communicate and distribute pertinent audio, MIDI and automation-related data throughout the connected network system. Digital timing (wordclock) and synchronization (SMPTE and/or MIDI timecode) should also be supported.
- *Speed and flexibility*: These are probably a workstation's greatest assets. After you've become familiar with a particular system, most production tasks can be tackled in far less time than would be required using similar analog equipment. Many of the extensive signal processing, automation and systems communications features would be far more difficult to accomplish in the analog domain.
- *Session recall*: Because all of the functions are in the digital domain, the ability to instantly save and recall a session and to instantly undo a performed action becomes a relatively simple matter.
- *Automation*: The ability to automate almost all audio, control and session functions allows a great degree of control over almost all DAW programs and session parameters.
- *Expandability*: Most DAWs are able to integrate new and important hardware and software components into the system with little or no difficulty.
- *User-friendly operation*: An important element of a DAW is its ability to communicate with its central interface unit: you! The operation of a workstation should be relatively intuitive and shouldn't obstruct the creative process by speaking too much "computerese".

I'm sure you've gathered from these points that a software system (and its associated hardware) that is capable of integrating audio, video and MIDI under a single, multifunctional umbrella can be a major investment, both in financial terms and in terms of the time that's spent learning to master the overall program environment. When choosing a system for yourself or your facility, be sure to take these considerations (as well as personal ones) into account. Each system has its own strengths, weaknesses and particular ways of working. When in doubt, it's always a good idea to research the system as much as possible before committing to it. Feel free to contact your local dealer for a salesroom test drive, or better yet, try the demo. As with a new car, purchasing a DAW and its associated hardware can be an expensive proposition that you'll probably have to live with for a while. Once you've taken the time to make the right choice for you, you can get down to the business of making music.

DAW HARDWARE

Keeping step with the modern-day truism "technology marches on", the hardware and software specs of a computer and the connected peripherals continue to change at an ever-increasing pace. This is usually reflected as general improvements in such areas as their:

- Need for speed (multiple processors and accelerated co-processors)
- Increased computing power
- Increased disk size and speed
- Increased memory size and speed
- Operating system (OS) and peripheral integration
- General connectivity (networking and the Web)

In this day and age, it's definitely important that you keep step with the ever-changing advances in computer-related production technology (Figure 7.4). That's not to say you need to update your system every time a new hard- or soft-whiz-bang comes onto the market. On the contrary, it's often a wise person who knows when a system is working just fine for his or her own personal needs and who does the research to update software and fine-tune the system (to the best of his or her ability). On the other hand, there will come a time (and you'll know all too well when it arrives) when this "march" of technology will dictate a system change to keep you in step with the times. As with almost any aspect of technology, the best way to judge what will work best for you and your system is to research any piece of hard- and software that you're considering – quite

FIGURE 7.4
Pro Tools HDX DAW for the Macor PC (courtesy of Avid Technology, Inc., www.avid. com).

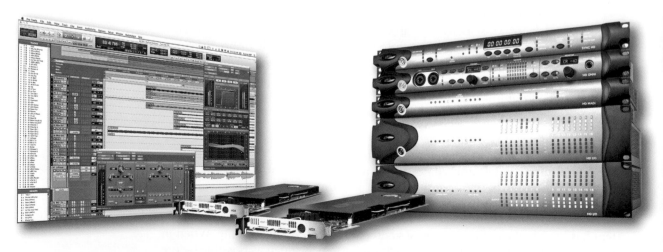

simply, read the specs, read the reviews, ask your friends and then make your best, most informed choice.

When buying a computer for audio production, one of the most commonly asked questions is "Which one – Mac or PC?" The answer as to which OS will work best for you will actually depend on:

- Your preference
- Your needs
- The kind of software you currently have
- The kind of computer platform and software your working associates or friends have
- Cost: The fact that you might already be heavily invested in either PC or Mac software, or that you are more familiar with a certain platform, will usually factor into your system choice
- OS: Even this particular question is being sidestepped with the advent of Apple's Boot Camp, which allows a Mac to boot up under the Mac or Windows OS, giving you freedom of choice to have either or both

The truth is, in this day and age, there isn't much of a functional difference between the two platforms. They both can do the job admirably and have their respective pluses and minuses.

Once you've decided which side of the platform tracks you'd like to live on, the more important questions that you should be asking are:

- Is my computer fast and powerful enough for the tasks at hand?
- Does it have enough hard-disk space that's large and fast enough for my needs?
- Is there enough random access memory (RAM)?
- Do I have enough monitor space (real estate) to see the important things at a single glance?

On the "need for speed" front, it's always a good idea to buy (or build) a computer at the top of its performance range at any given time. Keeping in mind that technology marches on, the last thing that you'll want to do is buy a new computer only to find out soon that it's underpowered for the tasks ahead.

The newer 8- and 12-core (multi-processor) systems allow faster calculations. Their tasks are spread across multiple CPUs; for example, a number of DAWs allow their odd/even track counts and/or effects processing to be split across multiple CPUs to increase the overall processing load for added track count and DSP capabilities.

With today's faster and higher-capacity hard drives, it's a simple matter to install cost-effective drives with terabyte capacities into a system. These drives can be internal, or they can be installed in portable drive cases that can be plugged into either a Thunderbolt®, FireWire® or universal serial bus (USB) port (preferably

USB 3 or C), making it easy to take your own personal drive with you to the studio or on stage.

The speed at which the disc platters turn will often affect a drive's access time. Modern drives with a high disc spin rate (7,200 rpm or higher) with large amounts of internal cache memory are often preferable for both audio and video production. SSDs (solid state drives) are probably the most commonly found drives in a modern production system, as they don't have moving parts at all and include solid state memory that can both read and write data at blazing speeds when compared with a hard-disk drive.

Within a media production system, it's always a wise precaution to have a dedicated drive that's strictly allocated for your audio, video and media files – this is generally recommended because the OS will often require access to the main drive in a way that can cause data interruptions and a slowed response should both the OS and media need to access data simultaneously.

Regarding random access memory, it's always good to install a more than adequate amount of high-speed RAM into the system (i.e., 16 Gb and higher). If a system doesn't have enough RAM, data will often have to be swapped to the system's hard drive, which can seriously slow things down and affect overall real-time DSP performance. When dealing with music sampling, video and digital imaging technologies, having a sufficient amount of RAM becomes even more of an issue.

With regard to system and application software, it's often wise to perform an update to keep your system, well, up to date. This holds true even if you just bought the software, because it's often hard to tell how long the original packaging has been sitting on the shelves – and even if it is brand spanking new, chances are that new revisions will still be available. Updates don't come without their potential downfalls, however; given the incredible number of hardware/software system combinations that are available, it's actually possible that an update might do as much harm as good. In this light, it's actually not a bad idea to do a bit of research before clicking that update button. Whoever said that all this stuff would be easy?

Just like there never seems to be enough space around the house or apartment, having a single, undersized monitor can leave you feeling cramped for visual "real estate". For starters, a sufficiently large monitor that's capable of working at higher resolutions will greatly increase the size of your visual desktop; however, if one is a good thing, two is always better! Both Windows® and Mac OS offer support for multiple monitors (Figure 7.5). By adding a commonly available "dual-head" video card, your system can easily double your working monitor space for fewer bucks than you might think. I've found that it's truly a joy to have your edit window, mixer, effects sections and transport controls in their own places – all in plain and accessible view over multiple monitors.

The Desktop Computer

Desktop computers are often (but not always) too large and cumbersome to lug around. As a result, these systems are most often found as a permanent installation in the professional, project and home studio (Figure 7.6). Historically, desktops have offered more processing power than their portable counterparts, but in recent times, this distinction has become less and less of a factor.

The Laptop Computer

One of the most amazing characteristics of the digital age is miniaturization. At the forefront of the studio-on-the-go movement is the laptop computer (Figure 7.7). From the creation of smaller, lighter and more powerful notebooks has come the technological Phoenix of the portable DAW and music performance machine. With the advent of USB, FireWire and Thunderbolt audio interfaces, controllers and other peripheral devices, these systems are now capable of handling most (if not all) of the edit and processing functions that can be handled in the studio. In fact, these AC/battery-powered systems are often powerful enough to handle advanced DAW edit/mixing functions, as well as happily handling a wide range of plug-in effects and virtual instruments, all in the comfort of – anywhere!

That's the good news! Now, the downside of all this portability is the fact that since laptops are optimized to run off a battery with as little power drain as possible, their:

FIGURE 7.5
You can never have too much visual "real estate"!

FIGURE 7.6
The desktop computer. (a) Creation Station 450 desktop PC (courtesy of Sweetwater, www.sweetwater.com). (b) The Mac Pro™ with Cinema display (courtesy of Apple Computers, Inc., www.apple.com).

(a)

(b)

(a)

(b)

- Processors *may* run slower so as to conserve battery power
- BIOS (the important subconscious brains of a computer) might be different (again, especially with regard to battery-saving features)
- Hard drives *might* not have spun as fast in previous times, although this has changed as most laptops are now fully equipped with high-speed SSD
- Video display capabilities are sometimes limited when compared with a desktop (video memory is often shared with system RAM, reducing graphic quality and refresh rate)
- Internal audio interface usually isn't so great (but that's why there are so many external interface options)

As the last option says, it's no secret that the internal audio quality of most laptops ranges from being quite acceptable to abysmal. As a result, the only true choice is to find an external audio interface that works best for you and your applications. Fortunately, there's a ton of audio interface choices for connecting via either USB or Thunderbolt, ranging from a simple on-the-go I/O device to those that include multichannel audio, MIDI and controller capabilities.

System Interconnectivity

In the not-too-distant past, installing a device into a computer or connecting between computer systems would have easily been a major hassle. With the development of the USB and other protocols (as well as the improved general programming of hardware drivers), hardware devices such as mice, keyboards, cameras, soundcards, modems, MIDI interfaces, CD and hard drives, MP3 players, portable fans, LED Christmas trees and cup warmers can be plugged into an available port, be installed and be up and running in no time – generally without a hassle. Additionally, with the development of a standardized network and Internet protocol, it's now possible to link computers together in a way that allows the fast and easy sharing of data throughout a connected system. Using such a system, artists and businesses alike can easily share and swap files on the other side of the world, and pro or project studios can swap sound files and video files over the Web with relative ease.

USB

In recent computer history, few interconnection protocols have affected our lives like the USB. In short, USB is an open specification for connecting external hardware devices to the personal computer, as well as a special set of protocols for automatically recognizing and configuring them. Here are the current USB specs:

- USB 2.0 (*up to 480 megabits/sec = 60 megabytes/sec*)
- USB 3.0 (*up to 5 gigabits/sec = 640 megabytes/sec*)
- USB 3.1 (*up to 10 gigabits/sec = 1.28 gigabytes/sec*)
- USB C (*up to 10 gigabits/sec = 1.28 gigabytes/sec*)

The basic characteristics of USB include:

- Up to 127 external devices can be added to a system without having to open up the computer. As a result, the industry has largely moved toward a "sealed case" or "locked-box" approach to computer hardware design.
- Newer operating systems will often automatically recognize and configure a basic USB device that's shipped with the latest device drivers.
- Devices are "hot pluggable", meaning that they can be added (or removed) while the computer is on and running.
- The assignment of system resources and bus bandwidth is transparent to the installer and end user.
- USB connections allow data to flow bidirectionally between the computer and the peripheral.
- USB cables can be up to 5 meters in length (up to 3 meters for low-speed devices) and include two twisted pairs of wires, one for carrying signal data and the other pair for carrying a DC voltage to a "bus-powered" device. Those that use less than 500 milliamps (1/2 amp) can get their power directly from the USB cable's 5-V DC supply, while those having higher current demands will need to be externally powered.
- Standard USB 1 through 3 cables have different connectors at each end. For example, a cable between the PC and a device would have an "A" plug at the PC (root) connection and a "B" plug for the device's receptacle.
- USB C is not a protocol but is an actual connector spec that includes a plug at each end that can be inserted in either orientation. Quite often (but not always), this connector can handle data rates up to the USB 3.1 data spec and when the Thunderbolt logo is present at the host port, is often capable of data speeds approaching that of Thunderbolt 3. In addition, this connector type can potentially supply up to 20 volts or 100 watts through the data cable for powering devices and charging systems such as laptops, interfaces, etc.

Cable distribution and "daisy-chaining" are done via a data "hub" (Figure 7.8). These devices act as traffic cops in that they cycle through the various USB inputs in a sequential fashion, routing the data into a single data output line.

FIGURE 7.8
USB hubs in action.

THUNDERBOLT

Originally designed by Intel (in collaboration with Apple) and released in 2011, Thunderbolt (Figure 7.9) combines the DisplayPort and PCIe bus into a single, serial data interface. A single Thunderbolt port can support a daisy chain of up to six Thunderbolt devices (two of which can be DisplayPort display devices), which can run at high speeds:

- Thunderbolt 1 (*up to 10 gigabits/sec = 1.28 gigabytes/sec*) makes use of the same connector as the Apple Mini DisplayPort (MDP)
- Thunderbolt 2 (*up to 20 gigabits/sec = 2.56 gigabytes/sec*) makes use of the same connector as the Apple Mini DisplayPort (MDP)
- Thunderbolt 3 (*up to 40 megabits/sec = 5.12 megabytes/sec*) makes use of a USB-C connector that can be plugged into any USB-C port (on either the Mac or the PC) that sports the Thunderbolt logo.

Obviously, the most amazing thing about this "hot pluggable" protocol is its speed. Another bonus is that it not only conforms to the USB spec (so that USB devices can be hot-plugged into it) but also conforms to the PCIe bus spec, allowing many hardware peripheral devices to be connected onto the Thunderbolt bus.

FIGURE 7.9
Intel's Thunderbolt3 protocol makes use of the USB-C type connector (courtesy of Apple Inc., www.apple.com).

One important fact when entering into the world of Thunderbolt are the requirements, limitations and special needs that are often required when using these specialized cables. TB1 and 2 cables are often very specialized in design and

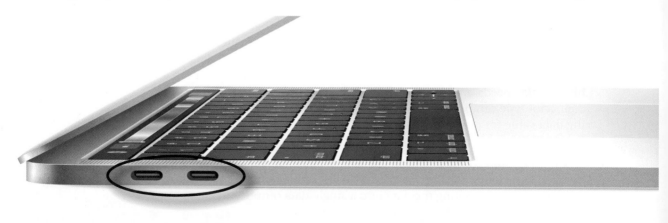

spec, often with few things that can go wrong in practice. Using TB3, on the other hand, requires that you carefully choose your cabling in order to make a proper TB spec connection.

For example, when connecting two TB devices with only a standard USB-C data cable, you stand an excellent chance that no connection will be made. Even though the cable "looks" right, it won't have the necessary spec to make the connection. You will need to actually buy a cable that conforms to the TB3 spec. Additionally, passive cables (not containing data conditioning circuitry) that are greater than half a meter will often not be able to pass data at the full 40 Mb/sec rate. Longer cables (up to 3 meters) must be high-quality active cables in order to get the job done. Passing the full TB3 40 Mb/sec rate at lengths greater than 3 meters can be done with the use of optical cables that conform to the TB3 spec. Basically, when setting up your system, it's wise to do your homework before buying just any ol' cable that might not work or might severely limit your system's capabilities.

FIREWIRE

Originally created in the mid-1990s as the IEEE-1394 standard, the FireWire protocol is similar to USB in that it uses twisted-pair wiring to communicate bidirectional, serial data within a hot-swappable, connected chain. Unlike USB (which can handle up to 127 devices per bus), up to 63 devices can be connected within a connected FireWire chain. FireWire most commonly supports two speed modes:

- FireWire 400 or IEEE-1394a (*400 megabits/sec*) is capable of delivering data over cables up to 4.5 meters in length. FireWire 400 is ideal for communicating large amounts of data to such devices as hard drives, video camcorders and audio interface devices.
- FireWire 800 or IEEE-1394b (*800 megabits/sec*) can communicate large amounts of data over cables up to 100 meters in length. When using fiber-optic cables, lengths in excess of 90 meters can be achieved in situations that require long-haul cabling (such as within sound stages and studios).

Unlike USB, compatibility between the two modes is mildly problematic, because FireWire 800 ports are configured differently from their earlier predecessor and therefore require adapter cables to ensure compatibility. It should be noted that this legacy data protocol isn't commonly found on newer devices.

AUDIO OVER ETHERNET

One of the more recent advances in audio and systems connectivity in the studio and on stage revolves around the concept of communicating audio over the Ethernet (AoE). Currently, there are several competing protocols that range from being open-source (no licensing fees) to those that require a royalty to be designed into a hardware networking system.

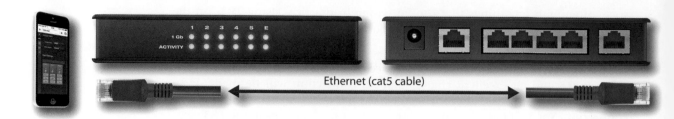

Ethernet (cat5 cable)

FIGURE 7.10
MOTU AVB (Audio Video Bridge) Switch and AVB Control app for communicating and controlling audio over Ethernet (courtesy of MOTU, Inc., www.motu.com).

By connecting hardware devices directly together via a standard cat5 Ethernet cable (Figure 7.10), it's possible for channel counts of up to 512 × 512 to be communicated over a single connected network. This straightforward system is designed to replace bulky snake cables and fixed wiring within large studio installations, large-scale stage sound reinforcement, convention centers and other complex audio installations. For example, instead of having an expensive, multi-cable microphone snake run from a stage to the main mixing console, a single Ethernet cable could be run directly from an A/D mic/line cable box to the mixer (as well as the on-stage monitor mixer, for that matter) – all under digital control that often can include a redundant cable/system in case of unforeseen problems or failures.

In short, AoE allows complex system setups to be interconnected, digitally controlled and routed in an extremely flexible manner, and since the system is connected to the Internet, wireless control via apps and computer software is often fully implemented.

The Audio Interface

An important device that deserves careful consideration when putting together a DAW-based production system is the *digital audio interface*. These devices can have a single, dedicated purpose, or they might be multifunctional in nature. In either case, their main purpose in the studio is to act as a connectivity bridge between the outside world of analog audio and the computer's inner world of digital audio (Figures 7.11 through 7.13). Audio interfaces come in all shapes, sizes and functionalities; for example, an audio interface can be:

- Built into a computer (although, more often than not, these devices are often limited in quality and functionality)
- A simple, two-I/O audio device
- A multichannel device, offering many I/Os and numerous I/O expansion options
- Fitted with one or more MIDI I/O ports
- One that offers digital I/O, wordclock and various sync options
- Fitted with a controller surface (with or without motorized faders) that provides direct DAW control integration
- Designed to include built-in DSP acceleration for offering additional plug-in processing

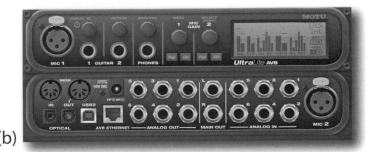

(a) (b)

These devices are most commonly designed as smaller stand-alone and/or 19″ rack mountable systems that plug into the system via USB, FireWire, Thunderbolt or AoE. An interface might have as few as two inputs and two outputs, or it might have more than 24. Recent units offer bit-depth/sample-rate options that range up to 24/96 or 24/192. In recent times, pretty much all interfaces will work with any DAW and platform (even Digidesign has dropped its use of proprietary hardware/software pairing).

Obviously, there are a wide range of options that should be taken into account when buying an interface. Near the top of this list (audio quality always being the top consideration) is the need for having an adequate number of inputs and outputs (I/O).

It's always important to fully research your needs and possible hardware options *before* you buy an interface. Anticipating your future needs is rarely an easy task … but it can save you from future heartaches, headaches and additional spending.

FIGURE 7.11

Portable audio interfaces, front/back: (a) Steinberg UR22C 2 × 2 audio interface (courtesy of Steinberg Media Technologies GmbH, a division of Yamaha Corporation, www.steinberg.net). (b) MOTU Ultralite ABV 18 × 18 USB/AVB audio interface with DSP mixing, Wi-Fi control and audio networking (courtesy of MOTU, Inc., www.motu.com).

Audio Driver Protocols

Audio driver protocols are software programs that set standards for allowing data to be communicated between the system's software and hardware. A few of the more common protocols are:

- *WDM*: This driver allows compatible single-client, multichannel applications to record and play back through most audio interfaces using Microsoft Windows. Software and hardware that conform to this basic standard can communicate audio to and from the computer's basic audio ports.

FIGURE 7.12

Presonus Quantum 26 × 32 Thunderbolt audio interface (courtesy of Presonus Audio Electronics, Inc., www.presonus.com).

(a)   (b)

FIGURE 7.13

Apollo audio interfaces with integrated UAD effects processing. (a) Apollo FireWire/Thunderbolt. (b) Apollo Twin USB (courtesy of Universal Audio, www.uaudio.com © 2022 Universal Audio, Inc. All rights reserved. Used with permission).

- *ASIO*: The Audio Stream Input/Output architecture (which was developed by Steinberg and offered free to the industry) forms the backbone of Virtual Studio Technology (VST). It does this by supporting variable bit depths and sample rates, multichannel operation and synchronization. This commonly used protocol offers low latency, high performance, easy setup and stable audio recording within VST.
- *MAS*: The MOTU Audio System is a system extension for the Mac that uses an existing central processing unit (CPU) to accomplish multitrack audio recording, mixer, bussing and real-time effects processing.
- *CoreAudio*: This driver allows compatible single-client, multichannel applications to record and play back through most audio interfaces using Mac OS X. It supports full-duplex recording and playback of 16-/24-bit audio at sample rates up to 96 kHz (depending on your hardware and CoreAudio client application).

In most circumstances, it won't be necessary for you to be familiar with the protocols – you just need to be sure that your software and hardware are compatible for use with the driver protocol that works best for you. Of course, further information can always be found on the respective companies' websites.

Latency

When discussing the audio interface as a production tool, it's important that we touch on the issue of latency. Quite literally, *latency* refers to the buildup of delays (measured in milliseconds) in audio signals as they pass through the audio circuitry of the audio interface, CPU, internal mixing structure and I/O routing chains. When monitoring a signal directly through a computer's signal path, latency can be experienced as short delays between the input and the monitored signal. If the delays are excessive, they can be unsettling enough to throw a performer off time. For example, when recording a synth track, you might actually hear the delayed monitor sound shortly after hitting the keys (not a happy prospect), and latency on vocals can be quite unsettling. However, by switching to a supported ASIO or CoreAudio driver and by optimizing the interface/DAW buffer settings to their lowest operating size (without causing the audio to stutter), these delay values can be reduced to an unnoticeable or barely noticeable range.

In response to this problem, most modern interface drivers include a function called *direct monitoring*, which allows the system to monitor inputs directly from the monitoring source in a way that bypasses the DAW's monitoring circuitry. The result is a monitor (cue) source that is free from latency, allowing the artist to hear themselves without the distraction of delays in the monitor path.

Need Additional I/O?

Obviously, there are a wide range of options that should be taken into account when buying an interface. Near the top of this list (audio quality always being the top consideration) is the need for having an adequate number of inputs and outputs (I/O).

Although a number of interface designs include a large number of I/O channels, by far the most have a limited I/O count, but instead offer access to additional I/O options should the need arise. This can include such options as:

- Lightpipe (ADAT) I/O, whereby each optical cable can give access to either 8 channels at sample rates of 44 or 48k or 4 channels at 96k (if this option is available) when used with an outboard lightpipe preamp (Figure 7.14).
- Connecting additional audio interfaces to a single computer. This is possible whenever several compatible interfaces can be detected by and controlled from a single interface driver.
- Using an audio over the Ethernet protocol and compatible interface systems, additional I/O can be added by connecting additional AoE devices onto the network and patching the audio through the system drivers.

It's important to fully research your needs and possible hardware options *before* you buy an interface. Anticipating your future needs is never an easy task, but it can save you from future heartaches, headaches and additional spending.

FIGURE 7.14

Outboard lightpipe (ADAT) preamp. (a) Warm Audio WA-412 (courtesy of Warm Audio LLC, www.warmaudio. com). (b) Universal Audio 4-710d Four-Channel Mic Preamplifier (courtesy of Universal Audio, www.uaudio .com, ©2022 Universal Audio, Inc. All rights reserved. Used with permission).

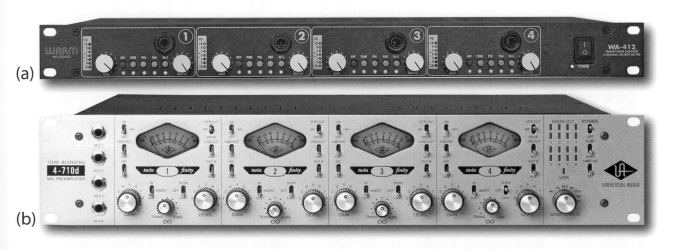

(a)

(b)

DAW Controllers

Originally, one of the more common complaints against most DAWs (particularly when relating to the use of on-screen mixers) was the lack of hardware control that gives the user direct, hands-on access. Over the years, this has been addressed by major manufacturers and third-party companies in the form of:

- Hardware DAW controllers
- MIDI instrument controller surfaces that can directly address DAW controls
- On-screen touch monitor surfaces
- iOS-based controller apps

It's important to note that there are a wide range of controllers from which to choose – and just because others feel that the mouse is cumbersome, this doesn't mean that you have to feel that way; for example, I have several controllers in my own studio, but the mouse is still my favorite tool. As always, the choice of what works best for you is totally up to you.

Hardware Controllers

Hardware controller types (Figure 7.15) generally mimic the design of an audio mixer in that they offer slide or rotary gain faders, pan pots, solo/mute and channel select buttons as well as full transport remote functions. A channel select button might be used to actively assign a specific channel to a section that contains a series of grouped pots and switches that relate to equalization (EQ), effects and dynamic functions, or the layout may be simple in form, providing only the most often used direct control functions in a standard channel layout.

Such controllers vary in the number of channel control strips that are offered at one time. They'll often (but not always) offer direct control over 8 input strips at a time, allowing channel groups to be switched in groups of 8 (1–8, 9–16, 17–24, etc.); any number of the grouped inputs can be accessed on the controller as well as on the DAW's on-screen mixer. These devices will also often include

FIGURE 7.15

Hardware controllers. (a) Mackie MCU Pro DAW controller (courtesy of Loud Technologies, Inc., www.mackie.com). (b) Presonus Faderport 16 DAW controller (courtesy of Presonus Audio Electronics, Inc., www.presonus.com).

(a)   (b)

(a)

(b)

software function keys that can be programmed to give quick and easy access to the DAW's more commonly used program keys.

Instrument Controllers

Since all controller commands are transmitted between the controller and audio editor via MIDI and device-specific MIDI SysEx messages (see Chapter 9), it makes sense that a wide range of MIDI instrument controllers (mostly keyboard controllers) offer a wide range of controls, performance triggers and system functionality that can directly integrate with a DAW (Figure 7.16). The added ability of controlling a mix, as well as remote transport control, is a nice feature should the keyboard controller be out of arm's reach of the DAW.

Touch Controllers

In addition to the wide range of hardware controllers that are available on the market, an ever-growing number of software-based touch-screen monitor controllers have begun to take over the market. These can take the form of standard touch-screen monitors that let you have simple, yet direct, control over any software commands, or they can include software that gives you additional commands and control over specific DAWs and/or recording-related software in an easy-to-use fashion (Figure 7.17a). Since these displays are computer-based devices themselves, they can change their form, function and entire way of working with a single, uh, touch.

In addition to medium-to-large touch control screens, a number of Wi-Fi-based controllers are available for the iPad (Figure 7.17b). These controller "apps"

FIGURE 7.16
Keyboard controllers will often provide direct access to DAW mixing and function controls. (a) Komplete Kontrol S49 keyboard controller (courtesy of Native Instruments GmbH, www.native-instruments.com). (b) KX49 keyboard controller (courtesy of Yamaha Corporation, www.yamaha.com).

FIGURE 7.17
Touch-screen controllers. (a) The Raven MTi Multi-touch Audio Production Console (courtesy of Slate Pro Audio, www.slateproaudio.com). (b) V-Control Pro DAW controller for the iPad (courtesy of Neyrinck, www.vcontrolpro.com).

(a)

(b)

FIGURE 7.18
Digidesign S6 integrated controller/console (courtesy of Avid Technology, Inc., www.avid.com).

offer direct control over many of the functions that were available on hardware controllers that used to cost hundreds or thousands of dollars but are now emulated in software and can be purchased from an app "store" for the virtual cost of a cup of coffee.

Large-Scale Controllers

Another controller type that is a different type of beastie is the large-scale controller (Figure 7.18). In fact, these controllers (which might or might not include analog hardware, such as mic preamps) are far more likely to resemble a full-sized music and media production console than a controller surface. They allow direct control and communication with the DAW but also offer a large number of physical and/or touch-screen controls for easy access during such tasks as mixing for film, television and music production.

SOUND FILE FORMATS

A wide range of sound file formats exist within audio and multimedia production. Here is a list of those used in professional audio that don't use data compression of any type:

- *Wave (.wav)*: The Microsoft Windows format supports both mono and stereo files at a variety of bit and sample rates. WAV files contain PCM coded audio (uncompressed pulse-code modulation formatted data) that follows the Resource Information File Format (RIFF) spec, which allows extra user information to be embedded and saved within the file itself.
- *Broadcast wave (.wav)*: In terms of audio content, broadcast wave files are the same as regular wave files; however, text strings for supplying additional information (most notably, timecode data) can be embedded in the file according to a standardized data format.
- *Apple AIFF (.aif or .snd)*: This standard sound file format from Apple supports mono or stereo, 8-, 16- and 24-bit audio at a wide range of sample rates. Like broadcast wave files, AIFF files can contain embedded text strings.

Sound File Sample and Bit Rates

While the sample rate of a recorded bit stream (samples per second) directly relates to the resolution at which a recorded sound will be digitally captured, the bit rate of a digitally recorded sound file directly relates to the number of quantization steps that are encoded into the bit stream. It's important that these rates be determined and properly set *before* starting a session. Further reading on sample- and bit-rate depths can be found in Chapter 6. Additional info on sound files and compression codecs can be found in Chapter 11.

Sound File Interchange and Compatibility Between DAWs

At the sound file level, most software editors and DAWs are able to read a wide range of uncompressed and compressed formats, which can then be saved into a new DAW session format. At the session level, there are several ways to exchange data for an entire session from one platform, OS or hardware device to another. These include the following:

- Open Media Framework Interchange (OMFI) is a platform-independent session file format intended for the transfer of digital media between different DAW applications; it is saved with an .omf file extension. OMF (as it is commonly called) can be saved in either of two ways: (1) "export all to one file", when the OMF file includes all of the sound files and session references that are included in the session (be prepared, this file will be extremely large) or (2) "export media file references", which does not contain the sound files themselves but will contain all of the session's region, edit and mix settings; effects (relating to the receiving DAW's available plug-ins and ability to translate effects routing); and I/O settings. This second type of file will be small by comparison; however, the original sound files must be transferred into the proper session folders.

- One audio-export-only option makes use of the broadcast wave sound file format. By using broadcast wave, many DAWs are able to directly read the time-stamped data that's embedded into the sound file and then automatically line them up within the session.

- Software options that can convert session data between DAWs also exist. Pro Convert from Solid State Logic, for example, is a stand-alone program that helps you tailor sound file information, level and other information and then transfer one DAW format into another format or readable XML file.

Although these systems for allowing file and session interchangeability between workstation types can be a huge time and work saver, it should be pointed out that they are, more often than not, far from perfect. It's not uncommon for files not to line up properly (or to load at all); plug-ins can disappear and/or lose their settings – Murphy's law definitely applies. As a result, the most highly recommended and surefire way to make sure that a session will load into any DAW platform is to make (print or export) a copy of each track, starting from the session's beginning (00:00:00:00) and going to the end of that particular track. Using this system, all you need to do is load each track into the new workstation at its respective track beginning point and get to work.

Of course, this method won't load any of the plug-in or mixer settings (often an interchange problem anyway). Therefore, it's extremely important that you properly document the original session, making sure that:

- All tracks have been properly named (supplying additional track notes and documentation, if needed).
- All plug-in names and settings are well documented (a screenshot can go a long way toward keeping track of these settings).
- Any virtual instrument or MIDI tracks are printed to an audio track. (Make sure to include the original MIDI files in the session, and to document all instrument names and settings – again, screenshots can help.)
- Any special effects or automation moves are printed to the particular track in question (make sure this is well documented), and you should definitely consider providing an additional copy of the track without effects or automation.

DAW SOFTWARE

By their very nature, DAWs (Figures 7.1 through 7.3 as well as 7.19 and 7.20) are software programs that integrate with computer hardware and functional applications to create a powerful and flexible audio production environment. These programs commonly offer extensive record, edit and mixdown facilities for such uses in audio production as:

- Extensive sound file recording, edit and region definition and placement
- MIDI sequencing and scoring
- Real-time, on-screen mixing
- Real-time effects
- Mixdown and effects automation
- Sound file import/export and mixdown export
- Support for video/picture playback and synchronization
- Systems synchronization
- Audio, MIDI and sync communications with other audio programs (e.g., ReWire)
- Audio, MIDI and sync communications with other effects and software instruments (e.g., VST technology)

This list is but a small smattering of the functional capabilities that can be offered by an audio production DAW.

Suffice it to say that these powerful software production tools are extremely powerful and varied in their form and function. As you can see, even with their inherent strengths, quirks and complexities, their basic look, feel and operational capabilities have, to some degree, become unified among the major DAW competitors. Having said this, there are enough variations in features, layout and basic operation that individuals (from aspiring beginner to seasoned professional) will have their favorite DAW make and model. With the growth of the DAW and computer industries, people have begun to customize their computers with features, added power and peripherals that rival their love for souped-up cars and motorcycles. In the end, though (as with many things in life), it doesn't matter which type of DAW you use – it's how you use it that counts!

Sound Recording and Editing

Most digital audio workstations are capable of recording sound files in mono, stereo, surround or multichannel formats (either as individual files or as a single interleaved file). These production environments graphically display sound file information within a main graphic window (left hand side of Figures 7.1 through 7.3 as well as 7.19 and 7.20), which contains drawn waveforms that graphically represent the amplitude of a sound file over time in a WYSIWYG (what you see is what you get) fashion. Depending on the system type, the sound file length and the degree of zoom, the entire waveform may be shown on the screen, or only a portion will show as it scrolls over the course of the song or program. Graphic editing differs greatly from the "razor blade" approach that's used to cut analog tape, in that the waveform gives us both visual and audible cues as to precisely where an edit point should be. Using this common display technique, any position, cut/copy/paste, gain or time changes will be instantly reflected in the waveforms on the screen. Almost always, these edits are nondestructive (a process whereby the original sound file isn't altered – only the way in which the region in/out points are accessed or the file is processed will be changed, undone, redone, copied and pasted – virtually without limit).

Only when a waveform is zoomed in fully is it possible to see the individual sample amplitude levels of a sound file (Figure 7.21). At this zoom level, it becomes simple to locate zero-crossing points (points where the level is at the 0, center-level line). In addition, when a sound file is zoomed in to this level,

FIGURE 7.19
Reaper DAW software (courtesy of Cockos Incorporated, www.reaper.fm).

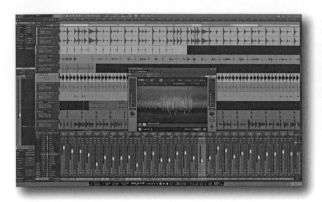

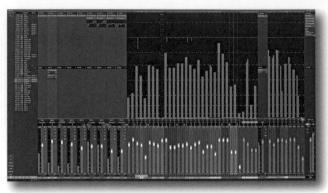

FIGURE 7.20
Presonus Studio One DAW software (courtesy of Presonus Audio Electronics, Inc., www.presonus.com).

the program might allow the sample points to be redrawn in order to remove potential offenders (such as clicks and pops) or to smooth out amplitude transitions between loops or adjacent regions.

The nondestructive edit capabilities of a DAW refer to a disk-based system's ability to edit a sound file without altering the data that was originally recorded to disk. This important capability means that any number of edits, alterations or program versions can be performed and saved to disk without altering the original sound file data.

Nondestructive editing is accomplished by accessing defined segments of a recorded digital audio file (often called *regions*) and allowing them to be reproduced in a user-defined order, defined segment in/out point or level in a manner that can be (and often is) different than the originally recorded sound file. In effect, when a specific region is defined, we're telling the program to access the sound file at a point that begins at a specific memory address on the hard disk and continues until a specified ending address has been reached (Figure 7.22). Once defined, these regions can be inserted into a program list (often called a playlist or edit list) in such a way that they can be accessed and reproduced in any order and any number of ti-ti-ti-times. For example, Figure 7.23 shows a snippet from *Gone With the Wind* that contains Rhett's immortal words "Frankly,

FIGURE 7.21
Zoomed-in edit window showing individual samples.

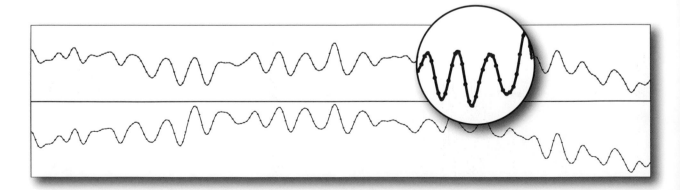

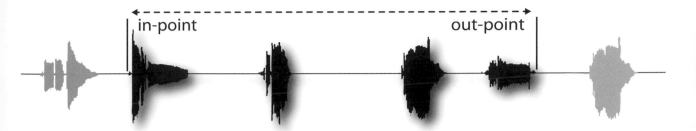

FIGURE 7.22
Nondestructive editing allows a region within a larger sound file to begin at a specific in-point and play until the user-defined end-point is reached.

my dear, I don't give a damn". By segmenting it into three regions, we could use a DAW editor to output the words in several ways.

When working in a graphic editing environment, regions can usually be defined by positioning the cursor over the waveform, pressing and holding the mouse or trackball button, and then dragging the cursor to the left or right, which highlights the selected region for easy identification. After the region has been defined, it can be edited, marked, named, maimed or otherwise processed.

As one might expect, the basic cut-and-paste techniques used in hard-disk recording are entirely analogous to those used in a word processor or other graphics-based programs:

- *Cut*: Places the highlighted region into clipboard memory and deletes the selected data (Figure 7.24a).
- *Copy*: Places the highlighted region into memory and doesn't alter the selected waveform in any way (Figure 7.24b).
- *Paste*: Copies the waveform data that's within the system's clipboard memory into the sound file beginning at the current cursor position (Figure 7.24c).

FIGURE 7.23
Example of how snippets from Rhett's famous Gone with the Wind dialogue can be easily rearranged using standard nondestructive editing.

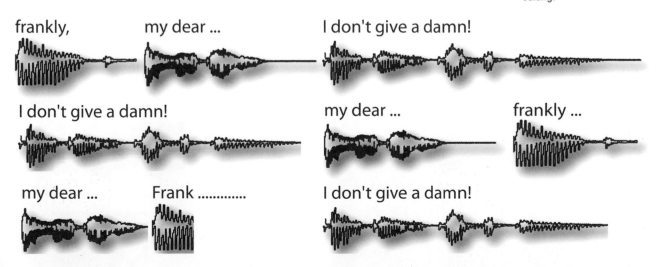

TRY THIS: RECORDING A SOUND FILE TO DISK

1. Download a demo copy of your favorite DAW (these are generally available off the company's website for a free demo period).
2. Download the workstation's manual and familiarize yourself with its functional operating basics.
3. Consult the manual regarding the recording of a sound file.
4. Assign a track to an interface input sound source.
5. Name the track! It's always best to name the track (or tracks) before going in to record. In this way,

the file will be saved to disk within the session folder under a descriptive name instead of an automatically generated file name (e.g., killerkick .wav instead of track16–01.wav).

6. Save the session and assign the input to another track, and overdub a track along with the previously recorded track.
7. Repeat as necessary until you're having fun!
8. Save your final results for the next tutorial.

FIGURE 7.24

Standard Cut, Copy and Paste commands. (a) Cutting inserts the highlighted region into memory and deletes the selected data. (b) Copying simply places the highlighted region into memory without changing the selected wave-form in any way. (c) Pasting copies the data within the system's clipboard memory into the sound file at the cur-rent cursor position.

Besides basic nondestructive cut-and-paste editing techniques, the amplitude processing of a signal is one of the most common types of change that are likely to be encountered. These include such processes as gain changing, normaliza-tion and fading.

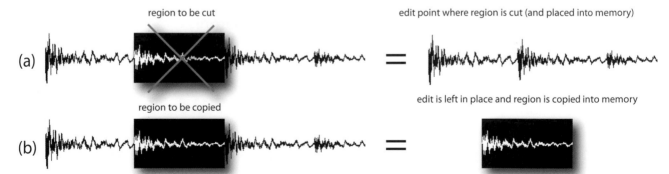

TRY THIS: COPY AND PASTE

1. Open the session from the preceding tutorial, "Recording a Sound File to Disk".
2. Consult your editor's manual regarding basic cut-and-paste commands (which are almost always the standard PC and Mac commands).
3. Open a sound file and define a region that includes a musical phrase, lyric or sentence.

4. Cut the region and try to paste it into another point in the sound file in a way that makes sense (musical or otherwise).
5. Feel free to cut, copy and paste to your heart's desire to create an interesting or totally wacky sound file.

Gain changing relates to the altering of a region or track's overall amplitude level such that a signal can be proportionally increased or reduced to a specified level (often in dB or percentage value). To increase a sound file or region's overall level, a function known as normalization can be used. *Normalization* (Figure 7.25) refers to an overall change in a sound file or defined region's signal level, whereby the file's greatest amplitude will be set to 100% full scale (or a set percentage level of full scale), with all other levels in the sound file or region being proportionally scaled up or down in gain level.

The fading of a region (either in or out, as shown in Figure 7.26) is accomplished by increasing or reducing a signal's relative amplitude over the course of a defined duration. For example, fading in a file proportionately increases a region's gain from infinity (zero) to full gain. Likewise, a fade-out has the opposite effect of creating a transition from full gain to infinity. These DSP functions have the advantage of creating a much smoother transition than would otherwise be humanly possible when performing a manual fade.

A cross-fade (or X-fade) is often used to smooth the transition between two audio segments that either are sonically dissimilar or don't match in amplitude at a particular edit point (a condition that would otherwise lead to an audible "click" or "pop"). This useful tool basically overlaps a fade-in and fade-out between the two waveforms to create a smooth transition from one segment to

FIGURE 7.25
Original signal and signal level normalized to 100% full scale.

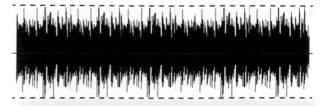

FIGURE 7.26
Examples of fade-in and fade-out curves.

the next (Figure 7.27). Technically, this process averages the amplitude of the signals over a user-definable length of time in order to mask the offending edit point.

Fixing Sound with a Sonic Scalpel

In traditional multitrack recording, should a mistake or bad take be recorded onto a new track, it's a simple matter to start over and record over the unwanted take. However, if only a small part of the take was bad, it's easy to go back and perform a punch-in (Figure 7.28). During this process, the recorder or DAW:

FIGURE 7.27
Example of cross-fade windows. (a) ProTools (courtesy of Avid Technology, Inc., www.avid.com). (b) Nuendo (courtesy of Steinberg Media Technologies GmbH, a division of Yamaha Corporation, www.steinberg.net).

- Silently enters into record at a predetermined point
- Records over the unwanted portion of the take
- Silently falls back out of record at a predetermined point.

A punch can be manually performed on most recording systems; however, DAWs and newer tape machines can be programmed to automatically go into and fall out of record at a predetermined time.

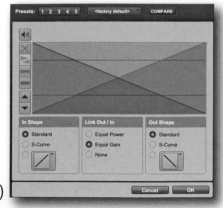
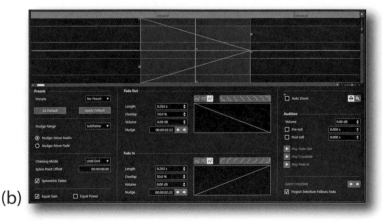

(a)   (b)

When punching-in, any number of variables can come into play. If a solo instrument is to be overdubbed, it's often easy to punch the track without fear of any consequences. If an offending musical section is within a group or ensemble, leakage from the original instrument could find its way into adjacent tracks, making a punch difficult or unwise. In such a situation, it's usually best to re-record the piece, pick up at a point just before the bad section and splice (edit insert) it back into the original recording, or attempt to punch the section using the entire ensemble.

From a continuity standpoint, it's often best to punch-in on a section immediately after the take has been recorded, because changes in mic choice, mic position or the session's general "vibe" can lead to a bad punch that simply doesn't match the original take's general sound. If this isn't possible, make sure that you match the sounds by carefully documenting the mic choice, placement, preamp type, etc. You'll be glad you did.

Naturally, performing a "punch" should always be done with care. In some analog or non-DAW cases, allowing the track to continue recording after the intended punch-out point could possibly cut off a section of the following, acceptable track and likewise require that the following section be redone. Stopping it short could cut off the natural reverb trail of the final note.

It needs to be pointed out that performing a punch using a DAW is often *far* easier than doing the same on an analog recorder. For example:

- If the overdub wasn't that great, you can simply click to "undo" it and start over!
- If the overdub was started early and cut into the good take (or went on too long), the leading and/or trailing edge of the punch can often be manually adjusted to expose or hide sections after the punch has been performed (a tape editor's dream).

These beneficial luxuries can go a long way toward reducing operator error (and its associated tensions) during a session.

FIGURE 7.28
Punch-ins let you selectively replace material and correct mistakes (courtesy of Steinberg Media Technologies GmbH, a division of Yamaha Corporation, www.steinberg.net).

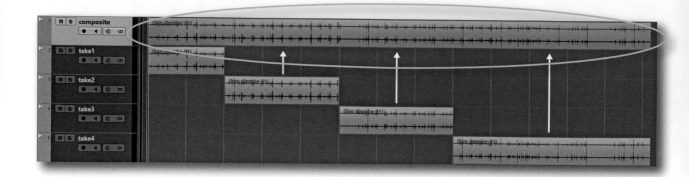

The Digital Audio Workstat

FIGURE 7.29
A single composite track can be created from several partially acceptable takes.

COMPING

When performing a musically or technically complex overdub, most DAWs will let you *comp* (short for composite) multiple overdubs together into a final master take (Figure 7.29). Using this process, a DAW can be programmed to automatically enter into and out of record at the appropriate points. When placed into record mode, the DAW will start laying down the overdub into a new and separate track (called a "lane"). At the end of the overdub, it'll loop back to the beginning and start recording the next pass onto a new and separate lane. This process of laying down consecutive takes will continue, until the best take is done or the artist simply gets tired of recording. Once done, an entire overdub might be chosen, or individual segments from the various takes can be assembled together into a final, master composite overdub. Such is the life of a digital micro-surgeon!

MIDI Sequencing and Scoring

FIGURE 7.30
MIDI edit windows with Steinberg's Cubase/Nuendo DAW. (a) Piano roll edit window. (b) Notation edit window (courtesy of Steinberg Media Technologies GmbH, a division of Yamaha Corporation, www.steinberg.net).

Most DAWs include extensive support for MIDI (Figure 7.30), allowing electronic instruments, controllers, effects devices and electronic music software to be integrated with multitrack audio and video tracks. This important feature often includes the full implementation for:

- MIDI sequencing, processing and editing
- Score editing and printing

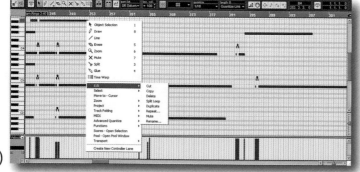

(a)

(b)

- Drum pattern and step note editing
- MIDI signal processing
- Support for linking the timing and I/O elements of an external music application (often via ReWire)
- Support for software instruments (VSTi and RTAS)

Further reading about the wonderful world of MIDI can be found in Chapter 9.

Support for Video and Picture Sync

Most high-end DAWs also include support for displaying a video track within a session, both as a video window that can be displayed on the desktop and in the form of a video thumbnail track (which often appears as a linear guide track within the edit window). Through the use of SMPTE timecode, MTC and wordclock, external video players and edit devices can be locked with the workstation's timing elements, allowing us to have full "mix to picture" capabilities (Figure 7.31).

Real-Time, On-Screen Mixing

In addition to their ability to offer extensive region edit and definition, one of the most powerful cost- and time-effective features of a digital audio workstation is the ability to offer on-screen mixing capabilities (Figure 7.32), known as mixing "in the box". Essentially, most DAWs include a digital mixer interface that offers most (if not more) of the capabilities that are offered by larger analog and/or digital consoles – without the price tag and size. In addition to the basic input strip fader, pan, solo/mute and select controls, most DAW software mixers offer broad support for EQ, effects plug-ins (offering a tremendous amount of DSP flexibility), routing, spatial positioning (pan and often surround-sound positioning), total automation (both mixer and plug-in automation), mixing and transport control from an external surface, support for exporting (bouncing)

FIGURE 7.31
Most high-end DAW systems are capable of importing a video file directly into the project session window.

(a)

(b)

0:03:35:09

FIGURE 7.32
DAW on-screen mixer. (a)
ProTools on-screen mixer
(courtesy of Avid Technology,
Inc., www.avid.com). (b)
Nuendo on-screen mixer
(courtesy of Steinberg Media
Technologies GmbH, a divi-
sion of Yamaha Corporation,
www.steinberg.net).

a mixdown to a file – the list goes on and on and on. Further reading on the mixers, consoles and the process of mixing audio can be found in Chapter 17.

DSP EFFECTS

In addition to being able to cut, copy and paste regions within a sound file, it's also possible to alter a sound file, track or segment using DSP techniques. In short, DSP works by directly altering the samples of a sound file or defined region according to a program algorithm (a set of programmed instructions) in order to achieve a desired result. These processing functions can be performed either in real time or non-real time (offline):

- *Real-time DSP*: Commonly used in most modern-day DAW systems, this process makes use of the computer's CPU or additional acceleration hardware to perform complex DSP calculations during actual playback. Because no calculations are written to disk in an offline fashion, significant savings in time and disk space can be realized when working with productions that involve complex or long processing events. In addition, the automation instructions for real-time processing are embedded within the saved session file, allowing any effect or set of parameters to be changed, undone and redone without affecting the original sound file.

- *Non-real-time DSP*: Using this method, signal processing (such as changes in level, L/R channel swapping, etc.) can be saved as a unique sound file in a non-real-time fashion. In this way, the newly calculated file (containing an effect, volume change, combined comp. tracks, sub-mix, etc.) will be played back without the need for additional, real-time CPU processing. It's good to know that DAWs will often have a specific term for tracks or processing functions that have been written to disk in order to save on processing – often called "locking" or "freezing" a file. These files can almost always be unlocked at a later time to revert to real-time DSP processing. When DSP is "printed" to a new file in non-real time, it's almost always wise to save both the original and the affected sound files, just in case you need to make changes at a later time.

- Most DAWs offer an extensive array of DSP options, ranging from options that are built into the basic I/O path of any input strip (e.g., basic EQ and gain-related functions) to DSP effects and plug-ins that come bundled with the DAW package, to third-party effects plug-ins that can be either inserted directly into the signal path (insert) or offered as a (send) that can be assigned to numerous tracks within a mix. Although the way in which effects are dealt with in a DAW will vary from one make and model to the next, the basic fundamentals will be much the same.

DSP PLUG-INS

Workstations often offer a number of stock DSP effects that come bundled with the program; however, a staggering range of third-party plug-in effects can be inserted into a signal path, which perform functions for any number of tasks, ranging from the straightforward to the wild-'n'-zany. These effects can be programmed to seamlessly integrate into a host DAW application that conforms to such plug-in platforms as:

- *DirectX*: A DSP platform for the PC that offers plug-in support for sound, music, graphics (gaming) and network applications running under Microsoft Windows (in its various OS incarnations)
- *AU (Audio Units)*: Developed by Apple for audio and MIDI technologies in OS X; allows a more advanced GUI and audio interface
- *VST*: A native plug-in format created by Steinberg for use on either a PC or a Mac; all functions of a VST effect processor or instrument are directly controllable and automatable from the host program
- *MAS (MOTU Audio System)*: A real-time native plug-in format for the Mac that was created by Mark of the Unicorn as a proprietary plug-in format for Digital Performer; MAS plug-ins are fully automatable and do not require external DSP in order to work with the host program
- *AudioSuite*: A file-based plug-in that destructively applies an effect to a defined segment or entire sound file, meaning that a new, affected version of the file is rewritten in order to conserve the processor's DSP overhead (when applying AudioSuite, it's often wise to apply effects to a copy of the original file so as to allow for future changes)
- *RTAS (Real-Time Audio Suite)*: A fully automatable plug-in format that was designed for various flavors of Digidesign's Pro Tools and runs on the power of the host CPU (host-based processing) on either the Mac or the PC
- *TDM (Time Domain Multiplex)*: A plug-in format that can only be used with Digidesign Pro Tools systems (Mac or PC) that are fitted with Digidesign Farm cards; this 24-bit, 256-channel path integrates mixing and real-time DSP into the system with zero latency and under full automation

These popular software applications (which are programmed by major manufacturers and smaller startup companies alike) have helped to shape the face

FIGURE 7.33

The UAD-2 DSP PCIe and Thunderbolt (Mac) or USB (Win) DSP processor and several plug-in examples (courtesy of Universal Audio, www.uaudio.com, © 2022 Universal Audio, Inc. All rights reserved. Used with permission).

of the DAW by allowing us to pick and choose the plug-ins that best fit our personal production needs. As a result, new companies, ideas and task-oriented products are constantly popping up on the market, literally on a monthly basis.

ACCELERATOR PROCESSING SYSTEMS

In most circumstances, the CPU of a host DAW program will have sufficient power and speed to perform all of the DSP effects and processing needs of a project. Under extreme production conditions, however, the CPU might run out of computing steam and choke during real-time playback. Under these conditions, there are a couple of ways to reduce the workload on a CPU. On the one hand, the tracks could be "frozen", meaning that the processing functions would be calculated in non-real time and then written to disk as a separate file. On the other hand, an accelerator card (Figure 7.33) that's capable of adding extra CPU power can be added to the system, giving it the necessary real-time power to perform the required effects calculations. Of course, as computers have gotten faster and more powerful, native processing packages have come onto the market, which make use of the computer's own multi-processor capabilities.

FUN WITH EFFECTS

The following effects notes describe only a few of the possible effects that can be plugged into the signal path of a DAW; however, further reading on effects processing can be found in Chapter 15 (Signal Processing).

EQUALIZATION

EQ is, of course, a feature that's often implemented at the basic level of a virtual input strip (Figure 7.34). Most DAW "strips" also include one that gives full parametric control over the entire audible range, offering overlapping control over several bands with a variable degree of bandwidth control (Q). Beyond the basic EQ options, numerous third-party EQ plug-ins are available on the market that vary in complexity, musicality and market appeal (Figure 7.35).

(a)   (b)

DYNAMICS

Dynamic range processors (Figures 7.36 and 7.37) can be used to change the signal level of a program. Processing algorithms are available that emulate a compressor (a device that reduces gain by a ratio that's proportionate to the input signal), limiter (reduces gain at a fixed ratio above a certain input threshold) or expander (increases the overall dynamic range of a program). These gain changers can be inserted directly into a channel or group master track or inserted into the final master output path.

In addition to the basic complement of stock and third-party dynamic range processors, wide assortments of multiband dynamic plug-in processors (Figures 7.38) are available for general and mastering DSP applications. These processors allow the overall frequency range to be broken down into various frequency bands. For example, a plug-in such as this could be inserted into a DAW's main output path, which allows the lows to be compressed while the mids are lightly limited and the highs are simultaneously de-essed to reduce harsh sibilance in the mix.

DELAY

Another important effects category that can be used to alter and/or augment a signal revolves around delays and regeneration of sound over time. These

FIGURE 7.34

DAWs offer a stock EQ on their channel strips. (a) 7-Band Digirack EQIII plug-in for Pro Tools (courtesy of Avid Technology, Inc., www. avid.com). (b) EQ plug-in for Cubase/Nuendo (courtesy of Steinberg Media Technologies GmbH, a division of Yamaha Corporation, www.steinberg.net).

FIGURE 7.35

EQ plug-ins. (a) FabFilter Pro-Q 24-band EQ plug-in for mixing and mastering (courtesy of FabFilter, www. fabfilter.com). (b) Lindell Audio TE-100 (courtesy of Lindell Plugins, www.lindell-plugins.com).

(a)   (b)

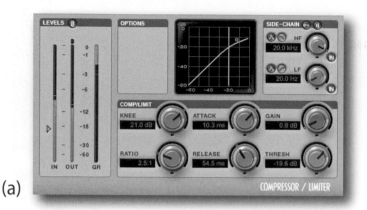

(a)

(b)

FIGURE 7.36
DAW stock compressor plug-ins. (a) Compressor/limiter plug-in for Pro Tools (courtesy of Avid Technology, Inc., www.avid.com). (b) Compressor plug-in for Cubase/Nuendo (courtesy of Steinberg Media Technologies GmbH, a division of Yamaha Corporation, www.steinberg.net).

FIGURE 7.37
Compressor plug-ins. (a) Summit Audio TLA-100A Tube Leveling Amplifier (courtesy of Avid Technology, Inc., www.avid.com). (b) API-2500 Compressor Plug-in (courtesy of Universal Audio, www.uaudio.com, © 2020 Universal Audio, Inc. All rights reserved. Used with permission).

time-based effects use delay (Figure 7.39) to add a perceived depth to a signal or change the way that we perceive the dimensional space of a recorded sound. A wide range of time-based plug-in effects exist that are all based on the use of delay (and/or regenerated delay) to achieve such results as:

- Delay
- Chorus
- Flanging
- Reverb

PITCH AND TIME CHANGE

Pitch change functions make it possible to shift the relative pitch of a defined region or track either up or down by a specific percentage ratio or musical interval. Most systems can shift the pitch of a sound file or defined region by determining a ratio between the current and the desired pitch and then adding (lower pitch) or dropping (raise pitch) samples from the existing region or sound file. In addition to raising or lowering a sound file's relative pitch, most systems can combine variable sample rate and pitch shift techniques to alter the duration of a region or track. These pitch- and time-shift combinations make it possible for such changes as:

- *Pitch shift only*: A program's pitch can be changed while recalculating the file so that its length remains the same.

(a)

(b)

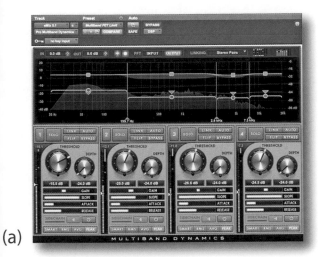

(a)

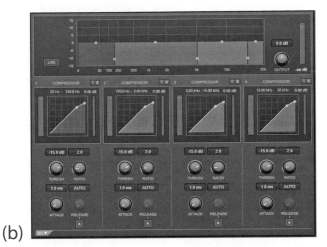

(b)

• *Change duration only*: A program's length can be changed while shifting the pitch so that it matches that of the original program.

• *Change in both pitch and duration*: A program's pitch can be changed while also having a corresponding change in length.

When combined with shifts in time (delay), changes in pitch make it possible for a world of time-based effects, alterations, tempo changes and more to be created. For example:

• Should a note be played that's out of pitch, instead of going back and doing an overdub, it's a simple matter to simply zoom in on that note and change its pitch up or down till it's right.

• Using pitch shift, it's a simple matter to perform time stretching to do any number of tasks. For example:

  • Should you be asked to produce a radio commercial that is 30 seconds long, and the producer tells you (after the fact) that it has to be 28 seconds, it's a simple matter to time stretch the entire commercial so as to trim the 2 seconds off.

  • The tempo of an entire song can be globally shifted in time or pitch, at the touch of a button, to change the entire key or tempo of a song.

FIGURE 7.38

Multiband compressor plug-ins. (a) Multiband compressor for Pro Tools (courtesy of Avid Technology, Inc., www.avid.com). (b) Multiband compressor for Cubase/Nuendo (courtesy of Steinberg Media Technologies GmbH, a division of Yamaha Corporation, www.steinberg.net).

FIGURE 7.39

Various delay plug-ins for Apollo and the UAD effects processing card (courtesy of Universal Audio, www.uaudio.com © 2022 Universal Audio, Inc. All rights reserved. Used with permission).

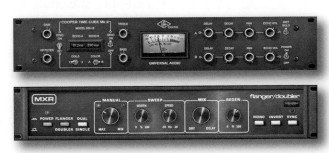

- Should you import a musical groove that's of a different tempo than your session tempo, most DAWs will let you slip the groove in time, so that its tempo fits perfectly, simply by dragging the boundaries. Changing the groove's pitch is likewise a simple matter.
- Dedicated plug-ins can also be used to automatically tune a vocal or instrumental track so that the intonation is corrected, smoothed out or exaggerated for effect.
- A process called "warping" can be used to apply micro changes in musical timing (using time and pitch shift processing) to fit, modify, shake up or otherwise mangle a section within a passage or groove. Definitely fun stuff!

If you're beginning to get the idea that there are few limitations to the wonderful world of pitch shifting – you're right. However, there are definitely limits and guidelines that should be adhered to or at least, experimented with. For starters:

- A single program will often have several algorithms that can be applied to a passage (depending on whether it's percussive, melodic or continuous in nature). Not all algorithms are created equal. Also, the algorithms of one program can easily sound totally different than those of another program. It isn't often straightforward or set in stone, as the processing is often simply too complex to predict … it often will require careful experimentation and artistry.
- Shifting in time or pitch (two sides of the same coin) by too great a value can cause audible side effects. You'll simply have to experiment.

REWIRE

ReWire and ReWire2 are special protocols that were co-developed by Reason Studios and Steinberg to allow audio to be streamed between two simultaneously running computer applications. Unlike a plug-in, where a task-specific application is inserted *into* a compatible host program, ReWire allows the audio and timing elements of an independent client program to be seamlessly integrated into another host program. In essence, ReWire provides virtual patch cords that link the two programs together within the computer. A few of ReWire's supporting features include:

- Real-time streaming of up to 64 separate audio channels (256 with ReWire2) at full bandwidth from one program into its host program application
- Sample accurate synchronization between the audio in the two programs
- The ability to allow the two programs to share a single soundcard or interface
- Linked transport controls that can be controlled from either program (provided it has some kind of transport functionality)

- An ability to allow numerous MIDI outs to be routed from the host program to the linked application (when using ReWire2)
- A reduction in the total number of system requirements that would be required if the programs were run independently

This useful protocol essentially allows a compatible program to be plugged into a host program in a tandem fashion. As an example, ReWire could allow Propellerhead's Reason (client) to be "ReWired" into Steinberg's Cubase DAW (host), allowing all MIDI functions to pass through Cubase into Reason while patching the audio outs of Reason into Cubase's virtual mixer inputs (Figure 7.40). For further information on this useful protocol, consult the supporting program manuals and Web videos.

Mixdown and Effects Automation

One of the great strengths of the "in the box" age is how easily all of the mix and effects parameters can be automated and recalled within a mix. The ability to change levels, pan and virtually control any parameter within a project makes it possible for a session to be written to disk, saved and recalled at a second's notice. In addition to grabbing a control and moving it manually (either virtually on screen or from a physical controller), another interface style for controlling automation parameters (known as *rubber band* controls) lets you view, draw and edit variables as a graphic line that details the various automation moves over time.

As with any automation moves, these rubber band settings can be undone, redone or recalled back to a specific point in the edit stage. Often (but not always), the fader volume moves within a mix can't be "undone" and reverted to any specific point in the mix. In any case, one of the best ways to save (and revert to) a particular mix version (or various alternate mix versions) is simply to save a specific mix under a unique, descriptive session file title (e.g., gamma _ultraviolet_radiomix01.ses) and then keep on working. By the way, it's always wise to save your mixes on a regular basis (many a great mix has been lost in a crash because it wasn't saved or the auto-save function didn't work properly); in

FIGURE 7.40
ReWire allows a client program to be inserted into a host program (often a DAW) so the programs can run simultaneously in tandem.

ReWire "client"

ReWire "host"

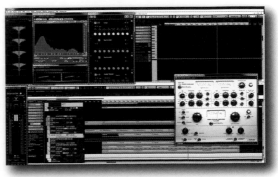

export/bounce
file to disk

FIGURE 7.41
Most DAWs can export
(bounce) session sound files,
effects and automation to a
final mixdown track.

addition, progressively saving your mixes under various name or version numbers (mix01.ses, mix02.ses, etc.) can come in handy if you need to revert to a past version. In short, save often and save regularly!

Exporting a Final Mixdown to File

Once your mix is ready, most DAW systems are able to export (bounce or print) part or all of a session to a single file or set of sound files (Figure 7.41).

An entire session or defined region can be exported as a single interleaved file (containing multiple channels that are encoded into a single L-R-L-R sound file) or can be saved as separate, individual (L.wav and R.wav) sound files. Of course, a surround or multichannel mix can be likewise exported as a single interleaved file or as separate files.

Often, the session can be exported in non-real time (a faster-than-real-time process that can include all mix, plug-in effects, automation and virtual instrument calculations) or in real time. Usually, a session can be mixed down in a number of final sound file and bit-/sample-rate formats.

POWER TO THE PROCESSOR ... UHHH, PEOPLE!

Speaking of having enough power and speed to get the job done, there are definitely some tips and tricks that can help you get the most out of your DAW. Let's take a look at some of the more important items. It's vital to keep in mind that keeping up with technology can have its triumphs and its pitfalls. No matter which platform you choose to work with, there's no substitute for reading, research and talking with your peers about your techno needs. It's generally best to strike a balance between our needs, our desires, the current state of technology and the relentless push of marketing to grab our money –and it's usually best to take a few big breaths (days, weeks, etc.) before making any important decisions.

Get a Computer That's Powerful Enough

With the increased demand for higher bit-/sample-rate resolution, more tracks, more plug-ins, more of everything, you'll obviously want to make sure that your computer is fast and powerful enough to get the job done in real time without spitting and sputtering digits. This often means getting the most up-to-date and powerful computer/processor system that your budget can reasonably handle. With the advent of 32- and 64-bit OS platforms and 8- or higher-core processors (chips that effectively contain multiple CPUs), you'll want to make sure that your hardware will support these features before taking the upgrade plunge.

The same goes for your production software and driver availability. If any part of this hardware, software and driver equation is missing, the system will not be able to make use of these advances. Therefore, one of the smartest things you can do is research the type and system requirements that would be needed to operate your production system and then make sure that your system exceeds these figures by a comfortable margin so as to make allowances for future technological advances and the additional processing requirements that are associated with them. If you have the budget to add some of the extra bells and whistles that go with living on the cutting edge, you should take the time to research whether or not your system will actually be able to deliver the extra goods when these advances actually hit the store shelves.

Make Sure You Have Enough Fast Memory

It almost goes without saying that your system will need to have an adequate amount of RAM and hard-disk storage in order for you to take full advantage of your processor's potential and your system's data storage requirements. RAM is used as a temporary storage area for data that is being processed and passed to and from the computer's CPU. Just as there's a "need for speed" within the computer's CPU, it's usually best that we install memory with the fastest possible transfer speed that can be supported by the computer. It's also important that you install as much memory as your computer and budget will allow. Installing too little RAM will force the OS to write this temporary data to and from the hard disk, a process that's much slower than transfer to RAM and causes the system's overall performance to slow to a crawl. For those who are making extensive use of virtual sampling technology (whereby samples are transferred to RAM), it's usually a wise idea to throw as much RAM into the system as possible.

Hard-disk requirements for a system are certainly an important consideration. The general considerations include:

- *Need for size*: Obviously, you'll want to have drives that are large enough to meet your production storage needs. With the use of numerous tracks within a session, often at sample rates of 24/96, data storage requirements can quickly become an important consideration.
- *Need for speed*: With the current track count and sample rate requirements that can commonly be encountered in a DAW session, it's easy to

understand how slower disk access times (the time that's required for the drive heads to move from one place to another on a disk and then output that data) becomes important. For these and other reasons, the blazingly high access speeds of SSDs have led to their common use for media production storage.

Keep Your Production Media Separate

Whenever possible, it's important that you keep your program and OS data on a separate hard or solid state drive from the one that holds your production media data. This is due to the simple fact that a computer periodically has to check in and interact with both the currently running program and the OS. Should the production media be on the same disk, interruptions in audio data can occur as the disk takes time to perform program-related tasks, resulting in a reduction in media and program data access and throughput time (not good).

Update Your Drivers ... With Caution!

In this day and age of software revisions, it's always a good idea to go on the Web and search for the latest update to a piece of software or a hardware driver. Even if you've just bought a new product out of the box, it might easily have been sitting on a music store shelf for over a year. By going to the company website and downloading the latest versions, you'll be assured that it has the latest and greatest capabilities. In addition, it's always wise to save these updates to disk in your backup directories. This way, if you're without Internet and there's a hardware or software problem, you'll be able to reload the software or drivers and should be on your way in no time.

I will say, however, sometimes it's wise to research an update (particularly if it's an important one) – it's a rare OS, software or app update that will rise up and bite you and your system in the butt, but it does happen (be especially careful of OS and program updates when traveling or when starting a new and important project).

Read Your Manuals

Unfortunately, DMH is not a big reader, but taking an occasional glance at my software manuals helps me get a better understanding of the finer points of my studio system. There are always new features, tips and tricks to be learned that you never would have thought of otherwise.

Going (at Least) Dual Monitor

Q: How do you fit the easy visual reference of multiple programs, documents and a DAW onto a single video monitor?

A: You often don't. Those of you who rely on your computer for recording and mixing, surfin', writing, etc., should definitely think about doubling your computer's visual real estate by adding an extra monitor to your computer system.

Folks who have never seen or thought much about adding a second monitor (see Figure 7.5 earlier in this chapter) might be skeptical and ask, "What's the big deal?" But, all you have to do is sit down and start opening programs onto a single screen just to see how fast your screen can get filled up. When using a complicated production program (such as a professional DAW or a high-end graphics app), getting the job done with a single monitor can be an exercise in frustration. There's just too much we need to see and not enough screen real estate to show it on.

Truth is, in this age of Mac and Windows, adding an extra monitor is a fairly straightforward proposition. Most systems can deal with two or more monitors with little or no fuss. Getting hold of a second monitor could be as simple as grabbing an unused one from the attic or buying a second monitor.

Once you've installed the hardware, the software side of building a dual-monitor system is relatively straightforward. Simply call up the resolution settings in the control panel or System Preferences and change the resolution settings and orientation for each monitor. Once you have extended your desktop across both monitors, you should be well on your way.

Those of you who use a laptop can also enjoy many of these benefits by plugging the second monitor into the video out and following the setup steps that are recommended by your computer's OS. You should be aware that many laptops are limited in the way they share video memory and might be restricted in the resolution levels that can be selected.

It's also worth noting that iOS software, such as Duet Display for the iPad, makes it possible for a Mac or Windows laptop or computer to use the iPad as an additional, fully functional monitor … a nice feature when traveling.

This might not seem much like a recording tip, but once you get a dual-monitor system going, your whole approach to producing content (of any type) on a computer will instantly change, and you'll quickly wonder how you ever got along without it!

Keeping Your Computer Quiet

Noise! Noise! Noise! It's everywhere! It's in the streets, in the car, and even in our studios. It seems like we spend all those bucks getting the best sound possible, only to gunk it all up by placing this big computer box that's full of noisy fans and whirring hard drives smack in the middle of a critical listening area. Fortunately, a number of companies have begun to find ways to reduce the problem. Here are a few solutions:

- Whenever possible, use larger, low-rpm fans to reduce noise.
- Certain PC motherboards come bundled with a fan speed utility that can monitor the CPU and case heat and adjust the fan speeds accordingly.
- Route your internal case cables carefully. They could block the flow of air, which can add to heat and noise problems.

- A growing number of hard-disk drives are available as quiet drives. Check the manufacturer's noise ratings.
- You might consider placing the computer in a well-ventilated area, just outside the production room. Always pay special attention to ventilation (both inside and outside the computer box), because heat is a killer that'll reduce the lifespan of your CPU. (*Note*: when building my own studio, I designed a special alcove/glass door enclosure that houses my main computer – no muss, no fuss and almost no noise.)
- Thanks to gamers and audio-aware buyers, a number of companies exist that specialize in quiet computer cases, fans and components. These are always fun to check out on the Web.

Backup, Archive and Networking Strategies

It's pretty much always true that it's not a matter of *if* an irreplaceable hard drive will fail, but *when*. At a time that we least expect it, disaster could strike. It's our job to be prepared for the inevitable. This type of headache can, of course, be partially or completely averted by backing up your active program and media files, as well as by archiving your previously created sessions and then making sure that these files are also backed up.

As previously stated, it's generally wise to keep your computer's OS and program data on a separate hard disk (usually the boot drive) and then store your session files on a separate media drive. Let's take this as a practical and important starting point. Beyond this premise, as most of you are quite aware, the basic rules of hard-disk management are extremely personal and will often differ from one computer user to the next (Figure 7.42). Given these differences, I'd still like to offer up some basic guidelines:

FIGURE 7.42
Data and hard-drive management (along with a good backup scheme) are extremely important facets of media production.

- It's important to keep your data (of all types) well organized, using a system that's both logical and easy to follow. For example, online updates of a program or hardware driver downloads can be placed into their own directories; data relating to your studio can be placed in the "studio" directory

SSD

- Operating System
- Programs
    - Music Programs
    - General Programs
- My Data
    - Documents
    - Graphic Files
    - MP3 Music
- Program Archives
- Driver Archives

SSD

- Current Project
  (faster access)

HDD (backup)

- Music Project #1
    - Session #1
    - Session #2
        - Song #1
        - Song #2
        - Song #3
- Music Project #2
    - Session #1
        - Song #1
        - Song #2

and subdirectories; documents, MP3s and all the trappings of day-to-day studio operations can be also placed on the disk, using a system that's easy to understand.

- Session data should likewise be logical and easy to find. Each project should reside in its own directory, and each song should likewise reside in its own subdirectory of that session's project directory.

- Remember to save various versions of a mix. If you just added the vocals to a song, go ahead and save the session under a new version name. This acts as an "undo" function that lets you go back to a specific point in a session. The same goes for mixdown versions. If someone likes a particular mix version or effect, go ahead and save the mix under a new name or version number ([my greatest song 1 ver15.ses] or [my greatest song 1 ver15 favorite effect.ses]). In fact, it's generally wise to save the various versions throughout the course of the mix. These session files are usually small and might save your butt at a later point in time. As a suggestion, you might want to create a "mix back" subdirectory in the session/song folder and move the older session files there so you don't end up being confused with 80 backup take names.

With regard to backup strategies, a number of options exist. In this day and age, hard drives are the most robust and cost-effective ways of backing up your precious data. Here are some options, although you may have better options that work for your own application and working scale:

- Primary data storage drive: drives (in addition to your main OS drive) that are within your computer can be used to store your primary (master) data files. It's often good to view a specific drive as a source where all information is held (Fort Knox) and that all data, sound files, etc. need to eventually make it to that drive.

- Backup drive or drives: external drives or portable high-capacity (2G and higher) drives can then be used to back up your program and media data. The latter portable drives are physically small and can be used with your laptop and other computers in a straightforward way.

- Off-site backup drive: it's almost always important to store a backup drive off site.

- Cloud: it's slow and cumbersome, but storing important data on another cloud network can be an effective backup scheme.

Having a relatively up-to-date backup that's stored in a bank vault safety box or at a second place (anywhere safe and secure) can literally save a crucial part of your personal life in case of theft or fire. Seriously – your hardware can be replaced, but your data can't.

COMPUTER NETWORKING

Beyond the concept of connecting external devices to a single computer, a larger concept hits at the heart of the connectivity age – *networking*. The ability to set up

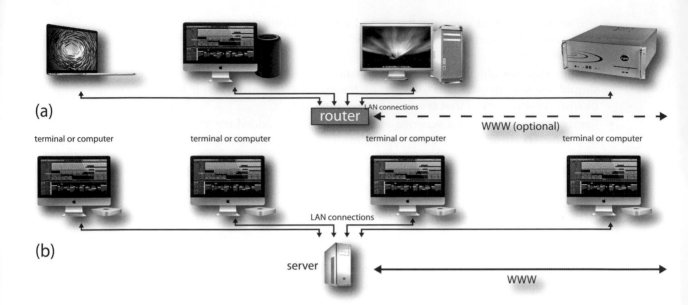

FIGURE 7.43

Local area network (LAN) connections. (a) Data can be shared between independent computers in a home or workplace LAN environment. (b) Computers or computer terminals may be connected to a centralized server, allowing data to be stored, shared and distributed from a central location and/or on the Web.

and make use of a *local area network* (*LAN*) can be extremely useful in the home, studio and/or office, in that it can be used to link multiple computers with various data, platforms and OS types. In short, a network can be set up in a number of different ways, with varying degrees of complexity and administrative levels. There are two common ways that data can be handled over a LAN (Figure 7.43):

- The first is a system whereby the data that's shared between linked computers resides on the respective computers and is communicated back and forth in a decentralized manner.
- The second makes use of a centralized computer (called a server) that uses an array of high-capacity hard drives to store all of the data that relates to the everyday production aspects of a facility. Often, such a system will have a redundant set of drives (RAID) that actually clones the entire system on a moment-to-moment basis as a safety backup procedure. In larger facilities where data integrity is highly critical, a set of backup tapes may be made on a daily basis for extra insurance and archival purposes.

No matter what level of complexity is involved, some of the more common uses for working with a network connection include:

- *Sharing files*: within a connected household, studio or business, a LAN can be used to share virtually anything (files, sound files, video images, etc.) throughout the connected facility. This means that various production rooms, studios and offices can simultaneously share and swap data and/or media files in a way that's often transparent to the users.
- *Shared Web connection*: one handy aspect of using a LAN is the ability to share an Internet connection over the network from a single, connected computer or server. The ability to connect from any computer with ease is

just another reason why you should strongly consider wiring your studio and/or house with LAN connections.

- *Archiving and backup*: in addition to the benefits of archiving and backing up data with a server system – even the simplest LAN can be a true life-saver. For example, let's say that we need to make a backup of a session. In this situation, we could simply run a backup to the main server that's connected to the system and continue working away on our DAW, without interruption – or the backups could automatically run in the background after work hours.
- *Accessing sound files and sample libraries*: it goes without saying that sound and sample files can be easily accessed from any connected computer. Actually, if you're wireless, you could go out to the pool, download or directly access the needed session files and soak up the sun while working on your latest project!

On a final note, those who are unfamiliar with networking are urged to learn about this powerful and easy-to-use data distribution and backup system for your pro or project studio. For a minimal investment in cables, hubs and educational reading, you might be surprised at the time-, trouble- and life-saving benefits that will be almost instantly realized.

Tips 'n' Tricks

In summary, the phrase "nothing lasts forever" is especially true in the digital domain of lost "1"s and "0"s, damaged media and dead hard drives … you know, the "Oh $@#%!" factor. It's a basic fact that you never quite know what lies around the techno bend, and it's extremely important that you protect yourself as much as is humanly possible against the inevitable. Of course, the answer to this digital dilemma is to back up your data in the most reliable (and redundant) way possible. Hardware and program software can (usually) be replaced; on the other hand, whenever un-backed, valuable session sound files are lost – they're lost!

Backing up a session can be done in several ways. Here are a few tips that can help you avoid data loss:

- As you might expect, the most straightforward backup system is to copy the session data, in its entirety, to the most appropriate media (most commonly onto one or more on- or off-site hard drives).
- In the case of a specific processing effect, you might want to save two copies of the track – one that contains the original, unaltered sound and

one that contains the affected signal (or simply the effect alone).

- Those who want additional protection against the degradation of unproven digital media may also want to back each track (or group of tracks) to the individual tracks of a multitrack analog recorder (often, major labels will stipulate that this be done in the band's contract for an important project).
- For those sessions that contain MIDI tracks, you should always keep these tracks within the session (i.e., don't delete them). These tracks might come in very handy during a remix or future mixdown.
- Whenever possible, make multiple backups and store them in separate locations. Having a backup copy in your home (or bank vault) as well as in the studio can save your proverbial butt in case of a fire or any other unforeseen situation. Remember the general backup rule of thumb: data is never truly backed up unless it's saved on three drives (and we would add – and in two physical places)!

Session Documentation

Most of us don't like to deal with housekeeping. But when it comes to recording and producing a project, documenting the creative process can save your butt after the session dust has settled – and help make your post-production life much easier (you never know when something will be reissued/remixed). So, let's discuss how to document the details that crop up before, during and after the session. After all, the project you save might be your own!

DOCUMENTING WITHIN THE DAW

One of the simplest ways to document and improve a session's workflow is to name a track before you press the record button, because most DAWs will use that as a basis for the file name. For example, by naming a track "Jenny's lead voc take 5", most DAWs will automatically save and place the newly recorded file into the session as "Jenny's lead voc take 5.wav" (or .aif). Locating this track later would be a lot easier than rummaging through sound files only to find that the one that you want is "Audio018-05".

Also, make use of your DAW's notepad (Figure 7.44). Most programs offer a scratchpad function that lets you fill in information relating to a track or the overall project; use this to name a specific synth patch, note the mic used on a vocal and include other information that might come in handy after the session's specifics have been long forgotten.

FIGURE 7.44
Cubase/Nuendo Notepad apps (courtesy of Steinberg Media Technologies GmbH, a division of Yamaha Corporation, www.steinberg .net).

Markers and marker tracks can also come in super-handy. These tracks can alert us to mix, tempo and other kinds of changes that might be useful in the production process. I'll often place the lyrics into a marker track so I can sing the track myself without the need for a lead sheet, or to help indicate phrasings to another singer.

MAKE DOCUMENTATION DIRECTORIES

The next step toward keeping better track of details is to create a "Song Name Doc" directory within the song's session and fill that folder with documents and files that relate to the session, such as:

- Your contact info
- Song title and basic production notes (composer, lyricist, label, business and legal contacts)
- Producer, engineer, assistant, mastering engineer, duplication facility, etc. (with contact info)
- Original and altered tempos, tempo changes, song key, timecode settings, etc.
- Original lyrics, along with any changes (changed by whom, etc.)
- Additional production notes
- Artist and supporting cast notes (including their roles, musician costs, address info, etc.)
- Lists of any software versions and plug-in types as well as any pertinent settings (you never know if they'll be available at a future time, and a description and screenshot might help you to duplicate it within another app)
- Lists of budget notes and production dates (billing hours, studio rates and studio addresses – anything that can help you write off the $$$)
- Scans of copyright forms, session contracts, studio contracts and billings
- Anything else that's even remotely important

In addition, I'll often take screenshots of some of my more complicated plug-in settings and place these into this folder. If I have to redo the track later for some reason, I refer to the screenshot so I can start reconstruction. Photos or movie clips can also be helpful in documenting which type of mic, instrument and specific placements were used within a setup. You can even use pictures to document outboard hardware settings and patch arrangements. Composers can use the "Doc" folder to hold original scratchpad recordings that were captured on your cell phone or message machine.

Furthermore, a "Song Name Graphics" directory can hold the elements, pictures and layouts that relate to the project's artwork … "Song Name Business" and "Project Name artwork" directory, etc. might also come in handy.

Accessories and Accessorizing

I know it seems like an afterthought, but there's an ever-growing list of hardware and travel accessories that can help you to take your portable rig on the road. Just a small listing includes:

- Laptop backpacks for storing your computer and gear in a safe, fun case
- Pad stands and cases

- Instrument cases and covers
- Flexible LED gooseneck lights that let you view your keyboard in the dark or on stage
- Laptop DJ stands for raising your laptop above an equipment-packed table

Protect Your Investment

When you've spent several years amassing your studio through hard-earned sweat-equity and bucks, it's only natural that you'll want to take the necessary precautions to protect your investment.

Obviously, the best way to protect your data is through a rigorous and straightforward backup scheme (the general rule is that something isn't backed up unless it's saved in three places – preferably with one of the backups being stored off site). However, you'll also want to take extra steps to protect your hardware and software investments as well by making sure that they're properly insured.

The best way to start the process of properly insuring your studio is to contact your trusted insurance agent or broker. If you don't have one, now's the time to get one. You might get some referrals from friends or people in your area and give them a call, set up some appointments and get several quotes. If you haven't already done so, sit down and begin listing your equipment, its serial numbers and replacement values. Next, you might consider taking pictures or a home movie of your listed studio assets. These steps will help your agent come up with an adequate replacement plan and will come in handy when filing a claim, should an unfortunate event occur. Being prepared isn't just for the Boy or Girl Scouts.

Protect Your Hardware

One of the best ways to protect against harmful line voltage fluctuations (both above and below their nominal power levels) is to use a quality power conditioner or an adequately powered uninterruptible power supply (UPS). In short, a quality power conditioner works by providing a regulated power voltage level that works within a specified tolerance that will protect your sensitive studio equipment (such as a computer, bank of effects devices, etc.) with a clean and constant voltage supply. It will also protect against power spikes in the mains line that could damage your system. Further info on this subject can be found in Chapter 14.

Protect Your Body

Producers, musicians, audio professionals and engineers spend a great deal of time in the control room and studio. It only makes sense that this environment should be laid out in a manner that's aesthetically, functionally and acoustically pleasing from a comfort, feng shui and cleanliness point of view. Creating a good working environment that's conducive to making good music is the goal of every professional and project studio owner. Beyond the basics of creating a well-designed facility from an acoustic and electronic standpoint, a number of basic concepts should be kept in mind when building or designing a recording facility, no matter how grand or humble. Here are a few helpful hints:

- Given the fact that a producer or engineer spends a huge amount of time sitting on his or her bum, it's always wise to invest in both your and your clients' posture and creature comforts by having comfortable, high-quality chairs around for both the production team and the musicians (Figure 7.45a). Of course, a functional workspace desk (Figure 7.45b) is always a big help to those of us who spend huge amounts of time sitting in front of the DAW. Posture, easy access and functionality, baby – it can't be stressed enough! Even something as simple as having an exercise ball in the room can really help when you've been sitting in a regular chair for too long.

- Velcro™ or tie-straps can be used to organize studio wiring bundles into groups that can be laid out in ways that reduce clutter, improve organization (by using color-coded straps) and make the studio look more professional.

- Most of us are guilty of cluttering up our workspace with unused gear, papers – you know, junk! I know it's hard, but a clean, uncluttered working environment tells your clients a lot about you, your facility and your work habits.

- Unused cables, adapters and miscellaneous stuff can be sorted into boxes and stacked for easy storage.

- Important tools and items that are used every day (such as screwdrivers, masking tape or markers) could be stored in a rack-mounted drawer that can be easily accessed without cluttering up your space – don't forget to pack a reliable LED flashlight (your phone's flash or screen display will also work in a pinch).

- Portable label printers can be used to identify cable runs within the studio, identify patch points, I/O strip instrumentation … you name it.

FIGURE 7.45
Functional and comfortable furniture is a must in the studio. (a) The venerable Herman Miller Aeron® chair (courtesy of Herman Miller, Inc., www.hermanmiller.com). (b) The Argosy Halo desk (courtesy of Argosy Console, www.argosyconsole.com).

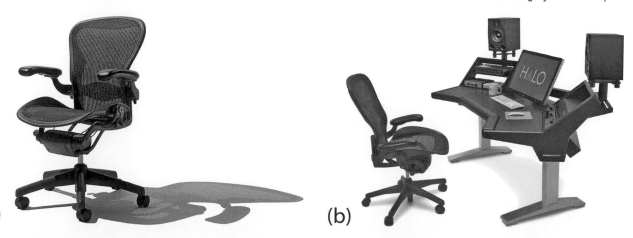

(a)   (b)

CHAPTER 8

Groove Tools and Techniques

The expression "getting into the groove" of a piece of music often refers to a feeling that's derived from the underlying foundation of music: rhythm. With the introduction and maturation of MIDI and digital audio, new and wondrous tools have made their way into the mainstream of music production. These tools (Figure 8.1) can help us to use technology to forge, fold, mutilate and create compositions that make direct use of rhythm and other building blocks of music through the use of looping technology. Of course, the cyclic nature of loops can be repeat-repeat-repetitive in nature, but new toys and technology and compositional techniques for looping can inject added flexibility, control and real-time processing into a project in wondrously expressive ways.

In this chapter, we'll be touching on many of the approaches and software packages that have evolved (and continue to evolve) into what is one of the fastest and most accessible facets of personal music production. It's literally impossible to hit on all of the finer operational points of these sample- and synth-based systems; for that, I'll rely on your motivation and ingenuity to:

- Download many of the software demos and apps that are readily available
- Delve into their manuals and working tutorials for both personal computers and iDevices

FIGURE 8.1
Probably the most widely used groove tool on the planet – GarageBand™ for the Mac and iDevices.

DOI: 10.4324/9781003260530-8

- Begin to create your own grooves and songs that can then be integrated into your music or that of collaborators

If you do these three things, you'll be shocked and astounded as to how much you'll quickly learn, and these experiences will directly translate into skills that'll widen your production horizons and possibly change your music.

THE BASICS

The basic idea behind groove-based tools rests with tempo matching, the idea that various rhythms, grooves, pads and any other imaginable sounds of various tempos, lengths and often musical keys can be artfully manipulated and crafted together into a single, working song.

Because groove-based tools often deal with rhythms and cyclic-based measures that are pulled from various musical sources, the technical factors that need to be generally managed are:

- Sync
- Tempo and length
- Time- and pitch-change techniques

The aspect of *sync* relates to the fact that the various loops in a groove project will need to sync up with each other (or with multiple lengths and timings of each other). It almost goes without saying that multiple loops that are successively or simultaneously triggered must have a synchronous timing relationship to one another – otherwise, it'll all end up being a jumbled mess of chaotic sound.

The next relationship relates to the aspect of tempo. Just as sync is imperative, it's also necessary for the files to be adjusted in length (time stretching) so that they precisely match the currently selected project tempo (or a relative multiple of its tempo).

A final aspect in groove production is associated with time- and pitch-change techniques. This is the process of altering a sound file (often one that's rhythmically repetitive and short in length) to match the current session tempo and then to synchronously align them within the software by using variable sample- and pitch-shifting techniques. Using these basic digital signal processing (DSP) tools, it's possible to alter a sound file's duration (varying the length of a program by raising or lowering its playback sample rate) and/or to alter its relative pitch (either up or down). In this way, loops can be matched up or musically combined by using any of the following time- and pitch-change combinations:

- *Time change*: A program's length can be altered without affecting its pitch.
- *Pitch change*: A program's length can remain the same while pitch is shifted either up or down.
- *Both*: Both a program's pitch and length can be altered using resampling techniques.

By setting the loop program to a master tempo (or a tempo at that point in the song), an audio segment or file can be imported, examined as to its sample rate and/or length and then recalculated to a new relative tempo that matches the current session tempo. Voilà! We now have a defined segment of audio that matches the tempo of all of the other segments within the project, allowing it to play and interact in relative sync with the other defined segments and/or loop files. (Note: more in-depth reading on pitch and time changing can be found in Chapter 15.)

Along the same lines, these pitch-changing techniques can be applied to change the relative pitch of the loop so that it matches (or best fits into) the musical key of the session. This final part of the musical jigsaw puzzle allows us to mix and match sounds or loops of various musical keys in ways that otherwise would never fit together, allowing new and interesting combinations of sounds, textures and rhythms to be created.

Pitch-Shift Algorithms

It's important to note that time-shifting processes are created according to a specific program algorithm. This means that the sound is shaped according to a set of basic mathematical/programming calculations and parameters. More often than not, the music program will let you choose between a basic set of algorithms according to the type of sound that's being processed (Figure 8.2). For example, a basic drum loop might sound really good when a "beats" detection algorithm is chosen, while a long, slow pad might sound totally unnatural with the same settings. In short, it's always a good idea to be aware of the various time-shift options that are available to you and to take the time to make processing choices that best match the music or selected segment at hand. It's also important to note that these algorithms will vary from one program to another. For example, the "pad" (a long, continuous series of chords) setting on one looping program might sound entirely different from a similar set of processes on another DAW or program – just something to be aware of (and possibly take advantage of).

FIGURE 8.2
Different DAWs will offer various pitch-shift algorithms that have their own sonic characteristics. (a) Ableton Live (courtesy of Ableton AG, www.ableton. com). (b) Cubase/Nuendo (courtesy of Steinberg Media Technologies GmbH, a division of Yamaha Corporation, www.steinberg.net).

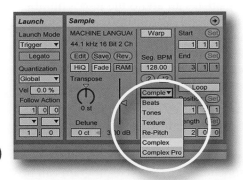
(a)

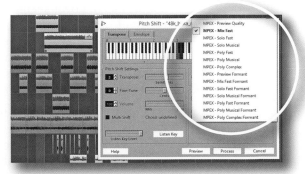
(b)

FIGURE 8.3

Hitpoint markers can be used to show and manipulate events and percussive transients in a sound file (courtesy of Ableton AG, www.ableton.com).

WARPING

In addition to basic pitch and time shifting, most looping tools make use of a tempo- and sound-processing technique called *warping*. This process uses various time-shift tools to match the timing elements of a sound file by detecting hitpoint markers (Figure 8.3) and entering them into the sound file. These markers are most often automatically detected at percussive transient points in a way that makes it easier for the time- and pitch-shifting process to best match the specific file to the session tempo (most often by adjusting the transient timings so they fall directly on the beat and its subdivisions).

In addition to helping the loops better match the session tempo, warp hitpoint markers can also be manually moved and manipulated to change the basic "feel" of a loop. For example, various hitpoints can be moved ahead or back in time to give the loop an entirely new feel or swing. In short, take the time to experiment and learn about how you can shape and create new sounds by using warp and hitpoint technology, as there are countless online videos on this subject that are both insightful and fun.

Beat Slicing

In addition to warp's use of pitch- and time-stretch techniques to alter the tempo and overall feel of a loop, another method, called *beat slicing*, makes use of an entirely different process to match the length and timings of a sound file segment to the session tempo. Rather than changing the speed and pitch of a sound file, the beat-slicing process actually breaks an audio file into a number of small segments by detecting the transient events within the loop and then automatically placing the new slices at their appropriate (and easily adjustable) tempo points according to automatic or user-definable sensitivity and detection controls. The process then simply changes the length and timing elements of a segment by adding or subtracting time between these slices.

The most universally used format for beat slicing is the REX file (Figure 8.4), which was created by Propellerhead. These files can be found in countless sample libraries as pre-formatted beat slice loops that can simply be loaded into most of the currently available DAWs. If you would like to edit and create your own loops from your own sound files, Propellerhead's ReCycle program can be used.

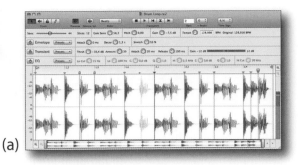

(a)

(b)

In short, REX files don't change the pitch, length or sound quality of the loop in order to match the current DAW's session tempo. That's to say, they are not altered in any way. Instead, these transient-detected hitpoint slices are physically moved to match the slices to the current session tempo. When the tempo is increased, the slices are pulled together in time and will simply overlap. When the tempo is slowed down, the slices are spread out in time. Slowing the tempo can easily cause audible gaps to fall between the cracks. These gaps are often masked by the other tracks within the session; however, the effect can be reduced by using a "stretch" algorithm that doesn't actually time stretch the loop file but instead, inserts a small sustain section from within the loop, reverses it and adds it to the end of the file, essentially smoothing out the gaps with quite acceptable results. From all of this, it's easy to understand why the beat-slicing process often works best on percussive sounds and not sustaining tracks.

FIGURE 8.4
Rex file beat slicing format.
(a) ReCycle Groove Editing Software (courtesy of Propellerhead Software, www.propellerheads.se).
(b) Rex file integration into Cubase/Nuendo (courtesy of Steinberg Media Technologies GmbH, a division of Yamaha Corporation, www.steinberg.net).

Audio to MIDI

Another ingenious way to allow your loops to be changed to match the session tempo or key or to be altered in almost any other way is to convert the audio loop into a musical instrument digital interface (MIDI) segment. This can be done using an *audio-to-MIDI* tool that can be found in many digital audio workstations (DAWs). Just as not all time-stretch algorithms are created equal, and they can often sound quite different from one program to the next, audio-to-MIDI algorithms can yield dramatically different results. Likewise, each software package will often offer different algorithms for different types of audio material (i.e., for converting harmony, melody, drum or sustained pad passages, as seen in Figure 8.5).

Once you get into the process of converting audio to MIDI, you'll quickly begin to notice that it can be a very hit-or-miss process. You might get totally lucky on your first try, and the musical line will translate perfectly into a MIDI file that can be altered in pitch, time, sound (or almost any other imaginable parameter), or you might find that it will need to be edited (notes corrected, added, removed, quantized, etc.) in order for it to be acceptable. With patience, however, the ability to alter a wimpy bass line into a powerful growler that perfectly fits your needs is an awesome and powerful tool that lets you continuously reinvent a loop.

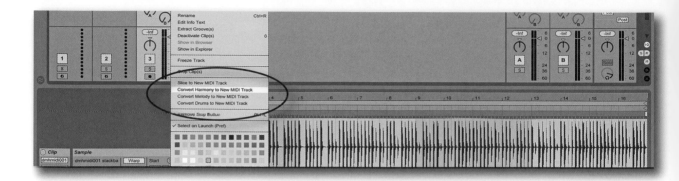

FIGURE 8.5
Certain DAWs are able to convert an audio track to a MIDI track (courtesy of Ableton AG, www.ableton.com).

As with so many other things we've discussed, the sky (and your imagination) is the limit when it comes to the tricks and techniques that can be used to match the tempos and various time-/pitch-shift elements between loops, MIDI and sound file segments. I strongly urge you to take the time to read the manuals and watch videos about the various loop and DAW software packages so as to learn more about the actual terms and procedures and then put them into practice. If you take the time, I guarantee that your production skills and your outlook on these tools will greatly expand.

For the remainder of this chapter, we'll be looking at several of the more popular groove tools and toys. This is by no means a complete listing, and I recommend that you keep reading the various trade magazines, websites and other resources for new and exciting toys and technologies that regularly come onto the market.

GROOVE HARDWARE

FIGURE 8.6
Groove hardware synths. (a) Older Roland MC-303 Groovebox. (b) Roland TR-8 Rhythm Performer (courtesy of Roland Corporation, www.roland.com).

Since the "groove" is so important to the foundation of modern dance and music production in general, groove-related hardware (Figure 8.6) has been around for a long time and continues to be used by musicians and producers alike. These instruments (and they truly are instruments) range in form and function from being strictly drum and percussion in nature (like the vintage Roland TR-808), to newer groove keyboards that are both rhythmic and melodic in nature, to the European models that dive deep into the dance culture (my

(a)

(b)

personal favorites). These devices not only offer up a wide range of press-n-play sounds and sequences but can be user-programmed to fit the tempo and pitch requirements of a special composition.

One of the best ways to get into the timing ballpark between your DAW, external groove or other MIDI hardware devices is through the use of MIDI timing messages (MIDI clock). In short, you can instruct your DAW to output MIDI timing clock data while slaving your external groove or MIDI hardware to this clock (look in your manual under "external MIDI clock"). In this way, when the DAW begins playback, MIDI timing messages will instruct the hardware to begin playing at the proper time and tempo – definitely an important feature.

For those who don't want to deal with an external MIDI hardware clock, there is another option. With the advent of powerful time- and pitch-shift processing within most DAWs, the sounds from these hardware devices can be pulled into a session without too much trouble. For example, a single groove loop (or multiple loops) could be recorded into a DAW (at a bpm that's fairly close to the session's tempo), edited and then imported into the session, at which time the loop could be easily stretched into the proper time sync, allowing it to be looped in exact sync with the song. Just remember, necessity is the mother of invention – patience and creativity are probably your most important tools in the looping process.

GROOVE SOFTWARE

Software is often the place where grooves come to life. This can take many forms, including looping capabilities that are (to varying degrees) available within most DAWs, plug-in instruments and various iOS applications. The possibilities are practically limitless, allowing combinations and manipulation of beats, patterns, sequences and pads in ways that can be addicting, fun and will help get the groove moving in your latest music track.

Looping Your DAW

Most digital audio workstations offer various features that make it possible to incorporate self-produced and/or imported loop grooves into a session. This can be as simple as making an edit cut on the beat at the beginning and end of a 4-, 8- or 16-bar segment and then repeating the newly created loop multiple times. Alternatively, a sample library loop can also be imported into a session, where it can be repeated and manipulated in order to fit the song. If the imported loop is at a different length and tempo than the song's session, most DAWs are able to use time-/pitch-shifting techniques to shift the loop length so that it matches the session's current tempo. Said another way, if a session has been set to a tempo of 115 bpm, and if a 100-bpm loop is imported at a specific measure, the segment can be time stretched until it fits snugly into the session's native tempo (Figure 8.7) by simply calling up the DAW's snap-to-grid

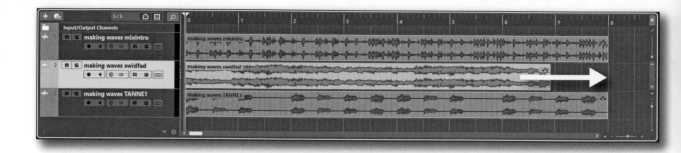

FIGURE 8.7

Most production workstations allow a loop or sound file to be resized to fit into the session's tempo. In this case, the middle loop track can be stretched to match the session's overall 8-bar, 120-bpm tempo.

and automatic time-stretch functions. Once done, the loop can be manually looped and manipulated to your heart's content.

As was mentioned, different DAWs and editing systems will have differing ways of tackling a time-stretch situation, with varying degrees of ease and accuracy. It's often a good idea to listen to the various stretch algorithms (beat, pad, vocal, etc.) and manually audition each option so as to get the best possible sound. For this very reason, DMH will make use of several DAW types in order to achieve the best results. Sometimes, when the main DAW falls short, a second DAW (such as Ableton Live) will actually end up achieving better results. Just remember, there are usually no hard and fast rules. With planning, ingenuity and your manual's help, you'll be surprised at the number of ways that a looping problem can be turned into an interesting sonic and learning opportunity. When a studio DAW is used in conjunction with groove loop toys, MIDI hardware and various instrument plug-ins, it's interesting to see and hear how each of these tools can work together to play an important role in the never-ending creative process of producing modern music.

Loop-based Audio Software

Loop-based audio systems are groove-driven DAWs, music programs or plug-ins that are designed to let you drag and drop pre-recorded or user-created loops and audio tracks into an easy-to-use, graphic production interface. At their basic level, these programs differ conceptually from their traditional DAW counterpart in that the pitch- and time-shift architecture is so variable and dynamic that even after the rhythmic, percussive and melodic grooves have been inserted into the project, their tempo, track patterns, pitch, session key, etc. can be quickly and easily changed at any time. With the help of custom, royalty-free loops (available from many manufacturer and third-party companies), users can quickly and easily experiment with setting up grooves, building backing tracks and creating a sonic ambience by simply dragging the loops into the program's main sound file view, where they can then be arranged, edited, processed, saved and exported.

One of the most interesting aspects of the loop-based editor is its ability to match the tempo of almost any programmed loop to the tempo of the current session. Amazingly enough, this process isn't that difficult to perform, because the program extracts the length, native tempo and pitch information from the imported files and (using the previously mentioned digital time- and pitch-change techniques) adjusts the loop to fit the native time and pitch parameters of the current session. This means that loops of various tempos and musical keys can be automatically adjusted to fit in time with previously existing session loops – just drag, drop and go!

Of course, the graphic user interfaces (GUIs) between looping software editors and tools can differ greatly.

Some layouts use a track-based system that lets you enter or drag a pre-programmed loop file into a track and then drag it to the right in a way that repeats the loop. Other loop programs make use of visual objects that can be triggered and combined into a real-time mix by clicking on an icon or track-based grid … then again, some make use of both visual feedback styles. Again, it's worth stressing that DAW editors will often include such looping functions that can be either basic in nature (requiring manual editing and/or sound file processing) or quite advanced (having any number of automated loop functions).

By far one of the most popular groove/looping programs is Live from Ableton (Figure 8.8). Live is an interactive loop-based program that's capable of recalculating the time, pitch and tempo structure of a sound file or easily defined segment, and then entering that loop into the session at a defined global tempo. This means that a segment of any length or tempo that's been pulled into the session grid will be recalculated to the master tempo and can be combined, mixed and processed in perfect sync with all other loops in the project session.

In Live's Arrangement View, as in all traditional sequencing programs, everything happens along a fixed song timeline, allowing the loops or tracks to be manipulated in a standard timeline. The program's Session View, however, breaks this traditional paradigm by allowing media files to be mapped onto a grid as buttons (called *clips*). Any clip can be played at any time and in any order in a random fashion that lends itself to interactive performance both in the studio and on stage. Functionally, each vertical column, or "track", can play

FIGURE 8.8
Ableton Live performance audio workstation: (a) Arrangement View. (b) Session View (courtesy of Ableton AG, www.ableton.com).

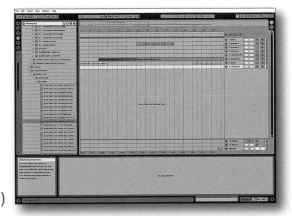

(a)   (b)

only one clip at a time. Any number of "clips", which are laid out in horizontal rows, can be played in a loop fashion by clicking on their launch button. By clicking on a "scene" launch at the screen's far right, every clip in that row will simultaneously begin playback. Of course, Live is fully capable of incorporating MIDI into a project or live interactive performance, along with a wide range of editing tools, effects and software instruments (both within the program and with external plug-in instruments and effects).

TRY THIS: HAVING FUN WITH A LOOP-BASED EDITOR

- Load Ableton Live, or go to www.ableton.com and download their 90-day trial version.
- Download a free set of loops from www .producerplanet.com (under "free loops and samplepacks") or simply use the included loops within the Abelton demo.
- Load the individual loops.
- Read the program's manual and/or begin to experiment with the loops.

- Mess around with the tempo and musical keys.
- Copy and duplicate the various loop tracks to your heart's content until you've made a really fun song!
- Import some of your own samples and incorporate them into the song.
- Save and export the session to play for your friends!

Another popular groove and production-related tool is Reason from the folks at Reason Studios. Reason differs from most sound file-based looping programs in that it's an overall music production environment, which includes a MIDI sequencer as well as a wide range of built-in software instrument modules (Figure 8.9) that can be played, mixed and integrated into a comprehensive music production environment that can be controlled from any external keyboard or MIDI controller. Reason also includes a large number of signal processors that can be applied to any instrument or instrument group under full automation control.

In essence, Reason is a combination of modeled representations of vintage analog synthesis gear mixed with the latest in digital synthesis and sampling technology (Figure 8.10). Combine these with a modular approach to signal and effects processing, add a generous amount of internal and remote mix and controller management (via external MIDI controllers), top this off with a quirky yet powerful sequencer, and you have an integrated software package that is powerful and convenient.

Once you've finished the outline of a track, the obvious idea is to create new instrument tracks (which can be combined in a traditional multitrack

(a)

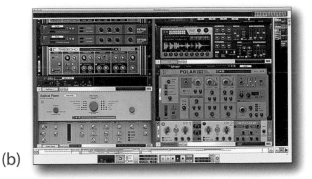

(b)

building-block approach) until a song begins to form. Additional sounds, loops and patches are widely available for sale or for free as "refills" that can be added to your collection to greatly expand the software's palette.

Feel free to download the demo and take it for a spin. You'll also want to check out the Reason Basics video clips. Due to its all-in-one production nature, this program might take a while to master, but the journey just might open you up to a whole new world of production possibilities.

FIGURE 8.9

Reason Music Production Software showing both the stand-alone program and Reason being "rewired" into a DAW: (a) Mixer/sequencer screen. (b) Instrument/effects screen (courtesy of Reason Studios, www.reasonstudios. com).

REWIRE

In Chapter 7, we learned that ReWire is a special protocol that was developed by Reason Studios and Steinberg, which allows audio to be streamed between two simultaneously running computer applications. Unlike a plug-in, where a task-specific application is inserted into a compatible host program, ReWire allows the audio and timing elements of a supporting program to be seamlessly integrated into another host program that also supports ReWire. In short, they are two independently running programs that have linked audio and timing elements. For example, DAWs such as Cubase/Nuendo and Pro Tools support ReWire, allowing production environment programs such as Ableton Live and Reason to be virtually "wired" into the inner workings of a DAW session. This allows audio to be routed through the DAW's virtual mixer or I/O while making it possible for the timing elements of the client application to be synchronously controlled from the host DAW. This capability allows greatly expanded instrument and production options within a studio or on-stage environment.

FIGURE 8.10

Examples of Reason's software instruments: (a) SubTractor polyphonic synth module. (b) NN-XT sampler module (courtesy of Reason Studios, www.reasonstudios .com).

(a)

(b)

(a)

(b)

FIGURE 8.11

Groove-based plug-ins:
(a) Stylus RMX real-time
groove module (courtesy of
Spectrasonics, www.spec-
trasonics.net). (b) Steinberg's
LoopMash2 Groove plug-in
(courtesy of Steinberg Media
Technologies GmbH, a division of Yamaha Corporation,
www.steinberg.net).

If you feel up to the task, download a few program demos, consult the involved program manuals, and try it out for yourself. The most important rule to remember when using ReWire is that the host program must always be opened first, and then you can open the client (the secondary program that's being ReWired into the DAW). When shutting down, the client program should always be closed first, and the primary DAW must be closed last..

GROOVE AND LOOP-BASED PLUG-INS

It's a sure bet that for every hardware looping tool, there are far more software plug-in groove tools and synths (Figure 8.11) that can be inserted into your DAW. These amazing software wonders often make life easier by:

- Automatically following the session tempo or tempo map
- Allowing I/O routing to plug directly into the DAW (without the need for ReWiring)
- Making use of the DAW's automation and external controller capabilities
- Allowing individual or combined groove loops to be easily imported as audio into a session

These software instruments come with a wide range of sounds (and/or added "sound packs") that can often be edited and effected using an on-screen user interface or be directly controlled from an external MIDI control editor.

Drum and Drum Loop Plug-Ins

Virtual software drum machines are also part of the present-day production landscape and can be used in a stand-alone, plugged-in or rewired production environment. These plug-ins (Figure 8.12) are capable of injecting a wide range of groove and sonic spice options into a digital audio and/or MIDI project at the session's current tempo. This allows accompaniment patterns that can range from being simple and non-varying over time to individually complex instrument parts that can be meticulously programmed into a session or performed on the fly. In addition, most of these tools include multiple signal paths that let

(a)

(b)

FIGURE 8.12
Virtual software drum machines: (a) Battery 3 virtual drum and loop module (courtesy of Native Instruments GmbH, www.native-instruments.com). (b) BFD2 acoustic drum library module (courtesy of FXpansion, www.fxpansion.com).

you route individual or grouped voices to a specific mixer or DAW input/track. This makes it possible for isolated voices to be individually mixed, panned or processed (using equalization, effects, etc.) in a session in new and interesting ways. From a practical standpoint, these "drummers" have gotten so good that it's sometimes difficult to tell that it's not a real drummer, and when working at slower or more complex tempos, finding a real, first-class drummer can be a real hit to the pocketbook.

PULLING LOOPS INTO A DAW SESSION

When dealing with loops in modern-day production, a concept that needs to be discussed is the various ways that grooves and loops can be managed within a DAW session – such as:

- It's certainly possible to ReWire a supported client program in conjunction with the host program. This, however, will often require special setup, making sure that the programs are opened in the right order and that each is opened with the right session files. You also should be aware that these applications might use up valuable computer program and memory resources.
- You might have a groove/loop plug-in inserted into a DAW session (a far more automated process, which will always open directly within the session).
- One of the better ways for freeing up resources, and for creating actual sound files that can be processed and backed up, is to export the groove/loop as a sound file track. This can be done in several ways (although with forethought, you might come up with a new way that works best for you):
  1. The instrument track can be soloed and exported to an audio track in a contiguous fashion from the beginning to the end of the session, or simply as defined regions during the instrument's performance.
  2. In the case of a repetitive loop, a defined segment (often of a precise length of 4, 8 or more bars that occurs on the metric boundaries) can be selected for export as a sound file loop. Once exported, it can be

dropped into the session and then looped, processed and easily manipulated in any number of straightforward ways.

3. In the case of an instrument or groove tool that has multiple parts or voices, each part can be soloed and exported to its own looping track. This will give you a far greater degree of control over variations, effects and any number of DAW parameters than otherwise would be possible during production and mixdown. In this way, you can actually "perform" the loops in a way that's not repetitive and adds life to the performance.

4. If you export a plug-in, loop tool, etc., it's always a good idea to save it within the session (and probably disable it, so that it doesn't take up resources) or to take a screenshot of the settings and place it in a "docs" or "pics" folder within the session (should you wish to manually recall it at a later time).

5. It also goes without saying that you'll want to properly document your export, just as you would with any recorded track. You might include the instrument type/name that was used directly within the sound file name or within the track's note area … along with any other identifying information. The confusion saved (not to mention lost time and frustration that you might save if a change is needed or a problem arises at a later time) will almost certainly be your own.

6. Obviously, the best approach to exporting loops and creating a loop library is to devise a system that makes sense to you and then stick with that personal organizational system.

IOS GROOVE APPS

As we'll see in Chapter 10, there are a wide range of free and low-cost apps for the iOS that range from being inventive to downright indispensable. By integrating an iPad or iOS device into your system, you could perform and capture a track in real time or under total MIDI control. Here, the sky is truly your only limit.

GROOVE CONTROLLERS

An increasing number of groove-based software/controller systems (Figure 8.13) are appearing on the market that control or work as stand-alone and/or plug-in based groove software systems. These controllers offer both hardware control over the various software parameters and direct performance playing surfaces (often mimicking basic drum machine pad control surfaces). Although some of these controllers are dedicated to playing and manipulating drum kits and percussion sounds, others allow any type of performance or loop-based sample data to be entered and edited in the system within an intuitive, hands-on performance environment.

(a)

(b)

In addition to using the hardware surface to directly control the system's groove software, many of these devices can be used as a controller for other third-party software plug-ins as well. For example, a hardware controller might be able to directly control sampler plug-ins, synths and other drum machine plug-ins from within the DAW. Sometimes, this will take ingenuity and a willingness to wade your way through any number of software integration options, but the fun and versatility can be worth it.

DJ Software

In addition to music production software, there is a growing number of software players, loopers, groovers, effects and digital devices that are available on the market for the twenty-first-century digital DJ. These hardware/software devices make it possible for digital grooves to be created from a laptop, iPad, controller, specially fitted turntable or digital turntable (jog/scratch CD player). Using such hardware-/software-based systems, it's possible to sync, scratch and perform with vinyl and/or digital media with an unprecedented amount of live performance and wireless interactivity that can be used on the floor, on stage or in the studio (Figure 8.14).

OBTAINING LOOP FILES FROM THE GREAT DIGITAL WELLSPRING

In this day and age, there's absolutely no shortage of pre-programmed loops that can be directly imported into a number of DAWs and groove editors. Some sound files might need to be manually edited, shaped or programmed to work with your DAW system, while others can be directly and straightforwardly entered into any loop-based program. Either way, these files can be easily obtained from any number of sources, such as:

- Downloads and CD-ROMs that are included for free with newly purchased software

FIGURE 8.13
Groove-based software/controller systems: (a) Maschine groove hardware/software production system (courtesy of Native Instruments GmbH, www.native-instruments.com). (b) Spark Drum Machine (courtesy of Arturia, www.arturia.com).

(a)   (b)

FIGURE 8.14
Digital DJ software: (a)
Traktor portable laptop
DJ rig (courtesy of Native
Instruments GmbH, www.
native-instruments.com). (b)
Serato DJ (courtesy of Serato
Audio Research, www.serato.
com).

- The Web (both free and for purchase)
- Commercial media
- Files within websites, promotions and CDs that are loaded as royalty-free demo content
- Rolling your own (creating your own loops can add a satisfying and personal touch to your project)

It's important to note that at any point during the creation of a composition, audio and MIDI tracks (such as vocals or played instruments) can be performed alongside a loop-based session in order to give the performance a fluid, alternative and interesting feel. Obviously, recording a live instrument into a session with a defined tempo and then editing these tracks into loops that can be dropped into the current and future sessions can also add a live performance feel to the track.

As with most music technologies, the field of looping and laying tracks into a groove-based project continues to advance and evolve at an amazing rate. It's almost a sure bet that your current system will support looping in one way or another. Take time to experiment, read the manuals, gather up some loops that fit your style and particular interests, and start working them into a session. It might take you some time to master the art of looping, or then again, you might be a natural Zen master. Either way, the journey will help your production style grow … not to mention that it's usually a ton of fun!

CHAPTER 9

MIDI and Electronic Music Technology

Today, professional and nonprofessional musicians alike are using the language of the *Musical Instrument Digital Interface* (MIDI) to perform an expanding range of music and automation tasks within audio production, audio for video, film post, stage production, etc. This industry-wide acceptance can, in large part, be attributed to the cost-effectiveness, power and general speed of MIDI production. Once a MIDI instrument or device comes into the production picture, there may be less need (if any at all) to hire outside musicians for a project. This alluring factor allows a musician/composer to compose, edit and arrange a piece in an electronic music environment that's extremely flexible. By this, we're not saying that MIDI replaces, or should replace, the need for acoustic instruments, microphones and the traditional performance setting. In fact, it's a powerful production tool that assists countless musicians to create music and audio productions in ways that are both innovative and highly personal. In short, MIDI is all about control, repeatability, flexibility, cost-effective production power and fun.

The affordable potential for future expansion and increased control over an integrated production system has spawned the growth of a production industry that allows an individual to cost-effectively realize a full-scale sound production, not only in his or her own lifetime but in a relatively short time. For example, much of modern-day film composition owes its very existence to MIDI. Before this technology, composers were forced to create without the benefits of hearing their work at all or by creating a reduction score that could only be played on a piano or small ensemble (due to the cost and politics of hiring a full orchestra). With the help of MIDI, composers can now hear their work in real time, make any necessary changes, print out the scores and take a full orchestra into the studio to record the final score version. At the other end of the spectrum, MIDI can be an extremely personal tool that lets us perform, edit and layer synthesized and/or sampled instruments to create a song that helps us to express ourselves to the masses – all within the comfort of the home or personal project studio. The moral of this story is that today's music industry would look and sound very different if it weren't for this powerful, four-letter production word.

DOI: 10.4324/9781003260530-9

THE POWER OF MIDI

In everyday use, MIDI can be thought of as a compositional tool for creating a scratch pad of sounds and then over time, helping to sculpt them into a final piece. It's an awesome compositional environment that, like the digital world, is extremely chameleon-like in nature.

- It can be used in straightforward ways, whereby sounds and textures are created, edited, mixed and blended into a composition.
- It can be used in conjunction with groove and looping tools to augment, control and shape a production in an endless number of ways and in a wide range of music genres.
- It can be used as a tool for capturing a performance (as a tip, if an instrument in the studio has a MIDI out jack, it's always wise to record it to a MIDI track on your digital audio workstation [DAW]). The ability to edit, change a sound or vary parameters after the fact is a helpful luxury that could save, augment and/or improve the track.
- MIDI, by its very nature, is a "reamp" beast; the ability to change a sound, instruments, settings and/or parameters in post-production is what MIDI is all about, baby. You could even play the instrument back in the studio, turn it up and re-record it acoustically – there are practically no limits.
- The ability to have real-time and post-production control over music and effects parameters is literally in MIDI's DNA. Almost every parameter can be mangled, mutilated and finessed to fit your wildest dreams – either during the composition phase or in post-production.

In short, the name of this game is editability, flexibility and individuality. There are so many ways of approaching and working with MIDI that it's very personal in nature. The ways that a system can be set up, and the various approaches by which the tools and toys are used to create music and sounds, can be extremely individualistic. How you use your tools to create your own style of music is literally up to you, both in production and in post-production. That's the true beauty of MIDI.

MIDI Production Environments

One of the more powerful aspects of MIDI production is that a system can be designed to handle a wide range of tasks with a degree of flexibility and ease that best suits an artist's main instrument, playing style and even personal working habits. By opening up almost any industry-related magazine, you'll easily see that a vast number of electronic musical instruments, effects devices, computer systems and other MIDI-related devices are currently available on the new and used electronic music market. MIDI production systems exist in all kinds of shapes and sizes and can be incorporated to match a wide range of production and budget needs. For example, working and aspiring musicians commonly install digital audio and MIDI systems in their homes (Figure 9.1).

(a)

(b)

These production environments range from ones that take up a corner of an artist's bedroom to larger systems that are integrated into a dedicated project studio. Systems such as these can be specially designed to handle a multitude of applications and have the important advantage of letting artists produce their music in a comfortable environment – whenever the creative mood hits. Newer, laptop-based systems allow you to make music "wherever and whenever" from the comfort of your trusty backpack. Such production luxuries, which would have literally cost an artist a fortune in the not-too-distant past, are now within the reach of almost every musician.

In effect, the true power of MIDI lies in its repeatability and ability to offer control and edit functions both during production and (even more so) after the fact in post-production. When combined with DAWs and modern-day recording technology, much of the music production process can be pre-planned and rehearsed before you even step into the studio. In fact, it's not uncommon for recorded tracks to be laid down before they ever see the hallowed halls of a pro studio (if they see them at all). In business jargon, this luxury has reduced the number of billable hours to the artist or label to a cost-effective minimum – this flexibility, editability and affordability have placed MIDI production and control squarely at the heart of modern-day music production.

Since its inception, electronic music has been an indispensable tool for the scoring and audio post-production of television and radio commercials, industrial videos and full-feature motion picture soundtracks (Figure 9.2). For productions

FIGURE 9.1

MIDI production rooms: (a) Gettin' it all going in the bedroom studio (courtesy of Steinberg Media Technologies GmbH, a division of Yamaha Corporation, www.steinberg. net). (b) MIDI-based Project Studio (courtesy of Ableton AG, www.ableton.com).

FIGURE 9.2
Skywalker Sound scoring stage with orchestra, Marin County, CA (courtesy of Skywalker Sound, www.skysound.com).

that are on a budget, an entire score can be created in the artist's project studio using MIDI, hard-disk tracks and digital recorders – all at a mere fraction of what it might otherwise cost to hire the musicians and rent a studio.

Electronic music production and MIDI are also very much at home on the stage. In addition to using synths, samplers, DAWs and drum machines on the stage, most or all of a MIDI instrument and effects device parameters can be controlled from a pre-sequenced or real-time source. This means that all the necessary settings for the next song (or section of a song) can be automatically called up before being played. Once underway, various instrument patch and controller parameters can also be changed during a live performance from a stomp box controller, DAW or other hardware controller.

MIDI also falls squarely under the multimedia banner. General MIDI (GM, which is discussed in greater detail within Chapter 11) is a standardized spec that allows any sound-card or GM-compatible device to play back a score using the originally intended sounds and program settings. A General MIDI sequence can therefore be played on any laptop, tablet or (last but not least) phone for use in playing back music and effects within multimedia games and websites.

FIGURE 9.3
One ringy-dingy … MIDI (as well as digital audio) helps us to reach out and touch someone through phone ringtones.

With the integration of the General MIDI standard into various media devices, one of the fastest-growing MIDI applications, surprisingly, is probably comfortably resting in your pocket or purse right now – the ringtone on your cell phone (Figure 9.3). The ability to use MIDI (or digital sound files) to let you know who is calling has spawned an industry that allows your cell to be personalized in a super-fun way. One of my favorite ringtone stories happened on Hollywood Boulevard in L.A. This tall, lanky man was sitting at a café when his cell phone

started blaring out an "If I Only Had a Brain" MIDI sequence from *The Wizard of Oz*. It wouldn't have been nearly as funny if the guy hadn't looked A LOT like the scarecrow character. Of course, everyone laughed.

WHAT IS MIDI?

Simply stated, the MIDI is a digital communications language and compatible specification that allows multiple hardware and software electronic instruments, performance controllers, computers and other related devices to communicate with each other over a connected network (Figure 9.4). MIDI is used to translate performance- or control-related events (such as playing a keyboard, selecting a patch number, varying a modulation wheel, triggering a staged visual effect, etc.) into equivalent digital messages and then transmit these messages to other MIDI devices, where they can be used to control sound generators and other performance/control parameters. The beauty of MIDI is that its data can be recorded into a DAW or hardware device (known as a sequencer), where it can then be edited and communicated between electronic instruments or other devices to create music or control any number of parameters in a performance or post-production setting.

In addition to composing and performing a song, musicians can also act as techno-conductors, having complete control over a wide palette of sounds, their timbre (sound and tonal quality), overall blend (level, panning) and other real-time controls. MIDI can also be used to vary the performance and control parameters of electronic instruments, recording devices, control devices and signal processors in the studio, on the road or on the stage.

FIGURE 9.4
Example of a typical MIDI system with the MIDI network connections highlighted in solid lines.

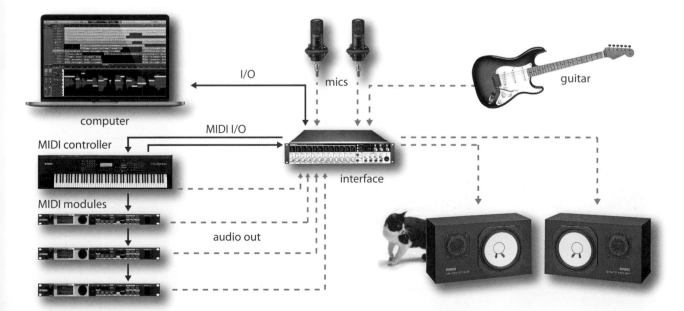

computer

I/O

mics

guitar

MIDI I/O

MIDI controller

interface

MIDI modules

audio out

The term *interface* refers to the actual data communications link and software/hardware systems in a connected MIDI and digital audio network. Through the use of MIDI, it's possible for all of the electronic instruments and devices within a network to be addressed through the transmission of real-time performance and control-related MIDI data messages throughout a system to multiple instruments and devices through one or more data lines (which can be chained from device to device). This is possible because a single data cable is capable of transmitting performance and control messages over 16 discrete channels. This simple fact allows electronic musicians to record, overdub, mix and play back their performances in a working environment that loosely resembles the multitrack recording process. Once mastered, MIDI surpasses this analogy by allowing a composition to be edited, controlled, altered and called up with complete automation and repeatability – all of this providing production challenges and possibilities that are well beyond the capabilities of the traditional tape-based multitrack recording process.

WHAT MIDI ISN'T

For starters, let's dispel one of MIDI's greatest myths: MIDI DOESN'T communicate audio, nor can it create sounds! It is strictly a digital language that instructs a device or program to create, play back or alter the parameters of sound or control function. It is a data protocol that communicates on/off triggering and a wide range of parameters to instruct an instrument or device to generate, reproduce or control audio or production-related functions. Because of these differences, the MIDI data path is entirely distinct and separate from the audio signal paths (Figure 9.5). Even when they digitally share the same transmission cable

FIGURE 9.5
Example of a typical MIDI system with the audio connections highlighted in solid lines.

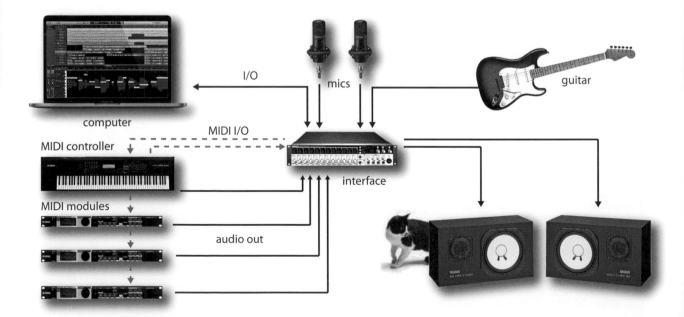

(such as FireWire, USB or Thunderbolt), the actual data paths and formats are completely separate in structure.

In short, MIDI communicates information that instructs an instrument to play or a device to carry out a function. It can be likened to the holes in a player-piano roll; when we put the paper roll up to our ears, we hear nothing, but when the cut-out dots pass over the sensors on a player piano, the instrument itself begins to make music. It's exactly the same with MIDI. A MIDI file or data stream is simply a set of instructions that pass down a wire in a serial fashion, but when an electronic instrument interprets the data, we begin to hear sound.

It's worth repeating here – of course, the power of MIDI lies in its ability to a capture performance, but its real strength squarely rests with the ability to edit that data, to manipulate it up, down and inside out, to control a wide range of musical and non-musical parameters as well as to alter the performance, individual notes and instrument sounds or parameters in almost an infinite way – all in the name of editing flexibility in post-production. That's where MIDI shines.

SYSTEM INTERCONNECTIONS

As a data transmission medium, MIDI is unique in the world of sound production in that it's able to transmit 16 discrete channels of performance, controller and timing information over a cable in one direction, using data densities that are economically small and easy to manage. In this way, it's possible for MIDI messages to be communicated from a specific source (such as a keyboard or MIDI sequencer) to any number of devices within a connected network over a single MIDI data chain. In addition, MIDI is flexible enough that multiple MIDI data lines can be used to interconnect devices in a wide range of possible system configurations; for example, multiple MIDI lines can be used to transmit data to instruments and devices over 32, 48, 128 or more discrete MIDI channels!

Of course, those of you who are familiar with the concept of MIDI know that over the years, the concept of interconnecting electronic instruments and other devices together has changed. These days, you're more likely to connect a device to a computer by using a USB, FireWire, Thunderbolt or network cable than by using standard MIDI cables. In recent times, these interconnections are made "under the virtual hood", where cable connectivity limitations are rarely an issue. However, it's still important that we have an understanding of how these data connections are made "at a basic level", because these older rules often still apply within a new connection environment – therefore, I present to you the MIDI cabling system.

The MIDI Cable

A MIDI cable (Figure 9.6) consists of a shielded, twisted pair of conductor wires that has a male 5-pin DIN plug located at each of its ends. The MIDI specification

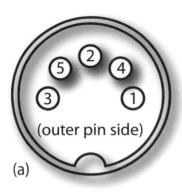

pin 1 - no connection
pin 4 - MIDI signal
pin 2 - ground
pin 5 - MIDI signal
pin 3 - no connection

(a)

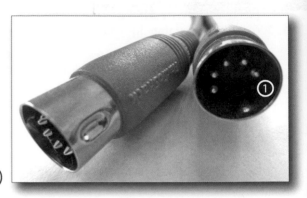

(b)

FIGURE 9.6
The MIDI cable: (a)
Wiring diagram. (b) Cable
connectors.

currently uses only 3 of the 5 pins, with pins 4 and 5 being used as conductors for MIDI data and pin 2 being used to connect the cable's shield to equipment ground. Pins 1 and 3 are currently not in use. The cables use twisted cable and metal shield groundings to reduce outside interference (such as electrostatic or radio-frequency interference), both of which can serve to distort or disrupt the transmission of MIDI messages.

MIDI Pin Description

- Pin 1 is not used in most cases; however, it can be used to provide the V– (ground return) of a MIDI phantom power supply.
- Pin 4 is a MIDI data line.
- Pin 2 is connected to the shield or ground cable, which protects the signal from radio and electro-magnetic interference.

- Pin 5 is a MIDI data line.
- Pin 3 is not used in most cases; however, it can be used to provide the +V (+9 to +15V) of a MIDI phantom power supply.

MIDI cables come prefabricated in lengths of 2, 6, 10, 20 and 50 feet and can commonly be obtained from music stores that specialize in MIDI equipment. To reduce signal degradation and external interference that tends to occur over extended cable runs, 50 feet is the maximum length specified by the MIDI spec.

It should be noted that in modern-day MIDI production, it's become increasingly common for MIDI data to be transmitted throughout a system network using USB, FireWire, Thunderbolt, network or Wi-Fi interconnections. Although the data isn't transmitted through traditional MIDI cabling, the data format still adheres to the MIDI protocol.

(a)   (b)

FIGURE 9.7
Wireless MIDI transmitter systems. (a) Yamaha Wireless MD-BT01 5-PIN DIN MIDI Adapter (courtesy of Yamaha Corporation of America; www.yamaha.com). (b) VORTEX WIRELESS 2 Wireless USB/MIDI Keytar Controller (courtesy of alesis; www.alesis.com).

MIDI PHANTOM POWER

In December 1989, Craig Anderton (musician and audio guru) submitted an article to *EM* proposing an idea that provides a standardized 12-V DC power supply to instruments and MIDI devices directly through pins 1 and 3 of a basic MIDI cable. Although pins 1 and 3 are technically reserved for possible changes in future MIDI applications (which never really came about), over the years, several forward-thinking manufacturers (and project enthusiasts) have begun to implement MIDI phantom power directly into their studio and on-stage systems.

WIRELESS MIDI

Wireless MIDI transmitters (**Figure 9.7**) also make it possible for a battery-operated MIDI guitar, wind controller, etc. to be footloose and fancy free on stage and in the studio. Working at distances of up to 500 feet, these battery-powered transmitter/receiver systems introduce very low delay latencies and can be switched over a number of radio channel frequencies. In recent times, however, dedicated wireless transmitters have given way to instruments, MIDI interface and portable iOS devices that communicate system-wide via Wi-Fi and network-based systems. The use of iOS devices for inputting, controlling and directly interfacing with MIDI-based systems has expanded into an important way to wirelessly communicate in the studio or on stage with relative ease (more on this in Chapter 10).

MIDI Jacks

MIDI is distributed from device to device using three types of MIDI jacks: MIDI In, MIDI Out and MIDI Thru (**Figure 9.8a**). These three connectors use 5-pin DIN jacks as a way to connect MIDI instruments, devices and computers into a music or production network system. As a side note, these ports (as strictly defined by the MIDI 1.0 spec) are optically isolated to eliminate possible ground loops that might occur when connecting numerous devices together.

- *MIDI In jack*: the MIDI In jack receives messages from an external source and communicates this performance, control and timing data to the device's internal microprocessor, allowing an instrument to be played or a device to be controlled. More than one MIDI In jack can be designed into

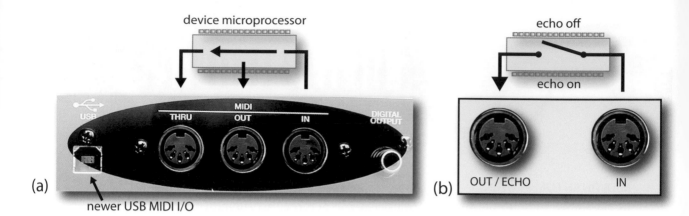

FIGURE 9.8
MIDI ports. (a) MIDI In, Out and Thru ports, showing the device's signal path routing. (b) MIDI echo configuration.

a system to provide for MIDI merging functions or for devices that can support more than 16 channels (such as a MIDI interface). Other devices (such as a controller) might not have a MIDI In jack at all.

• *MIDI Out jack*: The MIDI Out jack is used to transmit MIDI performance, control messages or SysEx data from one device to another MIDI instrument or device. More than one MIDI Out jack can be designed into a system, giving it the advantage of controlling and distributing data over multiple MIDI paths using more than 16 channels (i.e., 16 channels × N MIDI port paths).

• *MIDI Thru jack*: The MIDI Thru jack retransmits an exact copy of the data that's being received at the MIDI In jack. This process is important, because it allows data to pass directly through an instrument or device to the next device in the MIDI chain (more on this later). Keep in mind that this jack is used to relay an exact copy of the MIDI In data stream and isn't merged with the data being transmitted from the MIDI Out jack.

MIDI ECHO

Certain MIDI devices may not include a MIDI Thru jack at all. Some of these devices, however, may have the option of switching the MIDI Out between being an actual MIDI Out jack and a MIDI Echo jack (Figure 9.8b). As with the MIDI Thru jack, a MIDI Echo option can be used to retransmit an exact copy of any information that's received at the MIDI In port and route this data to the MIDI Out/Echo jack. Unlike a dedicated MIDI Out jack, the MIDI Echo function can often be selected to merge incoming data with performance data that's being generated by the device itself. In this way, more than one controller can be placed in a MIDI system at one time. Note that although performance and timing data can be echoed to a MIDI Out/Echo jack, not all devices are capable of echoing SysEx data.

Typical Configurations

Although electronic studio production equipment and setups are rarely alike (or even similar), there are a number of general rules that make it easy for MIDI

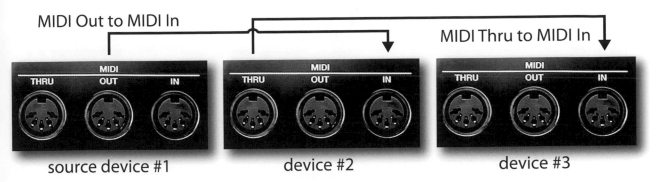

MIDI Out to MIDI In

MIDI Thru to MIDI In

source device #1   device #2   device #3

devices to be connected to a functional network. These common configurations allow MIDI data to be distributed in the most efficient and understandable manner possible.

FIGURE 9.9
The two valid means of connecting one MIDI device to another.

As a primary rule, there are only two valid ways to connect one MIDI device to another within a MIDI cable chain (Figure 9.9):

- The MIDI Out jack of a source device (controller or sequencer/computer) must be connected to the MIDI In of a second device in the chain.

- The MIDI Thru jack of the second device must be connected to the MIDI In jack of the third device in the chain, following this same Thru-to-In convention until the end of the chain is reached.

THE DAISY CHAIN

One of the simplest and most common ways to distribute data throughout a MIDI system is through a *daisy chain*. This method relays MIDI data from a source device (controller or sequencer/computer) to the MIDI In jack of the next device in the chain (which receives and acts on this data). This next device then relays an exact copy of the incoming data at its MIDI In jack to its MIDI Thru jack, which is then relayed to the next MIDI In within the chain, and so on through the successive devices. In this way, up to 16 channels of MIDI data can be chained from one device to the next within a connected data network – and it's precisely this concept of stringing multiple data lines/channels through a single MIDI line that makes the whole idea work! Let's try to understand this system better by looking at a few examples.

Figure 9.10a shows a simple (and common) example of a MIDI daisy chain whereby data flows from a controller (MIDI Out jack of the source device) to a synth module (MIDI In jack of the second device in the chain), and then, an exact copy of the data that flows into the second device is relayed to its MIDI Thru jack out to another synth (via the MIDI In jack of the third device in the chain). If our controller is set to transmit on MIDI channel 3, the second synth in the chain (which is set to channel 2) will ignore the messages and not play, while the third synth (which is set to channel 3) will be playing its heart out.

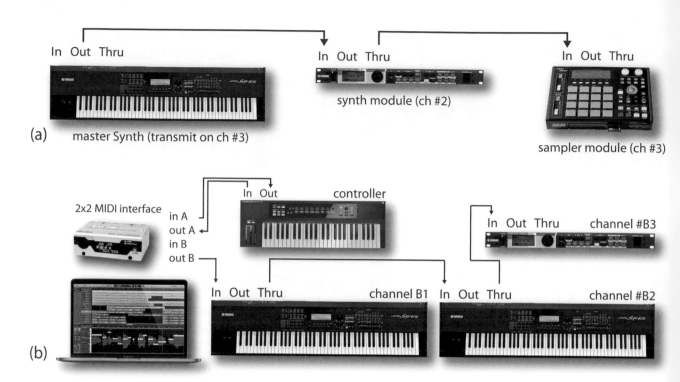

FIGURE 9.10

Example of a connected MIDI system using a daisy chain: (a) Daisy chain hookup. (b) Example of how a computer can be connected into a daisy chain.

The moral of this story is that although there's only one connected data line, a wide range of instruments and channel voices can be played in a surprisingly large number of combinations – all by using individual channel assignments along a daisy chain.

Another example (Figure 9.10b) shows how a computer can easily be designated as the master source within a daisy chain so that a sequencing program can be used to control the entire playback and channel routing functions of a daisy-chained system. In this situation, the MIDI data flows from a master controller/synth to the MIDI In jack of a computer's MIDI interface (where the data can be played into, processed and rerouted through a MIDI sequencer). The MIDI Out of the interface is then routed back to the MIDI In jack of the master controller/synth (which receives and acts on this data). The controller then relays an exact copy of this incoming data out to its MIDI Thru jack (which is then relayed to the next device in the chain) and so on, until the end of the chain is reached. When we stop and think about it, we can see that the controller is essentially used as a "performance tool" for entering data into the MIDI sequencer, which is then used to communicate this data out to the various instruments throughout the connected MIDI chain.

THE MULTIPORT NETWORK

Another common approach to routing MIDI throughout a production system involves distributing MIDI data through the multiple 2, 4 and 8 In/Out ports that are available on the newer multiport MIDI interfaces or through the use of multiple MIDI USB interface devices.

In larger, more complex MIDI systems, a multiport MIDI network (Figure 9.11) offers several advantages over a single daisy chain path. One of the most important is its ability to address devices within a complex setup that requires more than 16 MIDI channels. For example, a 2 × 2 MIDI interface that has 2 independent In/Out paths is capable of simultaneously addressing up to 32 channels (i.e., port A 1–16 and port B 1–16), whereas an 8 × 8 port is capable of addressing up to 128 individual MIDI channels.

> NOTE: Although the distinction isn't overly important, you might want to keep in mind that a MIDI "port" is a virtual data path that's processed through a computer, whereas a MIDI "jack" is the physical connection on a device itself.

This type of multiport MIDI network has a number of advantages. As an example, port A might be dedicated to three instruments that are set to respond to MIDI channels 1–6, 7 and finally channel 11, whereas port B might be transmitting data to two instruments that are responding to channels 1–4 and 510, and port C might be communicating SysEx MIDI data to and from a MIDI remote controller to a DAW. In this modern age of audio interfaces, multiport MIDI interfaces and controller devices that are each fitted with MIDI ports, it's a simple matter for a computer to route and synchronously communicate MIDI data throughout the studio in lots of ingenious and cost-effective ways. As you might remember, many of the newer devices that are MIDI capable might also be able to talk with the OS and host software via USB, thereby reducing the need for a multiport setup.

MIDI 1.0 AND 2.0

For decades, the MIDI 1.0 spec has remained unchanged. This is a true testament to the hard work and dedication that the original members of the MMA (MIDI Manufacturers Association, www.midi.org) and the AMEI (Association of Music Electronics Industry, its Japanese counterpart) put into making a spec that would withstand the test of time. However, at the 2019 Winter NAMM (National Association of Music Merchants) convention, it was announced that the spec

FIGURE 9.11
Example of a multiport network using two MIDI interfaces.

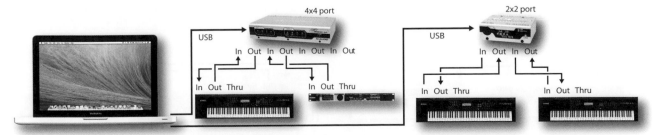

was getting an overhaul, and in the fall of 2019, MIDI 2.0 was ratified for general industry use.

Later within this chapter, we'll be taking a look into the changes and enhancements that went into the creation of MIDI 2.0, including bidirectionality, increased performance and controller resolution, expanded feature-sets, increased port/channel capabilities and so forth. Above all, the MMA and AMEI have made BACKWARDS COMPATIBILITY a key requirement, stating that users can expect MIDI 2.0 and the related newer systems to work seamlessly with MIDI devices that have been sold over the past 33 years.

For now, it's best to gain a basic understanding of MIDI 1.0, and then later in this chapter, we'll introduce many of the extra features and capabilities that MIDI 2.0 has to offer.

EXPLORING THE MIDI 1.0 SPEC

MIDI is a specified data format that must be strictly adhered to by those who design and manufacture MIDI-equipped instruments and devices. Because the format is standardized, you don't have to worry about whether the MIDI output of one device will be understood by the MIDI in port of a device that's made by another manufacturer. As long as the data ports say and/or communicate MIDI, you can be assured that the data (at least, most of the basic performance functions) will be transmitted and understood by all devices within the connected system. In this way, the user need only consider the day-to-day dealings that are involved with using electronic instruments, without having to be concerned with compatibility between devices.

The MIDI 1.0 Message

When using a standard MIDI 1.0 cable, it's important to remember that data can only travel in one direction from a single source to a destination (Figure 9.12a). In order to make two-way communication possible, a second MIDI data line must be used to communicate data back to the device, either directly or through the MIDI chain (Figure 9.12b).

FIGURE 9.12
MIDI 1.0 data can only travel in one direction through a single MIDI cable: (a) Data transmission from a single source to a destination. (b) Two-way data communication using two cables.

MIDI digitally communicates musical performance data between devices as a string of MIDI messages. These messages are made up of groups of 8-bit words (known as bytes), which are transmitted in a serial fashion (generally at a speed of 31,250 bits/sec) to convey a series of instructions to one or all MIDI devices within a system.

(a)

(b)

Only two types of bytes are defined by the MIDI 1.0 specification: the status byte and the data byte.

- A *status byte* is used to identify what type of MIDI function is to be performed by a device or program.

It is also used to encode channel data (allowing the instruction to be received by a device that's set to respond to the selected channel).

- A *data byte* is used to associate a value to the event that's given by the accompanying status byte.

Although a byte is made up of 8 bits, the most significant bit (MSB; the leftmost binary bit within a digital word) is used solely to identify the byte type. The MSB of a status byte is always 1, while the MSB of a data byte is always 0. For example, a 3-byte MIDI Note-On message (which is used to signal the beginning of a MIDI note) might read in binary form as a 3-byte Note-On message of (10010100) (01000000) (01011001). This particular example transmits instructions that would be read as: "Transmitting a Note-On message over MIDI channel #5, using keynote #64, with an attack velocity [volume level of a note] of 89".

MIDI Channels

Just as a public speaker might single out and communicate a message to one individual in a crowd, MIDI messages can be directed to communicate information to a specific device or range of devices within a MIDI system. This is done by embedding a channel-related nibble (4 bits) within the status/channel number byte. This process makes it possible for up to 16 channels of performance or control information to be communicated to a specific device or a sound generator through a single MIDI data cable (Figure 9.13).

FIGURE 9.13
Up to 16 channels can be communicated through a single MIDI cable or data port. Since this nibble is 4 bits wide, up to 16 discrete MIDI channels can be transmitted through a single MIDI cable or designated port.

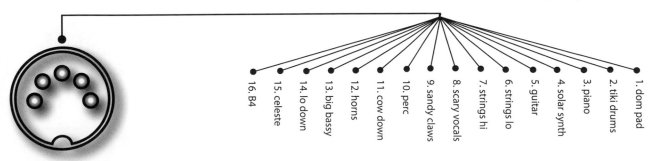

16. B4 15. celeste 14. lo down 13. big bassy 12. horns 11. cow down 10. perc 9. sandy claws 8. scary vocals 7. strings hi 6. strings lo 5. guitar 4. solar synth 3. piano 2. tiki drums 1. dom pad

Since this nibble is 4 bits wide, up to 16 discrete MIDI 1.0 channels can be transmitted through a single MIDI cable or designated port.

0000 = CH#1	0100 = CH#5	1000 = CH#9	1100 = CH#13
0001 = CH#2	0101 = CH#6	1001 = CH#10	1101 = CH#14
0010 = CH#3	0110 = CH#7	1010 = CH#11	1110 = CH#15
0011 = CH#4	0111 = CH#8	1011 = CH#12	1111 = CH#16

Whenever a MIDI device or sound generator within a device or program function is instructed to respond to a specific channel number, it will only react to messages that are transmitted on that channel (i.e., it ignores channel messages that are transmitted on any other channel). For example, let's assume that we're going to create a short song using a DAW that is connected to a MIDI controller and has been loaded with three MIDI tracks. Each of these tracks could be assigned to either hardware instruments (as shown in Figure 9.14) or virtual instruments (or any combination that you have at your disposal).

1. We could easily start off by downloading or producing our own simple drum pattern track into MIDI track #1, which could be assigned to a hardware or software synth, sampler or groove device that is assigned to channel #10. Once the track has been created, the sequence will thereafter transmit the notes and data over channel 10, allowing the device's percussion sounds to be played.

2. Next, we could route track #2 to a synthesizer or software synth to create a basic melody line. By performing the track on the hardware controller over Ch #2, the MIDI data will then be transmitted out to the second instrument in the chain (or to the soft synth itself). The sequence will then simultaneously transmit the melody line to the synth on ch #2 and to the perc voices on the first instrument assigned to ch#10. At this point, our song can hopefully begin to take shape.

FIGURE 9.14

MIDI setup showing a set of MIDI channel assignments.

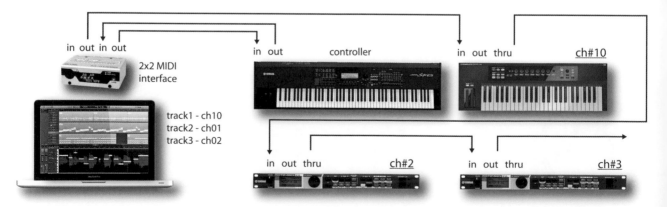

3. Now, we can set track #3 to route out to the third synth instrument that's set to respond to ch #3, which might be an underlying melody pad.

4. Now that the song's beginning to take shape, the sequencer can then play the musical parts to the instruments on their respective MIDI channels – all in a "multitrack" environment that gives us complete control over voicing, volume, panning and a wide range of edit functions. In short, we've created a true multichannel production environment that can be performed in a production environment and then saved, edited, embellished and mixed in a post-production environment.

It goes without saying that this example is just one of the infinite setup and channel possibilities that can be encountered in a production environment. It's often true, however, that even the most complex MIDI and production rooms will have a system strategy – a basic channel and overall layout that makes the day-to-day operation of making music easier to work with. The layout and basic decisions that you might make in your own room are, of course, up to you. Streamlining your system to work both efficiently and in a way that works best for you, as with all things, comes with time, experience and practice.

MIDI Modes

Electronic instruments often vary in the number of sounds and notes that can be simultaneously produced by their internal sound-generating circuitry. For example, certain instruments can only produce one note at a single time (known as a monophonic instrument), while others can generate 16, 32 and even 64 notes at once (these are known as polyphonic instruments). The latter type can easily play chords or more than one musical line on a single instrument at a time.

In addition, some instruments are only capable of producing a single generated sound patch (often referred to as a "voice") at any one time. Their generating circuitry could be polyphonic, allowing the player to lay down chords and bass or melody lines, but it can only produce these notes using a single, characteristic sound at any one time (e.g., an electric piano, a synth bass or a string patch). However, the vast majority of newer synths differ from this in that they're multitimbral in nature, meaning that they can generate numerous sound patches at any one time (e.g., an electric piano, a synth bass and a string patch, as can be seen in Figure 9.15). That's to say that it's common to run across electronic instruments that can simultaneously generate a number of voices, each offering its own control over a wide range of parameters. Best of all, it's also common for different sounds to be assigned to their own MIDI channels, allowing multiple patches to be internally mixed within the device to a stereo output bus or independent outputs.

The following list and figures explain the four modes that are supported by the MIDI spec:

Ch #01   Big Bass
Ch #02   Sub Stick
Ch #03   Brassman
Ch #04   Dyno Pad
....
Ch #16   Ice Vibes

internal mixer

FIGURE 9.15

Multitimbral instruments are virtual bands-in-a-box that can simultaneously generate multiple patches, each of which can be assigned to its own MIDI channel.

- *Mode 1 (Omni On/Poly):* in this mode, an instrument will respond to data that's being received on any MIDI channel and then redirect this data to the instrument's base channel. In essence, the device will play back everything that's presented at its input in a polyphonic fashion, regardless of the incoming channel designations. As you might guess, this mode is rarely used.

- *Mode 2 (Omni On/Mono):* as in Mode 1, an instrument will respond to all data that's being received at its input without regard to channel designations; however, this device will only be able to play one note at a time. Mode 2 is used even more rarely than Mode 1, as the device can't discriminate channel designations and can only play one note at a time.

- *Mode 3 (Omni Off/Poly):* in this mode, an instrument will only respond to data that matches its assigned base channel in a polyphonic fashion. Data that's assigned to any other channel will be ignored. This mode is by far the most commonly used, as it allows the voices within a multitimbral instrument to be individually controlled by messages that are being received on their assigned MIDI channels. For example, each of the 16 channels in a MIDI line could be used to independently play each of the parts in a 16-voice, multitimbral synth.

- *Mode 4 (Omni Off/Mono):* as with Mode 3, an instrument will be able to respond to performance data that's transmitted over a single, dedicated channel; however, each voice will only be able to generate one MIDI note at a time. A practical example of this mode is often used in MIDI guitar systems, where MIDI data is monophonically transmitted over six consecutive channels (one channel/voice per string).

Channel Voice Messages

FIGURE 9.16

Byte structure of a MIDI Note-On message.

Channel Voice messages are used to transmit real-time performance data throughout a connected MIDI system. They're generated whenever a MIDI

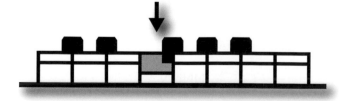

Status / Ch #	Note #	Attack Velocity
(0 - 15)	(0 - 127)	(0 - 127)
(1001 CCCC)	(NNNN NNNN)	(VVVV VVVV)

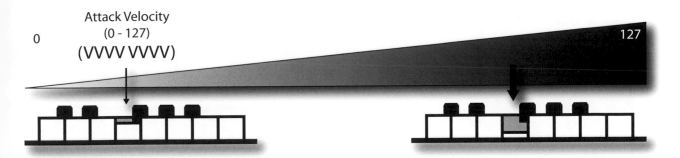

Attack Velocity
(0 - 127)
(VVVV VVVV)

0

127

FIGURE 9.17
Velocity is used to communicate the volume or loudness of a note within a performance.

instrument's controller is played, selected or varied by the performer. Examples of such control changes could be the playing of a keyboard, pressing of program selection buttons, or movement of modulation or pitch wheels. Each Channel Voice message contains a MIDI channel number within its status byte, meaning that only devices that are assigned to the same channel number will respond to these commands. There are seven Channel Voice message types: Note-On, Note-Off, Polyphonic Key Pressure, Channel Pressure, Program Change, Pitch Bend Change and Control Change.

- *Note-On* messages (Figure 9.16): indicate the beginning of a MIDI note. This message is generated each time a note is triggered on a keyboard, drum machine or other MIDI instrument (by pressing a key, striking a drum pad, etc.). A Note-On message consists of three bytes of information: a MIDI channel number, a MIDI pitch number and an attack velocity value (messages that are used to transmit the individually played volume levels [0–127] of each note, as shown in Figure 9.17).

- *Note-Off* messages: indicate the release (end) of a MIDI note. Each note played through a Note-On message is sustained until a corresponding Note-Off message is received. A Note-Off message doesn't cut off a sound; it merely stops playing it. If the patch being played has a release (or final decay) stage, it begins that stage upon receiving this message. It should be noted that many systems will actually use a Note-On message with a velocity 0 to denote a Note-Off message.

- *Polyphonic Key Pressure* messages (Figure 9.18): transmitted by instruments that can respond to pressure changes applied to the individual keys of a keyboard. A Polyphonic Key Pressure message consists of three bytes of information: a MIDI channel number, a MIDI pitch number and a pressure value.

FIGURE 9.18
Byte structure of a MIDI Polyphonic Key Pressure message (independently generated when additional pressure is applied to any key that's played).

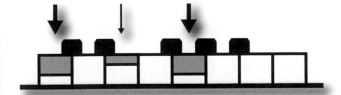

Status / Ch # (0 - 15)	Note # (0 - 127)	Pressure Value (0 - 127)
(1010 CCCC)	**(NNNN NNNN)**	**(VVVV VVVV)**

FIGURE 9.19
Program Change messages can be used to change sound patches from a DAW track, sequencer or remote controller.

- *Channel Pressure (or Aftertouch)* messages: are transmitted and received by instruments that respond to a single, overall pressure that's applied to all of the keys that are played. In this way, additional pressure on the keys can also be assigned to control such variables as pitch bend, modulation and panning.

- *Program Change* messages (Figure 9.19): change the active voice (generated sound) or preset program number in a MIDI instrument or device. Using this message format, up to 128 presets (a user- or factory-defined number that activates a specific sound-generating patch or system setup) can be selected. A Program Change message consists of two bytes of information: a MIDI channel number (1–16) and a program ID number (0 –127).

- *Pitch Bend Change* messages (Figures 9.20 and 9.21): transmitted by an instrument whenever its pitch bend wheel is moved in either the positive (raise pitch) or negative (lower pitch) direction from its central (no pitch bend) position.

- *Control Change* messages (Figures 9.22 and 9.23): transmit information that relates to real-time control over a MIDI instrument's performance parameters (such as modulation, main volume, balance and panning). Three types of real-time controls can be communicated through Control Change messages: continuous controllers, which communicate a continuous range of control settings, generally with values ranging from 0 to 127; switches (controls having an ON or OFF state with no intermediate settings); and data controllers, which enter data through either numerical keypads or stepped up/down entry buttons.

FIGURE 9.20
Byte structure of a Pitch Bend Change message.

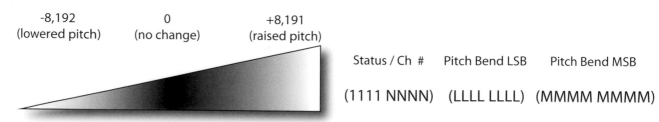

Minimum Value = 0   Mid Value = 64   Maximum Value = 127

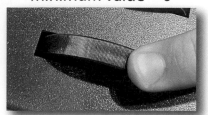  

FIGURE 9.21
Pitch bend wheel data value ranges.

Explanation of Controller ID Parameters

As you can see in Figure 9.22, the second byte of the Control Change message is used to denote the *controller ID number*. This all-important value is used to specify which of the device's program or performance parameters are to be addressed.

The following section details the general categories and conventions for assigning controller ID numbers to an associated parameter (as specified by the 1995 update of the MMA, www.midi.org). An overview of these controllers can be seen in Table 9.1. This is definitely an important table to earmark, because these numbers will be an important guide toward knowing and/or finding the right ID number that can help you on your path toward finding that perfect parameter for controlling a variable.

System Messages

As the name implies, System messages are globally transmitted to every MIDI device in the MIDI chain. This is accomplished because MIDI channel numbers aren't addressed within the byte structure of a System message. Thus, any device will respond to these messages, regardless of its MIDI channel assignment. The three System message types are:

- System Common messages
- System Real-Time messages
- System Exclusive messages

System Common messages are used to transmit MIDI timecode, song position pointer, song select, tune request and end-of-exclusive data messages throughout the MIDI system or 16 channels of a specified MIDI port.

- *MIDI timecode (MTC)* messages: provide a cost-effective and easily implemented way to translate SMPTE (a standardized synchronization timecode)

FIGURE 9.22
Control Change message byte structure.

0 ←--------------------→ 127

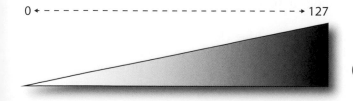

Status / Ch #	Controller ID#	Controller Value
(0 - 15)	(0 - 127)	(0 - 127)
(1011 NNNN)	(CCCC CCCC)	(VVVV VVVV)

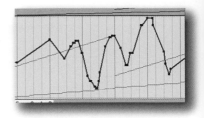

FIGURE 9.23
Control messages can be varied in real time or under automation using a number of input methods.

into an equivalent code that conforms to the MIDI 1.0 spec. It allows time-based codes and commands to be distributed throughout the MIDI chain in a cheap, stable and easy-to-implement way. MTC Quarter-Frame messages are transmitted and recognized by MIDI devices that can understand and execute MTC commands. A grouping of eight quarter frames is used to denote a complete timecode address (in hours, minutes, seconds and frames), allowing the SMPTE address to be updated every two frames. More in-depth coverage of MIDI timecode can be found in Chapter 12.

- *Song Position Pointer (SPP)* messages: allow a sequencer or drum machine to be synchronized to an external source (such as a tape machine) from any measure position within a song. This complex timing protocol isn't commonly used, because most users and design layouts currently favor MTC.

- *Song Select* messages: use an identifying song ID number to request a specific song from a sequence or controller source. After being selected, the song responds to MIDI Start, Stop and Continue messages.

- *Tune Request* messages: used to request that an equipped MIDI instrument initiate its internal tuning routine.

- *End of Exclusive (EOX)* messages: Indicate the end of a System Exclusive message.

System Real-Time messages provide the precise timing element required to synchronize all of the MIDI devices in a connected system. To avoid timing delays, the MIDI specification allows System Real-Time messages to be inserted at any point in the data stream, even between other MIDI messages.

- *Timing Clock* messages: the MIDI Timing Clock message is transmitted within the MIDI data stream at various resolution rates. It is used to synchronize the internal timing clocks of each MIDI device within the system and is transmitted in both the start and stop modes at the currently defined tempo rate. In the early days of MIDI, these rates (which are measured in pulses per quarter note [ppq]) ranged from 24 to 128 ppq; however, continued advances in technology have brought these rates up to 240, 480 or even 960 ppq.

- *Start* messages: upon receipt of a timing clock message, the MIDI Start command instructs all connected MIDI devices to begin playing from their internal sequences' initial start point. Should a program be in

Table 9.1	Listing of Controller ID Numbers, Outlining Both the Defined Format and Conventional Controller Assignments

Control #	Parameter
14-Bit Controllers Coarse/MSB (Most Significant Bit)	
0	Bank Select 0–127 MSB
1	Modulation Wheel or Lever 0–127 MSB
2	Breath Controller 0–127 MSB
3	Undefined 0–127 MSB
4	Foot Controller 0–127 MSB
5	Portamento Time 0–127 MSB
6	Data Entry MSB 0–127 MSB
7	Channel Volume (formerly Main Volume) 0–127 MSB
8	Balance 0–127 MSB
9	Undefined 0–127 MSB
10	Pan 0–127 MSB
11	Expression Controller 0–127 MSB
12	Effect Control 1 0–127 MSB
13	Effect Control 2 0–127 MSB
14	Undefined 0–127 MSB
15	Undefined 0–127 MSB
16–19	General Purpose Controllers 1–4 and 0–127 MSB
20–31	Undefined 0–127 MSB
14-Bit Controllers Fine/LSB (Least Significant Bit)	
32	LSB for Control 0 (Bank Select) 0–127 LSB
33	LSB for Control 1 (Modulation Wheel or Lever) 0–127 LSB
34	LSB for Control 2 (Breath Controller) 0–127 LSB
35	LSB for Control 3 (Undefined) 0–127 LSB
36	LSB for Control 4 (Foot Controller) 0–127 LSB
37	LSB for Control 5 (Portamento Time) 0–127 LSB
38	LSB for Control 6 (Data Entry) 0–127 LSB
39	LSB for Control 7 (Channel Volume, formerly Main Volume) 0–127 LSB
40	LSB for Control 8 (Balance) 0–127 LSB
41	LSB for Control 9 (Undefined) 0–127 LSB
42	LSB for Control 10 (Pan) 0–127 LSB

(*Continued*)

Table 9.1 (Continued)

Control #	Parameter
43	LSB for Control 11 (Expression Controller) 0–127 LSB
44	LSB for Control 12 (Effect control 1) 0–127 LSB
45	LSB for Control 13 (Effect control 2) 0–127 LSB
46–47	LSB for Control 14–15 (Undefined) 0–127 LSB
48–51	LSB for Control 16–19 (General Purpose Controllers 1–4) 0–127 LSB
52–63	LSB for Control 20–31 (Undefined) 0–127 LSB
7-Bit Controllers	
64	Damper Pedal On/Off (Sustain) <63 off, >64 on
65	Portamento On/Off <63 off, >64 on
66	Sostenuto On/Off <63 off, >64 on
67	Soft Pedal On/Off <63 off, >64 on
68	Legato Footswitch <63 Normal, >64 Legato
69	Hold 2 <63 off, >64 on
70	Sound Controller 1 (default: Sound Variation) 0–127 LSB
71	Sound Controller 2 (default: Timbre/Harmonic Intensity) 0–127 LSB
72	Sound Controller 3 (default: Release Time) 0–127 LSB
73	Sound Controller 4 (default: Attack Time) 0–127 LSB
74	Sound Controller 5 (default: Brightness) 0–127 LSB
75	Sound Controller 6 (default: Decay Time: see MMA RP-021) 0–127 LSB
76	Sound Controller 7 (default: Vibrato Rate: see MMA RP-021) 0–127 LSB
77	Sound Controller 8 (default: Vibrato Depth: see MMA RP-021) 0–127 LSB
78	Sound Controller 9 (default: Vibrato Delay: see MMA RP-021) 0–127 LSB
79	Sound Controller 10 (default undefined: see MMA RP-021) 0–127 LSB
80–83	General Purpose Controller 5–8 0–127 LSB
84	Portamento Control 0–127 LSB
85–90	Undefined
91	Effects 1 Depth (default: Reverb Send Level) 0–127 LSB

(Continued)

Table 9.1	(Continued)
Control #	**Parameter**
92	Effects 2 Depth (default: Tremolo Level) 0–127 LSB
93	Effects 3 Depth (default: Chorus Send Level) 0–127 LSB
94	Effects 4 Depth (default: Celesta [Detune] Depth) 0–127 LSB
95	Effects 5 Depth (default: Phaser Depth) 0–127 LSB
Parameter Value Controllers	
96	Data Increment (Data Entry +1)
97	Data Decrement (Data Entry –1)
98	Non-Registered Parameter Number (NRPN): LSB 0–127 LSB
99	Non-Registered Parameter Number (NRPN): MSB 0–127 MSB
100	Registered Parameter Number (RPN): LSB* 0–127 LSB
101	Registered Parameter Number (RPN): MSB* 0–127 MSB
102–119	Undefined
Reserved for Channel Mode Messages	
120	All Sound Off 0
121	Reset All Controllers
122	Local Control On/Off 0 off, 127 on
123	All Notes Off
124	Omni Mode Off (+ all notes off)
125	Omni Mode On (+ all notes off)
126	Poly Mode On/Off (+ all notes off)
127	Poly Mode On (+ mono off + all notes off)

mid-sequence, the start command will reposition the sequence to its beginning, at which point it will begin to play.

- *Stop* messages: upon receipt of a MIDI Stop command, all devices within the system will stop playing at their current position point.
- *Continue* messages: after receiving a MIDI Stop command, a MIDI Continue message will instruct all connected devices to resume playing their internal sequences from the precise point at which they were stopped.
- *Active Sensing* messages: when in the Stop mode, an optional Active Sensing message can be transmitted throughout the MIDI data stream

every 300 milliseconds. This instructs devices that can recognize this message that they're still connected to an active MIDI data stream.

- *System Reset* messages: a System Reset message is manually transmitted in order to reset a MIDI device or instrument back to its initial power-up default settings (commonly mode 1, local control on and all notes off).

System-exclusive (SysEx) messages allow MIDI manufacturers, programmers and designers to communicate customized MIDI messages between MIDI devices. The purpose of these messages is to give manufacturers, programmers and designers the freedom to communicate any device-specific data of an unrestricted length as they see fit. Most commonly, SysEx data are used for the bulk transmission and reception of program/patch data and sample data, as well as real-time control over a device's parameters. The transmission format of a SysEx message (Figure 9.24), as defined by the MIDI standard, includes a SysEx status header, manufacturer's ID number, any number of SysEx data bytes and an EOX byte. When a SysEx message is received, the identification number is read by a MIDI device to determine whether or not the following messages are relevant. This is easily accomplished by the assignment of a unique 1- or 3-byte ID number to each registered MIDI manufacturer and model. If this number doesn't match the receiving MIDI device, the subsequent data bytes will be ignored. Once a valid stream of SysEx data has been transmitted, a final EOX message is sent, after which the device will again begin to respond normally to incoming MIDI performance messages.

In actual practice, the general idea behind SysEx is that it uses MIDI messages to transmit and receive program, patch and sample data or real-time parameter information between devices. It's sort of like having an instrument or device that's a musical chameleon. One moment, it can be configured with a certain set of sound patches and setup data, and then, after it receives a new SysEx data dump, you could easily end up with an instrument that's literally full of new and hopefully exciting sounds and settings. Here are a few examples of how SysEx can be put to good use:

FIGURE 9.24
System-exclusive ID data and controller format.

- *Transmitting patch data between synths:* SysEx can be used to transmit patch and overall setup data between synths of identical make and (most often) model. Let's say that we have a Brand X Model Z synthesizer, and as it turns out, you have a buddy across town who also has a Brand X Model Z. That's

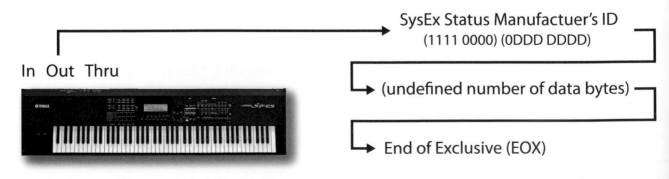

In   Out   Thru

SysEx Status Manufactuer's ID
(1111 0000) (0DDD DDDD)

(undefined number of data bytes)

End of Exclusive (EOX)

cool, except your buddy has a completely different set of sound patches loaded into her synth – and you want them! SysEx to the rescue! All you need to do is go over and transfer the patch data into your synth (to make life easier, make sure you take the instruction manual along).

- *Backing up your current patch data:* this can be done by transmitting a SysEx dump of your synth's entire patch and setup data to disk, to a SysEx utility program (often shareware) or to your DAW/MIDI sequencer. This is important: *back up your factory preset or current patch data before attempting a SysEx dump!* If you forget and download a SysEx dump, your previous settings will be lost until you contact the manufacturer, download the dump from their website or take your synth back to your favorite music store to reload the data.

- *Getting patch data from the Web:* one of the biggest repositories of SysEx data is on the Internet. To surf the Web for SysEx patch data, all you need to do is log on to your favorite search engine and enter the name of your synth. You'll probably be amazed at how many hits will come across the screen, many of which are chock-full of SysEx dumps that can be downloaded into your synth.

- *Varying SysEx controller or patch data in real time:* patch editors or hardware MIDI controllers can be used to vary system and sound-generating parameters in real time. Both of these controller types can ease the job of experimenting with parameter values or changing mix moves by giving you physical or on-screen controls that are often more intuitive and easier to deal with than programming electronic instruments, which'll often leave you dangling in cursor and 3-inch LCD screen hell.

Before moving on, I should also point out that SysEx data grabbed from the Web, disk, disc or any other medium will often be encoded using several SysEx file format styles (unfortunately, none of these are standardized). This can easily mean that sequencer Y might not recognize a SysEx dump that was encoded using sequencer Z. For this reason, dumps are often encoded using easily available, standard SysEx utility programs for the Mac or PC or as a standard MIDI file.

At last, it seems that a single unified standard has begun to emerge from the fray that's so simple that it's amazing it wasn't universally adopted from the start. This system simply records a SysEx dump as data on a single MIDI track within your DAW. Before recording a dump to a DAW MIDI track, you may need to consult the manual to make sure that SysEx filtering is turned off. Once this is done, simply place the track into record mode, initiate the dump, and save the track in your personal SysEx dump directory. Using this approach, it would also be possible to:

- Import the appropriate SysEx dump track (or set of tracks) into the current working session so as to automatically program the instruments before the sequence is played back.

- Import the appropriate SysEx dump track (or set of tracks) into separate MIDI tracks that can be muted or unassigned. Should the need arise, the track(s) can be activated and/or assigned in order to dump the data into the appropriate instruments.

EXPLORING THE MIDI 2.0 SPEC

The fact that MIDI 1.0 has been in place from 1983 till 2022 (almost 40 years!) is an amazing testament to the individuals and companies that joined forces to create a robust spec that has forever changed the face of music production. However, as with all things, the time has finally come to broaden the spec so as to more fully take advantage of the advances that have taken place in music hardware, music software, DAW capabilities and overall data distribution.

At the time of this writing, MIDI 2.0 has just been formally ratified and released by the MMA at the Winter NAMM, 2020, and was also finalized by the Association of Musical Electronics Industry (AMEI) of Japan. It was pointed out at this show by a good friend of mine with the Yamaha Corporation of America that the specifics of MIDI 2.0 could easily fill four books by themselves. For this reason, I feel it best that I present you with an overall summary of the capabilities, strengths and technology of the 2.0 spec … and not the full details of the spec itself. Of course, those of you who have a need to dig deeper into the spec can do so by logging onto the MMA website (www.midi.org) and downloading the current MIDI 2.0 or other related MIDI specs as they become available.

Introduction to MIDI 2.0

The result of a global, decade-long development, MIDI 2.0 is an effort to keep MIDI relevant in the future, using a new Universal MIDI Packet format that allows easier communication over any digital medium (such as USB, Thunderbolt, Ethernet or any other medium that's yet to be envisioned). Future growth and flexibility are also a big part of 2.0, in that additional space has also been reserved for other message and controller types that might be needed in the future.

Some of the more important additions to the MIDI 2.0 spec include:

- Bidirectional communication
- Backwards compatibility
- Both protocol specs are fully supported
- Higher resolution for velocity and control messages
- Tighter timing
- Sixteen channels become 256
- Built-in support for "per-note events"

THE THREE BS

To begin with, let's take a look at the basic foundation of the MIDI 2.0 spec, known as "The Three Bs":

- Bidirectional communication
- Backwards compatibility
- Both protocols (simultaneous support for both specs)

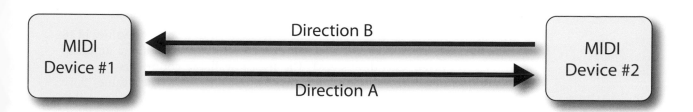

Bidirectional Communication

As we've read earlier in this chapter, MIDI 1.0 is only capable of being sent in one direction. It is only able to communicate from a source to a receiving destination, without the ability to know what the capabilities of the destination device are or whether that information can be properly interpreted. By contrast, MIDI 2.0 can both communicate bidirectionally (Figure 9.25) as well as set up a dialog that allows the involved devices to be automatically configured in ways that allow them to work together in better and/or faster ways.

FIGURE 9.25
MIDI 2.0 communication data in a bidirectional manner between compatible devices.

Some of you might remember that in the past, certain samplers and SCSI-capable devices made use of tiny jumper switches that had to be set with a screwdriver or a pen point so as to make sure that the receiver could properly receive the data that was being sent. Now, with 2.0, these compatibility settings can be automatically made with a digital handshake within the data stream itself. Just plug the devices in, and when the connection is made, the device capabilities are automatically negotiated and configured so as to best make use of their capabilities.

With 2.0's bidirectional capabilities, a device could ask: "Hey, who are you and what are you capable of?" … When it hears back, they can then agree to talk to each other in a way that makes communications faster, better and easier for the task at hand.

Backwards Compatibility

In this scenario, if a 2.0 capable device shouts out to another device asking: "Hey, who are you?" and hears nothing back, it'll assume that the destination device is older and will automatically revert to using the fully supported language of MIDI 1.0.

Since the overwhelming majority of electronic instruments, controllers and processors are built upon the 1.0 spec, it stands to reason that communication with the entire present-day music-making ecosystem must be fully and transparently supported.

Both Protocols

One of the core goals of MIDI 2.0 is to enhance the feature set of MIDI 1.0 whenever possible.

HIGHER RESOLUTION FOR VELOCITY AND CONTROL MESSAGES

One of the biggest developments in 2.0 is its expansion from 7-bit values (128 discrete steps) to 16-bit (which in the case of velocity, represents a jump up to 65,536 discrete encoding steps). The list of available controllers has been expanded to over 32,000 different controller options, while their values have been enhanced by offering a 32-bit range (representing 4,294,967,296 possible discrete steps). This can be thought of as going from having rough control over every aspect of MIDI 1.0 to having extremely precise control over such articulation and control aspects, such as volume, modulation, pitch bend, aftertouch, etc. Potentially, this increased degree of accuracy will allow greater control over performance values, making the process of music production more analog and fine-tunable in nature.

Since MIDI's inception, guitar, violin, woodwind and brass players often had to learn how to best express their instrument using the keyboard playing surface. Now, it should be possible to design newer, more sensitive controllers or translation pickups that would allow players to use their own instruments (or ones that are closer in form and playability to their own) in order to capture their sound into their favorite DAW, processing and/or printing software.

It could be argued that these higher degrees of resolution might be unnecessary for making music. However, newer types of instruments that are able to make use of micro-tonal performance characteristics to better mimic human performances would certainly benefit from the added control. In addition, MIDI is able to capture and control more than just musical data; its use in lighting and animatronics, for example, could certainly benefit from the added degree of resolution. In short, who knows what will be invented in the future that can best make use of the added capabilities and resolutions that MIDI 2.0 has to offer? … Time will tell.

TIGHTER TIMING

MIDI 2.0 has improved timing accuracy for making sure that notes and other event types occur at precisely the right time. Through the use of Jitter Reduction (JR) Timestamps, the precise time of an event can be directly encoded within the MIDI message itself. In this way, if a message gets delayed within the data stream, the system will be able to accurately keep track of the time that an event is supposed to take place.

Timing problems can often crop up when a great deal of MIDI data is passed through a single connection (data clogging). One situation where this might occur is when a great deal of data passes from a MIDI guitar to a sequenced track or synth instrument. In such a case, data might get clogged, and latency delays might cause sounds to stutter or be mistimed. By time stamping the actual MIDI data, it would actually be possible for the data to be returned to its proper position with a sequence … after it's been recorded.

SIXTEEN CHANNELS BECOME 256

The original 1.0 spec offers up 16 separate channels over a single MIDI cable or data line. Each message sent over the 2.0 spec includes a channel group (of which there are 16), which can address any of 16 channels, allowing the encoding of up to 256 distinct channels over a single cable or data line!

MIDI 1.0 data messages, such as System Exclusive messages, can also be transmitted along any one of these 256 channels, alongside MIDI 2.0 data, allowing full compatibility between the newer and older specifications.

BUILT-IN SUPPORT FOR "PER-NOTE" EVENTS

Another significant advancement within 2.0 is the addition of new Per-Note messages, which expand on the concepts adopted by the MMA in MPE (MIDI Polyphonic Expression). This means that 2.0 will directly encode performance data from specialized instruments that are designed to offer extended controller functions over pitch bend, expression, modulation, etc., all on a per-note basis. For example, newer-specialized keyboards and performance controllers (Figure 9.26) allow the artist to add pitch and other controller gestures to a single keyboard note or controller pad in a fully polyphonic fashion. In the future, the added capabilities of MIDI 2.0 over the encoding of controller data could take these and other performance/control functions even further.

Before the adoption of 2.0, the MPE-encoded messages were sent by rotating through a defined contiguous block of channels called Per-Note channels. These Per-Note messages were limited to Note-On, Note-Off, Channel Pressure (finger pressure), Pitch Bend (X-axis movement) and Controller #74 (Y-axis movement). All other messages (such as Program Change, volume and sustain) apply to all voices and are sent over a separate "Common" channel (which is usually CH 1 or 16). If channel 1 is used as the common channel, the Per-Note channels are 2–16. If 16 is used, the Per-Note channels will be 1–15. As an example, it's also possible, using MIDI 1.0, to split an MPE keyboard so that the left split makes use of CH 1 for the common channel, while 2 through 8 are used as the Per-Note channels, and the right split uses CH 16 at the common, and 9 through 15 are used for Per-Note data.

FIGURE 9.26
ROLI Seaboard Block Studio Edition 24-note USB/Bluetooth LE keyboard controller with MPE, Polyphonic Aftertouch/Pitch Bend and ROLI studio software (courtesy of ROLI, www.roli.com, photo © ROLI Ltd, 2022).

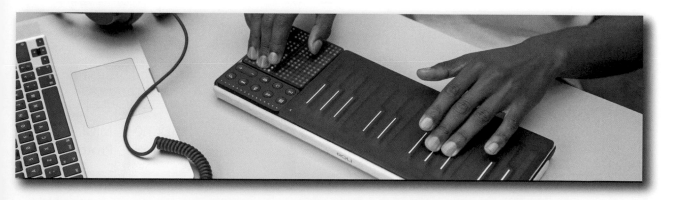

From this scenario, you can see why it's rather exciting to have Per-Note controllers built directly into MIDI 2.0 without having to make special channel assignments or other arrangements. The direct incorporation of MPE-like capabilities gives us an extreme amount of versatility and control over Per-Note polyphony, but with a higher degree of data and performance resolution. The implication of this is that not only can individual pitches be bent individually in a polyphonic fashion, but individual polyphonic controller data can be varied within a single data stream.

MIDI 2.0 MEETS VST 3

Steinberg, as a member of the MMA, was involved with MIDI 2.0's development. Given that it has been designed from the ground up to include higher resolutions, the long-established Virtual Studio Technologies (VST) plug-in protocol (in the form of VST3 instruments and other plug-in types) is designed to fully make use of 2.0's expanded resolution capabilities.

MIDI CAPABILITY INQUIRY (MIDI-CI)

The secret to the above-mentioned three Bs of MIDI 2.0 lies in its ability to have a bidirectional dialog between instruments, devices and software as to what the capabilities of the involved systems are and then to configure their communications accordingly so as to communicate most effectively. The protocol for this communication is known as MIDI Capability Inquiry (MIDI-CI) (**Figure** 9.27).

FIGURE 9.27
MIDI 2.0 MIDI-CI
environment.

The added functions that MIDI 2.0 brings to devices are made possible through the use of MIDI-CI. The basic concept of this bidirectional dialog is that devices can exchange information as to their basic capabilities. MIDI-CI is then used to negotiate or auto-configure any features that are common between the devices.

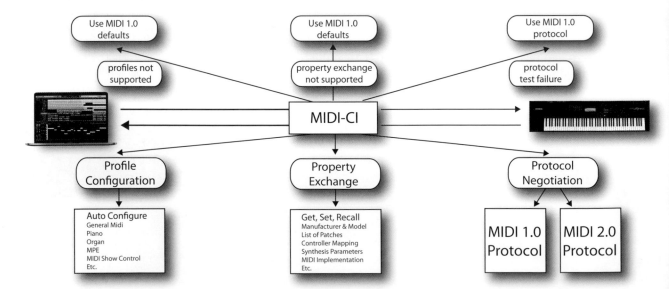

It then provides test mechanisms that can check compatibility when enabling these new features. If a test fails, the communication will revert to using the MIDI 1.0 protocol. In the end, the whole idea behind MIDI-CI is to expand MIDI's capabilities (such as higher resolution and Per-Note control) while protecting backwards compatibility.

THE THREE PS

MIDI-CI includes inquiries for three major areas of expanded MIDI functionality, known as "The Three Ps":

- Profile Configuration
- Property Exchange
- Protocol Negotiation

Profile Configuration

MIDI-CI Profile Configuration allows the use of agreed conventions within the electronic music industry and manufacturers, whereby a setup profile (an agreed-upon settings configuration) can be called up when needed to best configure connected hardware and software for automatic and intuitive use, thereby potentially eliminating the need for manual configuration of these devices by users.

Advanced users might be familiar with the "mapping" conventions that have been created by such companies as Native Instruments and Novation for pre-mapping the various controllers of an instrument or effects plug-in over a hardware/software layout. This mapping eliminates the need to manually configure the assignments for each controller, thereby making the process seamless and automatic.

MIDI-CI is likewise capable of natively mapping all of the controllers from one device to another in an equally seamless fashion. Profiles can be written for device types or for unique applications and/or devices that could be universally adopted by the industry for a wide range of hardware/software instruments, controllers or effects control layouts. Such profiles could, of course, be used in non-musical applications, such as with robotics, industrial machines or lighting controllers.

In one example, such a MIDI controller map could be configured for a grand piano sample setup, which could be programmed with standard configurations (such as Note-On/Off, sustain pedal, etc.). However, additional parameters could also be programmed into the profile that would allow for one or more specialized velocity curves, variable sustain pedals, variable open lid angles, tuning, decay and other parameters. Any device or software system that has been designed to conform to this specific and industry-standardized profile would be automatically "mapped" throughout the system (DAW, hardware, software and controllers) to this convention.

In another example, a workstation might use MIDI-CI to query a hardware device that can be used as a hardware mixer. Once the profile is confirmed, the faders, pan pots and other mixer parameters could be configured and made available to the user. Likewise, a drawbar controller could be instantly mapped to the various control parameters of an organ software plug-in ... or lighting control parameters could be adopted by the industry that would automatically map the needed parameters to create a software or hardware control surface in a way that can save time and eliminate tedious manual programming.

Once adopted, these profiles can be used to auto-configure the settings between DAW, hardware, software and controller systems in a standardized way, allowing MIDI map profiles to pave the way for system setups that can be automatically called up without any muss or fuss.

Property Exchange

Through the use of MIDI-CI, Property Exchange allows the access and sharing of configuration data between devices. This means that information such as parameter lists, controller auto-mapping, synth parameters and setup information about patch presets can be automatically shared. It can choose programs and patches by name and visually display relevant control and display data to DAWs without any prior device knowledge or specially crafted software.

Property Exchange makes use of JavaScript Object Notation (JSON) via System Exclusive messages for exchanging data set information and adding an extended degree of potential possibilities to MIDI 2.0. Using these messages, info such as patch and setup data can be instantly relayed between devices using their actual names and other meaningful info, which can be recognized by the user. Gone would be the use of generic numbers or abstract parameter lists that are hard to decipher in the heat of production. It could even be used to display everything you need to know about your hardware synthesizer on screen, effectively making hardware total synth parameter recall just as possible as with software synths. This could easily turn your recall and control over "Program #32" or "Controller #9" into control over your fave patch "Dreamstate" ... It's much better as a human being to know the actual name of what you're working with. Such is the power of Property Exchange.

Protocol Negotiation

MIDI-CI Protocol Negotiation allows devices to select between using the MIDI 1.0 Protocol and the MIDI 2.0 Protocol. Two devices that have established a two-way MIDI-CI session can negotiate a protocol and the various features of that protocol.

The MIDI 1.0 and the MIDI 2.0 Protocol have many messages in common, which are identical in both protocols. The MIDI 2.0 extends some of the 1.0 messages with a higher resolution and new set of features. Of course, some messages are exclusive to the MIDI 2.0 Protocol.

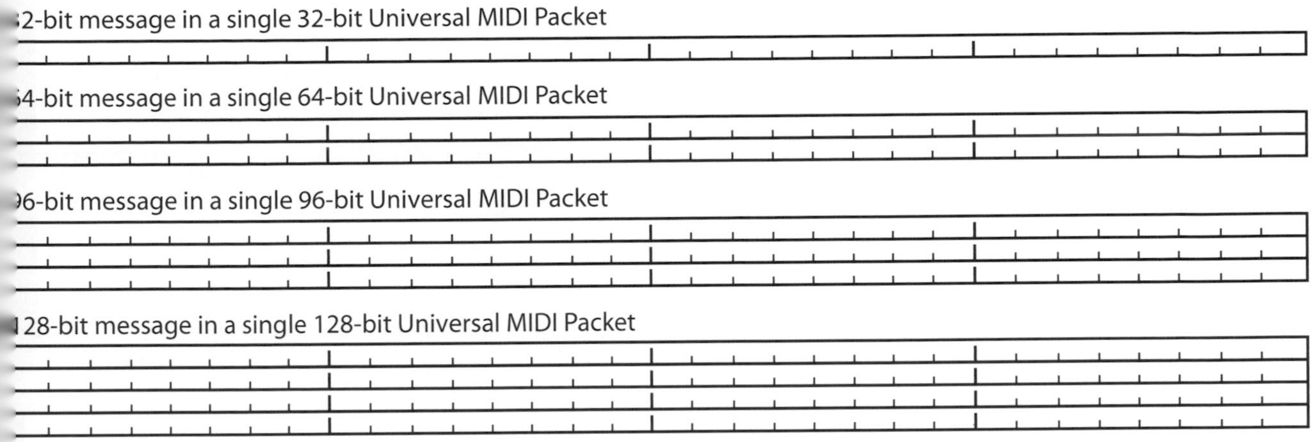

32-bit message in a single 32-bit Universal MIDI Packet

64-bit message in a single 64-bit Universal MIDI Packet

96-bit message in a single 96-bit Universal MIDI Packet

128-bit message in a single 128-bit Universal MIDI Packet

FIGURE 9.28
Universal MIDI packet message format.

THE UNIVERSAL MIDI PACKET

MIDI 2.0 has a new Universal MIDI Packet format for carrying MIDI 1.0 Protocol messages and MIDI 2.0 Protocol messages. A Universal MIDI Packet (Figure 9.28) contains a MIDI message that consists of one to four 32-bit words.

The Universal MIDI Packet format is suited to sending MIDI data over high-speed transports such as USB or a network connection or between applications running inside a personal computer OS.

The traditional 5-pin DIN transport from MIDI 1.0 uses a byte stream rather than packets. At the moment, there is no plan to use the Universal MIDI Packet on the 5-pin DIN cable. Unless/until that plan changes, 5-pin DIN will only support the MIDI 1.0 Protocol.

Message Types

The first 4 bits of every message contain a Message Type . The Message Type is used as a classification of message functions.

GROUPS

The Universal MIDI Packet carries 16 groups of MIDI messages, with each group containing an independent set of system messages and 16 MIDI channels. Therefore, a single connection using the Universal MIDI Packet carries up to 16 sets of system messages over up to 256 channels.

Each of the 16 groups can carry either MIDI 1.0 or MIDI 2.0 Protocol data. Therefore, a single connection can carry both protocols simultaneously. MIDI 1.0 Protocol and MIDI 2.0 Protocol messages, however, cannot be mixed together within one group.

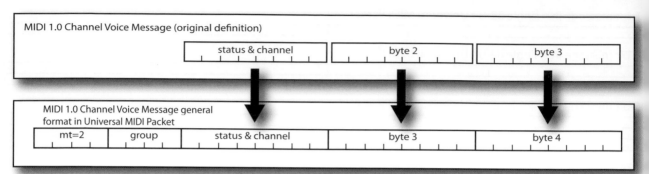

Note: MIDI 1.0 Channel Voice Messages are not all 3 bytes in length. When putting shorter MIDI 1.0 messages into the Channel Voice Message General Format packet, unused bytes are reserved and set to zero.

FIGURE 9.29
MIDI 1.0 Channel Voice Messages within a Univesa MIDI Packet.

Jitter Reduction Timestamps

The Universal MIDI Packet format adds a Jitter Reduction Timestamp mechanism. This Timestamp can be prepended to any MIDI 1.0 Protocol message or MIDI 2.0 Protocol message for improved timing accuracy.

MIDI 1.0 PROTOCOL INSIDE THE UNIVERSAL MIDI PACKET

All existing MIDI 1.0 messages are carried in the Universal MIDI 1.0. As an example, Figure 9.29 shows how MIDI 1.0 Channel Voice Messages are carried in 32-bit packets.

System messages, other than System Exclusive, are encoded similarly to Channel Voice Messages. System Exclusive messages vary in size; they can be very large and can span multiple Universal MIDI Packets.

MIDI 2.0 Protocol Messages

The MIDI 2.0 Protocol uses the architecture of the MIDI 1.0 Protocol to maintain backward compatibility and easy translation while offering expanded features:

- Extends the data resolution for all Channel Voice Messages.
- Makes some messages easier to use by aggregating combination messages into one atomic message.
- Adds new properties for several Channel Voice Messages.
- Adds several new Channel Voice Messages to provide increased Per-Note control and musical expression.
- Adds new data messages, including System Exclusive 8 and Mixed Data Set. The System Exclusive 8 message is very similar to MIDI 1.0 System Exclusive but with 8-bit data format. The Mixed Data Set Message is used to transfer large data sets, including non-MIDI data.
- Keeps all System messages the same as in MIDI 1.0.

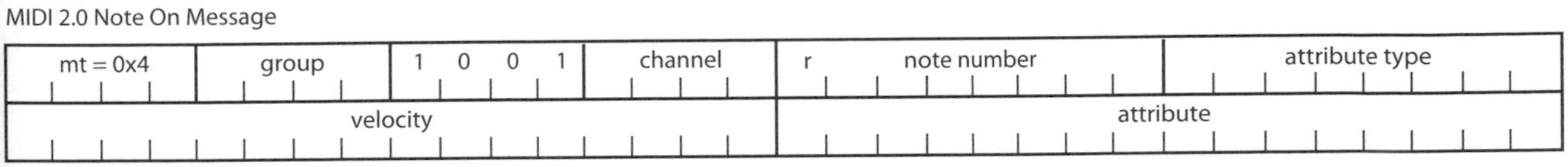

MIDI 2.0 Note On Message

mt = 0x4	group	1　0　0　1	channel	r	note number	attribute type

velocity	attribute

FIGURE 9.30
MIDI 2.0 Protocol Note
Message showing expanded
resolution and capabilities.

EXPANDED RESOLUTION AND EXPANDED CAPABILITIES

The example of a MIDI 2.0 Protocol Note message (Figure 9.30) shows the expansions that are used beyond their MIDI 1.0 Protocol equivalents. The MIDI 2.0 Protocol Note-On, for example, has higher velocity resolution. The two new fields, Attribute Type and Attribute data field, provide space for additional data such as articulation or tuning details.

The MIDI 2.0 Protocol replaces Registered Parameter Number (RPN) and Non-Registered Parameter Number (NRPN), having 16,384 Registered Controllers and 16,384 Assignable Controllers that are as easy to use as Control Change messages.

Creating and editing RPNs and NRPNs with the MIDI 1.0 Protocol requires the use of compound messages. These can be confusing or difficult for both developers and users. MIDI 2.0 Protocol replaces RPN and NRPN compound messages with single messages. The new Registered Controllers and Assignable Controllers are much easier to use.

Managing so many controllers might be cumbersome. Therefore, Registered Controllers are organized in 128 Banks, with each Bank having 128 controllers. Assignable Controllers are also organized into 128 Banks, with each Bank having 128 controllers.

MIDI 2.0 Program Change Message

The MIDI 2.0 Protocol combines the Program Change and Bank Select mechanisms from the MIDI 1.0 Protocol into one message. The MIDI 1.0 mechanism for selecting Banks and Programs requires sending three MIDI messages. MIDI 2.0 changes this mechanism by combining the Bank Select and Program Change into one new MIDI 2.0 Program Change message.

The MIDI 2.0 Program Change message always selects a Program. A Bank Valid bit (B) determines whether a Bank Select is also performed by the message. If Bank Valid = 0, then the receiver performs the Program Change without selecting a new Bank, and the receiver keeps its currently selected Bank. Bank MSB and Bank Least Significant Bit (LSB) data fields are filled with zeroes. If Bank Valid = 1, then the receiver performs both Bank and Program Change. Other option flags that are not yet defined and are Reserved for future use.

The Future of MIDI 2.0

Obviously, MIDI 2.0 is very much in its infancy as of this writing. For further information and updated information as the spec, instrument and usage of 2.0

begin to develop, it's best to follow the changes and expanded capabilities of the spec, which are posted on the MIDI Manufacturer's Association website (www .midi.org).

MIDI AND THE COMPUTER

Besides the coveted place of honor in which most electronic musicians hold their instruments, the most important device in a MIDI system is undoubtedly the personal computer. Through the use of software programs and peripheral hardware, the computer is often used to control, process and distribute information relating to music performance and production from a centralized, integrated control position.

Of course, two computer types dominate modern-day music production: the PC and the Mac. In truth, each brings its own particular set of advantages and disadvantages to personal computing, although their differences have greatly dwindled over the years. My personal take on the matter (a subject that's not even important enough to debate) is that it's a dual-platform world. The choice is yours and yours alone to make. Many professional software and hardware systems can work on either platform. As I write this, some of my music collaborators are fully Mac, some are PC and some (like me) use both, and it doesn't affect our production styles at all. Coexistence isn't much of a problem, either. Living a dual-platform existence can give you the edge of being familiar with both systems, which can be downright handy in a sticky production pinch.

Connecting to the Peripheral World

An important event in the evolution of personal computing has been the maturation of hardware and processing peripherals. With the development of the USB (www.usb.org), FireWire (en.wikipedia.org/wiki/IEEE_1394), Thunderbolt (www.thunderbolttechnology.net) and Dante (www.audinate.com) protocols, hardware devices such as mice, keyboards, cameras, audio interfaces, MIDI interfaces, CD and hard drives, MP3 players and even portable fans can be plugged into an available port without any need to change frustrating hardware settings or open up the box. External peripherals are generally hardware devices that are designed to do a specific task or range of production tasks. For example, an audio interface is capable of translating analog audio (and often MIDI, control and other media) into digital data that can be understood by the computer. Other peripheral devices can perform such useful functions as printing, media interfacing (video and MIDI), scanning, memory card interfacing, portable hard disk storage – the list could fill pages.

THE MIDI INTERFACE

Although computers and electronic instruments both communicate using the digital language of the 1s and 0s, computers simply can't understand the

(a)

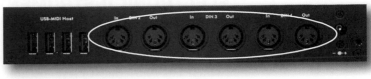

(b)

FIGURE 9.31
Device MIDI ports. (a) Many audio interface devices include MIDI I/O ports (courtesy of Native Instruments GmbH, www.native-instruments.com). (b) Mio xm 4 × 4 (64 channel) MIDI interface (courtesy of iConnectivity, www.iConnectivity.com).

language of MIDI without the use of a device that translates these serial messages into a data structure that computers can comprehend. Such a device is known as the *MIDI interface*. A wide range of MIDI interfaces currently exist that can be used with most computer system and OS platforms. For the casual and professional musician, interfacing MIDI into a production system can be done in a number of ways. Probably the most common way to access MIDI In, Out and Thru jacks is on a modern-day USB or Thunderbolt audio interface or keyboard controller surface (Figure 9.31a), although they usually only offer up a single I/O port.

The next option is to choose a USB MIDI interface. These can range from simpler devices that include a single port to multiple-port systems that can easily handle up to 64 channels over 4 I/O ports. The multiport MIDI interface (Figure 9.31b) is often the device of choice for most professional electronic musicians who require added routing and synchronization capabilities. These USB devices can easily be ganged together to provide eight or more independent MIDI Ins and Outs to distribute MIDI data through separate lines over a connected network.

In addition to distributing MIDI data, these systems often include driver software that can route and process MIDI data throughout the MIDI network. For example, a multiport interface could be used to merge together several MIDI Ins (or Outs) into a single data stream, filter out specific MIDI message types (used to block out unwanted commands that might adversely change an instrument's sound or performance) or re-channel data being transmitted on one MIDI channel or port to another channel or port (thereby allowing the data to be recognized by an instrument or device).

Another important function that can be handled by some multiport interfaces is synchronization. Synchronization (sync, for short) allows other, external devices (such as DAWs, video decks and other media systems) to be simultaneously played back using the same timing reference. Interfaces that include sync features will often read and write SMPTE timecode, convert SMPTE to MIDI timecode (MTC) and allow recorded timecode signals to be cleaned up when copying code from one analog device to another (jam sync). Further reading on synchronization can be found in Chapter 12.

In addition to these interface types, a number of MIDI keyboard controllers and synth instruments have been designed with MIDI ports and jacks built right into them. For those getting started, this useful and cost-saving feature makes it easy to integrate your existing instruments into your DAW and sequencing environment.

ELECTRONIC INSTRUMENTS

Since their inception in the early 1980s (www.midi.org/articles/the-history-of -midi), MIDI-based electronic instruments have played a central and important role in the development of music technology and production. These devices (which fall into almost every instrument category), along with the advent of cost-effective analog and digital audio recording systems, have probably been the most important technological advances to shape the industry into what it is today. In fact, the combination of hardware and newer software plug-in technologies has turned the personal project studio into one of the most important driving forces behind modern-day music production.

Inside the Toys

Although electronic instruments often differ from one another in looks, form and function, they almost always share a common set of basic building block components, including the following:

- *Central processing units (CPUs):* CPUs are one or more dedicated computing devices (often in the form of a specially manufactured microprocessor chip) that contain all of the necessary instructional brains to control the hardware, voice data and sound-generating capabilities of the entire instrument or device.
- *Performance controllers:* these include such interface devices as music keyboards, knobs, buttons, drum pads and/or wind controllers for inputting performance data directly into the electronic instrument in real time or for transforming a performance into MIDI messages. Not all instruments have a built-in controller. These devices (commonly known as modules) contain all the necessary processing and sound-generating circuitry; however, the idea is to save space in a cramped studio by eliminating redundant keyboards or other controller surfaces.
- *Control panel:* the control panel is the all-important human interface of data entry controls and display panels that let you select and edit sounds and route and mix output signals, as well as control the instrument's basic operating functions.
- *Memory:* digital memory is used for storing important internal data (such as patch information, setup configurations and/or digital waveform data). This digital data can be encoded in the form of either read-only memory (ROM; data that can only be retrieved from a factory-encoded chip, cartridge, or CD/DVD-ROM) or random access memory (RAM; memory that can be read from and stored to a device's resident memory, cartridge, hard disk or recordable media).
- *Voice circuitry:* depending on the device type, this section can chain together digital processing "blocks" to either generate sounds (voices) or process and reproduce digital samples that are recorded into memory for playback according to a specific set of parameters. In short, it's used to generate

or reproduce a sound patch, which can then be processed, amplified and heard via speakers or headphones.

- *Auxiliary controllers:* these are external controlling devices that can be used in conjunction with an instrument or controller. Examples of these include foot pedals (providing continuous-controller data), breath controllers, and pitch-bend or modulation wheels. Some of these controllers are continuous in nature, while others exist as a switching function that can be turned on and off. Examples of the latter include sustain pedals and vibrato switches.
- *MIDI communications ports:* these data ports and physical jacks are used to transmit and/or receive MIDI data.

Generally, no direct link is made between any of these functional blocks; the data from each of these components is routed and processed through the instrument's or device's CPU. For example, should you wish to select a certain sound patch from the instrument's control panel, the control panel could be used to instruct the CPU to recall all of the waveform and sound-patch parameters from memory that are associated with the particular sound. These instructional parameters would then be used to modify the internal voice circuitry so that when a key on the keyboard is pressed, or a MIDI Note-On message is received, the sound generators will output the desired patch's note and level values.

For the remainder of this section, we'll be discussing the various types of MIDI instruments and controller devices that are currently available on the market. These instruments can be grouped into such categories as keyboards, percussion, MIDI guitars and controlling devices.

Instrument and Systems Plug-Ins

Of course, one of the wonderful things about living in the digital age is that many (if not most) of our new toys aren't hardware at all – they exist as software synths, samplers, effects, manglers and musical toys of all types, features and genres. These systems (which can exist as an instrument or effects plug-in) include all known types of synths, samplers, and pitch- and sound-altering devices that are capable of communicating MIDI, audio, timing sync and control data between the software instrument/effect plug-in and a host DAW program.

Using an established plug-in communications protocol, it's possible for most or all of the audio and timing data to be routed through the host audio application, allowing the instrument or application I/O, timing and control parameters to be seamlessly integrated into the DAW or application. A few of these protocols include VST, AU, MAS, AudioSuite and RTAS.

ReWire is another type of protocol that allows audio, performance and control data of an independent audio program to be wired into a host program (usually a DAW) such that the audio routing and sync timing of the slave program are locked to the host DAW, effectively allowing them to work in tandem as a single production environment. Further reading on plug-in protocols and software can be found in Chapter 15.

KEYBOARDS

By far the most common instruments that you'll encounter in almost any MIDI production facility will probably belong to the keyboard family. This is due, in part, to the fact that keyboards were the first electronic music devices to gain wide acceptance; also, MIDI was initially developed to record and control many of their performance and control parameters. The two basic keyboard-based instruments are the synthesizer and the digital sampler.

The Synth

A synthesizer (or synth) is an electronic instrument that uses multiple sound generators, filters and oscillator blocks to create complex waveforms that can be combined into countless sonic variations. These synthesized sounds have become a basic staple of modern music and range from those that sound "cheesy" to ones that realistically mimic traditional instruments – and all the way to those that generate otherworldly, ethereal sounds that literally defy classification.

Synthesizers generate sounds using a number of different technologies or program algorithms. Examples of these include:

- *FM synthesis:* this technique generally makes use of at least two signal generators (commonly referred to as "operators") to create and modify a voice. It often does this by generating a signal that modulates or changes the tonal and amplitude characteristics of a base carrier signal. More sophisticated FM synths use up to four or six operators per voice, each using filters and variable amplifier types to alter a signal's characteristics.
- *Wavetable synthesis:* this technique works by storing small segments of digitally sampled sound into memory media. Various sample-based and synthesis techniques make use of looping, mathematical interpolation, pitch shifting and digital filtering to create extended and richly textured sounds that use a surprisingly small amount of sample memory, allowing hundreds if not thousands of samples and sound variations to be stored in a single device or program.
- *Additive synthesis:* This technique makes use of combined waveforms that are generated, mixed and varied in level over time to create new timbres that are composed of multiple and complex harmonics. Subtractive synthesis makes extensive use of filtering to alter and subtract overtones from a generated waveform (or series of waveforms).

Of course, synths come in all shapes and sizes and use a wide range of patented synthesis techniques for generating and shaping complex waveforms in a polyphonic fashion using 16, 32 or even 64 simultaneous voices (Figures 9.32

(a)   (b)

and 9.33). In addition, many synths often include a percussion section that can play a full range of drum and "perc" sounds in a number of styles. Reverb and other basic effects are also commonly built into the architecture of these devices, reducing the need for using extensive outboard effects when being played on stage or out of the box. Speaking of "out of the box", a number of synth systems are referred to as being "workstations". Such beasties are designed (at least in theory) to handle many of your basic production needs (including basic sound generation, MIDI sequencing, effects, etc.) all in one neat little package.

FIGURE 9.32
Hardware synths. (a) Mopho × 4 4-voice analog synth (courtesy of Dave Smith Instruments, www. davesmithinstruments.com). (b) Fantom-X7 Workstation Keyboard (courtesy of Roland Corporation US, www.rolan-dus.com).

Samplers

A sampler (Figure 9.34) is a device that can convert audio into a digital form that is then imported into, manipulated and output from internal RAM.

Once audio has been sampled or loaded into RAM, segments of audio can then be edited, transposed, processed and played in a polyphonic, musical fashion. Additionally, signal processing capabilities, such as basic editing, looping, gain changing, reverse, sample-rate conversion, pitch change and digital mixing, can also be easily applied to:

- Edit and loop sounds into a usable form
- Vary and modulate envelope parameters (e.g., dynamics over time)
- Apply filtering to alter the shape and feel of the sound
- Vary processing and playback parameters

A sample can then be played back according to the standard Western musical scale (or any other scale, for that matter) by altering the reproduced sample rate

FIGURE 9.33
Software synthesizers. (a) Reactor shown running "blocks" (courtesy of Native Instruments GmbH, www. native-instruments.com). (b) Omnisphere (courtesy of Spectrasonics, www.spectra-sonics.net).

(a)   (b)

(a)   (b)

FIGURE 9.34
Sampling systems. (a) Roland MC-707 hardware sampler (courtesy of Roland Corporation US, www.roland.com). (b) Kontakt Virtual Sampler (courtesy of Native Instruments GmbH, www.native-instruments.com).

over the controller's note range. In simple terms, a sample can be imported into the device, where it can be mapped (assigned) to a range of keys (or just a single key) on the keyboard. When the root key is played, the sample will play back at its original pitch; however, when notes above or below that key are played, they will be transposed upwards or downwards in a musical fashion.

By choosing the proper sample-rate ratios, these sounds can be polyphonically played (whereby multiple notes are sounded at once) at pitches that correspond to standard musical chords and intervals.

A sampler (or synth) with a specific number of voices (e.g., 64 voices) simply means that up to 64 notes can be simultaneously played on a keyboard at any one time. Each sample in a multiple-voice system can be assigned across a performance keyboard using a process known as splitting or mapping. In this way, a sound can be assigned to play across the performance surface of a controller over a range of notes, known as a zone (Figure 9.35). In addition to grouping samples into various zones, velocity can enter into the equation by allowing multiple samples to be layered across the same keys of a controller according to how soft or hard they are played. For example, a single key might be layered so that pressing the key lightly would reproduce a softly recorded sample, while pressing it harder would produce a louder sample with a sharp, percussive attack. In this way, mapping can be used to create a more realistic instrument or wild set of soundscapes that change not only with the played keys but with

FIGURE 9.35
Example of a sampler's keyboard layout that has been programmed to include zones. Notice that the upper register has been split into several zones that are triggered by varying velocities.

hard grand piano

loud honky piano

bass gong   upright bass   soft grand piano   soft honky piano

different velocities as well. Most samplers have extensive edit capabilities that allow the sounds to be modified in much the same way as a synthesizer, using such modifiers as:

- Velocity
- Panning
- Expression (modulation and user control variations)
- Low-frequency oscillation (LFO)
- Attack, delay, sustain and release (ADSR) and other envelope processing parameters
- Keyboard scaling
- Aftertouch

Many sampling systems will often include such features as integrated signal processing, multiple outputs (offering isolated channel outputs for added live mixing and signal processing power or for recording individual voices to a multitrack recording system) and integrated MIDI sequencing capabilities.

Sample Libraries and DIY Sampling

Just as patch data in the form of SysEx dump files can have the effect of breathing new life into your synth, a wide range of free or commercially available samples is commonly available online or as a purchase package that lets you experiment with loading new and fresh sounds into your production system. These files can exist as unedited sound file data (which can be imported into any sample system or DAW track) or as data that has been specifically programmed by a professional musician/programmer to contain all the necessary loops, system commands and sound-generating parameters, so that all you ideally need to do is load the sample and start having fun.

The mind boggles at the range of styles and production quality that has gone into producing samples that are just ready and waiting to give your project a boost. The general production level literally runs the entire amateur-to-pro gamut – meaning that whenever possible, it's wise to listen to examples to determine their quality and to hear how they might fit into your own personal or project style before you buy. As a final caveat, by now, you've probably heard of the legal battles that have been raging over sampled passages that have been "ripped" from recordings of established artists. In the fall of 2004, in the case of *Bridgeport Music et al. v. Dimension Films*, the 6th Circuit US Court of Appeals ruled that the digital sampling of a recording without a license is a violation of copyright, regardless of size or significance. This points to the need for tender loving care when lifting samples off a record, the Web or a CD.

It goes without saying that a great deal of satisfaction and individual artistry can be gained by making your own samples. Lifting sounds from your own instruments, music projects or simply walking around the house and recording "things" with a handheld recorder can be super fun and interesting. Editing them and then importing them into a software sampler is far easier than it used to be with hardware samplers, and playing the sounds of that Tibetan bell that you recorded in a cabin in the mountains of Austria can help make your music more personal. Get out, start recording and importing your own sounds, and have fun!

The MIDI Keyboard Controller

As computers, sound modules, virtual software instruments and other types of digital devices have come onto the production scene, it's been interesting to note that fewer and fewer instruments are being made that include a music keyboard in their design. As a result, the MIDI keyboard controller (Figure 9.36) has gained in popularity as a device that might include:

- A music keyboard surface
- Variable parameter controls
- Fader, mixing and transport controls
- Switching controls
- Trigger pads

FIGURE 9.36

MIDI keyboard controller. (a) Roland A-88mkII (courtesy of Roland Corporation US, www.roland.com). (b) Native Instruments S88 mk2 (courtesy of Native Instruments GmbH, www.native-instruments.com).

As was stated, these devices contain no internal tone generators or sound-producing elements. Instead they can be used in the studio or on the road as a simple and straightforward surface for handling MIDI performance, control and device-switching events in real time.

As you might imagine, controllers vary widely in the number of features that are offered. For starters, the number of keys can vary from the sporty, portable 25-key models to those having 49 and 61 keys and all the way to the full 88-key models

(a)

(b)

that can play the entire range of a full-size grand piano. The keys may be fully or partially weighted, and in a number of models, the keys might be much smaller than the full piano key size – often making a performance a bit difficult. Beyond the standard pitch and modulation wheels (or similar-type controller), the number of options and general features is up to the manufacturers. With the increased need for control over electronic instruments and music production systems, many model types offer up a wide range of physical controllers for varying an ever-widening range of expressive parameters.

A Word About Controllers

A MIDI controller is a device that's expressly designed to control other devices (be they for sound, light or mechanical control) within a connected MIDI system. As was previously mentioned, these devices contain no internal tone generators or sound-producing elements but often include a high-quality control surface and a wide range of controls for handling control, trigger and device-switching events.

Since controllers have become an integral part of music production and are available in many incarnations to control and emulate many types of musical instrument, don't be surprised to find controllers of various incarnations popping up all over this book for both recording and electronic music production applications.

One of the wonders of using MIDI to directly control any number of devices via MIDI control messages is that the overall communication between the host and the controller device is open to being easily configurable by the user. A controller message can easily be paired to a physical knob, button or other controller input type by a simple "learn" command. A few companies go a step further to make assignments between devices, plug-ins and software instruments that are much easier to control and operate by creating a unified system for automatically "mapping" a controller's hardware directly to the software's parameters in an easy-to-use, pre-programmed fashion.

One such system offered by Native Instruments is called the Native Kontrol Standard (NKS for short). NKS allows direct control over a growing number of software instruments and plug-ins when working within Native Instrument's Komplete 10 under the control of a Komplete Kontrol or Maschine hardware controller (Figure 9.37a). On the software side, newer DAWs are capable of having layout scripts that allow a controller to be integrated into a session, enabling it to have direct hardware control over instruments and effects in an open session (Figure 9.37b).

The Drum Machine

In its most basic form, the drum machine uses ROM-based, pre-recorded waveform samples to reproduce high-quality drum sounds from its internal memory. These factory-loaded sounds often include a wide assortment of drum sets, percussion sets and rare, wacky percussion hits, and effected drum sets

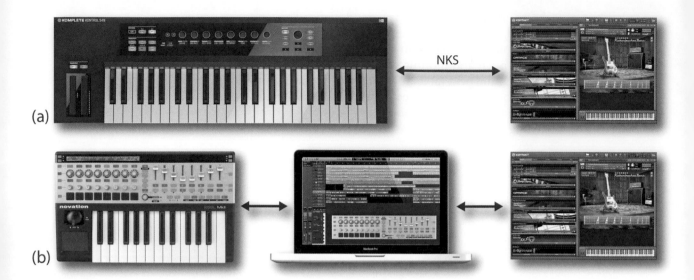

NKS

(a)

(b)

FIGURE 9.37
MIDI controller mapping. (a) Direct hardware mapping to a software plug-in or instrument. (b) Software mapping through a DAW for control over a software plug-in or instrument.

FIGURE 9.38
Examples of drum machines. (a) Alesis SR-18 stereo drum machine (courtesy of Alesis, www.alesis.com). (b) Roland 7X7-TR-8 Rhythm Performer (courtesy of Roland Corporation US, www.rolandus.com).

(e.g., reverberated, gated). Who knows, you might even encounter scream hits by the venerable King of Soul, James Brown. These pre-recorded samples can be assigned to a series of playable keypads that are generally located on the machine's top face, providing a straightforward controller surface that often sports velocity and aftertouch dynamics. Sampled voices can be assigned to each pad and edited using control parameters such as tuning, level, output assignment and panning position.

Because of new cost-effective technology, many drum machines (Figure 9.38) now include basic sampling technology, which allows sounds to be imported, edited and triggered directly from the box. As with the traditional "beat box", these samples can be easily mapped and played from the traditionally styled surface trigger pads. Of course, virtual software drum and groove machines are part of the present-day landscape and can be used in a stand-alone, plug-in and rewired production environment.

(a)

(b)

MIDI Drum Controllers

MIDI drum controllers are used to translate the voicing and expressiveness of a percussion performance into MIDI data. These devices are great for capturing the feel of a live performance while giving you the flexibility of automating or sequencing a live event. These devices range from having larger pads and trigger points on a larger performance surface to drum machine–type pads/buttons. Since the latter type of controller pads are generally too small and not durable enough to withstand drumsticks or mallets, they're generally played with the fingers. A few of the many ways to perform and sequence percussion include:

- *Drum machine button pads:* one of the most straightforward of all drum controllers is the drum button pad design that's built into most drum machines, portable percussion controllers (Figure 9.39), iOS devices and certain keyboard controllers. By calling up the desired setup and voice parameters, these small footprint triggers let you go about the business of using your fingers to do the walking through a performance or sequenced track.

- *The keyboard as a percussion controller:* since drum machines respond to external MIDI data, probably the most commonly used device for triggering percussion and drum voices is a standard MIDI keyboard controller. One advantage of playing percussion sounds from a keyboard is that sounds can be triggered more quickly because the playing surface is designed for fast finger movements and doesn't require full hand/wrist motions. Another advantage is its ability to express velocity over the entire range of possible values (0–127) instead of the limited number of velocity steps that are available on certain drum pad models.

- *Drum pad controllers:* In more advanced MIDI project studios or live stage rigs, it's often necessary for a percussionist to have access to a playing surface that can be played like a real instrument. In these situations, a dedicated drum pad controller would be better for the job. Drum controllers vary widely in design. They can be built into a single, semi-portable case, often having between six and eight playing pads, or the trigger pads can

FIGURE 9.39
Portable drum controllers. (a) Presonus ATOM Performance Controller (courtesy of Presonus Audio Electronics, Inc., www.presonus.com). (b) Roland Octapad spd-30 Digital Percussion Pad (courtesy of Roland Corporation US, www.rolandus.com).

(a)

(b)

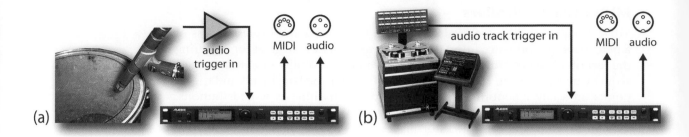

(a) audio trigger in   MIDI   audio

(b) audio track trigger in   MIDI   audio

FIGURE 9.40

By using a MIDI trigger device, a pickup can be either directly replaced or sent via MIDI to another device or sequenced track (a) Using a mic/instrument source. (b) Using a recorded track as a source.

be individual pads that can be fitted onto a special rack, traditional drum floor stand or drum set.

- *MIDI drums:* Another way to MIDI-fy an acoustic drum is through the use of trigger technology. Put simply, triggering is carried out by using a transducer pickup (such as a mic or contract pickup) to change the acoustic energy of a percussion or drum instrument into an electrical voltage. Using a MIDI trigger device (Figure 9.40), a number of pickup inputs can be translated into MIDI so as to trigger programmed sounds or samples from an instrument or recorded track for use on stage or in the studio.

- As was seen in Chapter 8, there are a number of very popular drum/percussion-based systems that combine a software plug-in (or a stand-alone app) with a connected hardware counterpart. These MPC pad-style systems (such as Native Instruments Maschine) are often beat and percussive driven in nature, and can be used to produce music and grooves in a vast number of ways.

Drum Replacement

FIGURE 9.41

Steven Slate Drums Trigger 2 drum replacement plug-in (courtesy of Steven Slate Drums, www.stevenslatedrums.com).

While we're still on the subject of triggering MIDI sounds from acoustic sources such as drums, another option is the use of triggering software to detect and replace drum sounds from an already recorded DAW track (Figure 9.41). Using drum replacement software, it's a relatively simple matter to replace a bad-sounding kick, snare or whatever with one of the many sounds that you might have in your existing drum sample library. These trigger plug-ins can also be

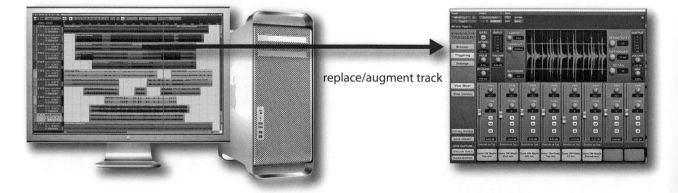

replace/augment track

used to simply augment or fatten up an existing percussion track – the options and possibilities are huge.

SEQUENCING

With regard to the modern-day project studio, one of the most important tools within MIDI production is the *MIDI sequencer*. A sequencer is a digital device or software application that's used to record, edit and output MIDI messages in a sequential fashion. These messages are generally arranged in a track-based format that follows the modern production concept of having separate instruments (and/or instrument voices) located on separate tracks. This traditional track environment makes it easy for us humans to view MIDI data as isolated tracks, most commonly on our DAW.

These sequenced tracks contain MIDI-related performance and control events that are made up of such channel and system messages as Note-On, Note-Off, Velocity, Modulation, Aftertouch and Program/Continuous Controller messages. Once a performance has been recorded into a sequencer's memory, these events can be graphically arranged and edited into a musical performance. The data can then be saved as a MIDI file or (most likely) within a DAW session and recalled at any time, allowing the data to be played back in its originally recorded or edited "sequential" order.

Integrated Hardware Sequencers

A type of keyboard synth and sampler system known as a *keyboard workstation* will often include much of the necessary production hardware that's required for music production, including effects and an integrated hardware sequencer. These systems have the advantage of letting you take your instrument and sequencer on the road without having to drag your whole system along. Similarly to the stand-alone hardware sequencer, a number of these sequencer systems have the disadvantage of offering few editing tools beyond basic transport functions, punch-in/out commands and other basic edit functions. Newer, more powerful keyboard systems include a larger, integrated LCD display and have extensive features that resemble their software sequencing counterparts. Other types of palm-sized sequencers offer such features as polyphonic synth voices, drum machine kits, effects, MIDI sequencing and in certain cases, facilities for recording multitrack digital audio in an all-in-one package that fits in the palm of your hand!

Software Sequencers

By far the most common sequencer type is the software sequencing section that exists within all major DAW programs (Figure 9.42). These tracks (which exist alongside audio tracks within a DAW) take advantage of the hardware and software versatility that only a computer can offer in the way of speed, hardware

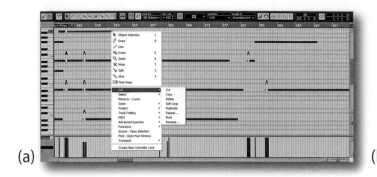

(a)

(b)

FIGURE 9.42
The MIDI edit window within a DAW. (a) Cubase audio production software (courtesy of Steinberg Media Technologies GmbH, a division of Yamaha Corporation, www.steinberg.net). (b) Pro Tools (courtesy of Digidesign, a division of Avid Technology, www.digidesign.com).

flexibility, memory management, signal routing, and digital signal and MIDI processing. These sequence tracks offer a multitude of advantages over their hardware counterparts, such as:

- Increased graphics capabilities (giving us direct control over track and transport-related record, playback, mix and processing functions)
- Standard computer cut-and-paste edit capabilities
- Ability to easily change note and controller values, one note at a time or over a defined range
- A graphic user interface environment that allows easy manipulation of program, controller and edit-related data
- Easy adjustment of performance timing and tempo changes within a session
- Powerful MIDI routing to multiple ports within a connected system
- Graphic assignment of instrument voices via Program Change messages
- Ability to save and recall files using standard computer memory media

Basic Introduction to Sequencing

When dealing with any type of sequencer, one of the most important concepts to grasp is that these devices don't store sound directly; instead, they encode MIDI messages that instruct instruments as to what note is to be played, over what channel, at what velocity and what, if any, optional controller values might be attributed to the messages. In other words, a sequencer simply stores command instructions that follow in a sequential order. This means that the amount of encoded data is a great deal less memory intensive than its digital audio or digital video media counterparts. Because of this, the data overhead that's required by MIDI is very small, allowing a computer-based sequencer to work simultaneously with the playback of digital audio tracks, video images, Internet browsing, etc., all without unduly slowing down the computer's CPU. For this reason, MIDI and the MIDI sequencer provide a media environment that plays well with other computer-based production media.

RECORDING

Commonly, a MIDI sequencer is an application within a DAW workspace for creating personal and commercial compositions in environments that range from the bedroom to more elaborate project and professional studios. As with audio tracks, these systems use a working interface that's roughly designed to emulate a traditional multitrack-based environment. A tape-like set of transport controls lets us move from one location to the next using standard Play, Stop, Fast Forward, Rewind and Record command buttons. Beyond using the traditional Record-Enable button to select the track or tracks that we want to record onto, all we need to do is select the MIDI input (source) port, output (destination) port, MIDI channel (although most DAWs are also able to select all MIDI inputs as a source for ease of use), instrument/plug-in patch and other setup requirements. Then, press the record button and begin laying down the track – it can be that easy.

TRY THIS: TUTORIAL: SETTING UP A SESSION AND LAYING DOWN A MIDI TRACK

1. Pull out a favorite MIDI instrument or call up a favorite instrument plug-in.
2. Route the instrument's MIDI and audio cables (or plug-in routing) to your DAW.
3. Create a MIDI track that can be recorded to.
4. Set the session to a tempo that feels right for the song.
5. Assign the track's MIDI input to the port that's receiving the incoming MIDI data.
6. Assign the track's MIDI output to the port and proper MIDI channel that'll be receiving the outgoing MIDI data during playback.
7. If a click track is desired, turn it on (more about this later).
8. Name the track (always a good idea, as this will make it easier to identify the MIDI instrument/ patch in the future).
9. Place the track into the Record-Ready mode.
10. Play the instrument or controller. Can you hear the instrument? Do the MIDI activity indicators light up on the sequencer, track and MIDI interface? If not, check your cables and run through the checklist again. If so, press Record and start laying down your first track.
11. Once you've finished a track, you can jump back to the beginning of the recorded passage and listen to it. From this point, you could then "arm" (a term used to denote placing a track into the Record-Ready mode) the next track and go about the process of laying down additional tracks (possibly with a different instrument or sound patch) until a song begins to form.

SETTING A SESSION TEMPO

When beginning a MIDI session, one of the first aspects to consider is the tempo and time signature. The beats-per-minute (bpm) value will set the general tempo speed for the overall session. This is important to set at the beginning of the session so as to lock the overall "bars and beats" timing elements to this initial speed, which is often essential in electronic and modern music production. This tempo/click element can then be used to lock the timing elements of other instruments and/or rhythm machines to the session (e.g., a drum machine plug-in can be pulled into the session that'll automatically lock to the session's speed and timing).

To avoid any number of unforeseen obstacles to a straight-forward production, it's often wise to set your session tempo (or at least, think about these options) before pressing the record button.

As with most things MIDI, when working strictly in this environment, it's extremely easy to change almost all aspects of your song (including the tempo) *provided* that you have initially worked to some form of an initial timing element (such as a base tempo). Although working without a set tempo can give a very human feel, it's easy to see how this might be problematic when trying to get multiple MIDI tracks to work together in any form of musical sync and timing control. In truth, there are usually ways to pull corrective timing rabbits out of your technical hat, but the name of the game (as with most things recording) is forethought and preparation.

CHANGING TEMPO

The tempo of a MIDI production can often be easily changed without worrying about changing the program's pitch or real-time control parameters. In short, once you know how to avoid potential conflicts and pitfalls, tempo variations can be made after the fact with relative ease. All you need to do is alter the tempo of a sequence (or part of a sequence) to best match the overall feel of the song. In addition, the tempo of a session can be dynamically changed over its duration by creating a tempo map that causes the speed to vary by defined amounts at specific points within a song. Care and pre-planning should be exercised when a sequence is to be synced to another media form or device.

CLICK TRACK

When musical timing is important (as is often the case in modern music and visual media production), a click track can be used as a tempo guide for keeping the performance as accurately on the beat as possible. A click track can be set to make a distinctive sound on the measure boundary or (for a more accurate timing guide) on the first beat boundary and on subsequent meter divisions (e.g., tock, tick, tick, tick, tock, tick, tick, tick …). Most sequencers can output a click track either by using a dedicated beep sound (often outputting from the device or main speakers) or by sending Note-On messages to a connected instrument in the MIDI chain. The latter lets you use any sound you want, and often at definable velocity levels. For example, a kick could sound on the beat, while a snare sounds out the measure divisions.

The use of a click track is by no means a rule. A strong reason for using a click track (at least initially) is that it serves as a rhythmic guide that can improve the timing accuracy of a performance. However, in certain instances, it can lead to a performance that sounds stiff. For compositions that loosely flow and are legato in nature, a click track can stifle the passage's overall feel and flow. As an alternative, you could turn the metronome down, have it sound only on the measure boundary, and then listen through your headphones. As with most creative decisions, the choice is up to you and your current circumstance.

> Care should be taken when setting the proper time signature at the session's outset. Listening to a 4/4 click can be disconcerting when the song is being performed in 3/4 time.

MULTITRACK MIDI RECORDING

Although only one MIDI track is commonly recorded at a time, most mid- and professional-level sequencers let us record multiple tracks at one time. This feature makes it possible for multiple instruments and performers to be recorded to a multitrack sequence in one live pass.

PUNCHING IN AND OUT

Almost all sequencing systems are capable of punching in and out of record while playing a sequence (Figure 9.43). This commonly used function lets you drop in and out of record on a selected track (or series of tracks) in real time, in a way that mimics the traditional multitrack overdubbing process. Although punch-in and punch-out points can often be manually performed on the fly from the transport or often from a convenient foot pedal, most sequencers can also automatically perform a punch by graphically or numerically entering in

FIGURE 9.43
Automated punch-in and punch-out points (courtesy of Steinberg Media Technologies GmbH, a division of Yamaha Corporation, www.steinberg.net).

the measure/beat points that mark the in and out location points. Once done, the sequence can be rolled back to a point a few measures before the punch-in point, and the artist can then play along while the sequencer automatically performs the necessary switching functions.

TRY THIS: PUNCHING DURING A TAKE

1. Create a MIDI track in your DAW and record a musical passage. Save/name the session.
2. Roll back to the beginning of the take and play along. Manually punch in and out during a few bars (simply by pressing the REC button). Now, undo or revert back to your originally saved session.
3. Read your DAW/sequencer manual and learn how to perform an automated punch (placing your punch-in and punch-out points at a logical place on the track).
4. Roll back to a point a few bars before the punch, and go into the record mode. Did the track automatically place itself into record? Was that easier than doing it manually?
5. Feel free to try other features, such as record looping or stacking.

STEP TIME ENTRY

In addition to laying down a performance track in real time, most sequencers will allow us to enter note values into a sequence one note at a time. This feature (known as step time, step input or pattern sequencing) makes it possible for notes to be entered into a sequence without having to worry about the exact timing. Upon playback, the sequenced pattern will play back at the session's original tempo. Fact is, step entry can be an amazing tool, allowing a difficult or a blazingly fast passage to be meticulously entered into a pattern and then be played out or looped with a degree of technical accuracy that would otherwise be impossible for most of us to play. Quite often, this data entry style is used with fast, high-tech musical styles where real-time entry just isn't possible or accurate enough for the song.

DRUM PATTERN ENTRY

In addition to real-time and step-time entry, most sequencers will allow drum and percussion notes to be entered into a drum pattern grid (Figure 9.44). This graphical environment is intuitive to most drummers and allows patterns to be quickly and easily programmed and then linked together as a string of patterns that can form the backbone of a song's beat.

For those who want to dive into a whole new world of experimentation and sonic surprises, here's a possible guideline that can be used when recording a live recording session: "If the instrument supports MIDI, record the performance data to a DAW MIDI track during each take". For example:

1. You might record the MIDI out of a keyboard performance to a DAW MIDI track. If there's a problem in the performance, you can simply change the note (just after it was played or later) without having to redo the performance – or if you want to change the sound, simply pick another sound.
2. Record the sequenced track from a triggered drum set or controller to a set of DAW MIDI tracks.

3. If a MIDI guitar riff needs some tweaking to fill out the sound, double the MIDI track with another sound patch or chord augmentation.
4. Acoustic drum recordings can benefit from MIDI by using a trigger device that can accept audio from a mic or recorded track and output the triggered MIDI messages to a sampler or instrument in order to replace the bad tracks with samples that rock the house. Of course, don't forget to record the trigger outputs to MIDI tracks on the DAW, just in case you want to edit or change the sounds at a later time.
5. Even if MIDI isn't involved, a drum replacement plug-in could be used to replace bad sounds or to fill out a sound at a later time.

It's all up to you – as you might imagine, surprises can definitely come from experiments like these.

MIDI to Audio

When mixing down a session that contains MIDI tracks, many folks prefer not to mix the sequenced tracks in the MIDI domain. Instead, they'll often export (bounce) the hardware or plug-in instruments to an audio track within the project. Here are a few helpful hints that can make this process go more smoothly:

- Set the main volume and velocities to a "reasonable" output level, much as you would with any recorded audio track.
- Solo the MIDI and instrument's audio input track and take a listen, making sure to turn off any reverb or other effects that might be on that instrument track. If you really like the instrument effect, of course, go ahead and

FIGURE 9.44
Drum pattern entry window in Cubase/Nuendo (courtesy of Steinberg Media Technologies GmbH, a division of Yamaha Corporation, www.steinberg.net).

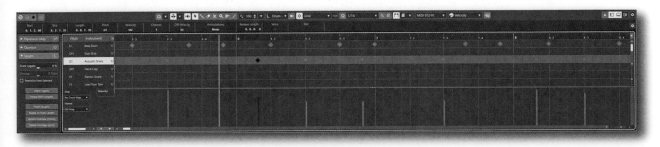

record it; however, you might consider recording the track both with and without effects, as you might want to make changes in mixdown.

- If any mix-related moves have been programmed into the sequence, you might want to strip out volume, pan and other controller messages before exporting the track. This is easily done by making a copy of the existing track, stripping the controller values from the copy and then exporting it to an audio track.

- If the instrument has an acoustic element to it (such as a MIDI acoustic grand piano or MIDI guitar room/amp setup), you might also consider recording the instrument in the studio and mixing these tracks in with the MIDI instrument tracks This will often allow a greater sense of acoustic "space" within the final mix.

It's worth mentioning again that it is always wise to save the original MIDI track or file within the session. This makes future changes in the composition infinitely easier. Failure to save your MIDI files could limit your future options or result in major production headaches down the road.

Various DAWs will offer different options for capturing MIDI tracks as audio tracks within a session. As always, you might consider consulting your DAW's manual for further details.

Audio to MIDI

Another tool that comes under the "You gotta try this!" category is the ability to take an audio file and extract relevant MIDI data from the track. The power in such an option is truly huge! For example, let's say that we have a really killer bass loop audio file that has just the right vibe, but it lacks punch, is too noisy or sounds wrong. A number of DAWs actually have algorithms for detecting and extracting MIDI note values that will result in a MIDI loop or sequence that can be saved as a separate MIDI file. As with all things MIDI, this new loop can be routed to a synth, sampler or anything to change the sound in a way that'll make it stand out. In short, you can now do "anything" with the sequence. You owe it to yourself to check this out!

There's one thing to keep in mind when extracting MIDI from an audio track or loop, however. Each DAW will usually use different detection algorithms for drum/percussion, complex melody patterns and sustained pads. You might experiment to find which will work best with your track, as different DAWs will often yield completely different results. Experimentation and experience is often the name of the game here. Lastly, the results might be close to the original (in pitch and rhythmic pattern), but it might require a bit of human editing to get the pattern and/or feel right. Once you're happy with the final results, the resulting sounds can be truly fun to experiment with.

Saving Your MIDI Files

Just as it's crucial that we carefully and methodically back up our program and production media, it's important that we save our MIDI and session files while we're in production. This can be done in two ways:

- Periodically save your files over the course of a production.
- At important points throughout a production, you might choose to save your session files under new and unique names (mysong001, mysong002, etc.), thereby making it possible to easily revert back to a specific point in the production. This can be an important recovery tool should the session take a wrong turn or for re-creating a specific effect and/or mix. Personally, I save these session versions under a "mysong_bak" sub-directory within the project session.

Again, We Can't Stress This Enough ...

- When working with MIDI within a DAW session, it's always a good idea to save the original MIDI tracks within the session. This makes it easy to go back and change a note, musical key or sounding voice or to make any other alterations you want. Not saving these files could lead to some major headaches or worse.
- Keeping your DAW's MIDI files within a folder called "MIDI" will make it easier to identify and even organizationally hide (collapse) these tracks when they're not needed.

MIDI files can also be converted to and saved as a standard MIDI file for use in exporting to and importing from another MIDI program or for distributing MIDI data for use on the Web, to cell phones, etc. These files can be saved in either of two formats:

- *Type 0:* saves all of the MIDI data within a session as a single MIDI track. The original MIDI channel numbers will be retained. The imported data will simply exist on a single track. Note that if you save a multi-instrument session as a Type 0, you'll lose the ability to save the MIDI data to discrete tracks within the saved sequence.
- *Type 1:* saves all of the MIDI data within a session onto separate MIDI tracks that can be easily imported into a sequencer in a multitrack fashion.

Documentation

When it comes to MIDI sequencing, one of the things to definitely keep in mind is the need for documenting any and all information that relates to your MIDI tracks. It's always a good idea to keep notes about:

- What instrument/plug-in is being used on the track?
- What is the name of the patch?

- What are the settings that are being used (if these settings are not automatically saved within the session – and even if they are)? In such cases, you might want to make a screenshot of the plug-in settings, or if it's a hardware device, get out your cell phone or camera and take a picture that can be placed in the documentation directory.

Placing these and any other relevant pieces of information into the session (within the track notes window or in a separate doc file) can really come in handy when you have to revisit the file a year or so later and you've totally forgotten how to rebuild the track. Trust me on this one; you'll eventually be glad you did.

Editing

One of the more important features that a sequencer (or MIDI track within a DAW) has to offer is its ability to edit sequenced tracks or blocks within a track. Of course, these editing functions and capabilities often vary from one DAW/sequencer to another. The main track window of a sequencer or MIDI track on a DAW is used to display such track information as the existence of track data, track names, MIDI port assignments for each track, program change assignments, volume controller values and other transport commands.

Depending on the sequencer, the existence of MIDI data on a particular track at a particular measure point (or over a range of measures) is indicated by the highlighting of track data in a way that's extremely visible. For example, back in Figure 9.42, you'll notice that the MIDI tracks contain graphical bar display information. This means that these measures contain MIDI messages, while the non-highlighted areas don't.

By navigating around the various data display and parameter boxes, it's possible to use cut-and-paste and/or edit techniques to vary note values and parameters for almost every facet of a musical composition. For example, let's say that we really screwed up a few notes when laying down an otherwise killer bass riff. With MIDI, fixing the problem is totally a no-brainer. Simply highlight each fudged note and drag it (them) to the proper note location – we can even change the beginning and endpoints in the process. In addition, tons of other parameters can be changed, including velocity, modulation and pitch bend, note and song transposition, quantization and humanizing (factors that eliminate or introduce human timing errors that are generally present in a live performance), in addition to full control over program and continuous controller messages … the list goes on and on.

PRACTICAL EDITING TECHNIQUES

When it comes to learning the Ins, Outs and Thrus of basic sequencing, absolutely nothing can take the place of diving in and experimenting with your own setup. Here, I'd like to quote Craig Anderton, who said: "Read the manual once when you get the program (or device), then play with the software and get to know it before you need it. Afterwards, reread the manual to pick up the system's finer operating points". Wise words – although I tend to take another route and simply start pressing buttons and icons until I learn what I need. I honestly think there's something to be said for both approaches.

In the following section, we'll be covering some of the basic techniques that'll help speed you on your way to sequencing your own music. Note that there are no rules to sequencing MIDI. As with all music production (and the arts, for that matter), there are as many right ways to perform and play with making music via MIDI as there are musicians. Just remember that there are no hard and steadfast rules to music production – but there are always guidelines, tools and tips that can speed and improve the process.

Transposition

As was mentioned earlier, a sequencer app is capable of altering individual notes in a number of ways, including pitch, start time, length and controller values. In addition, it's generally a simple matter for a defined range of notes in a passage to be altered in ways that could alter the overall key, timing and controller processes. Changing the pitch of a note or the entire key of a song is extremely easy to do on a MIDI track. Depending on the system, a song can be transposed up or down in pitch at the global or defined measure level, thereby affecting the pitch or musical key of a song. Likewise, a segment can be shifted in pitch from the main edit, piano roll or notation edit windows by simply highlighting the bars and tracks that are to be changed and then dragging them or by calling up the transpose function from the MIDI edit menu.

Quantization

By far most common timing errors begin with the performer. Fortunately, "to err is human", and standard performance timing errors often give a piece a live and natural feel. However, for those situations when timing goes beyond the bounds of nature, an important sequencing feature known as quantization can help correct these timing errors. Quantization allows timing inaccuracies to be adjusted to the nearest desired musical time division (such as a quarter, eighth, or sixteenth note). For example, when performing a passage where all involved notes must fall exactly on the quarter-note beat, it's often easy to make timing mistakes (even on a good day). Once the track has been recorded, the problematic passage can be highlighted, and the sequencer can recalculate each note's start and stop times so they fall precisely on the boundary of the closest time division.

Humanizing

The humanization process is used to randomly alter all of the notes in a selected segment according to such parameters as timing, velocity and note duration. The amount of randomization can often be limited to a user-specified value or percentage range, and parameters can be individually selected or fine-tuned for greater control. Beyond the obvious advantages of reintroducing human-like timing variations into a track, this process can help add expression by randomizing the velocity values of a track or selected tracks. For example, humanizing the velocity values of a percussion track that has a limited dynamic range can help bring it to life. The same type of life and human swing can be effective on almost any type of instrument.

Slipping in Time

Another timing variable that can be introduced into a sequence to help change the overall feel of a track is the slip time feature. Slip time is used to move a selected range of notes either forward or backward in time by a defined number of clock pulses. This has the obvious effect of changing the advance/retard times for these notes relative to the other notes or timing elements in a sequence.

Editing Controller Values

Almost every DAW sequencer allows controller message values to be edited or changed, often by using a simple, graphic window whereby a line or freeform curve can be drawn that graphically represents the effect that the controller messages will have on an instrument or voice. By using a mouse or other input device, it becomes a simple matter to draw a continuous stream of controller values that correspondingly change such variables as velocity, modulation, pan, etc. To physically change parameters using a controller, all you need to do is twiddle a knob, move a fader or graphically draw the variables on screen in a WYSIWYG ("what you see is what you get") fashion.

It almost goes without saying that a range of controller events can be altered on one or more tracks by allowing a range of MIDI events to be highlighted and then altered by entering in a parameter or processing function from an edit dialog box. This ability to define a range of events often comes in handy for making changes in pitch/key, velocity, main volume and modulation (to name a few).

Some of the more common controller values that can affect a sequence and/or MIDI mix values include (for the full listing of controller ID numbers, see Table 9.1 earlier in this chapter):

Control #	Parameter
1	Modulation Wheel
2	Breath Controller
4	Foot Controller
7	Channel Volume (formerly Main Volume)
8	Balance
10	Pan
11	Expression Controller
64	Damper Pedal on/off (Sustain) <63 off, >64 on

TRY THIS: TUTORIAL: CHANGING CONTROLLER VALUES

1. Read your DAW/sequencer manual and learn its basic controller editing features.
2. Open a MIDI track or create a new one.
3. Highlight a segment and reduce the overall Channel Volume levels by 50%. On playback, did the levels reduce?
4. Now, refer to your DAW/sequencer manual for how to scale MIDI controller events over time.
5. Select a range of measures and change their Channel Volume settings (controller 7) over time. Does the output level change over the segment when you play it back?
6. Highlight the segment and scale the velocity values so they have a minimum value of 64 and a maximum of 96. Could you see and hear the changes?
7. Again, highlight the segment and draw or instruct the software to fade it from its current value to an ending value of 0. Hopefully, you've just created a Channel Volume fade. Did you see the MIDI channel fader move?
8. Undo the last fade, and start a new one with an initial value of 0 and a current value of 100% toward the end of the section. Did the segment now fade in?
9. Feel free to make up your own edit moves and experiment with the track(s).

Playback

Once a sequence is composed and saved within a session, all of the sequence tracks can be transmitted through the various MIDI ports and channels to the instruments or devices to make music, create sound effects for film tracks, control device parameters in real time, etc. Because MIDI data exists as encoded real-time control commands and not as audio, you can listen to the sequence and make changes at any time. You could change the patch voices, alter the final mix, or change and experiment with such controllers as pitch bend or modulation – and as we've read, change the tempo and key signature. In short, this medium is infinitely flexible in the number of versions that can be created, saved, folded, spindled and mutilated until you've arrived at the overall sound and feel you want. Once done, you'll have the option of using the medium for live performance or mixing the tracks down to a final recorded product, either in the studio or at home.

During the summer, in a wonderful small-town tavern in the city where I live, there's a frequent performer who'll wail the night away with his voice, trusty guitar and a backup band that consists of several electronic synth modules and a DAW that's just chock-full of country-'n'-western sequences. His set of songs for the night is loaded into a song playlist that's pre-programmed for that night. Using this, he queues his sequences so that when he's finished one song, taken his bow, introduced the next song and complimented the lady in the red dress, all he needs to do is press the spacebar and begin playing the next song. Such is the life of a hard-workin', on-the-road sequencer.

Mixing a Sequence

Almost all DAW and sequencer types will let you mix a sequence in the MIDI domain using various controller message types. This is usually done by simply integrating the world of MIDI into the DAW's on-screen mixer. Instead of directly mixing the audio signals that make up a sequence, however, these controls are able to access such track controllers as Main Volume (controller 7), Pan (controller 10), and Balance (controller 8) in an environment that completely integrates into the workstation's overall mix controls. Therefore, even with the most basic DAW, you'll be able to mix and remix your sequences with complete automation and total settings recall whenever a new sequence is opened. As is usually the case with a DAW's audio and MIDI graphical user interface (GUI), the controller and mix interface will almost always have moving faders, controls and access to external hardware controllers.

Don't forget that you can export your MIDI tracks/instruments as audio tracks for easy use and manipulation within a session. Always remember to save (and disable) your MIDI track within the session just in case any changes need to be made later.

MUSIC PRINTING PROGRAMS

In recent times, the field of transcribing musical scores and arrangements has been strongly affected by both the computer and MIDI technology. This process has been greatly enhanced through the use of computer software that makes it possible for music notation data to be entered into a computer either manually (by placing the notes onto the screen via keyboard or mouse movements) or by direct MIDI input or by sheet music scanning technology. Once entered, these notes can be edited in an on-screen environment that lets you change and configure a musical score or lead sheet using standard cut-and-paste editing techniques. In addition, most programs allow the score data to be played directly from the score by electronic instruments via MIDI. A final and important program feature is their ability to quickly print out hard copies of a score or lead sheets in a wide number of print formats and styles.

A music printing program (also known as a music notation program) lets you enter musical data into a computerized score in a number of manual and automated ways (often with varying degrees of complexity and ease). Programs of this type (Figure 9.45) offer a wide range of notation symbols and type styles that can be entered from either a computer keyboard or a mouse. In addition to entering a score manually, most music transcription programs will accept MIDI input, allowing a part to be played directly into a score. This can be done in

real time (by playing a MIDI instrument/controller or finished sequence directly into the program), in step time (by entering the notes of a score one note at a time from a MIDI controller) or by entering a standard MIDI file into the program (which uses a sequenced file as the notation source).

In addition to dedicated music printing programs, most DAW or sequencer packages will include a basic music notation application that allows the sequenced data within a track or defined region to be displayed and edited directly within the program, from which it can be printed in a limited score-like fashion. However, a number of high-level workstations offer scoring features that allow sequenced track data to be notated and edited in a professional fashion into a fully printable music score.

As you might expect, music printing programs often vary widely in their capabilities, ease of use and offered features. These differences often center around the GUI, methods for inputting and editing data, the number of instrumental parts that can be placed into a score, the overall selection of musical symbols, the number of musical staves (the lines that music notes are placed onto) that can be entered into a single page or overall score, the ability to enter text or lyrics into a score, etc. As with most programs that deal with artistic production, the range of choices and general functionality reflects the style and viewpoints of the manufacturer, so care should be taken when choosing a professional music notation program to see which one would be right for your personal working style.

FIGURE 9.45
Notation 6 music notation program showing inserted video screen, as well as Notation 6 for the iOS (courtesy of Presonus Audio Electronics, Inc., www.presonus.com).

CHAPTER 10

The iOS in Music Production

One of the most awe-inspiring inventions in the modern age is mobile computing – specifically when it relates to all of the hundreds of millions of existing pads and phones (handys, mobiles, or whatever your country calls them). It has literally changed the way we communicate (or not communicate, if we don't occasionally look up) and the way that information is dealt with in an on-the-go, on-demand kind of way.

Of these mobile devices, the class of systems that has affected the audio production community in a very direct way is the iOS range of devices from Apple. Whether you are an Apple or Windows user, almost all of us in recording, live sound and music production have been directly touched in many ways by these devices, which offer:

- Mobility: as iOS devices are wireless and portable by their very nature, these mini-computers allow us to record, generate and play back audio; perform complex control functions over a production computer; be a mixer or production hardware device – our imagination is the only limit to what these devices can do from a remote, untethered location.

- Affordability: quite often, these multifunction devices will replace dedicated hardware systems that can cost a hundred or even a thousand times what an equivalent "app" might cost.

- Multifunctionality: given that these devices are essentially computers, their capabilities can be as far-reaching and as functional as the programming that went into them. Most importantly, they have the ability to be a chameleon – changing form from being a controller, to a synthesizer, to a mixer, to a calculator, to a flashlight, to a phone and social networking device – all in a fraction of a second.

- Touch capabilities: the ability to directly interact with portable devices through the sense of touch (and voice) is something that we've all come to take for granted. This direct interaction allows all of the above advantages to be available at the touch of a virtual button, which, once touched, can transform the device into another type of phoenix that can raise your

DOI: 10.4324/9781003260530-10

production system from the ashes into something beautiful, simple, cost-effective and functional.

Of course, it goes without saying that this chapter is simply a quick overview of a field that has expanded into a major industry and field of artistry in and of itself. The number of apps, options, and supporting connectivity and control hardware choices continues to grow on a monthly basis. To better find out what options are available and best suit your needs, I urge you to simply browse the App Store, search the web for an ever-expanding number of online resources (e.g., www.iosmidi.com) and wade your way through any number of YouTube videos and reviews on any iSubject that might be helpful to you and your studio scenario.

AUDIO INSIDE THE IOS

Apple has a long history of dealing with properly implementing audio into their operating systems (for the most part). One of the major advantages of dealing with the iOS for portable computing and media players comes down to two factors:

- The iOS has been developed to pass high-quality audio with a very low amount of latency (system delay within the audio path).
- Given the fact that the operating system (including its audio programming) is closed to outside third-party developers, audio applications can be developed in a standardized way that must make use of Apple's audio-friendly programming architecture.

The following sections offer a basic glimpse into how the iOS can integrate into an audio production system with professional, mobile, cost-effective and fun results.

Core Audio on the iOS

Audio on the Mac OS and the iOS is handled through an integrated programming service called Core Audio. This audio architecture (Figure 10.1) is broken into application-level services that include:

FIGURE 10.1
Core Audio's basic I/O architecture.

- Audio Cue Services: used to record, play back, pause, loop and synchronize audio.

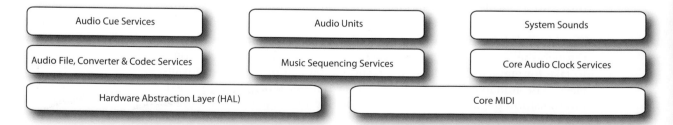

- Audio File, Converter and Codec Services: used to read and write from disk (or media memory) and to perform audio data format transformations (in OS X, custom codecs can also be created).
- Audio Units: used to host audio units (audio plug-ins) in your application.
- Music Sequencing Services: used to play (MIDI)-based control and music related data.
- Core Audio Clock Services: used for audio and MIDI synchronization and time format management.
- System Sounds: used to play system sounds and user-interface sound effects.

Core Audio on the iOS is optimized for the computing resources available in a battery-powered mobile platform.

AudioBus

AudioBus (Figure 10.2) is an iOS app that can be downloaded and used to act as a virtual patch cord for connecting together the ins, outs and throughs of various AudioBus compatible apps in a way that would otherwise not be possible. For example, an audio app that has been launched will obviously have its own input and an output. If we were to launch another audio app, there would be no way to "patch" the output of one app to the input of another within the iOS. Using AudioBus, it's now possible to connect a source device (input), route this through an effects device (which is not acting as "plug-in" but as a stand-alone, processing app) and then patch the output of the effects app through to a final destination app (output). In addition to AudioBus, you might want to be aware that other third-party apps (such as AUM [Audio Mixer]), have also come onto the market that allow audio and MIDI data to be routed between apps in various and innovative ways.

Audio Units for the iOS

With the release of the latest versions of the iOS, Apple has allowed effects and virtual instrument plug-ins to be directly inserted into an app in much the same way that plug-ins can be inserted into programs using the Mac OS. This protocol,

FIGURE 10.2
AudioBus showing input, effects and output routing. (a) Basic screen. (b) Showing audio apps that are available to its input.

(a)

(b)

which is universally known across all Apple platforms as Audio Units, lets us go to the App Store and download a plug-in that might work best in our situation, then insert it directly into the processing path of an iOS digital audio workstation (DAW), video editor or other host app for DSP processing.

CONNECTING THE IOS TO THE OUTSIDE WORLD

iOS devices can easily be connected to an audio or DAW system using cost-effective external hardware solutions that can be tailored to a wide range of studio and on-the-go applications.

Audio Connectivity

Audio passes through an iOS device using Core Audio. Connecting to the device can be as simple as using the internal mic and line/headphone out jack (or lightning/USB-C port) on the unit. Making a higher-quality, multichannel audio connection will often require an audio interface. The number of interface options for connecting an iPad or iPhone to a high-quality mic or integrating these devices into the studio system continues to grow. For example:

- A docking device that is specifically designed to connect an iOS device to the outside studio world of audio and possibly MIDI can be used.
- A standard audio interface that is Class Compliant (cc is a mode that allows the interface to be directly connected to an iOS device using a readily available Apple Camera Adapter, as shown in Figure 10.3). These interfaces can then be directly connected to the device, allowing MIDI and audio to pass in a normal I/O manner.
- Direct digital connection between an iOS device and a Mac through the use of Inter-Device Audio + MIDI (IDAM). IDAM lets us pass digital audio and MIDI through a standard lightning cable on Macs with iOS 11 and higher with an impressively low degree of latency. As an example, a DAW that is capable of communicating via Audio Units would be able to pass MIDI data from a track through the lightning cable, while audio could be passed from an instrument or groove-based app back to a track on the DAW.

A note on latency: in many situations involving the interconnection of audio and MIDI between an iOS and a computer device, it's always wise to be aware of any latency issues that might crop up, especially if the device is being performed and recorded in real time. This issue will often vary depending upon the connections and protocols that are being used.

MIDI Connectivity

Using Core Audio, all iOS devices are capable of receiving and outputting MIDI without the need for additional software or drivers. All that's needed (as always) is a compatible I/O device that is capable of communicating with it. Currently, this is possible in any of three ways:

- By way of a docking device that can serve as an interface for audio and/or MIDI
- By way of a MIDI interface that can act as a bridge to connect both MIDI and audio from an iOS device to a Mac or PC computer (Figure 10.4a) so as to integrate the two together through the use of "Audio PassThru Technology"
- Through the use of an Apple Camera Adapter that can connect the device to an audio interface with one or more MIDI I/O ports (Figure 10.4b)

RECORDING USING IOS

Given the fact that iOS devices are capable of recording, editing, processing and outputting audio, it stands to reason that they would also excel at giving us access to production tools on a bus, in an airplane, by the pool or on our way with Elon to Mars – all in a cost-effective and versatile fashion.

Handheld Recording Using iOS

Another way that an iOS device can come in handy for recording audio is as a handheld recording device. When used with a suitable mic accessory and recording app (Figure 10.5), an iPhone or iPad can be used to capture professional quality audio in an easy-to-use, on-the-go fashion, which can then be edited or transferred to a DAW for further editing, processing and integration into a project.

FIGURE 10.3
A camera adapter is used to connect a compliant device (such as an audio interface to an iOS device).

FIGURE 10.4
Class-compliant MIDI I/O for an iOS device. (a) iConnect MIDI4+ 4 × 4 MIDI interface (courtesy of iConnectivity, www.iConnectivity.com). (b) Rear of Steinberg's UR22 MK II interface showing MIDI I/O (courtesy of Steinberg Media Technologies GmbH, a division of Yamaha Corporation, www.steinberg.net).

(a)    (b)

Mixing With IOS

Another way that the iOS has integrated itself into the recording and mixing process is through its pairing with the console or mixer itself (Figure 10.6). This combination gives an engineer or producer unprecedented remote control over most mixing, panning, equalization (EQ), effects and monitor control, either in the studio or on stage.

iDAWs

Over the years, iPads have gotten powerful enough and apps have been written that allow an entire session to be transferred from our main workstation to a multitrack DAW on the pad (Figure 10.7). It's easy to see (and hear) how such a powerful and portable device would be of benefit to the on-the-go musician or producer. He or she could record, mix and process a mix virtually anywhere. Offering up a surprising number of DSP options that are compatible with and can be read by their main DAW system, anyone could put on a pair of in-ear monitors or noise-canceling headphones and begin mixing (as the iMantra goes) "virtually anywhere".

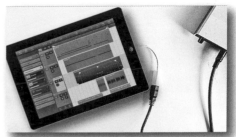

(a)

(b)

FIGURE 10.7

Several iDAW examples: (a) Auria Pro (courtesy of WaveMachine Labs, Inc., www.wavemachinelabs .com). (b) Steinberg's Cubasis (courtesy of Steinberg Media Technologies GmbH, a division of Yamaha Corporation, www.steinberg.net).

Taking Control of Your DAW Using the iOS

In addition to the many practical uses listed previously, another huge contribution that iOS and mobile computing technology have made to music production is the pad's ability to serve as a DAW remote control. For example, in the not-too-distant past, dedicated hardware DAW controllers would easily cost us over $1,000 and would need both a power and a wired USB connection. Now, with the introduction of iOS-based DAW controllers (Figure 10.8), the same basic functionality is available to us for less than $20 (if not for free). These apps allow the controller to fit in our hand and allow us to work wirelessly from any place in the studio, giving us the ability to:

- Mix from the studio, from your instrument stand, on stage, in an audience, virtually anywhere
- Remotely control all transport functions from anywhere within the facility
- Control studio monitor and headphone sub-mixes from within the DAW, allowing the musician to control their own monitor mix (Figure 10.9) simply by downloading the app, connecting their iOS device to the studio network and then creating their own personal sub-mix.

FIGURE 10.8

iOS-based DAW controllers. (a) V-Control Pro DAW controller (courtesy of Neyrinck, www.vcontrolpro .com). (b) Cubase iC Pro (courtesy of Steinberg Media Technologies GmbH, a division of Yamaha Corporation, www.steinberg.net).

The iOS doesn't stop with the idea of controlling a DAW by using simple transport, control and mixing functions – when used with a performance-based DAW (such as Ableton Live), a number of iOS applications can be used to wirelessly integrate with a DAW, allowing a performer to take control of their system to

(a)

(b)

FIGURE 10.9
Musicians can easily mix their headphone sends remotely via an iOS DAW controller app (courtesy of Steinberg Media Technologies GmbH, a division of Yamaha Corporation, www.steinberg.net).

create a live performance set in real time (Figure 10.10). Whereas in the past, a hardware controller was used with a DAW in a way that had limited mobility and reduced visual feedback, a modern iOS device can give the performer far better tactile and interactive response cues in a way that allows the performer to clearly see exactly which audio and MIDI loops, effects and any number of performance controls are available to them. All this is available in a way that can be interactively played, triggered and controlled live on stage. As an electronic musician who works with such tools in an on-stage environment, DMH can tell you that the iOS literally changes the performance game into one that's far easier to see, understand and interact with over its hardware counterpart.

FIGURE 10.10
Touchable Pro can be used to wirelessly control Ableton Live in a practice and performance setting (courtesy of Zerodebug, www.zerodebug.com).

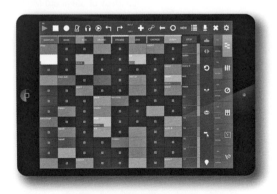

TRY THIS: INSTALLING A DAW CONTROLLER

Go ahead, go to the App Store, search under "DAW controllers" and choose one that will best work for you (as always, it's important to do a bit of research before buying – we all know there are always differences and quirks that will set one app apart from the rest for your particular DAW). Once you've followed the install and setup instructions, you'll be "remoting" around the studio from your pad and/or phone in no time.

THE IOS ON STAGE

Another huge advancement in wireless control comes in the form of the iOS-based live sound controller (Figure 10.11). These devices are literally changing the way live sound hands are able to do their jobs. By giving a live sound mixer the freedom to mix wirelessly from anywhere in the venue, he or she can walk around the place and make any adjustments that are needed from the middle of the audience, FOH (Front of House) position or virtually anywhere (Figure 10.12). Stage monitor mixing (one using a completely separate mix or sub-mix from the main FOH mixer) can now also be accomplished wirelessly from a mix app. Depending upon the size and scale of the performance and venue, these stage mixes can be performed by a dedicated monitor mix person or by the performers themselves. As with all things wireless, it's about freedom, mobility and flexibility.

iOS and the DJ

The modern DJ is definitely no stranger to the power and portability of the iOS. Full sets can be easily pre-programmed or performed on the fly from an iPad or iPhone, especially when used in conjunction with external control hardware. Offering up most or all of the control of a laptop DJ system, these devices give the DJ full freedom to strut their stuff anywhere and anytime.

FIGURE 10.11

iOS-based mixing systems. (a) Mackie DC16 mixing system (for iOS and Android) (courtesy of Loud Technologies, Inc., www .mackie.com). (b) StudioLive RML32AI wireless mixing system for live and recorded sound (courtesy of Presonus Audio Electronics, Inc., www. presonus.com).

(a)

(b)

FIGURE 10.12
Greg "Greedy" Williamson
mixing at Easy Street
Records in West Seattle
(courtesy of Greedtone, www.
greedtone.com).

FIGURE 10.13
iOS music software. (a)
GarageBand for the Mac,
Windows and iOS (courtesy
of Apple Inc., www.apple
.com). (b) Nanologue synth
and LoopMash HD Groove
app (courtesy of Steinberg
Media Technologies GmbH,
a division of Yamaha
Corporation, www.steinberg
.net).

IOS AS A MUSICAL INSTRUMENT

Electronic instruments and music production tools allow us to quickly save and work on musical ideas at the drop of a hat, or to integrate an instrument app into our on-the-go working environment, adding rich and complex musical expression to a track, at a mere fraction of the cost of its hardware equivalent.

All of this started with GarageBand (Figure 10.13a), a musical loop app that allows audio and MIDI to be dragged into a project timeline in a quick and easy way without extensive musical experience. This app allowed musicians to first grasp the concept that the iOS could act as a serious musical instrument. Offering up a wide range of electronic loops and beats, as well as a set of reasonably good-sounding virtual instruments (piano, guitar, bass, strings, etc.), these instruments could be sequenced from within the program itself or (with the use of a MIDI interface connection) could be played from an external MIDI sequencer or controller source.

After GarageBand, individual developers and electronic instrument manufacturers quickly realized that there was a huge market for the recreation of classic synths, new synth designs, groove synths, beat generators and other electronic instrument types that range from being super-simple in their operation to being sophisticated programs that equal or rival any hardware counterpart (Figure 10.13b).

(a)

(b)

When using an iOS-compatible audio and MIDI interface, IDAM over a lightning cable, or simply a camera adapter/MIDI interface setup and the unit's 1/8" headphone out jack, an iOS device can be integrated into a DAW-based MIDI setup to varying degrees, literally allowing the iOS instrument to fully integrate into your musical system.

THE ABILITY TO ACCESSORIZE

Naturally, there are tons of additional accessories that can be added to customize the look and functionality of your iOS device. These include:

- Desk stand adapters
- Mic stand adapters
- Docking stations
- Interface systems and adapters

Of course, the list could go on for quite some time. Suffice it to say, if you wanted blue flames to shoot out of your device in a way that would spell your name on the stage – just wait a few days, and you'll be able to get it from the store. It's all about adding the right tools and toys to your system and lifestyle in a way that'll make your production environment uniquely yours.

CHAPTER 11

Multimedia and the Web

It's no secret that modern-day computers, smartphones, gamestations and even smart televisions have gotten faster, sleeker, more touchable and sexier in their overall design. In addition to their ability to act as a multifunctional production workhorse, one of the crowning achievements of modern work and entertainment devices is their networking and media integration, which has come to be universally known by the household buzzword *multimedia*.

This combination of working and playing with multimedia has found its way into modern media and computer culture through the use of various hardware and software systems that combine in a multitasking environment to bring you an experience that seamlessly involves such media types as:

- Audio and music
- Video and video streaming
- Graphics and gaming
- Musical instrument digital interface (MIDI)
- Text and communication

The obvious reason for creating and integrating these media types is the human desire to share and communicate one's experiences with others. This has been done for centuries in the form of books and in relatively recent decades, through movies and television. Obviously, in the here and now, the powerful and versatile presence of the Internet can be placed at or near the top of this communications list. Nothing allows individuals and corporate entities to reach millions (or billions) so easily. Perhaps most importantly, the web is a multimedia experience that each individual can manipulate, learn from and even respond to in an interactive fashion. It has indeed unlocked the potential for experiencing events and information in a way that makes each of us a participant, and not just a passive spectator. To me, this is the true revolution that's occurring at the dawn of the twenty-first century!

DOI: 10.4324/9781003260530-11

THE MULTIMEDIA ENVIRONMENT

Although much of recording and music production has matured into a relatively stable industry, the web, multimedia and the music industry itself are in a full-speed-ahead tailspin of change. With online social media, on-demand video and audio streaming, network communications, computer gaming and hundreds of other media options entering into the marketplace on a weekly basis, it's no wonder that things are changing fast!

As with all things tech, I would say that the most important word in multimedia technology today is "integration". In fact, the perfect example of multimedia today is in your pocket or purse. Your cell phone (handy, mobile or whatever you call it) is a marvel of multimedia technology that can:

- Keep you in touch with friends
- Surf the web to find the best restaurant in the area
- Connect to a web-based and/or actual GPS to keep you from getting lost
- Play or stream your favorite music
- Let you watch the latest movie or YouTube video
- Take pictures
- Light your way to the bathroom late at night or remotely turn your home's lights on or off

Of course, that's just the short list. What more could you ever want from a mobile on-the-go device? I don't know, but we're surely to find out in the not too-distant-future; and because of these everyday tools, we've come to expect the same or similar experience from other media devices, such as:

- Computers
- Televisions
- Cars

The Computer

Obviously, the tool that started the multimedia revolution is the computer. The fact that it is a lean, mean multitasking machine makes it ideal for delivering all of the media that we want, all of the time. From multimedia, to gaming, to music, to video – if you have a good Internet connection – you can pretty much hold the world in your hands (or at least your lap).

Television and the Home Theater

As you might expect, newer generations of video and home theater systems incorporate more and more options for offering up "a rich multimedia experience". Newer TVs are able to directly connect to the Web, allowing us to stream our favorite movies or listen to Internet radio "on demand".

So, what holds all of these various media devices together? Media, data distribution and transmission formats!

Delivery Media

Although media data can be stored and/or transmitted on a wide range of storage devices, the most commonly found delivery media at the time of this writing are:

- Shared networks
- The Web
- Physical media

Networking

At its most basic level, a shared data network is a collection of computers and other hardware devices that are connected by protected data protocol links that enable the communication of shared resources, program apps and data. The most well-known network communications protocols are Ethernet (a standard for creating Local Area Networks [LANs]) and the Internet Protocol Suite (otherwise known as the World Wide Web).

Within media production, a LAN is a powerful tool that allows multiple computers to be linked within a production facility. For example, a central server (a dedicated data delivery and shared storage device) can be used in a facility to store and share large amounts of data throughout a facility. Simpler LAN connections could also be used in a project studio to link various computers so as to share media and backup data in a simple and cost-effective manner. For example, a single, remote computer within a connected facility could be used to store and share the large amounts of data that are required for video, music and sample library production, and then back up all of this data in a RAID (redundant array of independent disks) system, allowing the data and backups to be duplicated and stored on multiple hard drives (thereby reducing the chance of system data loss).

In short, it's always a good idea to become familiar with the strength, protection and power that a properly designed network can offer a production facility, no matter how big it is.

The Web

One of the most powerful aspects of multimedia is its ability to communicate experiences either to another individual or to the masses. For this, you need a very large network connection. The largest and most common network of all is the Internet (World Wide Web). Here's the basic gist of how this beast works:

- The Internet (Figure 11.1) can be thought of as a communications network that allows your computer (or connected network) to be connected to an

mac   server   WWW   server   pc

FIGURE 11.1
The Internet works by communicating requests and data from a user's computer to connected servers that are connected to other network access points around the world, which are likewise connected to other users' computers.

Internet Service Provider (ISP) server (a specialized computer or cluster of ISP computers that are designed to handle, pass and route data between other network user connections).

- These ISPs are then connected (through specialized high-speed connections) to a series of network access points (NAPs), which essentially form the connected infrastructure of the World Wide Web (www).

Therefore, in its most basic form, the Web can be simply thought of as a unified array of connected networks.

Internet browsers transmit and receive information on the web via a Uniform Resource Locator (URL) address. This address is then broken down into three parts: the protocol (e.g., http:// or https://), the server name (e.g., www.modrec .com) and the requested page or file name (e.g., index.htm). The connected server is able to translate the server name into a specific Internet Provider (IP) address, which is then used to connect your computer to the desired server, after which the requests to receive or send data are communicated, and the information is passed to your computer.

Email works in a similar data transfer fashion, with the exception that an email isn't sent to or requested from a specific server; rather, it's communicated through a worldwide server network from one specific email address (e.g., myname@ myprovider.com) directly to a destination email address (e.g., yourname@your-provider.com).

The Cloud

One of the more current buzz terms on the Web is "the cloud" or "cloud computing". Put simply, storing data on the cloud refers to data that's stored on a remote server system or Web-connected drive system. Most commonly, that server would be operated and maintained by a company that will store your data at a cost (although many services allow limited amounts of your personal data to be stored for free). For example, cloud storage companies can:

- Store backup media online in a manual or automated way to keep your files secure
- Store huge amounts of uploadable video data online in a social media context (i.e., YouTube)
- Store uploadable audio data online in a social media context for promoting artists and DJs (i.e., BandCamp, ReverbNation, SoundCloud)

- Store program-related data online, reducing the need for installing programs directly onto your computer
- Store application and program data online, allowing the user to "subscribe" to the use of their programs for a monthly or yearly fee

PHYSICAL MEDIA

Although online data distribution and management is increasingly becoming the primary way to get information from the distributor to the consumer, physical media (you know, the kind that we can hold in our hands) still have a very important role in delivering media to the consumer masses. Beyond the fun stuff – like vinyl records – the most common physical media formats are compact discs (CD), DVDs and Blu-ray discs.

The CD

Of course, one of the first and most important developments in the mass marketing and distribution of large amounts of digital media was the CD, in the form of both the CD-Audio and the CD-ROM. As most are aware, the CD-Audio disc is capable of storing up to 74 minutes of audio at a rate of 16 bits/44.1 kHz. Its close optical cousin, the CD-ROM, is most often capable of storing 700 MB of graphics, video, digital audio, MIDI, text and raw data. Consequently, these pre-manufactured and user-encoded media are still widely used to store large amounts of music, text, video, graphics, etc., largely due to the fact that you can hold them in your hand and store them away in a safe place. Table 11.1 details the various CD standards (often affectionately called the "rainbow book") that are currently in use.

It's important to note that Red Book CDs (audio CDs) are capable of encoding small amounts of user data that can be used to encode embedded metadata (user information). This metadata (called CD-Text) allows the media creator to enter and encode information such as artist, title, song number, track artist/title, etc. – all within the disc itself. Since most CD players, computers and media players are capable of reading this information, it's always a good idea to provide the listener with as much information as possible.

Another system for identifying CD- and disc-related data is provided by Gracenote (formerly CDDB or Compact Disc Data Base). In short, Gracenote maintains and licenses an Internet database that contains CD info, text and images. Many media devices that are connected to the Web are able to access this database and display the information on your player or device.

The DVD

Similar to their cousins, DVDs (which, after a great deal of industry deliberation, simply stands for "DVD"), can contain any form of data. These discs are capable of storing up to 4.7 gigabytes (GB) within a single-sided disc or 8.5 GB

Table 11.1	CD Format Standards
Format	**Description**
Red Book	Audio-only standard; also called CD-A (Compact Disc Audio)
Yellow Book	Data-only format; used to write/read CD-ROM data
Green Book	CD-I (Compact Disc Interactive) format; never gained mass popularity
Orange Book	CD-R (Compact Disc Recordable) format
White Book	VCD (Video Compact Disc) format for encoding CD-A audio and MPEG-1 or MPEG-2 video data; used for home video and karaoke
Blue Book	Enhanced Music CD format (also known as CD Extra or CD+); can contain both CD-A and data
ISO-9660	A data file format that's used for encoding and reading data from CDs of all types across platforms
Joliet	Extension of the ISO-9660 format that allows up to 64 characters in its file name (as opposed to the 8 file + 3 extension characters allowed by MS-DOS)
Romeo	Extension of the ISO-9660 format that allows up to 128 characters in the file name
Rock Ridge	Unix-style extension of the ISO-9660 format that allows long file names
CD-ROM/XA	Allows extended usage for the CD-ROM format – Mode-1 is strictly Yellow Book, while Mode-2 Form-1 includes error correction, and Mode-2 Form-2 doesn't allow error correction; often used for audio and video data
CD-RFS	Incremental packet writing system from Sony that allows data to be written and rewritten to a CD or CD-RW (in a way that appears to the user much like the writing/retrieval of data from a hard drive)
CD-UDF	UDF (Universal Disc Format) is an open incremental packet writing system that allows data to be written and rewritten to a CD or CD-RW (in a way that appears to the user much like the writing/retrieval of data from a hard drive) according to the ISO-13346 standard
HDCD	The High-Definition Compatible Digital system adds 6 dB of gain to a Red Book CD (when played back on an HDCD-compatible player) through the use of a special companion mastering technique
Macintosh HFS	An Apple file system that supports up to 31 characters in a file name; includes a data fork and a resource fork that identify which application should be used to open the file

Table 11.2	DVD Video/Audio Formats					
Format	Sample	Bit Rate (kHz)	Bit/s	Ch	Common Format	Compression
PCM	48, 96	16, 20, 24	Up to 6.144 Mbps	1 to 8	48 kHz, 16 bit	None
AC3	48	16, 20, 24	64 to 448 kbps	1 to 6.1	192 kbps, stereo	AC3 and 384 kbps, 448 kbps
DTS	48, 96	16, 20, 24	64 to 1,536 kbps	1 to 7.1	377 or 754 kbps	DTS coherent acoustics for stereo and 754.5 or 1,509.25 kbps for 5.1
MPEG-2	48	16, 20	32 to 912 kbps	1 to 7.1	Seldom used	MPEG
MPEG-1	48	16, 20	384 kbps	2	Seldom used	MPEG
SDDS	48	16	Up to 1,289 kbps	5.1, 7.1	Seldom used	ATRAC

on a double-layered disc. This capacity makes the DVD a good delivery medium for encoding video (generally in the MPEG-2 encoding format), data-intensive games, DVD-ROM titles and program installation discs. The increased demand for multimedia games, educational products, etc. has spawned the computer-related industry of CD and DVD-ROM authoring. The term *authoring* refers to the creative, design and programming aspects of putting together a CD or DVD project. At its most basic level, a project can be authored, mastered and burned to disc from a single commercial authoring program. Whenever the stakes are higher, trained professionals and expensive systems are often called in to assemble, master and produce the final disc for mass duplication and packaging. Table 11.2 details the various DVD video/audio formats that are currently in use.

Blu-ray

Although similar in size to the CD and DVD, a Blu-ray disc can store up to 25 GB of media-related data onto each data layer (50 GB for a dual-layer disc). In addition to most of the standard video formats that are commonly encoded onto a DVD, the Blu-ray format can play back both compressed and uncompressed Pulse-Code Modulation (PCM) audio (Table 11.3) in a multichannel, high-resolution environment.

The Flash Card and Memory USB Stick

In our on-the-go world, another useful media device is the flash memory card. More specifically, the SD (secure digital) card typically ranges in size up to 128 Gb in capacity and comes in various physical sizes. These media cards can be used for storing audio, video, photo, app and any other type of digital info that can be used with your laptop, phone, car player, Blu-ray player – you name it!

Table 11.3	Specification of BD-ROM Primary Audio Streams							
	LPCM	**Dolby Digital**	**Dolby Digital Plus**	**Dolby TrueHD (Lossless)**	**DTS Digital Surround**	**DTS-HD Master Audio (Lossless)**	**DRA**	**DRA Extension**
Max. Bit-rate	27.648 Mbit/s	640 kbit/s	4.736 Mbit/s	18.64 Mbit/s	1.524 Mbit/s	24.5 Mbit/s	1.5 Mbit/s	3.0 Mbit/s
Max. Channel	8 (48 kHz, 96 kHz), 6 (192 kHz)	5.1	7.1	8 (48 kHz, 96 kHz), 6 (192 kHz)	5.1	8 (48 kHz, 96 kHz), 6 (192 kHz)	5.1	7.1
Bits/ sample	16, 20, 24	16, 24	16, 24	16, 24	16, 20, 24	16, 24	16	16
Sample Freq.	48 kHz, 96 kHz, 192 kHz	48 kHz	48 kHz	48 kHz, 96 kHz, 192 kHz	48 kHz	48 kHz, 96 kHz, 192 kHz	48 kHz	48 kHz, 96 kHz

Media Delivery Formats

Now that we've taken a look at the various delivery media, the next most important aspect of delivering the multimedia experience rests with the data distribution formats (the coding and technical aspects of data delivery) themselves.

When creating content for the various media systems, it's extremely important that the media format and bandwidth be matched with the requirements of the content delivery system that's being used. In other words, it's always smart to maximize the efficiency of the message (media format and required bandwidth) to match (and not alienate) your intended audience. The following section outlines many standard and/or popular formats for delivering media to a target audience.

UNCOMPRESSED SOUND FILE FORMATS

Digital audio is obviously a component that adds greatly to the multimedia experience. It can augment a presentation by adding a dramatic music soundtrack, help us to communicate through speech or give realism to a soundtrack by adding sound effects. Because of the large amounts of data required to pass video, graphics and audio from a disc, the Internet or other media, the bit- and sample-rate structure of an uncompressed audio file is usually limited compared with that of a professional-quality sound file. At the "lo-fi" range, the generally accepted sound file standard for older multimedia production is either 8-bit or 16-bit audio at a sample rate of 11.025 or 22.050 kHz. This standard came about mostly because older CD drive and processor systems generally couldn't pass the professional rates of 44.1 kHz and higher. With the introduction of faster processing systems and better hardware, these limitations have generally been lifted to include 16/44.1 (16 bit/44.1 kHz), 24/44.1 and as high as 24/192. Obviously, there are limitations to communicating uncompressed professional-rate sound files over the Internet or from an optical disc that's also streaming full-motion video. Fortunately, with improvements in codec (encode/decode) techniques, hardware speed and design, the overall sonic and production quality of compressed audio data has greatly improved.

PCM AUDIO FILE FORMATS

Although several formats exist for encoding and storing sound file data, only a few have been universally adopted by the industry. These standardized formats make it easier for files to be exchanged between compatible media devices.

In audio, PCM is the standard system for encoding, storing and decoding audio. Within a PCM stream, the amplitude of the analog signal is sampled at precise intervals, with each sample being quantized to the nearest value within a range of digital steps. This level (amplitude) is then sampled at precise time intervals (frequency) so as to represent analog audio in a numeric form.

Probably the most common file type is the Wave (or .wav) format. Developed for the Microsoft Windows format, this universal file type supports both mono and stereo files at a wide range of uncompressed resolutions and sample rates. Wave files contain PCM coded audio that follows the Resource Information File Format (RIFF) spec, which allows extra user information to be embedded and saved within the file itself. The newly adopted Broadcast Wave format, which has been adopted by the Producers and Engineers wing (www.grammypro.com/producers-engineers-wing) as the preferred sound file format for DAW production and music archiving, allows timecode-related positioning information to be directly embedded within the sound file's data stream.

In addition to the .wav format, the Audio Interchange File (AIFF; .aif) format is commonly used to encode digital audio within Apple computers. Like Wave files, AIFF files support mono or stereo, 8-bit, 16-bit and 24-bit audio at a wide range of sample rates – and like Broadcast Wave files, AIFF files can also contain embedded text strings. Table 11.4 details the differences between uncompressed file sizes as they range from the 24-bit/192-kHz rates all the way down to low–voice quality 8-bit/10-kHz files.

DIRECT STREAMING DIGITAL (DSD) AUDIO

DSD was a joint venture between Sony and Phillips for encoding audio onto the Super Audio CD (SACD). Although the SACD has fallen out of favor (with the Blu-ray format's wide acceptance), the audio format itself survives as a high-resolution audio format. Unlike PCM, DSD makes use of Pulse-Density Modulation (Figure 11.2) to encode audio. That's to say that it doesn't follow

Table 11.4	Audio Bit Rate and File Sizes				
Sample Rate	Word Length	No. of Channels	Date Rate (kbps)	MB/min	MB/hour
192	24	2	1152	69.12	4,147.2
192	24	1	576	34.56	2,073.6
96	32	2	768	46.08	2,764.8
96	32	1	384	23.04	1,382.4
96	24	2	576	34.56	2,073.6
96	24	1	288	17.28	1,036.8
48	32	2	384	23.04	1,382.4
48	32	1	192	11.52	691.2
48	24	2	288	17.28	1,036.8
48	24	1	144	8.64	518.4
48	16	2	192	11.52	691.2
48	16	1	96	5.76	345.6
44.1	32	2	352	21.12	1,267.2
44.1	32	1	176	10.56	633.6
44.1	24	2	264	15.84	950.4
44.1	24	1	132	7.92	475.2
44.1	16	2	176	10.56	633.6
44.1	16	1	88	5.28	316.8
32	16	2	128	7.68	460.8
32	16	1	64	3.84	230.4
22	16	2	88	5.28	316.8
22	16	1	44	2.64	158.4
22	8	1	22	1.32	79.2
11	16	2	44	2.64	158.4
11	16	1	22	1.32	79.2
11	8	1	11	0.66	39.6

PCM's system of the periodic sampling of audio at a specific rate; rather, the level and change of relative gain levels over time are a result of the density of the bits within the stream. A stream that has all 0s will have no level at that point in time, while one having all 1s will have a maximum voltage level. The density of 1s to 0s will determine the overall change in gain over time at a sampling rate

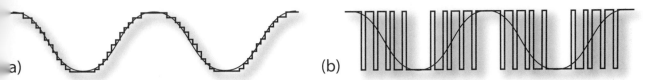

(a)   (b)

FIGURE 11.2
Uncompressed audio coding. (a) PCM encodes the absolute level at that sample period and stores that number within memory. (b) DSD, on the other hand, does not encode the level within samples and words, but encodes the "density" of 1s to 0s within a period of time to determine the signal's level … there is no coding per se.

of 2.8224 MHz (or 64 times the 44.1-kHz sample rate), 5.6448 MHz (DSD128) or higher. Currently, only a few DAWs are able to work natively in the DSD modulation code.

COMPRESSED CODEC SOUNDFILE FORMATS

As was mentioned earlier, high-quality uncompressed sound files often present severe challenges to media delivery systems that are restricted in terms of bandwidth, download times or memory storage. Although the streaming of audio data from various media and high-bandwidth networks (including the Web) has improved over the years, memory storage space and other bandwidth limitations have led to the popular acceptance of compressed audio data formats known as codecs. These formats can encode audio in a manner that reduces data file size and bandwidth requirements and then decode the information upon playback using a system known as perceptual coding.

PERCEPTUAL CODING

The central idea behind perceptual coding is the psychoacoustic principle that the human ear will not always be able to hear all of the information that's present in a recording. This is largely due to the fact that louder sounds will often mask sounds that are both lower in level and relatively close in frequency to another louder signal. These perceptual coding schemes take advantage of this masking effect by filtering out noises and sounds that can't be detected by our ears and removing them from the encoded audio stream.

The perceptual encoding process is said to be "lossy" because once the filtered data has been taken away, it can't be replaced or introduced back into the file. For the purposes of audio quality, the amount of perceived data compression reduction can be selected by the user during the encoding process. Higher bandwidth compression rates will remove less data from a stream (resulting in a reduced amount of filtering and higher audio quality), while low bandwidth rates will greatly reduce the data stream (resulting in smaller file sizes, increased filtering, increased artifacts and lower audio quality). The amount of filtering that's to be applied to a file will depend on the intended audio quality and the delivery medium's bandwidth limitations. Due to the lossy character of these encoded files, it's always a good idea to keep a copy of the original, uncompressed sound file in a data archive backup in case changes in content or future technologies occur (never underestimate Murphy's law).

Many of the listed codecs are capable of encoding and decoding audio using a constant–bit rate (CBR) and a variable–bit rate (VBR) structure:

- CBR encoding is designed to work effectively in a streaming scenario where the end user's bandwidth is a consideration. With CBR encoding, the chosen bit rate will remain constant over the course of the file or stream.
- VBR encoding is designed for use when you want to create a downloadable file that has a smaller file size and bit rate without sacrificing sound and video quality. This is carried out by detecting which sections will need the highest bandwidth and adjusting the encode process accordingly. When lower rates will suffice, the encoder adjusts the processing to match the content. Under optimum conditions, you might end up with a VBR-encoded file that has the same quality as a CBR-encoded file but is only half the file size.

Perceptual coding schemes that are in most common use include:

- MP3
- MP4
- Windows Media Audio (WMA)
- Advanced Audio Coding (AAC)
- RealAudio
- FLAC

MP3

MPEG (which is pronounced "M-peg" and stands for the Moving Picture Experts Group; www.mpeg.org) is a standardized format for encoding digital audio into a compressed format for the storage and transmission of various media over the web. As of this writing, the most popular format is the ISO-MPEG Audio Level-2 Layer-3, commonly referred to as MP3. Developed by the Fraunhofer Institute (www.iis.fraunhofer.de) and Thomson Multimedia in Europe, MP3 has advanced the public awareness and acceptance of compressing and distributing digital audio by creating a codec that can compress audio by a substantial factor while still maintaining quality levels that approach those of a CD (depending on which compression levels are used). Although a wide range of compression rates can be chosen to encode/decode an MP3 file, the most common rate for the general masses is 128 kbps (kilobits per second). Although this rate is definitely "lossy" (containing increased distortion, reduced bandwidth and sideband artifacts), it allows us to literally put thousands of songs on an on-the-go player. Higher rates of 160, 192 and 320 kbps offer higher "near-CD sound quality", with the obvious tradeoff being larger file sizes.

Although faster web connections are capable of streaming MP3 (and higher rates) in real time, this format is most often downloaded to the end consumer for storage to disk, disc and SD media for the storage and playback of down-loaded songs. Once saved, the data can then be transferred to playback devices (such as phones, pads, etc.). In fact, billions of music tracks are currently being downloaded every month on the Internet using MP3, practically every personal computer contains licensed MP3 software and virtually every song has been encoded into this format – it's actually hard to imagine how many players are out there on the global market. This makes it the Web's most popular audio compression format by far … although, as this book goes to press, Fraunhofer has decided to stop its support and licensing for this codec.

MP4

Like MP3, the MPEG-4 (MP4) codec is largely used for streaming media data over the web or for viewing media over portable devices. MP4 is largely based on Apple's QuickTime "MOV" format and can be used to encode A/V and audio-only content over a wide range of bit-rate qualities with both stereo and multichannel (surround) capabilities. In addition, this format can employ DRM (copy protection) so as to restrict copying of the downloaded file.

WMA

Developed by Microsoft as their corporate response to MP3, Windows Media Audio (WMA) allows compression rates to encode high-quality audio at low bit-rate and file-size settings. Designed for ripping (extracting audio from a CD) and sound file encoding/playback from within the popular Windows Media Player (Figure 11.3), this format has grown and then fallen in general acceptance and popularity. In addition to its high quality at low bit rates, WMA also allows for a wide range of bit-rate qualities with both stereo and multichannel (surround) capabilities while being able to imbed DRM (copy protection) so as to restrict copying of the downloaded file.

FIGURE 11.3
Windows Media Player.

AAC

Jointly developed by Dolby Labs, Sony, ATT and the Fraunhofer Institute, the AAC scheme is touted as a multichannel-friendly format for secure digital music distribution over the Internet. Stated as having the ability to encode CD-quality audio at lower bit rates than other coding formats, AAC not only is capable of encoding 1, 2 and 5.1 surround sound files but can also encode up to 48 channels within a single bit stream at bit/sample rates of up to 24/96. This format is also SDMI (Secure Digital Music Initiative) compliant, allowing copyrighted material to be protected against unauthorized copying and distribution. AAC is the default or standard audio format for YouTube, iPhone, iPod, iPad, iTunes (it's the backbone of Apple's music and audio media distribution), Nintendo DSi, Nintendo 3DS, DivX Plus Web Player and PlayStation 3.

FLAC

FLAC (Free Lossless Audio Codec) is a format that makes use of a data compression scheme that's capable of reducing an audio file's data size by 40% to 50% while playing back in a lossless fashion that maintains the sonic integrity of the original stereo and multichannel source audio (up to 8 channels with bit depths of up to 32 bits at sample rates that range up to 96k or 192k). As the name suggests, FLAC is a free, open-source codec that can be used by software developers in a royalty-free fashion.

With the increase in memory storage size and higher download speeds, many enthusiasts in audio are beginning to demand higher playback quality. As such, FLAC is growing in popularity as a medium for playing back high-quality, lossless audio, both in stereo and in various surround formats.

Tagged Metadata

Within most types of multimedia file formats, it's possible to embed a wide range of content identifier data directly into the file itself or within a Web-related page or event. This "tagged" data (also known as metadata) can identify and provide extensive and extremely important information that relates to the content of the file. For example, let's say that little Sally is looking to find a song that she wants to download from her favorite artist. Now, Sally consumes a lot of music, and she goes to iTunes to download that song into her phone. So, how can she find her favorite needle in a digital haystack? By searching for songs under "Mr. Right", the site is able to find several of his latest songs. and BOOM – she's groovin' to the beat – all thanks to metadata.

Now that she knows the name of the song, she can find it under "Mr. Right" in her player, and she's groovin' on the underground or heading to school. On the flip side, if the song name hadn't been entered (or was incorrectly entered) into the metadata database, poor Sally's song would literally get lost in the

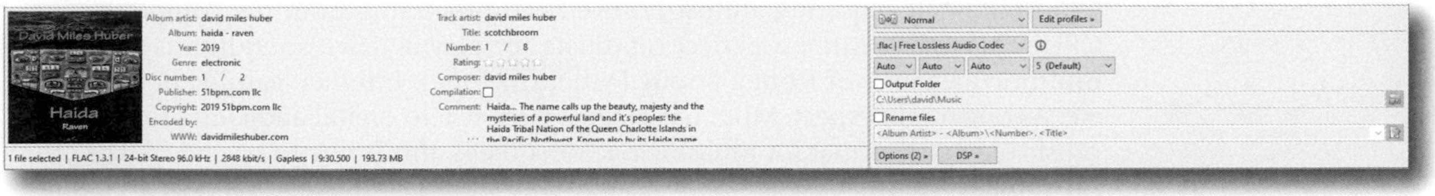

FIGURE 11.4
Embedded metadata file tags can be added to a media file via various media libraries, rippers or editors and then viewed by a media player or file manager.

shuffle. Sally would be bummed, and the record label/artist would lose the sale. On another day, let's say that Sally really wanted to buy a new song. She could enter her favorite music genre into the field and look through the songs that have been properly tagged with that genre. By clicking the "sounds like" button, songs that have been tagged in the same genre could pop up that might completely flip her out – BOOM, again – a sale and (even better) a fan of a new artist is born, all because the "sounds like" metadata was properly tagged by her new, favorite band. Are you getting the idea that "tagging" a song, project or artist band data with the proper metadata can be a make or break deal in the new digital age?

Metadata in all its media and website glory tells the world who it is, the song title, what genre type, etc. It's the gateway to telling the world: "Hey, I'm here! Listen to me!" Metadata can also be extracted from an audio using a central music database and then be automatically entered into a music copy (ripping) program (Figure 11.4).

Due to the fact that there is no existing set of rules for filling out metadata, folks who make extensive use of iTunes, Discogs and other music playlist services often go nuts when the tags are incorrect, conflicting or non-standard. For example, a user might download an album that might go by the artist name of "XYZ Band" and then download another album that might have it listed as "The Band XYZ". One would show up in a player under "X", while the same band would also show up under "T" – and it can get A LOT worse than that. In fact, many people actually go so far as to manually fix their library's metadata to their own liking. The moral of this story is to research your metadata and stick to a single, consistent naming scheme.

It's worth mentioning here that the Producers and Engineers wing of the Grammys have been doing extensive work on the subject of metadata (www .grammy.com/credits) so that tags can be more uniform and extensive, allowing producer, engineer and other project-related data to be entered into the official metadata. It is hoped that such credit documentation will help to get royalties to those who legally deserve to be recognized and paid for their services in the here and now or in the future (when new payment legislations might be passed).

MIDI

One of the unique advantages of MIDI as it applies to multimedia is the rich diversity of musical instruments and program styles that can be played back in real time while requiring almost no overhead processing from the computer's CPU. This makes MIDI a perfect candidate for playing back soundtracks from multimedia games or from a phone (MIDI ringtone), Internet, gaming devices, etc. As one might expect, MIDI has taken a back seat to digital audio as a serious music playback format for multimedia. Most likely, this is due to several factors, including:

- A basic misunderstanding of the medium
- The fact that producing MIDI content often requires a fundamental knowledge of music
- The frequent difficulty of synchronizing digital audio to MIDI in a multimedia environment
- The fact that soundcards, phones, etc. often include poorly designed FM synthesizers (although most operating systems now include higher-quality software synths)

Fortunately, a number of companies have taken up the banner of embedding MIDI within their media projects, and Google's Chrome now includes integrated MIDI support within the browser itself. All of these factors have helped push MIDI a bit more into the Web mainstream. As a result, it's becoming more common for your PC to begin playing back a MIDI score on its own or perhaps in conjunction with a game or more data-intensive program.

The following information relates to MIDI as it functions within the multimedia environment. Of course, more in-depth information on the spec and its use can be found within Chapter 9: MIDI.

STANDARD MIDI FILES

The accepted format for transmitting music-related data and real-time MIDI information within multimedia (or between sequencers from different manufacturers) is the standard MIDI file. This file type (which is labeled with a .mid or .smf extension) is used to distribute MIDI data, song, track, time signature and tempo information to the general masses. Standard MIDI files can support both single and multichannel sequence data and can be loaded into, edited and then directly saved from almost any sequencer package. When exporting a standard MIDI file, keep in mind that they can come in two basic flavors – type 0 and type 1:

- Type 0 is used whenever all of the tracks in a sequence need to be merged into a single MIDI track. All of the notes will have a channel number attached to them (i.e., will play various instruments within a sequence); however, the data will have no definitive track assignments. This type might be the best choice when creating a MIDI sequence for a standard device or the Internet (where the sequencer or MIDI player application might not know or care about dealing with multiple tracks).

- Type 1, on the other hand, will retain its original track information structure and can be imported into another sequencer type with its basic track information and assignments left intact.

GENERAL MIDI

One of the most interesting aspects of MIDI production is the absolute uniqueness of each professional and even semi-pro project studio. In fact, no two studios will be even remotely alike (unless they've been specifically designed to be the same, or there's a very unlikely coincidence). Each artist will have his or her own favorite equipment, supporting hardware, assigned patches and way of routing channels/tracks. The fact that each system setup is unique and personal has placed MIDI at odds with the need for complete compatibility in the world of multimedia. For example, if you import a MIDI file over the Net that's been created in another studio, the song will most likely attempt to play with a totally irrelevant set of sound patches (it might sound interesting, but it won't sound anything like it was originally intended). If the MIDI file is loaded into completely different setups, the sequence will again sound completely different, and so on.

To eliminate (or at least reduce) the basic differences that exist between systems, a standardized set of patch settings, known as General MIDI (GM), was created. In short, General MIDI assigns a specific instrument patch to each of the 128 available program change numbers. Since all electronic instruments that conform to the GM format must use these patch assignments, placing GM program change commands at the header of each track will automatically instruct the sequence to play with its originally intended sounds and general song settings. In this way, no matter what synth, sequencer and system setup is used to play the file back, as long as the receiving instrument conforms to the GM spec, the sequence will be heard using its intended instrumentation.

Tables 11.5 and 11.6 detail the program numbers and patch names that conform to the GM format. These patches include sounds such as synthesizer sounds, ethnic instruments and sound effects that have been derived from early Roland synth patch maps. Although the GM spec states that a synth must respond to all 16 MIDI channels, the first 9 channels are reserved for instruments, while GM restricts the percussion track to MIDI channel 10.

MULTIMEDIA IN THE "NEED FOR SPEED" ERA

The household phrase "surfin' the Web" has become synonymous with jumping onto the Net, browsing the sites and grabbing all of those hot songs, videos and graphics that might wash your way. With improved audio and video codecs and

Table 11.5	GM Non-percussion Instrument (Program Change) Patch Map	
1. Acoustic Grand Piano	44. Contrabass	87. Lead 7 (fifths)
2. Bright Acoustic Piano	45. Tremolo Strings	88. Lead 8 (bass + lead)
3. Electric Grand Piano	46. Pizzicato Strings	89. Pad 1 (new age)
4. Honky-tonk Piano	47. Orchestral Harp	90. Pad 2 (warm)
5. Electric Piano 1	48. Timpani	91. Pad 3 (polysynth)
6. Electric Piano 2	49. String Ensemble 1	92. Pad 4 (choir)
7. Harpsichord	50. String Ensemble 2	93. Pad 5 (bowed)
8. Clavichord	51. SynthStrings 1	94. Pad 6 (metallic)
9. Celesta	52. SynthStrings 2	95. Pad 7 (halo)
10. Glockenspiel	53. Choir Aahs	96. Pad 8 (sweep)
11. Music Box	54. Voice Oohs	97. FX 1 (rain)
12. Vibraphone	55. Synth Voice	98. FX 2 (soundtrack)
13. Marimba	56. Orchestra Hit	99. FX 3 (crystal)
14. Xylophone	57. Trumpet	100. FX 4 (atmosphere)
15. Tubular Bells	58. Trombone	101. FX 5 (brightness)
16. Dulcimer	59. Tuba	102. FX 6 (goblins)
17. Drawbar Organ	60. Muted Trumpet	103. FX 7 (echoes)
18. Percussive Organ	61. French Horn	104. FX 8 (sci-fi)
19. Rock Organ	62. Brass Section	105. Sitar
20. Church Organ	63. SynthBrass 1	106. Banjo
21. Reed Organ	64. SynthBrass 2	107. Shamisen
22. Accordion	65. Soprano Sax	108. Koto
23. Harmonica	66. Alto Sax	109. Kalimba
24. Tango Accordion	67. Tenor Sax	110. Bagpipe
25. Acoustic Guitar (nylon)	68. Baritone Sax	111. Fiddle
26. Acoustic Guitar (steel)	69. Oboe	112. Shanai
27. Electric Guitar (jazz)	70. English Horn	113. Tinkle Bell
28. Electric Guitar (clean)	71. Bassoon	114. Agogo
29. Electric Guitar (muted)	72. Clarinet	115. Steel Drums
30. Overdriven Guitar	73. Piccolo	116. Woodblock
31. Distortion Guitar	74. Flute	117. Taiko Drum
32. Guitar Harmonics	75. Recorder	118. Melodic Tom
33. Acoustic Bass	76. Pan Flute	119. Synth Drum

(Continued)

Table 11.5	(Continued)	
34. Electric Bass (finger)	77. Blown Bottle	120. Reverse Cymbal
35. Electric Bass (pick)	78. Shakuhachi	121. Guitar Fret Noise
36. Fretless Bass	79. Whistle	122. Breath Noise
37. Slap Bass 1	80. Ocarina	123. Seashore
38. Slap Bass 2	81. Lead 1 (square)	124. Bird Tweet
39. Synth Bass 1	82. Lead 2 (sawtooth)	125. Telephone Ring
40. Synth Bass 2	83. Lead 3 (calliope)	126. Helicopter
41. Violin	84. Lead 4 (chiff)	127. Applause
42. Viola	85. Lead 5 (charang)	128. Gunshot
43. Cello	86. Lead 6 (voice)	

Table 11.6	GM Percussion Instrument (Program Key Number) Patch Map (Channel 10)	
35. Acoustic Bass Drum	51. Ride Cymbal 1	67. High Agogo
36. Bass Drum 1	52. Chinese Cymbal	68. Low Agogo
37. Side Stick	53. Ride Bell	69. Cabasa
38. Acoustic Snare	54. Tambourine	70. Maracas
39. Hand Clap	55. Splash Cymbal	71. Short Whistle
40. Electric Snare	56. Cowbell	72. Long Whistle
41. Low Floor Tom	57. Crash Cymbal 2	73. Short Guiro
42. Closed Hi-Hat	58. Vibraslap	74. Long Guiro
43. High Floor Tom	59. Ride Cymbal 2	75. Claves
44. Pedal Hi-Hat	60. Hi Bongo	76. Hi Wood Block
45. Low Tom	61. Low Bongo	77. Low Wood Block
46. Open Hi-Hat	62. Mute Hi Conga	78. Mute Cuica
47. Low-Mid Tom	63. Open Hi Conga	79. Open Cuica
48. Hi-Mid Tom	64. Low Conga	80. Mute Triangle
49. Crash Cymbal 1	65. High Timbale	81. Open Triangle
50. High Tom	66. Low Timbale	

Note: In contrast to Table 11.5, the numbers in Table 11.6 represent the percussion keynote numbers on a MIDI keyboard, not program change numbers.

Table 11.7	Internet Connection Speeds	
Connection	**Speed**	**Description**
56k dial-up	56 kbps (usually less)	Common modem connection
ISDN	128 kbps (older technology)	
ISDN PRI/E1	1.5 Mbps/1.9 Mbps	
DSL	256 kbps to 20 Mbps	
Cable	Up to 85 Mbps	
OC-1	52 Mbps	Optical fiber
OC-3	155 Mbps	Optical fiber
OC-12	622 Mbps	Optical fiber
OC-48	2.5 Gbps	Optical fiber
Ethernet	10 Mbps (older technology) up to 100 Gbps	Local Area Network (LAN), not an Internet connection

ever-faster data connections (Table 11.7), the ability to search on any subject, download files, and stream audio or radio stations from any point in the world has definitely changed our perception of modern-day communications.

Streaming Audio over the Internet

A streaming encoding system can be used to capture, compress and broadcast audio and/or video over the Internet as a real-time, streaming event. On the audio front, it can encode audio as MP3, AAC, Ogg and other audio protocols and then stream this audio out to your host Internet streaming provider. This can be done through the use of streaming software that can integrate your system's hardware and software to help get your media message out to the general public in the form of Web conferences, making tutorials, social media streaming, live music shows, meetings and more.

By making use of a properly designed audio interface, high-quality audio can be presented in the best possible way to the public. Additionally, newer interfaces will often include a "loopback" function that will allow the audio that's being sent out of the interface to be routed back directly to your streaming software so that the program audio will be as high quality as is possible.

ON A FINAL NOTE

One of the most amazing things about multimedia, cyberspace and their related technologies is the fact that they're ever-changing. By the time you read this book, many changes will have occurred. Old concepts will have faded away, and new – possibly better – ones will take over and then begin to take on a new life of their own. Although I've always had a fascination with crystal balls and have often had a decent sense about new trends in technology, there's simply no way to foretell the many amazing things that lie ahead in the fields of music, music technology, gaming, visual media, multimedia and especially, cyberspace. As with everything techno, I encourage you to read the trades and surf the Web to keep abreast of the latest and greatest tools that have recently arrived, or are on the horizon.

CHAPTER 12

Synchronization

Of course, it's a safe bet to say that music and audio itself, in all its various forms, is an indispensable part of almost all types of media production. In video post-production, digital video editors, digital audio workstations (DAWs), audio and video transports, automated console systems, electronic musical instruments and stage lighting routinely work together to help create a finished soundtrack (Figure 12.1). The underlying technology that allows multiple audio and visual media to operate in tandem (so as to maintain a direct time relationship) is known as *synchronization* or *sync*.

Strictly speaking, synchronization occurs when two or more related events happen at precisely the same relative time. With respect to analog audio and video systems, sync is achieved by interlocking the transport speeds of two or more machines. For computer-related systems (such as digital video, digital audio and musical instrument digital interface [MIDI]), synchronization between devices is often achieved through the use of a timing clock that can be fed through a separate line or one that is directly embedded within the digital data line itself.

FIGURE 12.1
Example of an audio production system that's been integrated together via sync.

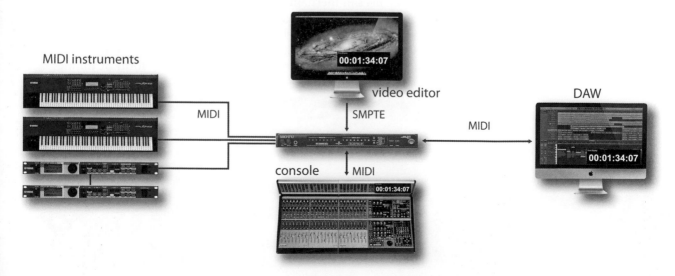

MIDI instruments

MIDI

video editor

SMPTE

MIDI

DAW

console   MIDI

DOI: 10.4324/9781003260530-12

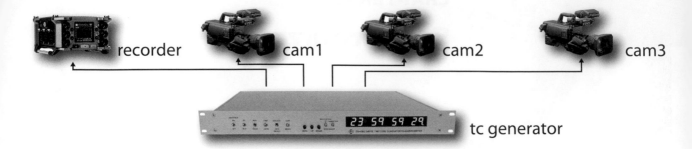

tc generator

FIGURE 12.2

Example of a video production shoot, whereby a master timecode generator is used to timestamp all of the various devices so that they can be synchronized together later during post-production.

Within such an environment, it's often necessary for analog and digital devices to be synchronized together, resulting in a number of ingenious forms of communication and data translation systems. In this chapter, we'll explore the various forms of synchronization used for both digital and analog devices, as well as current methods for maintaining sync between media types (Figure 12.2).

Of course, in modern-day, computer-based systems, audio sync can be easily placed onto a single digital timeline, where audio, video and all sorts of time-related information can live in complete lockstep without the need for maintaining a lock with any external devices (Figure 12.3). The purpose of this chapter, on the other hand, is to introduce you to the concept of sync and to help prepare you for when you DO need to interlock media between various devices.

TIMECODE

FIGURE 12.3

Example of an integrated digital audio or video timeline where no external sync is needed.

Maintaining relative sync between media devices doesn't require that all transport speeds involved in the process be constant; however, it's critical that they maintain the same relative speed and position over the course of a program. Physical analog devices, for example, have a particularly difficult time achieving this. Due to differences in mechanical design, voltage fluctuations and tape slippage, it's a simple fact of life that analog tape devices aren't able to maintain a constant playback speed, even over relatively short durations. For this reason, accurate sync between analog and digital machines would be nearly impossible to achieve over any reasonable program length without some form of timing

Time Display

00:01:34:07

FIGURE 12.4
Readout of a SMPTE
timecode address in
HH:MM:SS:FF.

lock. It therefore quickly becomes clear that if production is to utilize multiple forms of media and record/playback, sync is essential.

The standard method of interlocking audio, video and film transports makes use of a code that was developed by the Society of Motion Picture and Television Engineers. This timecode (SMPTE timecode or TC for short) identifies an exact position within recorded media or onto tape by assigning a digital address that increments over the course of a program's duration. This address code can't slip in time and always retains its original location, allowing the continuous monitoring of tape position to an accuracy of between 1/24th and 1/30th of a second (depending on the media type and frame rates being used). These divisional segments are called *frames*, a term taken from film production. Each audio or video frame is tagged with a unique identifying number, known as a "timecode address". This eight-digit address is displayed in the form 00:00:00:00, whereby the successive pairs of digits represent hours:minutes:seconds:frames or HH:MM:SS:FF (Figure 12.4).

The recorded timecode address is then used to locate a position on the hard disk, magnetic tape or any other recorded medium in much the same way that a letter carrier uses a written address to match up, locate and deliver a letter to a specific, physical residence (i.e., by matching up the address, you can then find the desired physical location point, as shown in Figure 12.5a). For example, let's suppose that a time-encoded analog multitrack tape begins at time 00:01:00:00, ends at 00:28:19:00 and contains a specific cue point (such as a glass shattering) that begins at 00:12:53:19 (Figure 12.5b). By monitoring the timecode readout, it's a simple matter to locate the precise position that corresponds to the cue

FIGURE 12.5
Location of relative
addresses: (a) Postal address
analogy. (b) Timecode
addresses and a cue point on
longitudinal tape.

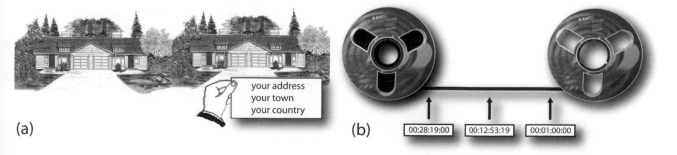

(a)

your address
your town
your country

(b)

| 00:28:19:00 | 00:12:53:19 | 00:01:00:00 |

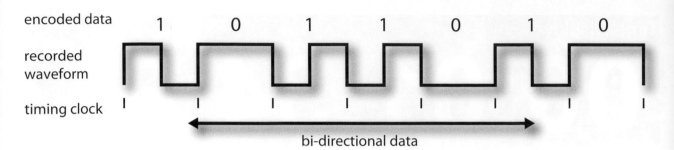

bi-directional data

FIGURE 12.6
Bi-phase modulation
encoding.

point on the media and then perform whatever function is necessary, such as inserting an effect into the soundtrack at that specific point – CRAAAASH!

Timecode Word

The total of all time-encoded information that's encoded within each audio or video sync frame is known as a *timecode word*. Each word is divided into 80 equal bits, which are numbered consecutively from 0 to 79. One word covers an entire audio or video frame, such that for every frame, there is a unique and corresponding timecode address. Address information is contained in the digital word as a series of bits that are made up of binary 1s and 0s.

In the case of an analog medium, an SMPTE signal is electronically encoded in the form of a modulated square wave. This method of encoding information is known as bi-phase modulation. Using this code type, a voltage or bit transition in the middle of a half-cycle of a square wave represents a bit value of 1, while no transition within this same period signifies a bit value of 0 (Figure 12.6). The most important feature about this system is that detection relies on shifts within the pulse and not on the pulse's polarity or direction. Consequently, timecode can be read in either the forward or reverse play mode, as well as at fast or slow shuttle speeds.

TRY THIS: SMPTE TIMECODE

1. Go to youtube.com/modernrecordingtechniques and search for "smpte timecode". SMPTE in its LTC form is not my favorite musical tune, but it's a useful one! Take a listen, just so you'll know what it sounds like, but make sure that your monitor volume control is turned down! … SMPTE Timecode isn't musical, quiet or particularly pleasing!

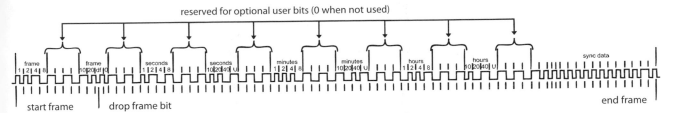

reserved for optional user bits (0 when not used)

| frame | frame | seconds | seconds | minutes | minutes | hours | hours | sync data |

start frame | drop frame bit

end frame

The 80-bit timecode word is subdivided into groups (Figure 12.7), whereby each grouping represents a specific coded piece of information. Each segment represents a binary-coded decimal (BCD) number that ranges from 0 to 9. When the full frame is scanned, all of these groupings are read out as a single SMPTE frame number (in hours, minutes, seconds and frames).

FIGURE 12.7
Bi-phase representation of the SMPTE timecode word.

SYNC INFORMATION DATA

An additional form of information that's encoded into the timecode word is sync data. This information exists as 16 bits at the end of each timecode address word. These bits are used to define the end of each frame. Because timecode can be read in either direction, sync data is also used to tell the device in which direction the tape or digital device is moving.

TIMECODE FRAME STANDARDS

In productions using timecode, it's important that the readout display be directly related to the actual elapsed time of a program, particularly when dealing with the exacting time requirements of broadcasting. Due to historical and technical differences between countries, timecode frame rates may vary from one medium, production house or region of origin to another. Here are the most common frame rates:

- *30 fr/sec (monochrome US video):* in the case of a black-and-white (monochrome) video signal, a rate of exactly 30 frames per second (fr/sec) is used. If this rate (often referred to as non-drop code) is used on a black-and-white program, the timecode display, program length and actual clock-on-the-wall time will all be in agreement.

- *29.97 fr/sec (drop-frame timecode for color NTSC US video):* the simplicity of 30 fr/sec was eliminated, however, when the National Television Standards Committee (NTSC) set the frame rate for the color video signal in the US and Japan at 29.97 fr/sec. Thus, if a timecode reader that's set up to read the monochrome rate of 30 fr/sec were used to read a color program, the timecode readout would pick up an extra 0.03 frame for every second that passes. Over the duration of an hour, the timecode readout would unfortunately differ from the actual elapsed time by a total of 108 frames (or 3.6 seconds). To correct for this difference and bring the timecode readout and the actual elapsed time back into agreement, a series of frame adjustments was introduced into the code. Because the goal is to drop 108 frames over the course of an hour, the code used for color has come to be known as drop-frame code. In this system, two frame counts for every minute of operation are omitted from the code, with the exception of minutes 00, 10, 20, 30, 40 and 50. This has the effect of adjusting the frame count so that it agrees with the actual elapsed duration of a program.

- *29.97 fr/sec (non-drop-frame code):* in addition to the color 29.97 drop-frame code, a 29.97 non-drop-frame color standard can also be found in video production. When using non-drop timecode, the frame count will always advance one count per frame without any drops. As you might expect, this mode will result in a disagreement between the frame count and the actual clock-on-the-wall time over the course of the program. Non-drop, however, has the distinct advantage of easing the time calculations that are often required in the video editing process (because no frame compensations need to be taken into account).

- *25 fr/sec EBU (standard rate for PAL European video):* another frame rate format that's used throughout Europe is the European Broadcast Union (EBU) timecode. EBU utilizes SMPTE's 80-bit code word but differs in that it uses a 25-fr/sec frame rate. Because both monochrome and color video EBU signals run at exactly 25 fr/sec, an EBU drop-frame code isn't necessary.

- *24 fr/sec (international standard rate for film work):* the medium of film differs from all of these in that it makes use of an SMPTE timecode format that runs at 24 fr/sec.

From this, it's easy to understand why confusion often exists as to which frame rate should be used on a project. If you are working on an in-house project that doesn't incorporate time-encoded material that comes from the outside world, you should choose a rate that both makes sense for you and is likely to be compatible with an outside facility (should the need arise).

For example, electronic musicians who are working in-house in the US will often choose to work at 30 fr/sec. Those in Europe have it easy, because 25 fr/sec is the logical choice for all music and video production. On the other hand, those who work with projects that come through the door from other production houses will need to take special care to reference their timecode rates to those used by the originating media house. This can't be stressed enough: if care isn't taken to keep your timecode references at the proper rate and relative address times (while keeping degradation to a minimum from one generation to the next), the various media might have trouble syncing up when it comes time to put the final master together – and that could spell BIG trouble.

3:2 PULLDOWN RATE

From the preceding list, we see that film runs at a framerate of 24 fr/sec, while the NTSC color frame rate runs at 29.97 fr/sec. Whenever there is a need to retain timecode lock when transferring film to color video, a rate conversion is needed to accurately retain the media's TC address locations. NTSC video makes use of two video fields (a series of still images that are sequentially played to create the illusion of motion) for every timecode frame. Using a process known as 3:2 (three to two) pulldown rate, a telecine film playback device is used to add a third video field to these existing two fields. This added field can't be seen by the untrained eye of the vast majority of viewers, but when combined with a speed slowdown of 1/1,000th, the TC translation process can be as accurate as possible.

Timecode Within Digital Media Production

Given that SMPTE exists in a digitally encoded data form, current-day digital professional media devices are able to accept and communicate SMPTE directly without too much trouble. Professional camera, film, controllers and editing systems are able to directly chain and synchronize SMPTE using a multitude of complicated yet standardized methods that make use of both digital- and analog-style timecode data streams.

Of course, there is a wide range of approaches that can be taken when media devices (cameras, video editing software and field audio recorders) are to be synchronized together. These can range from "shoots" that make use of multiple cameras and separate field recorders, which are "locked" to a single timecode source on a set, all the way down to a simple camera and digital hand recorder with audio that can be manually synced up within the digital editor, without the need for timecode at all. The types of equipment and the ways that they deal with the technology of sync are ever-changing. Therefore, it's important to keep abreast of current technology, read the manuals (about how connections and settings can best be made) and dive into the study of various types of media production.

BROADCAST WAVE FILE FORMAT

Although digital media devices and software are able to import, convert and communicate using the language of SMPTE timecode (in all its various flavors), the folks at the EBU saw the need to create a universal audio file format that would include timecode data within all forms of audio and visual media production. The result was the birth of the Broadcast Wave Format (BWF). Broadcast Wave is in most ways completely compatible with its Microsoft Wave (.WAV) counterpart, with the exception that it is able to embed metadata (information about the recorded content – photo, take number, date, technical data, etc.) as well as SMPTE timecode address data. The inclusion of such important content and timecode information means that the time-related information will actually be embedded within the media file itself, allowing sound files that are imported into a video or audio editor to automatically snap to their appropriate timecode position. Obviously, this means that Broadcast Wave can be a huge time saver in the production and post-production process.

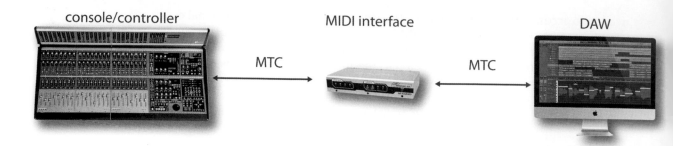

console/controller — MTC — MIDI interface — MTC — DAW

FIGURE 12.8
Many time-based media devices in the studio can be cost-effectively connected via MIDI timecode (MTC).

MIDI Timecode

In earlier times, the synchronization of audio devices to other video and/or audio devices was a very expensive proposition, far beyond the budget of most projects or independent production houses. Today, however, an easy-to-use and inexpensive standard makes use of MIDI to transmit sync and timecode data throughout a connected production system (Figure 12.8). This has made it possible for even the most budget-minded project studios to synchronize media devices and software using timecode.

MIDI timecode (MTC) was developed to allow electronic musicians, project studios, video facilities and virtually all other production environments to cost-effectively and easily translate timecode into time-stamped messages that can be transmitted over MIDI data lines. Created by Chris Meyer and Evan Brooks, MIDI timecode allows SMPTE-based timecode to be distributed throughout the MIDI chain to devices or instruments that are capable of synchronizing to and executing MTC commands. MIDI timecode is an extension of the MIDI standard, making use of existing SysEx message types that were either previously undefined or were being used for other, non-conflicting purposes.

Since most modern recording, music and stage lighting systems include MIDI in their design, there's often no need for external hardware when making direct connections. Simply chain the MIDI data lines from the master to the appropriate slaves within the system (via physical cables, USB or virtual internal routing). Although MTC uses a reasonably small percentage of MIDI's available bandwidth (about 7.68% at 30 fr/sec), it's customary (but not at all necessary) to separate these lines from those that are communicating performance data when using physical MIDI cables. As with conventional SMPTE, only one master can exist within an MTC system, while any number of slaves can be assigned to follow, locate and chase to the master's speed and position. Because MTC is easy to use and is often included free in many systems and program designs, this technology has grown to become the most straightforward and commonly used way to lock together such devices as DAWs, external devices, basic analog/video and lighting setups.

MIDI TIMECODE MESSAGES

The MIDI timecode format can be divided into two parts:

- Timecode
- MIDI cueing

The timecode capabilities of MTC are relatively straightforward and allow devices to be synchronously locked or triggered to SMPTE timecode. MIDI cueing is a format that informs a MIDI device of an upcoming event that's to be performed at a specific time (such as load, play, stop, punch-in/out, reset). This latter protocol envisions the use of intelligent MIDI devices that can prepare for a specific event in advance and then execute the command on cue.

MIDI timecode is made up of three message types:

- *Quarter-frame messages:* these are transmitted only while the system is running in real or variable-speed time, in either forward or reverse direction. True to its name, four quarter-frame messages are generated for each timecode frame. Since eight quarter-frame messages are required to encode a full SMPTE address (in hours, minutes, seconds and frames: 00:00:00:00), the complete SMPTE address time is updated once every two frames (in other words, MIDI timecode actually has half the resolution accuracy of its SMPTE timecode counterpart). Each quarter-frame message contains 2 bytes. The first byte is F1, the quarter-frame common header; the second byte contains a nibble (four hits) that represents the message number (0 through 7) and a nibble for encoding the time field digit.

- *Full messages:* quarter-frame messages are not sent in the fast-forward, rewind or locate modes, because this would unnecessarily clog a MIDI data line. When the system is in any of these shuttle modes, a full message is used to encode a complete timecode address. After a fast shuttle mode is entered, the system generates a full address message and then places itself in a pause mode until the time-encoded slaves have located to the correct position. Once playback has resumed, MTC will again begin sending incremental quarter-frame messages.

- *MIDI cueing messages:* MIDI cueing messages are designed to address individual devices or programs within a system. These 13-bit messages can be used to compile a cue or edit decision list, which in turn instructs one or more devices to play, punch in, load, stop and so on, at a specific time. Each instruction within a cueing message contains a unique number, time, name, type and space for additional information. At the present time, only a percentage of the possible 128 cueing event types have been defined.

SMPTE/MTC CONVERSION

Although MIDI timecode connections can be directly made between compatible MIDI devices, a SMPTE-to-MIDI converter is required to read incoming longitudinal (LTC) SMPTE timecode and convert it into MIDI timecode (and vice versa) for other device types. These conversion systems are available as a stand-alone device or as an integrated part of certain audio interfaces or multiport MIDI interface/patch bay/synchronizer systems (Figure 12.9).

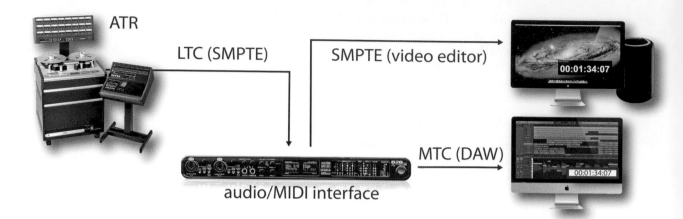

ATR

LTC (SMPTE)

SMPTE (video editor)

00:01:34:07

MTC (DAW)

audio/MIDI interface

00:01:34:07

FIGURE 12.9
SMPTE timecode can often be generated throughout a production system, either as LTC or as MTC via a capable MIDI or audio interface.

TIMECODE PRODUCTION IN THE ANALOG AUDIO AND VIDEO WORLDS

Fortunately for us, most digital editing systems (such as digital video and audio workstations) are able to communicate timecode in a relatively seamless and straightforward manner (at least at a basic level, often only requiring that the various systems be set to the same frame rates, etc.). Synching analog-to-analog or analog-to-digital devices, on the other hand, is often far less straightforward and needs to be understood, at least at a fundamental level.

Timecode that's recorded onto an analog audio or video cue track of an older-style video tape recorder is known as longitudinal timecode (LTC). LTC encodes a bi-phase timecode signal onto an analog track in the form of a modulated square wave at a bit rate of 2,400 bits/sec. The recording of square wave onto a magnetic audio track is difficult even under the best of conditions. For this reason, the SMPTE standard has set forth an allowable rise time of 25 ± 5 microseconds for the recording and reproduction of valid code. This tolerance requires a signal bandwidth of at least 15 kHz, which is well within the range of most professional audio recording devices. Variable-speed timecode readers are often able to decode timecode information at shuttle rates ranging from 1/10th to 100 times normal playing speed, which is often necessary when monitoring videotape at slow or near-still speeds.

Whenever a video tape recorder (VTR) is used (which is uncommon these days) at speeds slower than 1/10th to 1/20th normal play speed, a character generator is used to burn timecode addresses directly into the video image of a work tape copy. This superimposed readout allows the timecode to be easily seen and identified, even at very slow or still picture shuttle speeds.

LTC Refresh and Jam Sync

Longitudinal timecode operates by recording a series of square-wave pulses onto magnetic tape. As was said, it's somewhat difficult to record a square waveform

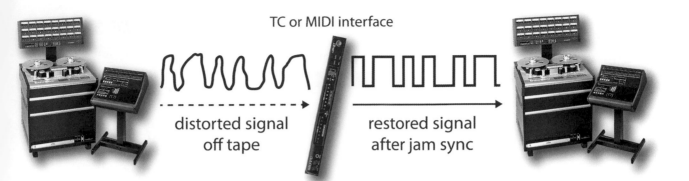

TC or MIDI interface

distorted signal
off tape

restored signal
after jam sync

onto analog magnetic tape without having the signal suffer moderate to severe waveform distortion. Although timecode readers are designed to be relatively tolerant of waveform amplitude fluctuations, such distortions are severely compounded when code is copied from one analog recorder to another over one or more generations. Should the quality of a copied SMPTE signal degrade to the point where the synchronizer can't differentiate between the pulses, the code will disappear, and the slaves will come to a stop. For this reason, a timecode refresher (Figure 12.10) has been incorporated into most timecode synchronizers and certain MIDI interface devices that have sync capabilities. Basically, this process (known as jam sync) reads the degraded timecode information from a previously recorded track and then regenerates the square wave back into its original shape. In this way, the timecode can be cleanly recorded to a new track and accurately read by another device.

Jam sync can also refer to the synchronizer's ability to predict the next timecode value even though the next valid value hasn't appeared at its input. The generator is then said to be working in a freewheeling fashion, since the generated code may not agree with the actual recorded address values. If a short-term dropout occurs, jam sync can often detect or refresh the lost signal. This process is often useful when dealing with dropouts or undependable code from audio tracks on an analog video machine. Two forms of jam sync options are available:

- Freewheeling: in the freewheeling mode, the receipt of timecode causes the generator's output to initialize when a valid address number is detected. The generator then begins to count in an ascending order on its own, ignoring any deterioration or discontinuity in code and producing fresh, uninterrupted SMPTE address numbers.
- Continuous: continuous jam sync is used in cases where the original address numbers must remain intact and shouldn't be regenerated as a continuously ascending count. After the reader has been activated, the generator updates the address count for each frame in accordance with incoming address numbers and outputs an identical, regenerated copy.

FIGURE 12.11
Example of timecode sync production using a simple MIDI interface synchronizer (possibly one that's already designed into an audio interface) within a studio setting.

Synchronization Using SMPTE Timecode

In order to achieve a frame-by-frame timecode lock between multiple audio, video or film analog transports, it's necessary to use a device or integrated system that's known as a synchronizer. The basic function of a synchronizer is to control one or more tape, computer-based or film transports (designated as slave machines) so their speeds and relative positions are made to accurately follow one specific transport (designated as the master).

The use of a synchronizer within a project studio environment (Figure 12.11) often involves a multiport MIDI interface that includes provisions for locking an analog audio or video transport to a digital audio, MIDI or electronic music system by translating LTC SMPTE code into MIDI timecode. In this way, one simple device can cost-effectively serve multiple purposes to achieve lock with a high degree of accuracy. Systems that are used in video production and in higher levels of production will often require a greater degree of control and remote-control functions throughout the studio or production facility. Such a setup will often require a more sophisticated device, such as a control synchronizer or an edit decision list (EDL) controller.

SMPTE Offset Times

In the real world of audio production, programs or songs don't always begin at 00:00:00:00 (although this can easily happen when using a video or audio workstation within an in-house project). Let's say that you were handed a recording that needed a synth track to be laid down onto track 7 of a song that goes from 00:11:24:03 to 00:16:09:21. Instead of inserting more than 11 minutes of empty bars into a MIDI track on your synched DAW, you could simply insert an offset start time of 00:11:24:03. This means that the sequenced track will begin to increment from measure 1 at 00:11:24:03 and will maintain relative offset sync throughout the program.

Offset start times are also useful when synchronizing devices to an analog or videotape source that doesn't begin at 00:00:00:00. As you're probably aware, it always takes a bit of time for an analog audio transport to settle down and begin playing (this wait time often quadruples whenever an analog videotape transport is involved). If a program's timecode were to begin at the head of the tape, it's extremely unlikely that you would want to start a program at 00:00:00:00, since playback would be delayed and extremely unstable at points near this begin time. Instead, most programming involving an analog audio or video media is striped with an appropriate pre-roll of anywhere from 10 seconds to 2 minutes. Such a pre-roll gives any analog transports ample time to begin playback and sync up to the master timecode source.

In addition, it's often wise to begin the actual production or first song at an offset or SMPTE start time of 00:01:00:00 (some facilities set the start offset at 01:00:00:00). This minimizes the possibility that rolling over at midnight will confuse the synchronizer. That's to say, if the content starts at 00:00:00:00 (midnight), the pre-roll would be in the 23:59:00:00 range, and the synchronizer would try to rewind the tape in the wrong direction to find 00:00:00:00 (rolling it backwards off the reel) instead of rolling forward. Not fun in the heat of a production!

Distribution of SMPTE Signals

Generally, when analog media devices are synced together, connections will need to be made between each transport and the synchronizer. These include lines for the LTC timecode track and the control interface (which often uses the Sony nine-pin remote protocol for giving the synchronizer full logic transport and speed-related feedback information). LTC signal lines can be distributed throughout the production system in much the same way that any other audio lines are distributed. They can be routed directly from machine to machine or patched through audio switching systems via balanced, shielded cables or unbalanced cables. It should be noted that because the timecode signal is bi-phase or symmetrical, it's immune to cable polarity problems.

TIMECODE LEVELS

One problem that can plague systems using timecode is crosstalk. This happens when a high-level signal leaks into adjacent signal paths or analog tape tracks. Although no industry standard levels exist for the recording of timecode onto magnetic tape or digital tape track, the levels shown in Table 12.1 can help you get a good signal level while keeping distortion and analog crosstalk to a minimum.

Table 12.1	Optimum Timecode Recording Levels	
Tape Format	**Track Format**	**Optimum Recording Level**
ATR	Edge track (highest number)	−5 to −10 VU
Digital device	Highest number track or dedicated timecode I/O ports	−20 dB

Note: if the VTR is equipped with automatic gain compensation (AGC), override the AGC and adjust the signal gain controls manually.

Again, it's nice to keep in mind that in most modern-day settings, LTC code almost certainly won't be necessary when syncing between digital video and audio devices.

However a basic understanding of sync will help keep you out of trouble.

REAL-WORLD APPLICATIONS USING TIMECODE AND MIDI TIMECODE

Before we delve into the many possible ways that a system can be set up to work in a timecode environment, it needs to be understood that each system will often have its own particular personality. Quite commonly, the connections, software and operation of one system might totally differ from those of another. This is often due to factors such as system complexity and the basic hardware types that are involved, as well as the type of hardware and software systems that are installed in a DAW. Larger, more expensive setups that are used to create television and film soundtracks will often involve extensive timecode and system interconnections that can easily get more complicated.

Fortunately, the use of MIDI timecode and digital systems has greatly reduced the cost and complexity of connecting and controlling a synchronous pro and project studio system down to levels that can be easily managed by both experienced and novice users. Having said these things, I'd still like to stress that solving synchronization problems will often require as much intuition, perseverance, insight and art as it will technical skill. For the remainder of this chapter, we'll be looking into some of the basic concepts and connections that can be used to get your system up and running. Beyond this, the next best course of action will be to consult your manuals, seek help from an experienced friend or call a company's tech department about the particular hardware or software that's giving both you and your system the willies.

Master/Slave Relationship

Since synchronization is based on the timing relationship between two or more devices, it follows that the logical way to achieve sync is to have one or more devices (known as slaves) follow the relative movements of a single transport or device (known as the master). The basic rule to keep in mind is that there can be only one master in a connected system; however, any number of slaves can be set to follow the relative movements of a master transport or device (Figure 12.12).

Generally, the rule for deciding which device will be the master in a production system (during the pre-planning phase) can best be determined by asking a few questions:

- What type of media is the master timecode media recorded on?
- Which device will provide the most stable timing reference?
- Which device will most easily and cost-effectively serve as the master?

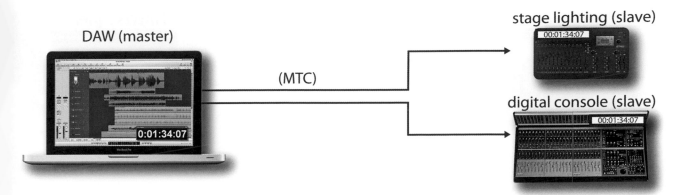

DAW (master)

0:01:34:07

(MTC)

stage lighting (slave)

00:01:34:07

digital console (slave)

00:01:34:07

If the master comes to you from an outside source, asking lots of questions about the source specs could help solve many of your problems. If the project is in-house, and you have total say in the matter, you might want to research your options more fully to make the best choice for your facility. The following sections can help give you insights into which devices will best serve as the master within a particular system.

FIGURE 12.12
There can be only one master in a synchronized system; however, there can be any number of slaves.

Video's Need for a Stable Timing Reference

Whenever a video signal is copied from one machine to another, it's essential that the scanned data (containing timing, video and user information) be copied in perfect sync from one frame to the next. Failure to do so will result in severe picture breakup or at best, the vertical rolling of a black line over the visible picture area. Copying video from one machine to another generally isn't a problem (because the video recording device that's doing the copying normally provides sync from the playback machine within the picture itself). Video post-production houses, however, often simultaneously use any number of video and audio workstations, switchers and edit controllers during the production and editing of a single program. Mixing and switching between these combined sources without a stable sync source would almost certainly result in chaos – with the end result being a very unhappy client.

Fortunately, referencing all of the video, audio and timing elements to an extremely stable timing source (using black burst or a house sync generator) will generally resolve this sync nightmare. This accurate reference clock serves to synchronize the video frames and timecode addresses that are received or transmitted by nearly every video-related device in a production facility, so the leading frame edge of every video signal occurs at exactly the same instant in time (Figure 12.13). By resolving all video and audio devices to a single black burst reference, you're assured that relative frame transitions, speeds and TC addresses throughout the system will be consistent and stable.

DAW/sequencer

blackburst generator (master)

00:01:34:07

00:01:34:07

00:01:34:07

video editor

FIGURE 12.13
Example of a system whose overall timing elements are locked to a black burst reference signal.

Video Workstation or Recorder

Since video is often an extremely stable timing source, a digital video editor (or even an analog video device) would be a stable timing source within a connected production system. This process still shouldn't be taken lightly, because the timecode must (in most cases) conform to the timecode addresses on the original video or working master. The rule of thumb is: if you're working on a project that was created out of house, always use the code that was provided by the original production team. Stripping your own code or erasing over the original code with your own would render the original timing elements useless, because the new code wouldn't relate to the original addresses or include any timing variations that might be a part of the original master source. In short, make sure that your working copy includes an SMPTE track that is a regenerated copy of the original code! Should you overlook this, you might run into timing and sync troubles, either immediately or later in the post-production phase – factors that will definitely lead to the premature loss of both your hair and client.

Digital Audio Workstations

A computer-based DAW can often be set to act as either a master or a slave. This will ultimately depend on the software and the situation, because most professional workstations can be set to chase (to follow or be triggered by) a stable master timecode source as well as to generate timecode (often in the form of MIDI or SMPTE timecode within a higher-end system). Should an analog system be used in conjunction with a DAW, most often, the DAW would be set to be the master, as the DAW's timing source is infinitely more stable than most analog tape machines (for example).

The process of maintaining a synchronous lock between digital audio devices differs fundamentally from the process of maintaining relative speed between analog transports. This is due to the fact that a digital system generally achieves synchronous lock by adjusting its playback sample rate (and thus its speed and/or pitch ratio) so as to precisely match the relative playback speed of the

FIGURE 12.14
Most high-end DAW systems
are capable of importing a
video file directly into the
project session window.

master transport. Therefore, whenever a digital system is synchronized to a time-encoded master, a stable timing source is relatively important in order to keep jitter (in this case, an increased distortion due to rapid pitch shifts) to a minimum. In other words, the source's program speed should vary as little as possible to prevent any degradation in the digital signal's quality. As such, a digital audio system that's working within a video production environment would also benefit from the above-mentioned house sync timing source.

Most modern DAWs include support for displaying a video track (Figure 12.14) within a session (both as a separate video screen that can be displayed on the monitor desktop and in the form of a video thumbnail track that appears within the track view). Of course, the video track provides important visual cues for tracking live music, accurately placing automation moves and effects (sfx) at specific hit points within the scene or for adding music sweetening. This feature allows audio to be built up within a DAW environment without the need for syncing to an external device at all. As you might expect, with the use of recorded tracks, software instruments and internal mixing capabilities, tracks can easily be built up, spotted and mixed – all inside the box.

Routing Timecode to and From Your Computer

From a connections standpoint, most DAW, MIDI and audio application software packages are flexible enough to let you choose from any number of available sync sources (whether connected to a hardware port, MIDI interface port or virtual sync driver). All you have to do is assign all of the slaves within the system to the device driver that's generating the system's master code (Figure 12.15). In many cases, it's best to have your DAW or editor generate the master code for the system with the appropriate settings and timecode address times.

FIGURE 12.15
Cubase/Nuendo sync setup and project sync dialog boxes (courtesy of Steinberg Media Technologies GMBH, www. steinberg.net).

Analog Audio Recorders

In many audio production situations, whenever an analog tape recorder is connected in a timecode environment, this machine will most often want to act as the master in a basic LTC environment. Although this might be counterintuitive, it's far easier and less expensive for an analog recorder to output master SMPTE code than to be controlled in an external slave relationship. This is because special and expensive equipment is generally required to continuously adjust the transport regulator's speed (using a DC capstan servo) so as to maintain a synchronous relationship to the master SMPTE address.

A SIMPLE CAVEAT

The preceding guidelines are just that – guidelines. As you might expect, each and every setup will be slightly different and might require that you come up with a novel solution to a quirky problem. Again, the Internet is full of insights and solutions from those who have already gone down that long and treacherous path. Be warned, though. It's important that you prepare and make your decisions wisely, lest a problem raise its ugly head at a later and crucial time.

Keeping out of Trouble

Here are a few guidelines that can help save your butt when using SMPTE and other timecode translations during a project:

- Familiarize yourself with the hardware and software involved in a project *before* the session starts.
- When in doubt about frame rates, special requirements or anything else, for that matter, ask! You (and your client) will be glad you did.
- Fully document your timecode settings, offsets, start times, etc.
- If the project isn't to be used in-house, ask the producer what the proper frame rate should be. Don't assume or guess it.

- When beginning a new session (when using a tape-based device), always stripe the master contiguously from the beginning to the end before the session begins. It never hurts to stripe an extra tape, just in case.

- Whenever analog machines are involved, start generating new code at a point after midnight (i.e., 00:01:00:00 or 01:00:00:00 to allow for a safety pre-roll). If the project isn't to be used in-house, ask the producer what the pre-roll/start times should be. Don't assume or guess.

- Never dub (copy) timecode directly. Always make a refreshed (jam synced) copy of the original timecode (from an analog master) before the session begins.

- Disable noise reduction and AGC (Automatic Gain Control) on analog audio tracks (on both audio and video devices).

- Work with copies from the original production video, and make a new one when sync troubles appear.

- It's not unusual for the timecode to be read incorrectly (when short drop-outs occur on the track, usually on videotape). When this happens, you might set the synchronizer to freewheel once the transports have initially locked.

In closing, I'd like to point out that synchronization can be a simple procedure, or it can be an extremely complex one, depending on your requirements and the type of equipment that's involved. A number of books and articles have been written on this subject. If you're serious about production, I suggest that you do your best to keep up on it. Although the fundamentals often remain the same, new technologies and techniques are constantly emerging. For further reading, go to the SMPTE website (www.smpte.org). As always, the best way to learn is simply by reading and then jumping in and doing it.

CHAPTER 13

Amplifiers

In the world of audio, amplifiers have many applications. They can be designed to amplify, equalize, combine, distribute or isolate a signal. They can even be used to match signal impedances between devices. At the heart of any *amplifier* (*amp*) system is either a vacuum tube or a semiconductor-type transistor series of devices. Everyone has heard of these regulating devices, but few have a grasp of how they operate, so let's have a basic look into these electronic wonders.

AMPLIFICATION

To best understand how the theoretical process of amplification works, let's draw on an analogy. The original term for the tube used in early amplifiers is *valve* (a term that's still used in England and other Commonwealth countries). If we hook up a physical water valve to a high-pressure hose, large amounts of water pressure can be controlled with very little effort simply by turning the valve (Figure 13.1). By using a small amount of expended energy, a trickle of water can be turned into a high-powered gusher and back down again. In practice, both the vacuum tube and the transistor work much like this valve. For example, a vacuum tube operates by placing a DC current across its plate and a heated cathode element (Figure 13.2). A wire mesh grid separating these two elements acts like a control valve, allowing electrons to pass from the plate to the cathode. By introducing a small and varying signal at the input onto the tube's

FIGURE 13.1
The current through a vacuum tube or transistor is controlled in a manner that's similar to the way that a valve tap can control water pressure through a water pipe: (a) Open valve. (b) Closed valve.

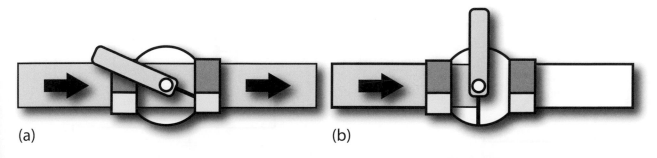

(a)   (b)

DOI: 10.4324/9781003260530-13

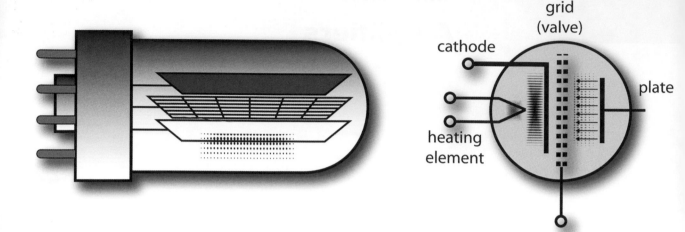

FIGURE 13.2

An example of a triode vacuum tube.

FIGURE 13.3

A simple amplifier schematic: (a) Showing how small changes in voltage at the tube's grid can produce much larger, corresponding amplitude changes between its cathode and plate. (b) Showing how small changes in current at the transistor's base can produce much larger, corresponding amplitude changes through the emitter and collector to the output.

grid, a much larger electrical signal can be used to correspondingly regulate the flow of electrons between the plate and the cathode (Figure 13.3a).

The transistor (a term originally derived from "trans-resistor", meaning a device that can easily change resistance) operates under a different electrical principle than a tube-based amp, although the valve analogy still applies. Figure 13.3b shows a basic amplifier schematic with a DC power source that's placed across the transistor's collector and emitter points. As with the valve analogy, by presenting a small control signal at the transistor's base, the resistance between the collector and emitter will correspondingly change. This allows a much larger analogous signal to be passed through to the device's output.

As a device, the transistor isn't inherently linear; that is, applying an input signal to the base won't always produce a corresponding output change. The linear operating region of a transistor lies between the device's lower-end cutoff region and an upper saturation point (Figure 13.4a). Within this operating region, however, changes at the input will produce a corresponding (linear) change in the collector's output signal. When operating near these cutoff or saturation

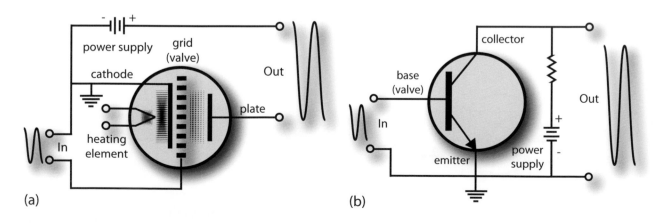

(a)

(b)

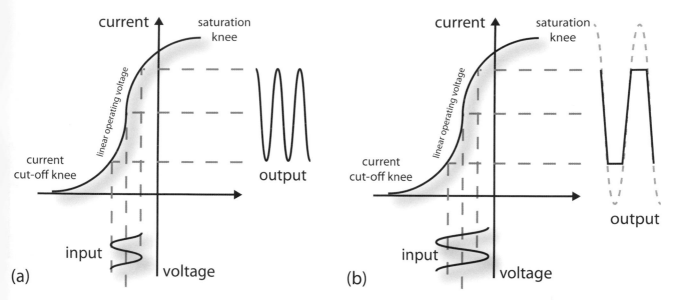

FIGURE 13.4
Output curves of a transistor:
(a) Proper operating region.
(b) A clipped waveform.

points, the base current lines won't be linear, and the output will become distorted. In order to keep the signal within this linear operating range, a DC bias voltage signal is applied to the base of the transistor (for much the same reason, a high-frequency bias signal is applied to an analog recording head). After a corrective voltage has been applied, and sufficient amplifier design characteristics have been met, the amp's dynamic range will be limited by only two factors: noise (which results from thermal electron movement within the transistor and other circuitry) and saturation.

Amplifier *saturation* results when the input signal is so large that its DC output supply isn't large enough to produce the required, corresponding output signal. Overdriving an amp in such a way will cause a mild to severe waveform distortion effect known as *clipping* (Figure 13.4b). For example, if an amp having a supply voltage of +24 volts (V) is operating at a gain ratio of 30:1, an input signal of 0.5 V will produce an output of 15 V. Should the input be raised to 1 V, the required output level would have to be increased to 30 V. However, since the maximum supply voltage is limited to 24 V, levels above this point will be chopped off or "clipped" at the upper and lower edges of the output waveform. Whenever a transistor and integrated circuit design clips, severe odd-order harmonics are often introduced that are immediately audible as distortion. Tube amp designs, on the other hand, tend to lend a more musical-sounding, even-order harmonic aspect to a clipped signal. I'm sure you're aware that clipping distortion can be a sought-after part of a tube instrument's sound (electric guitars thrive on it); however, it's rarely a desirable effect in quality studio and monitoring gear. The best way to avoid undesirable distortion from either amp type is to be aware of the various device gain stage levels throughout the studio's signal chain.

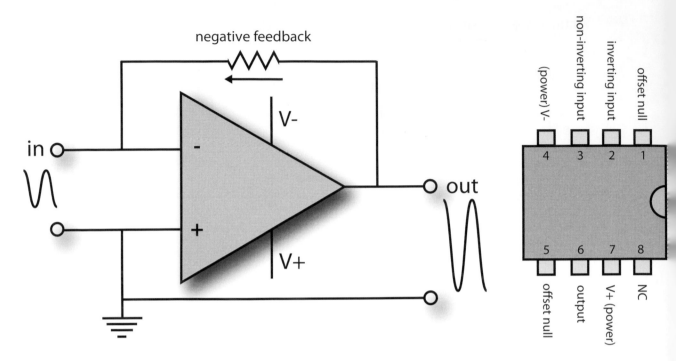

negative feedback

V-

in

out

+

V+

(power) V-

non-inverting input

inverting input

offset null

4 3 2 1

5 6 7 8

offset null

output

V+ (power)

NC

FIGURE 13.5

(a) Basic op-amp circuit and (b) 741-type pin configuration.

The Operational Amplifier

An *operational amplifier (op-amp)* is a stable, high-gain, high-bandwidth amp that has a high input impedance and low output impedance. These qualities allow op-amps (Figure 13.5) to be used as a basic building block for a wide variety of audio and video applications in both older and newer circuit designs, simply by adding components onto the basic circuit in a building-block fashion to fit the design's needs. To reduce an op-amp's output gain to more stable, workable levels, a negative feedback loop is often required. *Negative feedback* is a technique that applies a portion of the output signal through a limiting resistor back into the negative or phase-inverted input terminal. By feeding a portion of the amp's output back into the input out of phase, the device's output signal level is reduced. This has the effect of controlling the gain (by varying the negative resistor value) in a way that also serves to stabilize the amp's performance and further reduce distortion.

Op-amps are commonly used in current-day analog systems, often performing tasks such as amplification, equalization (EQ), filtering and signal processing. They're also used in such digital circuits as analog-to-digital, digital-to-analog converters and buffering circuits. As a matter of fact, hundreds of thousands of op-amp blocks are commonly designed into modern-day computers.

Preamplifiers

One of the mainstay amplifier types found at the input section of most professional mixer, console, audio interface and outboard devices is the *preamplifier* (*preamp*). This amp type is often used in a wide range of applications, such as boosting a mic's signal to line level, providing variable gain for various signal types, isolating input signals and equalization, just to name a few. Preamps are an important component in audio engineering because they often set the "tone" of how a device or system will sound. Just as a microphone has its own sonic character, a preamp design will often have its own "sound". Questions such as "Are the op-amps designed from quality components?" "Do they use tubes or transistors?" and "Are they quiet or noisy?" are considerations that can greatly affect the overall sound of a device.

Equalizers

You might be surprised to know that basically, an *equalizer* is nothing more than a frequency-discriminating amplifier. In most analog designs, EQ is achieved through the use of resistor/capacitor networks that are located in an op-amp's negative feedback loop (Figure 13.6) in order to boost (amplify) or cut (attenuate) certain frequencies in the audible spectrum. By changing the circuit design, complexity and parameters, any number of EQ curves can be achieved.

Summing Amplifiers

A *summing amp* (also known as an active combining amplifier) is designed to combine any number of discrete inputs into a single output signal bus while providing a high degree of isolation between them (Figure 13.7). The summing amplifier is an important component in analog console/mixer design because the large number of internal signal paths requires a high degree of isolation in order to prevent signals from inadvertently leaking into other audio paths.

FIGURE 13.6
Basic equalizer circuit:
(a) Low-frequency. (b)
High-frequency.

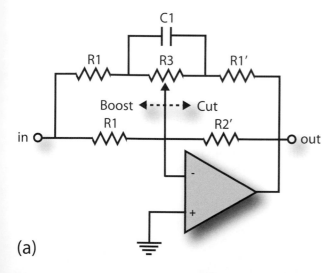

(a)

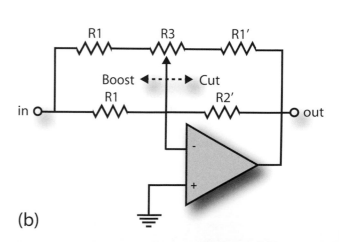

(b)

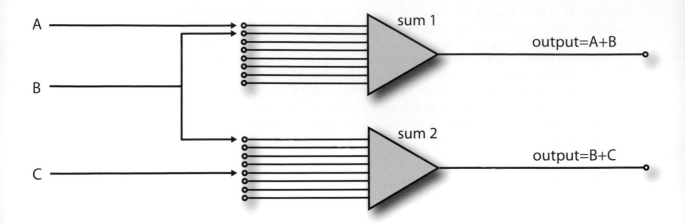

FIGURE 13.7
A summing amp is used to provide isolation between various inputs and/or outputs in a signal chain.

Distribution Amplifiers

Often, it's necessary for audio signals to be distributed from one device to several other devices or signal paths within a recording console or music studio. In this situation, a *distribution amp* isn't used to provide gain but instead, will amplify the signal's current (power) that's being delivered to one or more loads (Figure 13.8). Such an amp, for example, could be used to boost the overall signal power so that a single feed could be distributed to a large number of headphones during a string or ensemble session.

POWER AMPLIFIERS

As you might expect, *power amplifiers* (Figure 13.9) are used to boost the audio output to a level that can drive one or more loudspeakers at their rated volume levels. Although these are often reliable devices, power amp designs have their own special set of problems. These include the fact that transistors don't like to work at the high temperatures that can be generated during continuous, high-level operation. Such temperatures can also result in changes in the unit's response and distortion characteristics or outright failure. This often requires that protective measures (such as fuse and thermal protection) be taken. Fortunately, many of the newer amplifier models offer protection under a wide range of circuit conditions (such as load shorts, mismatched loads and even

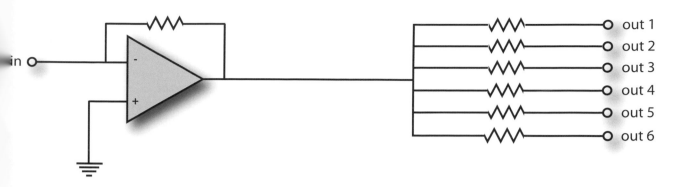

FIGURE 13.8
Distribution amp.

open "no-load" circuits) and are usually designed to work with speaker impedance loads ranging between 4 and 16 ohms (most speaker models are designed
to present a nominal load of 8 ohms). When matching an amp to a speaker set,
the amp should be capable of delivering sufficient power to properly drive the
speakers. If the speaker's sensitivity rating is too low or the power rating too
high for what the amp can deliver, there could be a tendency to "overdrive" the
amp at levels that could cause the signal to be clipped. In addition to sounding
distorted, clipped signals can contain a high-level DC component that could
potentially damage the speaker's voice coil drivers.

Power Amplifier Types

As one might suspect, not all amplifier designs are made using the same operation principle and output stage design. The main design goal in amp design is
to create a device that boosts or controls gain in a linear fashion over its frequency range in an efficient power-handling manner. It must be able to supply
the needed current during the peak transient times so as to properly drive the
speaker(s).

There are several amp design types, which are designated according to their circuit classification or "class".

FIGURE 13.9
Crown XLi800 200W (at
8 ohms) two-channel Power
Amplifier (courtesy of Crown
Audio, Division of HARMAN
Professional, www.crownau-
dio.com).

CLASS A AMPLIFIER

The Class A amp is the most common design type, which employs only a single output power transistor (most commonly of the bipolar, insulated-gate bipolar transistor or field-effect transistor [FET] type). This single transistor is biased (central reference crossover point) in the middle of its power "S" curve so as to be driven into its cutoff or saturation points. This simple design allows the transistor to output its current over the full 360° of an audio input signal.

This design is considered to be excellent due to its high degree of linearity, low distortion and high signal gain. These advantages, however, come at the price of higher power requirements that can result in high-heat problems in the power supply.

CLASS B AMPLIFIER

This design was created as a solution to the problems of efficiency and heat that are often associated with Class A. It uses two inverted transistors (bipolar or FET), whereby one operates in the positive range of the audio cycle, while the other takes care of the negative range, in a push–pull arrangement. Thus, each transistor only works in half of the waveform range.

Since there's no DC bias required to offset the zero crossover point, this type of amp tends to be far more efficient, though sometimes at the expense of linearity and overall distortion.

CLASS AB AMPLIFIER

As the name suggests, this design type draws from both A and B types. It is generally considered to be a good compromise in that music levels are often low enough to operate easily within the Class A linear region of the amp design (where distortion and noise are low). Once the signal becomes great enough, the Class B design takes over, allowing high volume levels without sacrifices in efficiency or excessive heat.

CLASS D AMPLIFIER

A Class D or switching amplifier commonly makes use of metal-oxide-semiconductor FET (MOSFET) transistors, which act as pulse width modulation gates that switch at a high frequency between the positive and negative rails of the amp's supply voltage. This modulated pulse width or pulse density is used to encode the amplified audio signal level over time. A simple low-pass filter is then used to smooth out the digital component of the signal, and the resulting audio is fed to the speaker system. In recent times, Class D has become so efficient, with low-distortion and high-output specs, that this lightweight and low-heat system has become a favorite in recent home and professional active speaker designs.

CHAPTER 14

Power- and Ground- Related Issues

Throughout this book, we've seen how a production facility, in all its forms, involves the interconnection of various digital and analog devices to create a common task – to capture and produce good music and audio without adding clicks, pops and spurious noises to our tracks. This brings us to two aspects that are often overlooked in the overall design of a facility:

1. The need for proper grounding techniques (the way that devices interconnect without introducing electrical noises from other devices)
2. The need for proper power conditioning (keeping the purity and isolation of a room's power from the big, bad outside world)

GROUNDING CONSIDERATIONS

Proper grounding is essential to maintaining equipment safety; however, within an audio facility, small AC voltage potentials between various devices in an audio system can leak into a system's grounding circuit. Although these potentials are small, they are sometimes large enough to induce noise in the form of hums, buzzes or radio-frequency (RF) reception that can be injected (and amplified) directly into the audio signal path. These unwanted signals generally occur whenever improper grounding allows a piece of audio equipment to detect two or more different paths to ground.

Because grounding problems arise as a result of electrical interactions between any number of equipment combinations, the use of proper grounding techniques and troubleshooting within an audio production facility are by their very nature situational and often frustrating. As such, the following procedures are simply a set of introductory guidelines for dealing with this age-old problem. There are a great number of technical papers, books, methods and philosophies on grounding, and it's recommended that you carefully research the subject further before tackling any major ground-related problems. When in doubt, an experienced professional should be contacted, and care should always be taken not to sacrifice safety. Here are some general guidelines that might help:

DOI: 10.4324/9781003260530-14

- Keep all studio electronics on the same AC electrical circuit – most stray hums and buzzes occur whenever parts of the sound system are plugged into outlets from different AC circuits. Plugging into a circuit that's connected to such noise-generating devices as air conditioners, refrigerators, light dimmers, neon lights, etc. will often invite stray noise problems. Because most project studio devices don't require a great deal of current (with the possible exception of power amplifiers), it's often safe to run each of these devices from its own, properly grounded line from the electrical circuit panel.

- Try to keep audio wiring away from AC wiring – whenever AC and audio cables are laid side by side, portions of the 50/60-Hz signal might be induced into a high-gain, unbalanced circuit as hum. If this occurs, check to see if separating or shielding the lines helps reduce the noise. Of course, the use of balanced audio lines between devices will greatly help.

- When all else fails: if you only hear hum coming from one particular input channel, check that source device for ground-related problems. If the noise still exists when the console or mixer inputs are turned down, check the amp circuit or any device that follows the mixer. If the problem continues to elude you, then …

- Disconnect all of the devices (both power and audio) from the console, mixer or audio interface, then methodically plug them back in one at a time (it's often helpful to carefully monitor through a pair of headphones).

- Check the cables for bad connections or improper polarity. It's also wise to keep the cables as short as possible (especially in an unbalanced circuit).

- Another common path for ground loops is through a chassis into a 19-inch rack that provides not only a mount for multiple devices but often a common ground path. Test this by removing devices from the rack one at a time. If needed, a device can be electrically isolated from the rack by using special nylon mounting washers.

- Investigate the use of a balanced power source if traditional grounding methods don't work.

- Lastly, it's definitely not a good (and potentially dangerous) idea to get rid of your ground noise by removing the ground pin leads from your power connectors. These connections are there to make sure that a problematic power voltage is routed to earth (ground) and not through you. Any grounding problems in the system should be able to be corrected without resorting to improper wiring techniques.

Troubleshooting a ground-related problem can be tricky, and finding the problem's source might be a needle-in-a-haystack situation. When putting on your troubleshooting hat, it's usually best to remain calm, be methodical and consult with others who might know more than you do (or might simply have a fresh perspective).

POWER CONDITIONING

As was said earlier, one of the best ways to ensure that the power that's being delivered to your production system is as "clean" as possible is to get power from a single circuit source. If you have the luxury of building your facility from the ground up, one of the best ways to ensure that you have clean and reliable power is to take care to deliver power directly from the circuit box (over one or two separate lines) directly to the studio. If this isn't possible, then pulling your power from a single circuit or power socket might be your best bet. The latter scenario is possible for many home project studios, as the combined power requirements are well below the maximum load of a wall socket before the breaker trips (15 A × 120 V = 1,800 watts or 20 A × 120 V = 2,400 watts).

One of the best and simplest ways to connect all of your studio electronics to the same AC electrical circuit is by using a power conditioner (Figure 14.1). In addition to obtaining power from a single source, such a device can regulate, isolate and protect the voltage supply that's feeding one of your studio's most precious investments (besides you and your staff) – the equipment! Other added benefits to using power conditioning can be broken down into two important topics:

- Voltage regulation (fluctuations and surges)
- Keeping the lines quiet (reducing noise and interference)

In an ideal world, the power that's being fed to your studio outlets should be very close to the standard reference voltage of the country where you are working (e.g., 120 V, 220 V, 240 V). The real fact of the matter is that these line voltages regularly fluctuate from this standard level, resulting in voltage sags (a condition that can seriously under-power your equipment), surges (rises in voltage that can harm or reduce the working life of your equipment), transient spikes (sharp, high-level energy surges from lightning and other sources that can do serious damage) and brown-outs (long-term sags in the voltage lines). Through the use of a voltage regulator, high-level, short-term spikes and surge conditions can be clamped (stopped or reduced), thereby reducing or eliminating the chance that the main voltage will rise above a standard, predetermined level.

FIGURE 14.1
Furman M-8Lx power conditioner (courtesy of Furman Sound, Inc., www.furman-sound.com).

FIGURE 14.2

Furman SS-6B power conditioner/strip (courtesy of Furman Sound, Inc., www. furmansound.com).

While most power conditioners are meant to work best in a rack-mounted scenario, a number of quality power strips on the market offer protection against surges as well as RF and electromagnetic interference (Figure 14.2). These corded power strips can offer better protection over cheaper power strips and can be used in portable settings that can go into tight spots where many larger conditioners simply can't fit.

Certain devices that are equipped with voltage regulation circuitry are able to deal with power sags, long-term surges and brown-outs by electronically switching between the multiple voltage level taps of a transformer so as to match the output voltage to the ideal mains level (or as close to it as possible). One of the best approaches for regulating voltage fluctuations both above and below nominal power levels is to use an adequately powered uninterruptible power supply (UPS). In short, a quality UPS works by using a regulated power supply to constantly charge a rechargeable battery or bank of batteries. This battery supply is again regulated and used to feed sensitive studio equipment (such as a computer, bank of effects devices, etc.) with a clean and constant voltage supply.

Multiple-Phase Power

Another good defense against noise, power drops and other equipment-related power problems is to make use of multiple circuits in your overall studio power design. This approach reduces the amount of potential interference between power systems by physically placing the major system groups on their own power circuit. For example, an ideal scenario would place each of the following groups on their own power phase (separate circuit):

Phase 1: Equipment, computer (UPS isolated) and studio power
Phase 2: Lighting
Phase 3: Air conditioning and heating

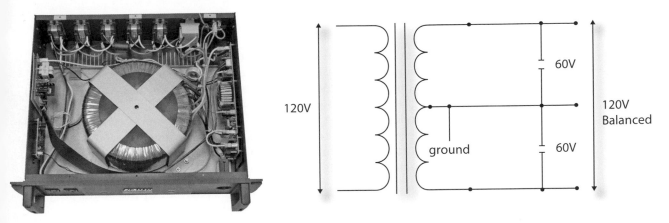

FIGURE 14.3
Furman P-2400 IT 20A
Prestige Symmetrically
Balanced Power Conditioner
(courtesy of Furman Sound,
Inc., www.furmansound.com).

The use of separate circuits can greatly help to guard against the everyday surges, noise and any other number of gremlins that might creep into your beloved production system.

Balanced Power

For those facilities that are located in areas where power lines are overtasked by heavy machinery, air conditioners and the like, a balanced power source might be considered. Such a device makes use of a power transformer (Figure 14.3) that has two secondary windings with a potential to ground on each side of 60 V. Because each side of the circuit is 180° out of phase with the other, a 120-V supply is maintained. Also, since the two 60-V legs are out of phase, any hum, noise or RF that's present at the device's input will be canceled at the transformer's center tap (a null point that's tied to ground).

A few important points relating to a balanced power circuit include:

- A balanced power circuit is able to reduce line noise if all of the system's gear is plugged into it. As a result, the device must be able to deliver adequate power.
- Balanced power will not eliminate noise from gear that's already sensitive to outside induced hums and buzzes.
- Choosing when to use balanced power is often open to interpretation, depending on who you talk to. For example, some feel that a balanced power conditioner should be used only after all other options to eliminate noise have been explored, while others believe it is a starting point from which to build a noise-free environment.

HUM, RADIO FREQUENCY (RF) AND ELECTROMAGNETIC INDUCTION (EMI)

In our modern world, where electrical and electronic gear are everywhere and where radio frequencies literally surround us as urban densities increase, the need for keeping noise out of our production system becomes more and more problematic.

The equipment in our production system usually has lots of analog amplifiers, and most of them have one task – to provide gain to make lower-level signals LOUDER. Unfortunately, whenever noises, buzzes, pops and ticks make their way into the signal before this gain stage, they, too, are amplified and quickly become a nuisance in the audio chain. These pests can be introduced into a system as hums (noise that occurs as a result of improper grounding or shielding) and electromagnetic interference (noise from radio and cell phone transmissions, lights or other electrically induced signals).

Several options for keeping electrically induced signal out of the audio path include:

- Use a clean power source (see the previous section on power conditioning).
- Make sure that the devices and audio cables are properly shielded to ground. This can be best done by using quality cables that have metal shielding.
- Whenever possible, use balanced cable connections (cables that have two audio conductors *plus* a ground connection – see the wiring explanations in Chapter 4 for more info).
- Whenever it's practical and possible, keep your audio and power signals physically separate.
- Some believe that the use of ferrite beads (you know, those heavy lumps that you'll find at the end of USB cables, etc.) can be useful for reducing high-frequency (MHz) transmission through cable lines. Since there are definitely debates on this subject, I'll let you do the research as to whether this might be helpful in your situation or not.

When talking about power, grounding and interference, one thing's for certain – there's usually no one-size-fits-all answer to problems that might creep up in an audio signal's path. You will probably be called upon to be patient, methodical, insightful and even a bit psychic so as to best understand and correct any of the innumerable problems that you might encounter. Oh, I almost forgot the last (but not least) element in helping to get this elusive job done – luck!

CHAPTER 15

Signal Processing

Over the years, signal processing has become an increasingly important part of audio and music production. It's the function of a *signal processor* to change, augment or otherwise modify an audio signal in either the analog or the digital domain. This chapter offers some insight into the basics of effects processing and how they can be integrated into a recording or mixdown in ways that sculpt sound using forms that are subtle, lush, extreme, or just plain whimsical and wacky.

Of course, the processing power of an effects system can be harnessed in either the hardware or the software plug-in domain. Regardless of how you choose to work with sound, the important rule to remember is that there are no rules; however, there are a few general guidelines that can help you get the sound that you want. When using effects, the most important asset you can have is experience and your own sense of artistry. The best way to learn the art of processing, shaping and augmenting sound is through experience – and gaining experience takes time, a willingness to learn and lots of patience.

THE WONDERFUL WORLD OF ANALOG, DIGITAL OR WHATEVER

Signal processing devices and their applied practices come in all sizes, shapes and flavors. These tools and techniques might be analog, digital or even acoustic in nature. The very fact that early analog processors have made a serious comeback (in the form of reissued hardware and software plug-in emulations, as seen in Figure 15.1) points to the importance of embracing past tools and techniques while combining them with the technological advances of the day to make the best possible production.

The Whatever

Although these aren't the first thoughts that come to mind, the use of acoustics and ambient mic techniques are often the first line of defense when dealing with

DOI: 10.4324/9781003260530-15

(a)

(b)

FIGURE 15.1

1176LN limiter: (a) Hardware version. (b) Powered software plug-in (courtesy of Universal Audio, www.uaudio.com. © 2017 Universal Audio, Inc. All rights reserved. Used with permission).

the processing of an audio signal. For example, as we saw in Chapter 4, changing a mic or its position might be the better option for changing the character of a pickup over using equalization (EQ). Placing a pair of mics out into a room or mixing a guitar amp with a second, distant pickup might fill the ambience of an instrument in a way that a device just might not be able to duplicate. In short, never underestimate the power of your acoustic environment and your ingenuity as an effects tool.

Analog

For those wishing to work in the world of analog (Figure 15.2), an enormous variety of devices can be put to use in a production. Although these generally relate to devices that alter a source's relative volume levels (e.g., equalization and dynamics), there are also a number of analog devices that can be used to alter effects that are time based. For example, an analog tape machine can be patched so as to make an analog delay or regenerative echo device. Although they're not commonly used, spring and plate reverb units that can add their own distinctive sound can still be found on the used and sometimes new market.

ANALOG RECALL

FIGURE 15.2

Analog hardware in the studio (courtesy of EastWest Studios, www.eastweststudios.com).

Of course, we simply can't overlook the reason why many seek out analog hardware devices, especially tube hardware – their sound! Many audio professionals are on a continual quest for that warm, smooth sound that's delivered by tube mics, preamps, amps and even older analog tape machines. It's part of what makes it all fun.

The downside of all those warm and fuzzy tools rests with their inability to be saved and instantly recalled with a session (particularly a digital audio workstation [DAW] session). If that analog sound is to be part of the recording, then you can use your favorite tube preamp to "print" just the right sound to a track. If the entire session is to be mixed in real time on an analog console, plugging your favorite analog toys into the channels will also work just fine. If, however, you're going to be using a DAW with an analog console (or even in the box, under special circumstances), then you'll need to do additional documentation in order to manually "recall" the session during the next setup. Of course, you could write the settings into the track notepads; however, another excellent way to save your analog settings is to take a picture of the device's settings and save it within the song's session documentation directory.

A number of forward-thinking companies have actually begun to design analog gear that can be digitally controlled, directly from within the DAW's session. The settings on these devices can then be remotely controlled from the DAW in real time and then instantly recalled from within the session.

Digital

The world of digital audio, on the other hand, has definitely set signal processing on fire by offering an almost unlimited range of effects that are available to the musician, producer and engineer. One of the biggest advantages to working in the digital signal processing (DSP) domain is the fact that software can be used to configure a processor in order to achieve an ever-growing range of effects (such as reverb, echo, delay, equalization, dynamics, pitch shifting, gain changing and signal re-synthesis).

The task of processing a signal in the digital domain is accomplished by combining logic or programming circuits in a building-block fashion. These logic blocks follow basic binary computational rules that operate according to a special program algorithm. When combined, they can be used to alter the numeric values of sampled audio in a highly predictable way. After a program has been configured (from either internal ROM, RAM or system software), complete control over a program's setup parameters can be altered and inserted into a chain as an effected digital audio stream. Since the process is fully digital, these settings can be saved and precisely duplicated at any time upon recall (often using musical instrument digital interface [MIDI] program change messages to recall a programmed setting). Even more amazing is how the overall quality and functionality have steadily increased while at the same time becoming more cost effective. It has truly brought an overwhelming amount of production power to the audio production table.

PLUG-INS

In addition to both analog and digital hardware devices, an ever-growing list of signal processors are available for the Mac and PC platforms in the form of software *plug-ins*. These software applications offer virtually every processing function that's imaginable (often at a fraction of the price of their hardware counterparts and with little or no reduction in quality, capabilities or automation features). These programs are, of course, designed to be "plugged" into an editor or DAW production environment in order to perform a particular real-time or non-real-time processing function.

Currently, several plug-in standards exist, each of which functions as a platform that serves as a bridge to connect the plug-in, through the computer's operating system (OS), to the digital audio production software. This means that any plug-in (regardless of its manufacturer) will work with an OS and DAW that's compatible with that platform standard, regardless of its form, function and/or manufacturer. As of this writing, the most popular standards are VST (PC/Mac), AudioSuite (Mac), Audio Units (Mac), MAS (MOTU for PC/Mac), as well as TDM and RTAS (Digidesign for PC/Mac).

By and large, effects plug-ins operate in a native processing environment (Figure 15.3a). This means that the computer's main central processing unit (CPU) carries the processing load for both the DAW and plug-in DSP functions. With the ever-increasing speed and power of modern-day CPUs, this has become less and less of a problem; however, hardware cards and external systems can be added to your system to "accelerate" the DSP processing power of your computer (Figure 15.3b) by adding additional hardware CPUs that are directly dedicated to handling the signal processing plug-in functions.

FIGURE 15.3

Signal processing plug-ins. (a) Collage of the various screens within the Cubase media production DAW (courtesy of Steinberg Media Technologies GmbH, www .steinberg.net). (b) Producer/ engineer Billy Bush mixing Garbage's "Not Your Kind of People" with Universal Audio's Apollo and UAD2 accelerated plug-ins (courtesy of Universal Audio/David Goggin, www.uaudio.com).

Plug-In Control and Automation

A fun and powerful aspect of working with various signal processing plug-ins on a DAW platform is the ability to control and automate many or all of the various effects parameters with relative ease and recall. These controls can be manipulated on screen (via hands-on or track parameter controls) or from an external hardware controller (Figure 15.4), allowing the parameters to be physically controlled in real time.

(a)

(b)

(a)

(b)

SIGNAL PATHS IN EFFECTS PROCESSING

Before diving into the process of effecting and/or altering sound, we should first take a quick look at an important signal path concept – the fact that a signal processing device can be inserted into an analog, digital or DAW chain in several ways. The most common of these are:

- Insert routing
- Send routing

Insert Routing

Insert routing is often used to alter the sonic or effects characteristics of a single track or channel signal. It occurs whenever a processor is directly inserted into a signal path in a serial (pass thru) fashion. Using this approach, the audio source enters into the input path, passes through the inserted signal processor and then continues on to carry out the record, monitor and/or mix function.

This method of inserting a device is generally used for the processing of a single instrument, voice or grouped set of signals that are present on a particular hardware or virtual input strip. Often, but not always, the device tends to be an amplitude-based processing function (such as an equalizer, compressor or limiter). In keeping with the "no rules" concept, however, time- and pitch-changing devices can also be used to tweak an instrument or voice as an insert. Here are but a few examples of how an insert can be used:

- A device can be plugged into an input strip's insert (direct send/return) point. This approach is used to insert an outboard device directly into the input signal path of an analog, digital or DAW strip (Figure 15.5a).
- A processor (such as a stereo compressor, limiter, EQ, etc.) could be inserted into a mixer's main output bus to effect an overall mix.
- A processor (such as a stereo compressor, limiter, EQ, etc.) could be inserted into a grouping to affect a sub-mix.

FIGURE 15.4
External EFX hardware controllers. (a) Novation's Automap 4 automatically maps its controls to various parameters on the target plug-in (courtesy of Novation Digital Music Systems, Ltd., www.novationmusic.com). (b) Native Instruments Komplete Kontrol system maps instruments and plug-ins to its hardware controls (courtesy of Native Instruments GmbH, www.native-instruments.com).

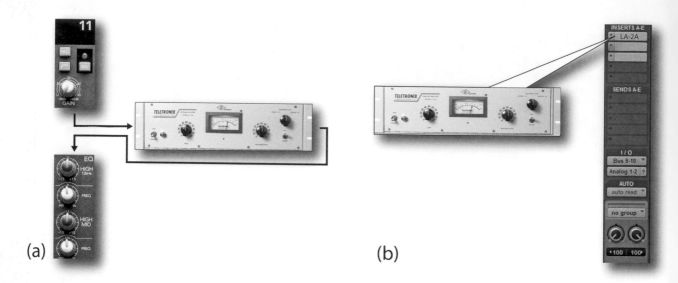

(a)

(b)

FIGURE 15.5

Inserting an effect into a channel strip. (a) Analog insert. (b) DAW insert.

- An effects stomp box could be inserted between a mic preamp and console input to create a grungy distortion effect.
- A DAW plug-in could be inserted into a strip to process only the signal on that channel (Figure 15.5b).

An insert is used to "insert" an effect or effects chain into the signal path of a single track or group. It tends (but not always) to be amplitude-based in nature, meaning the processor is often used to effect amplitude levels (i.e., compressor, limiter, gate, etc.)

EXTERNAL CONTROL OVER AN INSERT EFFECT'S SIGNAL PATH

FIGURE 15.6

Diagram of a key sidechain input to a noise gate. (a) The signal is passed whenever a signal is present at the key input. (b) No signal is passed when no signal is present at the key input.

Certain insert effects processors allow an external audio source to act as a control for affecting a signal as it passes from the input to the output of a device (Figure 15.6). Devices that offer an external "key" or "sidechain" input can be quite useful, allowing a signal source to be used as a control for varying another audio path. For example:

- A gate (an infinite expander that can control the passing of audio through a gain device) might take its control input from an external "key" signal that will determine when a signal will or will not pass in order to reduce leakage, tighten up an instrument decay or create an effect.

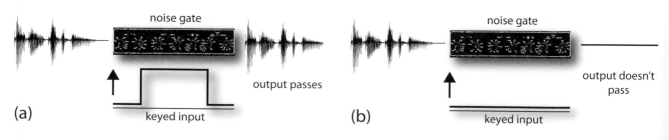

(a)

(b)

- A vocal track could be inserted into a vocoder's control input so as to synthetically add a robot-like effect to a track.
- A voice track could be used for vocal ducking at a radio station, as a control to fade out the music or crowd noise when a narrator is speaking.
- An external keyed input can be used to make a mix "pump" or "breathe" in a dance production.

It's important to note that there is no set standard for providing a sidechain key in software. Some software packages provide native sidechain capability, others support sidechaining via "multiple input" plug-ins and complex signal routing, and many don't support sidechaining at all.

Send Routing

Effects "sends" are often used to augment a signal (generally being used to add reverb, delay or other time-based effects). This type differs from an insert in that instead of inserting a signal-changing device directly into the signal path, a portion of the signal (which is essentially a combined mix of the desired channels) is then "sent" to one or more effects devices. Once effected, the signal can then be proportionately mixed back into the monitor or main out signal path so as to add an effects blend of the selected tracks to the overall output mix.

> A send is used to "send" a mix of multiple channel signals to a single effect or effects chain, after which, it can be routed to the monitor or main mix bus. It tends (but not always) to be time-based in nature, meaning the processor affects time functions (i.e., reverb, echo, etc.)

As an example, Figure 15.7 shows how sends from any number of channel inputs can be mixed together and then be sent to an effects device. The effected output signal is then sent back either into the effects return section, to a set of spare mixer inputs or directly to the main mixing bus outputs.

FIGURE 15.7
An "aux" sends path flows in a horizontal fashion to a send bus. The combined sum can be effected (or sent to another destination) and then returned back to the monitor or main output bus. (a) Analog aux send. (b) Digital aux send.

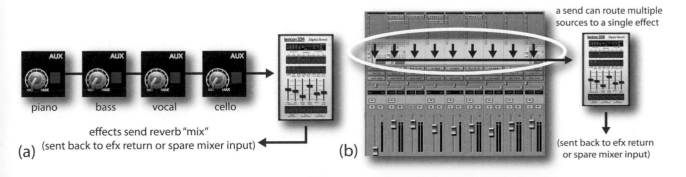

a send can route multiple sources to a single effect

piano   bass   vocal   cello

effects send reverb "mix"
(a) (sent back to efx return or spare mixer input)

(b)

(sent back to efx return or spare mixer input)

Vive la Difference

There are a few insights that can help you to understand the difference between insert and send effects routing. Let's start by reviewing the basic ways that their signal paths function:

- First, from Figure 15.5, we can see that the signal flow of an effects insert is basically "vertical", moving from the channel input through an inserted device and then back into the input strip's signal chain (for mixing, recording or whatever).

- From Figure 15.7, we can see that an effects send basically functions in a "horizontal" direction, allowing portions of the various input signals to be mixed together and sent to an effects device, which can then be routed back into the mixer signal chain for further effects and mix blending.

- Finally, it's important to grasp the idea that when a device is "inserted" into a signal chain, it's usually a single, dedicated hardware/plug-in device (or sometimes, a chain of devices) that is used to perform a specific function. If a large number of tracks/channels are used in a session or mix, inserting numerous DSP effects into multiple strips could take up too much processing power (or is simply unnecessary and hard to manage). Setting up an effects send can save a great deal of DSP processing overhead by creating a single effects send "mix" that can then be routed to a single effects device (or device chain) – it works like this: "why use a whole bunch of EFX devices to do the same job on every channel, when sending a signal mix to a single EFX device/chain will do?"

On a final note, each effects routing technique has its own set of strengths and weaknesses – it's important to play with each so as to know the difference and when to make best use of each of them.

Sidechain Processing

Another effects processing tool that can be used to alter the processing that's applied to a track or overall mix is the idea of *sidechain processing*. In short, a sidechain takes the signal from one track and uses that source signal to trigger an effect or gain change for a target audio signal on another track.

As an example, Figure 15.8 shows a kick drum track. Here, we'll take a sidechain signal from this track to feed a compressor that has been inserted into a synth pad track. Using this sidechain setup, the compressor won't be taking its feed from the synth signal but instead, will be triggered by the kick drum. In this way, whenever the kick repeatedly sounds, it will cause the compressor to momentarily reduce the level of the synth (an effects process known as "ducking"), so that the affected track will duck in level, thereby making the source signal seem louder.

This is only one of the countless ways that a sidechain circuit can be used to alter the signal of a target track, for example:

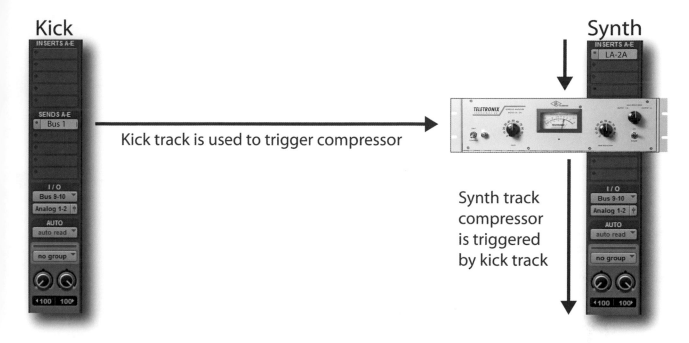

Kick

Synth

Kick track is used to trigger compressor

Synth track compressor is triggered by kick track

- Backing vocals can be ducked by the lead vocal to make the lead shine through.
- A bass track can be ducked by the kick so as to make the kick appear louder.
- The kick can be used to duck the entire main bus signal so as to make it pulse with the beat.
- A fast repeating snare track can be used to feed a gate on another track, so that the target track will repeatedly mute with the fast beat of the source track, as an effect.
- A source vocal track could be sent to an echo delay that is being used to effect the vocal track, such that when the vocalist is singing, the delay will be ducked or muted (depending upon whether a compressor or gate is being used), allowing the delay to repeat only the ending parts of the vocal phrases.

These are only a few examples of how sidechaining can be used to spice up a track. Go ahead, check out the wide range of YouTube videos on the subject, then read the sidechaining section of your DAW's manual and begin experimenting with this unique feature on your own.

FIGURE 15.8
A sidechain is set up such that the kick drum signal is used to trigger compression on a synth track, causing the synth signal to "duck" whenever the kick is present.

Parallel Processing

In short, parallel processing is the result of summing two different instances of the same audio signal through different processing paths so as to create a unique, combined sound. Most commonly, this method is used to create one path that has been processed in any number of ways (distortion, compression, delay, or a combination of these and other processes) along with the unaltered

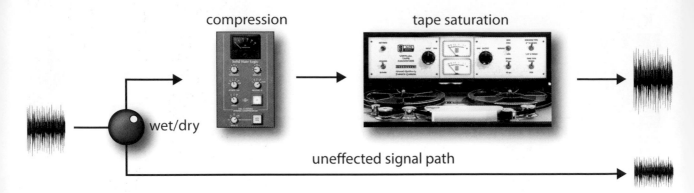

compression

tape saturation

wet/dry

uneffected signal path

FIGURE 15.9
Parallel processing occurs when an effected path and another effected (or uneffected) path are proportionately mixed together.

audio signal (Figure 15.9). Combining the two together will often yield a cleaner, more present sound than if you were to simply use the affected track alone in the mix.

The options and possible effects combinations are endless. By adding compression and saturation to an effects track and naming it "glue", "dirt", "grit" or anything you want, a degree of character that is uniquely yours can be added to the mix. I'd like to add a bit of caution here, however: adding grit to your mix can be all well and good (especially if it's a rock track that just begs for some extra dirt). However, it should be kept in mind that too much of a good thing can sometimes simply be TOO MUCH, especially when you're just starting your journey into the mix process.

There is one more parallel processing control that you should be aware of. A number of hardware and plug-in devices commonly have an internal "wet/dry mix" control that serves as an internal sidechain mix control for varying the amount of "dry" (original) signal that's to be mixed with the "wet" (effected) signal (usually varied in percentage). This control can have an effect on your settings (when used in either an insert or send setting), and careful awareness is advised, as many such controls are set to 50/50 by default ... not necessarily the best setting when inserting a reverb into an effects track.

EFFECTS PROCESSORS

From this point on, this chapter will be taking an in-depth look at many of the signal processing devices, applications and techniques that have traditionally been the cornerstone of music and sound production, including systems and techniques that exert an ever-increasing degree of control over:

- *The spectral content of a sound:* in the form of equalization and bandpass filtering
- *Amplitude-based processing:* in the form of volume and dynamic range processing
- *Time-based effects:* augmentation or re-creation of room ambience, delay, time/pitch alterations and tons of other special effects that can range from being sublimely subtle to "in yo' face".

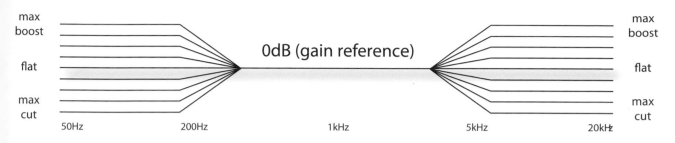

FIGURE 15.13
High/low, boost/cut curves of a shelving equalizer.

Shelving Filters

Another type of equalizer is the *shelving filter*. Shelving refers to a rise or drop in frequency response at a selected frequency, which tapers off to a preset level and continues at that level to the end of the audio spectrum. Shelving can be inserted at either the high or the low end of the audio range and is the curve type that's commonly found on home stereo bass and treble controls (Figure 15.13).

High-Pass and Low-Pass Filters

Equalizer types also include *high-pass* and *low-pass filters*. As their names imply, this EQ type allows certain frequency bandwidths to be passed at full level while other sections of the audible spectrum are attenuated. Frequencies that are attenuated by less than 3 dB are said to be inside the *passband*; those attenuated by more than 3 dB are located outside, in the *stopband*. The frequency at which the signal is attenuated by exactly 3 dB is called the *turnover* or *cutoff frequency* and is used to name the filter frequency.

Ideally, attenuation would become infinite immediately outside the passband; however, in practice, this isn't always attainable. Commonly, attenuation is carried out at rates of 6, 12 and 18 dB per octave. This rate is called the *slope* of the filter. Figure 15.14a, for example, shows a 700-Hz high-pass filter response curve with a slope of 6 dB per octave, and Figure 15.14b shows a 700-Hz low-pass filter response curve having a slope of 12 dB per octave. High- and low-pass filters differ from shelving EQ in that their attenuation doesn't level off outside the passband. Instead, the cutoff attenuation continues to increase. A high-pass filter in combination with a low-pass filter can be used to create a *bandpass filter*, with the passband being controlled by their respective turnover frequencies and the Q by the filter's slope (Figure 15.15).

FIGURE 15.14
A 700-Hz filter: (a) High-pass filter with a slope of 6 dB per octave. (b) Low-pass filter with a slope of 12 dB per octave.

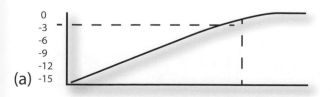

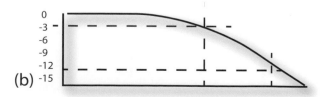

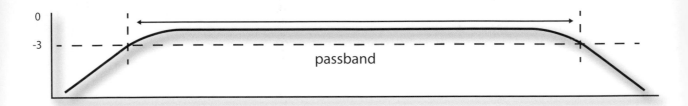

0

-3

passband

FIGURE 15.15

A bandpass filter is created by combining a high- and low-pass filter with different cutoff frequencies.

EQUALIZER TYPES

The four most commonly used equalizer types that can incorporate one or more of the previously described filter types are the:

- Selectable frequency equalizer
- Parametric equalizer
- Graphic equalizer
- Notch filter

The *selectable frequency equalizer* (Figure 15.16), as its name implies, has a set number of frequencies from which to choose. These equalizers usually allow a boost or cut to be performed at a number of selected frequencies with a predetermined Q. They are most often found on older console designs, certain low-cost production consoles and outboard gear.

The *parametric equalizer* (Figure 15.17a) lets you adjust most or all of its frequency parameters in a continuously variable fashion. Although the basic design layout will change from model to model, each band will often have an adjustment for continuously varying the center frequency. The amount of boost or cut is also continuously variable. Control over the center frequency and Q can be either selectable or continuously variable, although certain manufacturers might not have provisions for a variable Q.

Generally, each set of frequency bands will overlap into the next band section so as to provide smooth transitions between frequency bands or allow multiple curves to be placed in nearby frequency ranges. Because of its flexibility and performance, the parametric equalizer has become the standard design for most input strips, digital equalizers and workstations.

FIGURE 15.16

The Warm Audio EQP-WA selectable frequency tube equalizer (courtesy of Warm Audio LLC, www.warmaudio.com).

A *graphic equalizer* (Figure 15.17b) provides boost and cut level control over a series of center frequencies that are equally spaced (ideally, according to music intervals). An "octave band" graphic equalizer might, for example, have

(a)

(b)

FIGURE 15.17

Equalizer types: (a) The EQF-100 full range, parametric vacuum tube equalizer (courtesy of Summit Audio, Inc., www.summitaudio.com). (b) Rane GE 130 single-channel, 30-band, 1/3-octave graphic equalizer (courtesy of Rane Corporation, www.rane.com).

12 equalization controls spaced at the octave intervals of 20, 40, 80, 160, 320 and 640 Hz and 1.25, 2.5, 5, 10 and 20 kHz, while 1/3-octave equalizers could have up to 36 center frequency controls. The various EQ band controls generally use vertical sliders, which are arranged side by side so that the physical positions of these controls can provide a "graphic" readout of the overall frequency response curve at a glance. This type is often used in applications that can help fine-tune a system to compensate for the acoustics in various types of rooms, auditoriums and studio control rooms.

Notch filters are often used to zero in on and remove 60- or 50-Hz hum or other undesirable discrete-frequency noises. They use a very narrow bandwidth to fine-tune and attenuate a particular frequency in such a way as to have little effect on the rest of the audio program. Notch filters are used more in film location sound and broadcast than in studio recording, because severe narrowband problems aren't often encountered in a well-designed studio – hopefully.

APPLYING EQUALIZATION

When you get right down to it, EQ is all about compensating for deficiencies in a sound pickup or about reducing extraneous sounds that make their way into a pickup signal. To start our discussion on how to apply EQ, let's again revisit the all-important "Good Rule" from Chapter 4.

The "Good Rule"

Good musician + good instrument + good performance + good acoustics + good mic + good placement = good sound.

Whenever possible, EQ should not be used as a Band-Aid. By this, I mean that it's often a good idea to correct for a problem on the spot rather than to rely on the hope that you can "fix it in the mix" at a later time using EQ and other methods.

When in doubt, it's often better to deal with a problem as it occurs. This isn't always possible, however – therefore, EQ is best used in situations where:

- There's no time or money left to redo the track
- The existing take was simply magical and shouldn't be re-recorded
- The track was already recorded during a previous session and is in need of being fixed

EQ IN ACTION!

Although most equalization is done by ear, it's helpful to have a sense of which frequencies affect an instrument in order to achieve a particular effect. On the whole, the audio spectrum can be divided into four frequency bands: low (20 to 200 Hz), low–mid (200 to 1,000 Hz), high–mid (1,000 to 5,000 Hz) and high (5,000 to 20,000 Hz). When the frequencies in the 20- to 200-Hz (low) range are modified, the fundamental and the lower harmonic range of most bass information will be affected. These sounds often are felt as well as heard, so boosting in this range can add a greater sense of power or punch to music. Lowering this range will weaken or thin out the lower frequency range.

The fundamental notes of most instruments lie within the 200- to 1,000-Hz (low–mid) range. Changes in this range often result in dramatic variations in the signal's overall energy and add to the overall impact of a program. Because of the ear's sensitivity in this range, a minor change can result in an effect that's very audible. The frequencies around 200 Hz can add a greater feeling of warmth to the bass without loss of definition. Frequencies in the 500- to 1,000-Hz range

Table 15.1	Instrumental Frequency Ranges of Interest
Instrument	**Frequencies of Interest**
Kick drum	Bottom depth at 60–80 Hz, slap attack at 2.5 kHz
Snare drum	Fatness at 240 Hz, crispness at 5 kHz
Hi-hat/cymbals	Clank or gong sound at 200 Hz, shimmer at 7.5 kHz to 12 kHz
Rack toms	Fullness at 240 Hz, attack at 5 kHz
Floor toms	Fullness at 80–120 Hz, attack at 5 kHz
Bass guitar	Bottom at 60–80 Hz, attack/pluck at 700–1000Hz, string noise/pop at 2.5 kHz
Electric guitar	Fullness at 240 Hz, bite at 2.5 kHz
Acoustic guitar	Bottom at 80–120 Hz, body at 240 Hz, clarity at 2.5–5 kHz
Electric organ	Bottom at 80–120 Hz, body at 240 Hz, presence at 2.5 kHz
Acoustic piano	Bottom at 80–120 Hz, presence at 2.5–5 kHz, crisp attack at 10 kHz, honky-tonk sound (sharp Q) at 2.5 kHz
Horns	Fullness at 120–240 Hz, shrill at 5–7.5 kHz
Strings	Fullness at 240 Hz, scratchiness at 7.5–10 kHz
Conga/bongo	Resonance at 200–240 Hz, presence/slap at 5 kHz
Vocals	Fullness at 120 Hz, boominess at 200–240 Hz, presence at 5 kHz, sibilance at 7.5–10 kHz

Note: These frequencies aren't absolute for all instruments but are meant as a subjective guide.

could make an instrument sound hornlike, while too much boost in this range can cause listening fatigue.

Higher-pitched instruments are most often affected in the 1,000- to 5,000-Hz (high–mid) range. Boosting these frequencies often results in an added sense of clarity, definition and brightness. Too much boost in the 1,000- to 2,000-Hz range can have a "tinny" effect on the overall sound, while the upper mid-frequency range (2,000 to 4,000 Hz) affects the intelligibility of speech. Boosting in this range can make music seem closer to the listener, but too much of a boost can also cause listening fatigue.

The 5,000- to 20,000-Hz (high-frequency) region is composed almost entirely of instrument harmonics. For example, boosting frequencies in this range will often add sparkle and brilliance to a string or woodwind instrument. Boosting too much might produce sibilance on vocals and make the upper range of certain percussion instruments sound harsh and brittle. Boosting at around 5,000 Hz has the effect of making music sound louder. A 6-dB boost at 5,000 Hz, for example, can sometimes make the overall program level sound as though it's been doubled in level; conversely, attenuation can make music seem more distant. Table 15.1 provides an analysis of how frequencies and EQ settings can interact with various instruments. (For more information, refer to the Microphone Placement Techniques section in Chapter 4.)

TRY THIS: EQUALIZATION

1. Solo an input strip on a mixer, console or DAW. Experiment with the settings using the previous frequency ranges. Can you improve on the original recorded track, or does it take away from the sound?

2. Using the input strip equalizers on a mixer, console or DAW, experiment with the EQ settings and relative instrument levels within an entire mix using the previous frequency ranges as a guide. Can you bring an instrument out without changing the fader gains? Can you alter the settings of two or more instruments to increase the mix's overall clarity?

3. Plug an outboard or plug-in equalizer into the main output buses of a mixer, console or DAW, and change the program's EQ settings using the previous frequency range discussions as a guide. How does it change the mix?

One way to zero in on a particular frequency using an equalizer (especially a parametric one) is to accentuate or attenuate the EQ level and then vary the center frequency until the desired range is found. The level should then be scaled back until the desired effect is obtained. If boosting in one instrument range causes you to want to do the same in another frequency range , it's likely that you're simply overdoing it. It's easy to get caught up in the "bigger! Better!

MORE!" syndrome of wanting an instrument to sound louder. If this continues to happen on a mix, it's likely that one of the frequency ranges of an instrument or ensemble is too dominant and requires attenuation. On the subject of laying down a recorded track with EQ, there are a number of situations and differing opinions regarding them:

- Some will "track" (record) the sound of the mic/instrument/room sound directly to tape or DAW, so that little or no EQ (or any other changes) will be needed in mixdown.

- Some use EQ liberally to make up for placement and mic deficiencies, whereas others might use it sparingly, if at all. One example where EQ might be used sparingly is when an engineer knows that someone else will be mixing a particular song or project. In this situation, the engineer who's doing the mix might have a very different idea of how an instrument should sound. If large amounts of EQ were recorded to a track during the session, the mix engineer might have to work very hard to counteract the original EQ settings.

- If everything was recorded flat, the producer and artists might have difficulty passing judgment on a performance or hearing the proper balance during the overdub phase. Such a situation might call for equalization in the monitor mix while leaving the recorded tracks alone.

- In situations where several mics are to be combined onto a single track or channel, the mics can only be individually equalized (exchanged, altered or moved) during the recording phase. In situations where a project is to be engineered, mixed and possibly even mastered by the same person, the engineer might want to discuss in advance the type and amount of EQ that the producer and/or artist might want.

- Above all, it's wise that any "sound-shaping" should be determined and discussed with the producer and/or artist before the sounds are committed to a track.

In the end, there's no getting around the fact that an equalizer is a powerful tool. When used properly, it can greatly enhance or restore the musical and sonic balance of a signal. Experimentation and experience are the keys to proper EQ usage, and no book can replace the trial-and-error process of "just doing it!"

Before moving on, it's important to keep one age-old viewpoint in mind – that an equalizer shouldn't be regarded as a cure-all for improper mic, playing or instrument technique; rather, it should be used as a tool for correcting problems that couldn't be easily fixed on the spot through mic and/or performance adjustments. If an instrument is poorly recorded during an initial recording session, it's often far more difficult and time consuming to "fix it in the mix" at a later time. Getting the best possible sound down onto tape or DAW will definitely improve your chances for attaining a sound and overall mix that you can be proud of in the future.

Sound-Shaping Effects Devices and Plug-Ins

Another class of effects devices that aren't equalizers but instead, affect the overall tonal character of a track or mix come under the category of sound-shaping devices. These systems can be either hardware or plug-in in nature and are used to alter the tonal and/or overtone balance of a signal. For example, a device that's been around for decades is the Aphex Aural Exciter. This device is able to add a sense of presence to a sound by generating additional overtones that are subdued or not present in the program signal. Other such devices are able to modify the shape of a sound's transient envelope (Figure 15.18a) or to filter the sound in unique ways (Figure 15.18b).

Another class of sound-shaper comes in the form of virtual tape machine plug-ins that closely model and emulate analog tape recorders – right down to their sonic character, tape noise, distortion, changes with virtual tape formulation, and bias settings (Figure 15.19).

DYNAMIC RANGE

Like most things in life that get out of hand from time to time, the level of a signal can vary widely from one moment to the next. For example, if a vocalist gets caught up in the moment and lets out an impassioned scream following a soft whispery passage, you can almost guarantee that the mic and preamp will push the recording chain from its optimum recording level into severe distortion – OUCH! Conversely, if you set an instrument's mic to properly accommodate the loudest level, its signal might be buried in the mix during the rest of the song. For these and other reasons, it becomes obvious that it's sometimes necessary to exert some form of control over a signal's dynamic range by using various techniques and dynamic controlling devices. In short, the dynamics of an audio program's signal resides somewhere in a continuously varying realm between three level states:

FIGURE 15.18
Sound-shaping plug-ins. (a) Oxford Envolution. (b) Moog multimode filter (courtesy of Universal Audio, www.uaudio.com. © 2022 Universal Audio, Inc. All rights reserved. Used with permission).

- Saturation
- Average signal level
- System/ambient noise

(a)

(b)

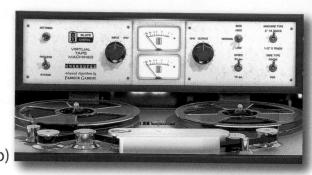

(a)   (b)

FIGURE 15.19

Analog tape emulation plug-in. (a) Oxide Tape Recorder for the Apollo and the UAD effects processing card (courtesy of Universal Audio, www.uaudio.com © 2022 Universal Audio, Inc. All rights reserved. Used with permission). (b) Slate Digital Virtual Tape Machine (courtesy of Slate Digital, www.slatedigital.com).

As you may remember from various chapters in this book, *saturation* occurs when an input signal is so large that an amp's supply voltage isn't large enough to produce the required output current or is so large that a digital converter reaches full scale (where the A/D output reads as all 1s). In either case, the results generally don't sound pretty and should be avoided in the channel's audio chain. The *average signal level* is where the overall signal level of a mix often likes to reside. Logically, if an instrument's level is too low, it can get buried in the mix – if it's too high, it can unnecessarily stick out and throw the entire balance off. It is here that the art of creating an average mix level that's high enough to stand out in any playlist, while still retaining enough dynamic "life", truly becomes an applied balance of skill and magic.

DYNAMIC RANGE PROCESSORS

The overall dynamic range of music is potentially on the order of 120 to 140 dB, whereas the overall dynamic range of a compact disc is often 80 to 90 dB, and analog magnetic tape is on the order of 60 dB (excluding the use of noise reduction systems, which can improve this figure by 15 to 30 dB). However, when working with 24-bit digital word lengths, a system's, processor's or channel's overall dynamic range can actually approach or exceed the full range of hearing. Even with such a wide dynamic range, unless the recorded program is played back in a noise-free environment, either the quiet passages will get lost in the ambient noise of the listening area (35 to 45 dB sound pressure level [SPL] for the average home and much worse in a car) or the loud passages will simply be too loud to bear. Similarly, if a program of wide dynamic range were to be played through a medium with a limited dynamic range (such as the 20- to 30-dB range of an AM radio or the 40- to 50-dB range of FM), a great deal of information would get lost in the general background noise. To prevent such problems, the dynamics of a program can be restricted to a level that's appropriate for the reproduction medium (theater, radio, home system, car, etc.) as shown in Figure 15.20. This gain reduction can be accomplished either by manually riding the fader's gain or through the use of a *dynamic range processor* that can alter the range between the signal's softest and loudest passages.

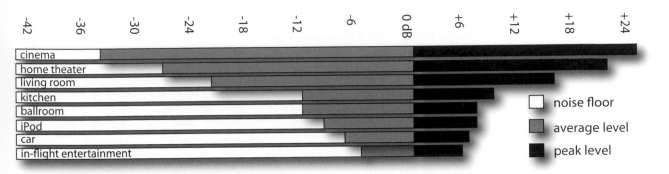

| -42 | -36 | -30 | -24 | -18 | -12 | -6 | 0 dB | +6 | +12 | +18 | +24 |

cinema
home theater
living room
kitchen
ballroom
iPod
car
in-flight entertainment

☐ noise floor
▨ average level
■ peak level

The concept of automatically changing the gain of an audio signal (through the use of compression, limiting and/or expansion) is perhaps one of the most misunderstood aspects of audio recording. This can be partially attributed to the fact that a well-done job won't be overly obvious to the listener. Changing the dynamics of a track or overall program will often affect the way in which it will be perceived (either consciously or unconsciously) by making it "seem" louder, thereby reducing its volume range to better suit a particular medium, or by making it possible for a particular sound to ride at a better level above other tracks within a mix.

FIGURE 15.20
Dynamic ranges of various audio media, showing the noise floor (black), average level (white) and peak levels (gray) (courtesy of Thomas Lund, tc electronic, www .tcelectronic.com).

Compression

A *compressor* (Figure 15.21), in effect, can be thought of as an automatic fader. It is used to proportionately reduce the dynamics of a signal that rises above a user-definable level (known as the *threshold*) to a lesser volume range. This process is done so that:

- The dynamics can be managed by the electronics and/or amplifiers in the signal chain without distorting the signal chain
- The range is appropriate to the overall dynamics of a playback or broadcast medium

FIGURE 15.21
Universal Audio 1176LN limiting amplifier (courtesy of Universal Audio, www.uaudio .com. © 2022 Universal Audio, Inc. All rights reserved. Used with permission).

input gain   output gain   attack   slope ratio   meter select

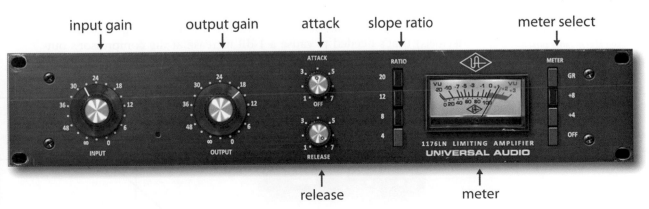

release   meter

- An instrument or vocal better matches the dynamics of other recorded tracks within a song or audio program

Since the signals of a track, group or program will be automatically turned down (hence the terms *compressed* or *squashed*) during a loud passage, the overall level of the newly reduced signal can now be amplified upwards to better fit into a mix or to match the required dynamics of the medium. In other words, once the dynamics have been reduced downward, the overall level can be boosted such that the range between the loud and soft levels is less pronounced (Figure 15.22). We've not only restored the louder signals back to a prominent level but have also turned up the softer signals that would otherwise be buried in the mix or ambient background noise.

The most common controls on a compressor (and most other dynamic range devices) include input gain, threshold, output gain, slope ratio, attack, release and meter display:

- *Input gain*: this control is used to determine how much signal will be sent to the compressor's input stage.
- *Threshold*: this setting determines the level at which the compressor will begin to proportionately reduce the incoming signal. For example, if the threshold is set to –20 dB, all signals that fall below this level will be unaffected, while signals above this level will be proportionately attenuated, thereby reducing the overall dynamics. On some devices, varying the input gain will correspondingly control the threshold level. In this situation, raising the input level will lower the threshold point and thus reduce the overall dynamic range. Most quality compressors offer hard and soft knee threshold options. A soft knee widens or broadens the threshold range, making the onset of compression less obtrusive, while the hard knee setting causes the effect to kick in quickly above the threshold point.
- *Output gain*: this control is used to determine how much signal will be sent to the device's output. It's used to boost the reduced dynamic signal into a range where it can best match the level of a medium or be better heard in a mix.
- *Slope ratio*: this control determines the slope of the input-to-output gain ratio. In simpler terms, it determines the amount of input signal (in decibels) that's needed to cause a 1-dB increase at the compressor's output. For example, a linear amplifier has an input-to-output ratio of 1:1

FIGURE 15.22

A compressor reduces input levels that exceed a selected threshold by a specified amount. Once reduced, the overall signal can then be boosted in level, thereby allowing the softer signals to be raised above other program or background sounds.

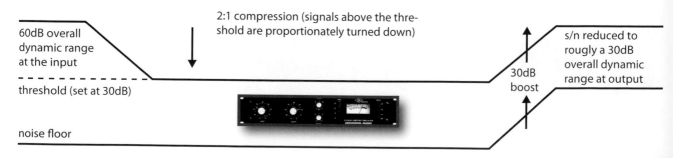

60dB overall dynamic range at the input

2:1 compression (signals above the threshold are proportionately turned down)

threshold (set at 30dB)

30dB boost

s/n reduced to rougly a 30dB overall dynamic range at output

noise floor

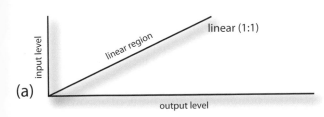

(a)

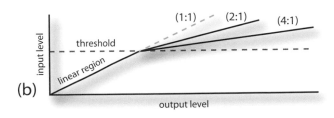

(b)

(one-to-one), meaning that for every 1-dB increase at the input, there will be a corresponding 1-dB increase at the output (Figure 15.23a). When using a 2:1 compression ratio, below the threshold, the signal will be linear (1:1); however, above this level, an increase of 2 dB will result in an increase of only 1 dB at the output (Figure 15.23b). An increase of 4 dB above the threshold will result in an output gain of 1 dB when a 4:1 ratio (slope) is selected. Get the idea?

- *Attack*: this setting (which is calibrated in milliseconds: 1 msec = 1 thousandth of a second) determines how fast or how slowly the device will turn down signals that exceed the threshold. It is defined as the time it takes for the gain to decrease to a percentage (usually 63%) of its final gain value. In certain situations (such as might occur with instruments that have a long sustain, such as the bass guitar), setting a compressor to instantly turn down a signal might be audible (possibly creating a sound that pumps the signal's dynamics). In this situation, it would be best to use a slower attack setting. On the other hand, such a setting might not give the compressor time to react to sharp, transient sounds (such as a hi-hat). In this case, a fast attack time would probably work better. As you might expect, you'll need to experiment to arrive at the fastest attack setting that won't audibly color the signal's sound.

- *Release*: similar to the attack setting, release (which is calibrated in milliseconds) is used to determine how slowly or quickly the device will restore a signal to its original dynamic level once it has fallen below the threshold point (defined as the time required for the gain to return to 63% of its original value). Too fast a setting will cause the compressor to change dynamics too quickly (creating an audible pumping sound), while too slow a setting might affect the dynamics during the transition from a loud to a softer passage. Again, it's best to experiment with this setting to arrive at the slowest possible release that won't color the signal's sound.

- *Meter display*: this control changes the compressor's meter display to read the device's output or gain reduction levels. In some designs, there's no need for a display switch, as readouts are used to simultaneously display output and gain reduction levels.

FIGURE 15.23
The output ratios of a compressor. (a) A linear slope results in an increase of 1 dB at the output for every 1 dB increase at the input. (b) A compression slope follows an input/output gain reduction ratio above the threshold, proportionately reducing signals that fall above this point.

As was previously stated, the use of compression (and most forms of dynamics processing) is often misunderstood, and compression can easily be abused. Generally, the idea behind these processing systems is to reduce the overall dynamic range of a track, music or sound program or to raise its overall perceived level without adversely affecting the sound of the track itself. It's a well-known fact that over-compression can actually squeeze the life out of a performance by limiting the dynamics and reducing the transient peaks that can give life to a performance. For this reason, it's important to be aware of the general nuances of the controls that have been discussed.

During a recording or mixdown session, compression can be used in order to balance the dynamics of a track to the overall mix or to keep the signals from overloading preamps, the recording medium and your ears. Compression should be used with care for any of the following reasons:

- Minimize changes in volume that might occur whenever the dynamics of an instrument or vocal are too wide for the mix. As a tip, a good starting point might be a 0-dB threshold setting at a 4:1 ratio, with the attack and release controls set at their middle positions.
- Smooth out momentary changes in source-to-mic distance.
- Balance out the volume ranges of a single instrument. For example, the notes of an electric or upright bass often vary in volume from string to string. Compression can be used to "smooth out" the bass line by matching their relative volumes (often using a slow attack setting). In addition, some instruments (such as horns) are louder in certain registers because of the amount of effort that's required to produce these notes. Compression is often useful for smoothing out these volume changes. As a tip, you might start with a ratio of 5:1 with a medium-threshold setting, medium attack and slower release time. Over-compression should be avoided to prevent any pumping effects.
- Reduce other frequency bands by inserting a filter into the compression chain that causes the circuit to compress frequencies in a specific band (multiband compression). A common example of this is a de-esser, which is used to detect high frequencies in a compressor's circuit so as to suppress those "SSSS", "CHHH" and "FFFF" sounds that can distort or stand out in a recording.
- Reduce the dynamic range and/or boost the average volume of a mix so that it appears to be significantly louder (as occurs when a song's volume sticks out in a playlist, or a television commercial seems louder than your favorite show).

Although it may not always be the most important, this last application often gets a great deal of attention, because many producers strive to cut their recordings as "hot" as possible. That is, they want the recorded levels to be as far above the normal operating level as possible without blatantly distorting. In this competitive business, the underlying logic behind the concept is that louder recordings (when placed into a Top 40, podcast, phone or MP3 playlist) will stand out from the softer recordings and get noticed. In fact, reducing the dynamic range of a song or program's dynamic range will actually make the overall levels appear to be louder. By using a slight (or not-so-slight) amount of compression and limiting to squeeze an extra 1- or 2-dB gain out of a song, the increased gain will also add to the perceived bass and highs because of our ears' increased sensitivity at louder levels (remember the Fletcher–Munson curve discussed in

Chapter 2). To achieve these hot levels without distortion, multiband compressors and limiters often are used during the mastering process to remove peaks and to raise the average level of the program. You'll find more on this subject in Chapter 20 (Mastering).

Compressing a mono mix is done in much the same way as one might compress a single instrument – although greater care should be taken. Adjusting the threshold, attack, release and ratio controls is more critical in order to prevent "pumping" sounds or loss of transients (resulting in a lifeless mix). Compressing a stereo mix gives rise to an additional problem: if two independent compressors are used, a peak in one channel will only reduce the gain on that channel and will cause sounds that are centered in a stereo image to shift (or jump) toward the channel that's not being compressed (since it will actually be louder). To avoid this center shifting, most compressors (of the same make and model) can be linked as a stereo pair. This procedure of ganging the two channels together interconnects the signal-level sensing circuits in such a way that a gain reduction in one channel will cause an equal reduction in the other.

Before moving on, let's take a look at a few examples of the use of compression in various applications. Keep in mind, these are only beginning suggestions – nothing can substitute for experimenting and finding the settings that work best for you and the situation:

- *Acoustic guitar*: a moderate degree of compression (3 to 8 dB) with a medium compression ratio can help to pull an acoustic forward in a mix. A slower attack time will allow the string's percussive attack to pass through.
- *Bass guitar*: the electric bass is often a foundation instrument in pop and rock music. Due to variations in note levels from one note to another on an electric bass guitar (or upright acoustic, for that matter), a compressor can be used to even out the notes and add a bit of presence and/or punch to the instrument. Since the instrument often (but not always) has a slower attack, it's often a good idea to start with a medium attack (4:1, for example) and threshold setting, along with a slower release time setting. Harder compression of up to 10:1 with gain reductions ranging from 5 to 10 dB can also give a good result.
- *Brass*: the use of a faster attack (1 to 5 ms) with ratios that range from 6:1 to 15:1 and moderate to heavy gain reduction can help keep the brass in line.
- *Electric guitar*: in general, an electric guitar won't need much compression, because the sound is often evened out by the amp, the instrument's natural sustain character and processing pedals. If desired, a heavier compression ratio with 10 or more decibels of compression can add to the instrument's "bite" in a mix. A faster attack time with a longer release is often a good place to start.
- *Kick drum and snare:* these driving instruments often benefit from added compression. For the kick, a 4:1 ratio with an attack setting of 10 ms or slower can help emphasize the initial attack while adding depth and presence. The snare attack settings might be faster so as to catch the initial transients. Threshold settings should be set for a minimum amount of reduction during a quiet passage, with larger amounts of gain reduction happening during louder sections.

- *Synths:* these instruments generally don't vary widely in dynamic range and thus, won't require much (or any) compression. If needed, a 4:1 ratio with moderate settings can help keep synth levels in check.

- *Vocals:* singers (especially inexperienced ones) will often place the mic close to their mouths. This can cause wide volume swings that change with small moves in distance. The singer might also shout out a line just after delivering a much quieter passage. These and other situations lead to the careful need for a compressor so as to smooth out variations in level. A good starting point would be a threshold setting of 0 dB, with a ratio of 4:1 and with attack and release settings set at their midpoints. Gain reductions that fall between 3 and 6 dB will often sit well in a mix (although some rock vocalists will want greater compression); be careful of over-compression and its adverse pumping artifacts. Given digital's wide dynamic range, you might consider adding compression later in the mixdown phase rather than during the actual session.

- *Final mix compression:* it's often a common practice to compress an entire mix during mixdown. If the track is to be professionally mastered, you should consult with the mastering engineer before the deed is done (or you might provide him or her with both a compressed and an uncompressed version). When applying bus compression, it is usually a good idea to start with medium attack and release settings, with a light compression ratio (say, 4:1). With these or your preferred settings, reduce the threshold detection until a light amount of compression is seen on the meter display. Levels of between 3 and 6 dB will provide a decent amount of compression without audible pumping or artifacts (given that a well-designed unit or plug-in is used).

TRY THIS: COMPRESSION

1. Go to youtube.com/modernrecordingtechniques, search for "compression" and listen to the tutorial sound files that relate to compression (which include instrument/music segments in various dynamic states).

2. Listen to the tracks.

3. If you'd like to DIY, then record or obtain an uncompressed bass guitar track and monitor it through a compressor or compression plug-in. Increase the threshold level until the compressor begins to kick in. Can you hear a difference?

Can you see a difference on the console or mixer meters?

4. Set the levels and threshold to a level you like and then set the attack time to a slow setting. Now, select a faster setting and continue until it sounds most natural. Try setting the release to its fastest setting. Does it sound better or worse? Now, select a slower setting. Does it sound more natural?

5. Repeat this routine and settings using a snare drum track. Were your findings any different?

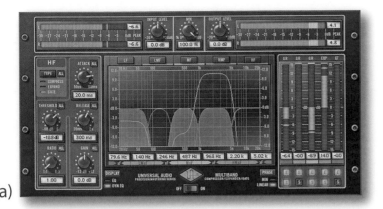

(a)

(b)

Multiband Compression

Multiband compression (Figure 15.24) works by breaking up the audible spectrum into various frequency bandwidths through the use of multiple bandpass filters. This allows each of the bands to be isolated and processed in ways that strictly minimize the problems or maximize the benefits in a particular band. Although this process is commonly done in the final mastering stage, multiband techniques can also be used on an instrument or grouping. For example:

- The dynamic upper range of a slap bass could be lightly compressed, while heavier amounts of compression could be applied to the instrument's lower register.
- An instrument's high end can be brightened simply by adding a small amount of compression. This can act as a treble boost while reducing any sharp attacks that might jump out in a mix.

Limiting

If the compression ratio is made large enough, a compressor will actually become a limiter. A limiter (Figure 15.25) is used to keep signal peaks from exceeding a specified level in order to prevent the overloading of amplifier signals, recorded signals onto tape or disc, broadcast transmission signals, and so on. Most limiters have ratios of 10:1 (above the threshold, for every 10 dB increase at the input,

FIGURE 15.24

Multiband compressors. (a) Universal Audio's UAD Precision Multiband plug-in (courtesy of Universal Audio, www.uaudio.com. © 2022 Universal Audio, Inc. All rights reserved. Used with permission). (b) DyneOne multiband compressor (courtesy of Leapwing, www.leapwingaudio.com).

FIGURE 15.25

Limiter plug-ins. (a) Fabfilter Pro L-2 limiter plug-in (courtesy of fabfilter, www.fabfilter.com). (b) Universal Audio's UAD Precision Limiter plug-in (courtesy of Universal Audio, www.uaudio.com. © 2017 Universal Audio, Inc. All rights reserved. Used with permission).

(a)

(b)

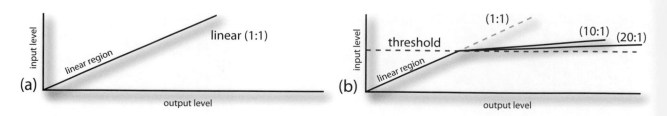

FIGURE 15.26

The output ratios of a limiter. (a) A linear slope results in an increase of 1 dB at the output for every 1 dB increase at the input. (b) A limiter slope follows a very high input/output gain reduction ratio (10:1, 20:1 or more) above the threshold, proportionately "limiting" signal level, so that they do not increase above a set point.

there will be a gain of 1 dB at the output) or 20:1 (Figure 15.26), although some have ratios that can range up to 100:1. Since a large increase above the threshold at the input will result in a very small increase at its output, the likelihood of overloading any equipment that follows the limiter will be greatly reduced. Limiters have three common functions:

- *To prevent signal levels from increasing beyond a specified level:* certain types of audio equipment (often those used in broadcast transmission) are often designed to operate at or near their peak output levels. Significantly increasing these levels beyond 100% would severely distort the signal and possibly damage the equipment. In these cases, a limiter can be used to prevent signals from significantly increasing beyond a specified output level.

- *To prevent short-term peaks from reducing a program's average signal level:* should even a single high-level peak exist at levels above the program's rms average, the overall level can be significantly reduced. This is especially true whenever a digital audio file is normalized at any percentage value, because the peak level will become the normalized maximum value and not the average level. Should only a few peaks exist in the file, they can easily be zoomed in on and manually reduced in level. If multiple peaks exist, then a limiter should be considered.

- *To prevent high-level, high-frequency peaks from distorting analog tape:* when recording to certain media (such as cassette and videotape), high-energy, transient signals actually don't significantly add to the program's level; however, if allowed to pass, these transients can easily result in distortion or tape saturation.

TRY THIS: LIMITING

1. Go to youtube.com/modernrecordingtech niques, search for "limiting" and listen to the tutorial sound files that relate to compression (which include instrument/music segments in various dynamic states).

2. Listen to the tracks. If you have access to an editor or DAW, import the files and look at the waveform amplitudes for each example. If you'd like to DIY, then:

3. Feed an isolated track or entire mix through a limiter or limiting plug-in.
4. With the limiter switched out, turn the signal up until the meter begins to peg (you might want to turn the monitors down a bit).
5. Now, reduce the level and turn it up again – this time with the limiter switched in. Is there a point where the level stops increasing, even though you've increased the input signal? What does the gain reduction meter show? Decrease and increase the threshold level, and experiment with the signal's dynamics. What did you find out?

Unlike the compression process, extremely short attack and release times are often used to quickly limit fast transients and to prevent the signal from being audibly pumped. Limiting a signal during the recording and/or mastering phase should only be used to remove occasional high-level peaks, as excessive use would trigger the process on successive peaks and would be noticeable. If the program contains too many peaks, it's probably a good idea to reduce the level to a point where only occasional extreme peaks can be detected.

Expansion

Expansion is the process by which the dynamic range of a signal is proportionately increased. Depending on the system's design, an *expander* (Figure 15.27) can operate either by decreasing the gain of a signal (as its level falls below the threshold) or by increasing the gain (as the level rises above it). Most expanders are of the first type, in that as the signal level falls below the expansion threshold, the gain is proportionately decreased (according to the slope ratio), thereby increasing the signal's overall dynamic range (Figure 15.28). These devices can also be used as noise reducers. You can do this by adjusting the device so that the noise is downwardly expanded during quiet passages, while louder program levels are unaffected or only moderately reduced. As with any dynamics device, the attack and release settings should be carefully set to best match the program material. For example, choosing a fast release time for an instrument that has a long sustain can lead to audible pumping effects. Conversely, slow release times on a fast-paced, transient instrument could cause the dynamics to return to its

FIGURE 15.27
Expander. (a) The Aphex Model 622 Logic-Assisted Expander/Gate (courtesy of Aphex Systems, Inc., www .aphex.com). (b) Cubase/ Nuendo Expander plug-in (courtesy of Steinberg Media Technologies GmbH, a division of Yamaha Corporation, www.steinberg.net).

(a)

(b)

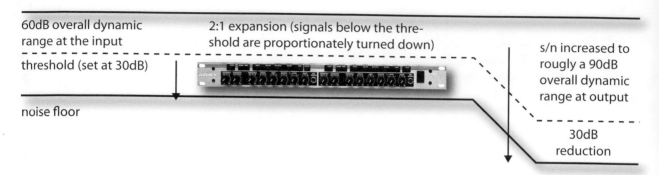

60dB overall dynamic range at the input

- - - - - - - -

threshold (set at 30dB)

noise floor

2:1 expansion (signals below the threshold are proportionately turned down)

s/n increased to rougly a 90dB overall dynamic range at output

30dB reduction

FIGURE 15.28

Commonly, the output of an expander is linear above the threshold and follows a low input/output gain expansion ratio below this point.

linear state more slowly than would be natural. As always, the best road toward understanding this and other dynamics processes is through experimentation.

The Noise Gate

One other type of expansion device is the noise gate (Figure 15.29). This device allows a signal above a selected threshold to pass through to the output at unity gain (1:1) and without dynamic processing; however, once the input signal falls below this threshold level, the gate acts as an infinite expander and effectively mutes the signal by fully attenuating it. In this way, the desired signal is allowed to pass, while background sounds, instrument buzzes, leakage or other unwanted noises that occur between pauses in the music are muted. Here are a few examples of where a noise gate might be used:

FIGURE 15.29

Noise gates are commonly included within many dynamic plug-in processors. (a) Noise Gate Plug-in (courtesy of Avid Technology, Inc., www.avid.com). (b) Noise Gate Plug-in (courtesy of Steinberg Media Technologies GmbH, a division of Yamaha Corporation, www.steinberg .net). (c) Noise Gate found on Duality console (courtesy of Solid State Logic, www.solid -state-logic.com).

- To reduce leakage between instruments. Often, parts of a drum kit fall into this category; for example, a gate can be used on a high-tom track in order to reduce excessive leakage from the snare.

- To eliminate tape or system noise from an instrument or vocal track during silent passages.

When "gating" a specific track, it will often be necessary to take time out to fine-tune the device's attack and release controls. This important step is done to

(a)

(b)

(c)

reduce or eliminate any unwanted signal "cutouts, pumping or breathing" as the signal, noise or leakage signal falls and rises around the threshold point.

The general rules for these settings are the same as those that apply to gain-change processors (see the compressor, limiter and expander settings section earlier in this chapter). Fortunately, these settings are often audibly more obvious than with any other dynamic tool. Setting the threshold, attack and release times at inappropriate levels will often be immediately audible because the sound will cut in and out at inappropriate times. For this reason, it's important that care be taken with this very useful tool when adjusting the settings by both listening to the track on its own (solo track) and listening to it within the context of the full mix.

TRY THIS: NOISE GATE

1. Go to youtube.com/modernrecordingtechniques and search for "noise gate"

and/or...

1. Find a set of live multitrack drum tracks from the web that have been recorded with excessive leakage.
2. Insert a gate onto one or more tracks, using its initial settings at first.

3. Solo a track, and then play with the settings to reduce any background noises, rings or hums.
4. Bypass and insert the gate on that track. What's the difference?
5. Do this for any or all tracks that need help, and then, bypass/insert the gate on all of the gated tracks. How does it change the overall presence of the drums?

Commonly, a key input (as previously shown in Figure 15.6) is included as a sidechain path for triggering a noise gate. A key input is an external control that allows an external analog signal source (such as a miked instrument or signal generator) to trigger the gate's audio output path. For example, a mic or recorded track of a kick drum could be used to key a low-frequency oscillator. Whenever the kick sounds, the oscillator will be passed through the gate. By combining the two, you can have a deep kick sound that'll make the room shake, rattle and roll.

Noise Reduction

With the advent of newer and better digital audio technologies, low-noise systems, hi-res audio and surround-sound home theaters, an increase in dynamic range and a demand for better quality sound have steadily been on the rise. Because of this, it's more important than ever that those in audio production pay close attention to the background noise levels that are produced by preamp and amplifier self-noise, synths, analog magnetic tape and the like. Although the overall dynamic range of human hearing roughly encompasses a full 140 dB, and well-designed digital systems are capable of much wider ranges in everyday practice, such dynamics often won't be fully captured, played back or appreciated for several reasons:

- An acoustic or electronic weak link in the chain might introduce noise and/or restrict the program's dynamic range.
- The medium or device itself might be incapable of capturing a wide dynamic range.
- Background noises in the environment might mask the subtleties of the sound.

Not all the blame for added noise can be placed on our older technology friends. Even though a 16-bit digital recording has a theoretical dynamic range of 96 dB, and a 24-bit system can actually encode 144 dB, noises can (and often will) crop up from such modern-day gremlins as mic preamps, effects and outboard gear, analog communication lines and poorly designed digital audio converters. Honestly, though, one of the biggest noise problems that you'll often encounter is the need for restoring tracks that were poorly recorded or were made under adverse and/or noisy acoustic conditions.

DIGITAL NOISE REDUCTION

As you might expect, in modern music production, DSP is most commonly used to reduce noise levels within a recorded sound file. These noises might include artifacts such as tape hiss, hum, obtrusive background ambience, needle ticks, pops and even certain types of distortion that are present in the original recording. Although stand-alone digital noise reduction processors definitely exist, most of these processors exist as plug-ins that can be introduced at multiple points into the signal chain of a DAW. For example, an "NR" plug-in can be inserted into a single track to reduce the amp noise on a guitar, synth or other type of track to instantly clean up a mix that might otherwise be problematic. Likewise, an NR plug-in can be inserted into a group or final output mix stage to clean up an overall mix that's overly noisy.

ADAPTIVE FILTERING

Modern-day noise reduction systems, particularly real-time processors and plug-ins, make use of adaptive filtering to reduce noise. This is done by dynamically reducing the bandwidth of an audio signal to downwardly expand (reduce) those signals that aren't considered to be important to the overall signal content. For example, if we listen to a bass track that contains a great deal of high-end amp or tape noise, this noise can be reduced by turning down parts of the audio bandwidth that don't contain valuable information … in this case, the offensive noise at the upper end of the spectrum. In short, it acts as an intelligent, dynamic equalizer, which can reduce an offensive noise that falls below a set threshold and leave important information in that upper bandwidth alone whenever it rises above that threshold.

These systems can be then used to extract noise from an audio source by combining a downward dynamic range expander with a variable low-pass filter. These real-time processing devices, which can be analog or digital in nature, are used to dynamically analyze and process an existing program to reduce the unwanted noise content with few or no audible effects (or giving us a best possible compromise, in extreme cases).

In addition, the emphasis or de-emphasis curve on a number of these systems can be used to shape the program's noise spectrum, allowing differing frequency spectrums or artifacts to be reduced (Figure 15.30). Such noise shaping can truly be a useful tool towards bringing out the qualities of a sound to best suit the program content.

These systems often work by breaking up the audio spectrum into a number of frequency bands, such that whenever the signal level within each band falls below a user-defined threshold, the signal will be attenuated. This downward expansion/filtering process accomplishes noise reduction by taking advantage of two basic psychoacoustical principles:

- Music is capable of masking noise (covering up a lower-level noise by a louder signal) that exists within the same bandwidth
- Reducing the bandwidth of an audio signal will reduce the perceived noise

It's a psychoacoustical fact that our ears are more sensitive to noises that contain a greater number of frequencies than to those containing fewer frequencies. Thus, whenever the sounds within a certain bandwidth are reduced or restricted, the dynamic filtering process will sense this and reduce by downward expansion (or cut by gating) this unused bandwidth accordingly, thereby reducing its overall perceived noise content. When the program's signal returns, the filter will again pass the frequency bandwidth (allowing the increased program content to mask the background noise). Such a dynamic filter can also be built using a multiband dynamics processor (Figure 15.31), whereby the high-end (or offending) band can be set to reduce or gate the signal in that band whenever

FIGURE 15.30
The use of frequency-selective emphasis to shape the nature of a noise reduced soundfile can be a very helpful tool (courtesy of Acon AS, www.acondigital.com).

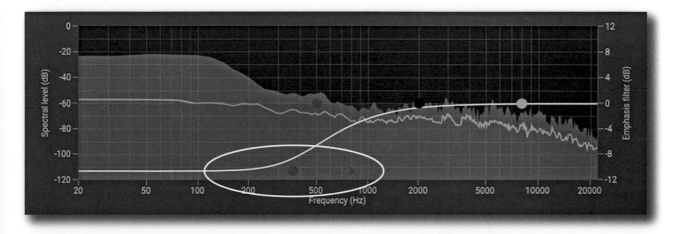

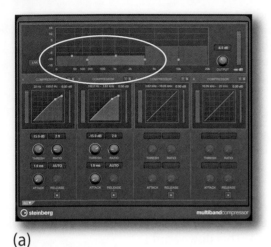

(a)

(b)

FIGURE 15.31

A Multiband Compressor/ Expander/Gate can effectively be programmed to become a single-ended noise reduction unit. (a) Steinberg Multiband Compressor (courtesy of Steinberg Media Technologies GmbH, a division of Yamaha Corporation, www.steinberg.net). (b) Universal Audio Precision Multiband Compressor/Expander/Gate plug-in (courtesy of Universal Audio, www.uaudio.com, © 2022 Universal Audio, Inc. All rights reserved. Used with permission).

the signal falls below the set threshold and pass the signal when it returns in this frequency range.

FAST FOURIER TRANSFORM

Another noise reduction process makes use of a mathematically intense algorithm known as Fast Fourier Transform (FFT). These applications and plug-ins (Figure 15.32) are able to analyze the amplitude/frequency domain of an audio signal in order to reduce hum, tape hiss and other extraneous noises from your recordings. This digital analysis generally begins by taking a digital "snapshot" of a short snippet of the offending noise (a brief section that contains only the noise to be eliminated will yield the best results). This noise template can then be digitally subtracted from the original sound file or segment in varying amounts (and under the control of various program parameters), such that only the footprint noise is reduced, while (under the best of conditions) the original program material is left intact and unaffected.

Although FFT algorithms for reducing noise have greatly improved over the years, it's still important that we briefly discuss a few of the unfortunate artifacts that can occur when using (and overusing) FFT-based noise reduction. The most notable of these is "chirping". This audible artifact most often occurs when too much FFT processing is applied. It literally sounds like a flock of small chirping birds that can be heard either in the background or in an obnoxious way that sounds like a bad Alfred Hitchcock movie. If you find yourself running for cover, it's best to pull back on the FFT settings (and/or increase the processing quality level) until the artifacts are less noticeable.

Should chirping and/or bandwidth limitations become a problem, you might consider using equalization or a single-ended noise reduction device/plug-in instead. Because single-ended noise reduction (to be discussed later) uses an adaptive filter to intelligently change the program's bandwidth, no chirping artifacts will be introduced.

(a)   (b)

FIGURE 15.32
Noise reduction plug-ins.
(a) Digidesign DINR plug-in
(courtesy of Digidesign, a
division of Avid Technology,
Inc., www.digidesign.com).
(b) Antares SoundSoap 5
Audio restoration plug-in
(courtesy of Antares Audio
Technologies, www.antares-
ech.com).

Finally, it's a misconception that an FFT-based noise reduction application can only be used for reducing noise. Literally any sound can be used as a sonic removal footprint, and as a result, vocal formants, snare hits or any sound that you can imagine can be pulled from a sound file to create unique and interesting effects. The sky's literally the limit!

SPECTRAL ANALYSIS NOISE REDUCTION

In addition to using your ears and the on-screen standard plug-in controls to analyze and reduce noise, a number of FFT-based noise reduction systems are additionally able to display frequencies in an easy-to-detect graphic form that lets the user view the overall spectral analysis of a recorded section (Figure 15.33). These various frequencies, noises, pops, etc. can then be easily seen and effectively "drawn out" in a way that allows the offending frequencies to be mathematically removed from the audio signal. Such a useful tool can also be used to remove coughs, squeaky chairs, sirens and any other unfortunate noises that might make their way into a recording.

SNAP, CRACKLE AND POP

In addition to removing noise, programs and DAW plug-ins also exist for removing clicks and pops from vinyl and older recordings (Figure 15.34). Although FFT analysis is often involved in the process, click removal differs slightly from FFT noise reduction. This multistep process begins by detecting high-level clicks (or those exceeding a user-defined threshold) that exist within a sound file or defined segment. Once the offending noises are detected, the program performs a frequency analysis (both before and after the click) and then goes about the business of making a best plausible "guess" as to what the damaged amplitude/frequency content should actually sound like. Finally, the calculated sound is pasted over the nasty offender (ideally rendering it less noticeable or gone) and then moves on to the next click and restarts the detection/replacement process.

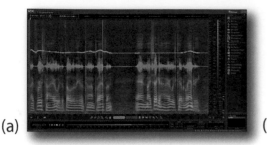

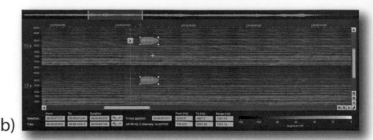

(a)　　　　　　　　　　　　　　(b)

FIGURE 15.33

Spectral Analysis Software is capable of displaying the spectral (frequency) content of a passage, allowing the user to selectively "draw out" any unwanted noises from a program. (a) iZotope's RX9 Audio Repair Software (courtesy of iZotope Inc., www.izotope.com). (b) Acon Digital's Acoustica Digital Audio Editor (courtesy of Acon AS, www.acondigital.com).

Because click and pop noises can be different in nature from each other (in both duration and frequency makeup), noise reduction plug-ins might offer applications that are specifically suited to reducing each type.

TIME-BASED EFFECTS

Another important effects category that can be used to alter or augment a signal revolves around delays and regeneration of sound over time. These time-based effects often add a perceived depth to a signal or change the way we perceive the dimensional space of a recorded sound. Although a wide range of time-based effects exist, they are all based on the use of delay (and/or regenerated delay) to achieve such results as:

- Time-delay or regenerated echoes, chorus and flanging
- Reverb

FIGURE 15.34

Click/pop eliminator applications: (a) Within Adobe Audition CC (courtesy of Adobe Systems, Inc., www .adobe.com). (b) Within iZo-tope RX9 (courtesy of iZotope Inc., www.izotope.com). (c) DeClick2 (courtesy of Acon AS, www.acondigital.com).

Delay

One of the most common effects used in audio production today alters the parameter of time by introducing various forms of delay into the signal path. Creating a delay circuit is a relatively simple task to accomplish digitally. Although dedicated delay devices (often referred to as digital delay lines, or DDLs) are readily available on the market, most multifunction signal processors and time-related plug-ins are capable of creating this straightforward effect (Figure 15.35). In its basic form, digital delay is accomplished by storing sampled audio directly into RAM. After a defined length of time (usually measured in milliseconds),

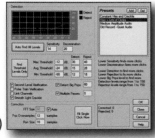

(a)　　　　　　　　　　(b)　　　　　　　　　　(c)

(a) (b)

the sampled audio can be read out from memory for further processing or direct output (Figure 15.36a). Using this basic concept, a wide range of effects can be created simply by assembling circuits and program algorithms into blocks that can introduce delays or regenerated echo loops. Of course, these circuits will vary in complexity as new processing blocks are introduced.

DELAY IN ACTION: LESS THAN 15 MS

Probably the best place to start looking at the delay process is at the sample level. By introducing delays downward into the microsecond (one millionth of a second) range, control over a signal's phase characteristics can be introduced to the point where selective equalization actually begins to occur. In reality, controlling very short-term delays is actually how EQ is carried out in both the analog and digital domains!

Whenever delays that fall below the 15-ms range are slowly varied over time and then are mixed with the original undelayed signal, an effect known as *combing* is created. Combing is the result of changes that occur when equalized peaks and dips appear in the signal's frequency response. By either manually or automatically varying the time of one or more of these short-term delays, a constantly shifting series of effects known as *flanging* can be created. Depending on the application, this effect (which makes a unique "swishing" sound that's often heard on guitars or vocals) can range from being relatively subtle to having moderate to wild shifts in time and pitch. It's interesting to note the differences between the effects of phasing and flanging. Phasing uses all-pass filters to create uneven peaks and notches, whereas flanging uses delay lines to create even peaks and notches, although the results are somewhat similar.

FIGURE 15.35

Delay plug-ins. (a) Pro Tools Mod Delay II (courtesy of Avid Technology, Inc., www.avid.com). (b) Galaxy Tape Echo Plug-in (courtesy of Universal Audio, www.uaudio.com. © 2022 Universal Audio, Inc. All rights reserved. Used with permission).

FIGURE 15.36

Digital delay. (a) A ddl stores sampled audio into RAM, where it can be read out at a later time. (b) In certain instances, ddl (doubling or double delay) can fool the brain into thinking that more instruments are playing than actually are.

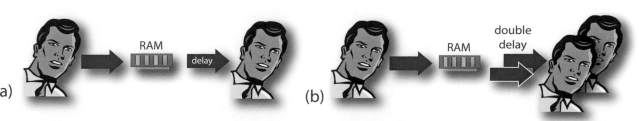

(a) (b)

DELAY IN ACTION: 15 TO 35 MS

By combining two identical (and often slightly delayed) signals that are slightly detuned in pitch from one another, an effect known as *chorusing* can be created. Chorusing is an effects tool that's often used by guitarists, vocalists and other musicians to add depth, richness and harmonic structure to their sound. Increasing delay times into the 15- to 35-ms range will create signals that are spaced too closely together to be perceived by the listener as being discrete delays. Instead, these closely spaced delays create a *doubling effect* when mixed with an instrument or group of instruments (Figure 15.36b). In this instance, the delays actually fool the brain into thinking that more instruments are playing than actually are – subjectively increasing the sound's density and richness. This effect can be used on background vocals, horns, string sections and other grouped instruments to make the ensemble sound as though it has doubled (or even tripled) its actual size. This effect also can be used on foreground tracks, such as vocals or instrument solos, to create a larger, richer and fuller sound. Some "chorus" delay devices introduce slight changes in delay and pitch shifting, allowing detunings that can create an interesting, humanized sound.

Should time or budget be an issue, it's also possible to create this doubling effect by actually recording a second pass to a new set of tracks. Using this method, a 10-piece string section could be made to sound like a much larger ensemble. In addition, this process automatically gives vocals, strings, keyboards and other legato instruments a more natural effect than the one you get by using an electronic effects device. This having been said, these devices can actually go a long way toward duplicating the effect. Some delay devices even introduce slight changes in delay times in order to create a more natural, humanized sound. As always, the method you choose will be determined by your style, your budget and the needs of your particular project.

TRY THIS: DELAY

1. Go to youtube.com/modernrecordingtechniques, search for "delay", and listen to the tracks (which include segments with varying degrees of delay).

2. If you'd like to DIY, then insert a digital delay unit or plug-in into a program channel and balance the dry track's output mix so that the input signal is set equally with the delayed output signal. (Note: if there is no mix control, route the delay unit's output to another input strip and combine delayed/undelayed signals at the console.)

3. Listen to the track with the mix set to listen equally to the dry and affected signal.

4. Vary the settings over the 1- to 10-ms range. Can you hear any rough EQ effects?

5. Manually vary the settings over the 10- to 35-ms range. Can you simulate a rough phasing effect?

6. Increase the settings above 35 ms. Can you hear the discrete delays?

7. If the unit has a phaser setting, turn it on. How does it sound different?

8. Now, change the delay settings a little faster to create a wacky flange effect. If the unit has a flange setting, turn it on. Try playing with the time-based settings that affect its sweep rate. Fun, huh?

DELAY IN ACTION: MORE THAN 35 MS

When the delay time is increased beyond the 35- to 40-ms point, the listener will begin to perceive the sound as being a discrete echo. When mixed with the original signal, this effect can add depth and richness to an instrument or range of instruments that can really add interest to an instrument within a mix.

Adding delays to an instrument that are tied to the tempo of a song can go even further toward adding a degree of depth and complexity to a mix. Most delay-based plug-ins make it easy to insert tempo-based delays into a track. For hardware delay devices, it's usually necessary to calculate the tempo math that's required to match the session. Here's the simple math for making these calculations:

$$60,000/tempo = time \text{ (in ms)}$$

For example, if a song's tempo is 100 bpm (beats per minute), then the amount of delay needed to match the tempo at the beat level would be:

$$60,000/100 = 600 \text{ ms}$$

Using divisions of this figure (300, 150, 75, etc.) would insert delays at 1/2, 1/4, 1/8 measure intervals.

Caution should be exercised when adding delay to an entire musical program, because the program could easily begin to sound muddy and unintelligible. By feeding the delayed signal back into the circuit, a repeated series of echo … echo … echoes can be made to simulate the delays of yesteryear – you'll definitely notice that Elvis is still in the house.

Reverb

In professional audio production, natural acoustic reverberation is an extremely important tool for the enhancement of music and sound production. A properly designed acoustical environment can add a sense of space and natural depth to a recorded sound that'll often affect the performance as well as its overall sonic character. In situations where there is little, no or substandard natural ambience, a high-quality reverb device or plug-in (Figure 15.37) can be extremely helpful in filling the production out and giving it a sense of dimensional space and perceived warmth. In fact, reverb consists of closely spaced and random multiple echoes that are reflected from one boundary to another within a determined space (Figure 15.38). This effect helps give us perceptible cues as to the size, density and nature of a space (even though it might have been artificially generated). These cues can be broken down into three subcomponents:

- Direct signal
- Early reflections
- Reverberation

(a)   (b)

FIGURE 15.37

Digital reverb effects processors. (a) Bricasti M7 Stereo Reverb Processor (courtesy of Bricasti Design Ltd, www.bracasti.com). (b) Steinberg Reverence reverb plug-in (courtesy of Steinberg Media Technologies GmbH, a division of Yamaha Corporation, www.steinberg.net).

The *direct signal* is heard when the original sound wave travels directly from the source to the listener. *Early reflections* is the term given to those first few reflections that bounce back to the listener from large, primary boundaries in a given space. Generally, these reflections are the ones that give us subconscious cues as to the perception of size and space. The last set of reflections makes up the signal's *reverberation* characteristic. These sounds are comprised of zillions of random reflections that travel from boundary to boundary within the confines of a room. These reflections are so closely spaced in time that the brain can't discern them as individual reflections, so they're perceived as a single, densely decaying signal.

REVERB TYPES

By varying program and setting parameters, a digital reverb device can be used to simulate a wide range of acoustic environments, reverb devices and special effects. A few popular categories include:

- *Hall:* simulates the acoustics of a concert hall. This is often a diffuse, lush setting with a longer RT60 decay time (the time that's required for a sound to decay by 60 dB).
- *Chamber:* simulates the acoustics of an echo chamber. Like a live chamber, these settings often simulate the brighter reflectivity of tile or cement surfaces.
- *Room:* as you might expect, these settings simulate the acoustics of a mid- to large-sized room. It's often best suited to intimate solo instruments or a chamber atmosphere.
- *Live (stage):* simulates a live performance stage. These settings can vary widely but often simulate long early-delay reflections.

FIGURE 15.38

Signal level versus reverb time.

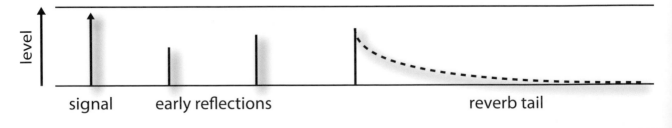

- *Spring:* simulates the low-fidelity "boingyness" of yesteryear's spring reverb devices.
- *Plate:* simulates the often-bright diffuse character of yesteryear's metallic plate reverb devices. These settings are often used on vocals and percussion instruments.
- *Reverse:* these backward-sounding effects are created by reversing the decay trail's envelope so that the decay increases in level over time and is quickly cut off at the tail end, yielding a sudden break effect. This can also be real-istically created in a DAW by reversing a track or segment, applying reverb and then reversing it again to yield a true backward reverb trail.
- *Gate:* cuts off the decay trail of a reverb signal. These settings are often used for emphasis on drums and percussion instruments.

TRY THIS: REVERB TYPES

Open up a DAW session that contains various instruments (such as a snare, guitar and vocal).

Insert (or send) a reverb plug-in into each track.

Listen to each track while changing the plug-in presets between "hall", "plate", "small room", etc. How do they sound different? Do they sound different for each instrument?

Psychoacoustic Enhancement

A number of signal processors rely on psychoacoustic cues in order to fool the brain into perceiving a particular effect. The earliest and most common of these devices are those that enhance the overall presence of a signal or entire recording by synthesizing upper-range frequency harmonics and inserting them into a mix in order to brighten the perceived sound. Although the additional harmonics won't significantly affect the program's overall volume, the effect is a marked increase in its perceived presence. Other psychoacoustic devices that make use of complex harmonic, phase, delay and equalization parameters have become standard production tools in the field of mastering in order to shape the final sound into one that's interesting, with a sonic character all their own.

In addition to synthesizing harmonics in order to change or enhance a recording or track, other digital psychoacoustic processors deal exclusively with the sub-ject of spatialization (the placement of an audio signal within a three-dimen-sional acoustic field), even though the recording is being played back over stereo speakers. By varying the parameters of a stereo or multiple input source, this processing function creates phase and amplitude paths that can fool the brain into perceiving that the stereo image is actually emanating from a sound field that's wider than the physical speaker positions. In practice, care should be taken when using these devices, because the effect is often carried off with degrees of success that vary from system to system. In addition, the use of phase relationships to expand the stereo sound field can actually cause obvious cancellation problems when the program is listened to in mono.

PITCH SHIFTING

Ever had a perfectly good vocal take that was spoiled by just one or two flat notes? Or had a project come in the door with a guitar track that was out of tune? Or needed to change the key on a 30-second radio spot? It's at times like these that pitch shifting can save your day! *Pitch shifting* can be used to vary the pitch of a signal or sound file (either upward or downward) in order to transpose the relative pitch of an audio program without affecting its duration. This process can take place in either real time or non-real time. Pitch shifting works by writing sampled audio data to a temporary memory, where it's resampled to either a higher or a lower sample rate (according to the desired final pitch). Once this is done, the processor either adds interpolated samples to (lowers the pitch) or subtracts them from (raises the pitch) the resampled data to return it back to the original output rate while keeping the altered pitch intact. Figure 15.39 gives two basic examples of how this is often carried out.

A degree of caution should be used when changing the pitch of a program or audio segment. Whenever uneven or minute interval changes are made, the interpolation of samples doesn't always fall perfectly into place. This can lead to digital artifacts, which add amounts of harmonic distortion that can range from slightly noticeable to unacceptable. If the track is in the background, there shouldn't be a problem; however, care should be taken with upfront instruments and vocals. It's important to keep in mind that large pitch changes might be more noticeable. As always, your ears are the best judge.

TIME AND PITCH CHANGES

By combining variable sample rates and pitch-shifting techniques, it's possible to create three different variations:

- *Time change:* a program's length can be altered, without affecting its pitch, by raising or lowering its playback sample rate.

FIGURE 15.39
Two pitch-shift examples with an initial 1-kHz digital signal and a sample rate of 44.1 kHz. (a) The signal can be halved in pitch (to 500 Hz) by internally downsampling to a new rate of 22.05k. To return the output rate to 44.1 (while retaining the 500-Hz pitch), new sample points must be added into each dropped position. (b) The signal can be doubled in pitch (to 2 kHz) by internally upsampling to a new rate of 88.2k. To return the output rate to 44.1 (while retaining the 2-kHz pitch), every other sample point must be dropped.

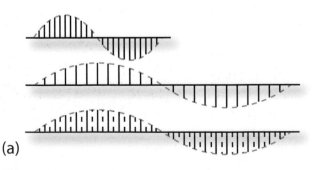

(a)

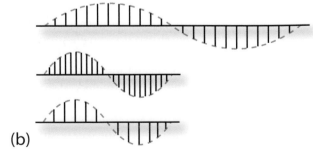

(b)

- *Pitch change:* a program's length can remain the same while pitch is shifted either up or down.
- *Both:* both a program's pitch and its length can be altered by means of simple resampling techniques.

These functions have become an important part of the signal processing and music production arsenals that are used by the audio-for-video, film and broadcast industries. They help give producers control over the running time of film video and audio soundtracks while maintaining the original, natural pitch of voice, music and effects. For example, using a DAW, we could add 5 seconds onto the end of an existing 25-second public service radio spot simply by time shifting the 25-second spot to 30 seconds (while keeping the pitch intact).

In addition to the basic time/pitch techniques that are commonly used in music production (most often by electronic musicians), this technology has enabled the huge explosion in loop-based music composition and production. These popular programs and music plug-ins involve the use of recorded sound files that are encoded with headers that include information on their native tempo and length (in both samples and beats). When you set the loop program to a master tempo (or a specific tempo at that point in the score), a loop segment, once imported, can go about the process of recalculating its pitch and tempo to match the current session tempo, and – voilà! The file's in sync with the song! Further info on loop-based production tools can be found in Chapter 8 (Groove Tools and Techniques).

It's important to realize that the algorithms that control how time and pitch are to be calculated differ between programs (as well as having various algorithm options within the same program). It would be wise to read up on how your favorite programs deal with these shifting techniques (or better yet, listen to various percussion and pad-based loops) and then familiarize yourself with how the different algorithms sound. The quality of your sound files will thank you.

Automatic Pitch Correction

One other pitch correction category makes use of *auto-tuning* software or plug-ins that are able to automatically detect off-tune or slight pitch-bend segments of a track and then correct these inaccuracies (according to user parameters). These systems (Figure 15.40) are commonly used in modern music production in situations that include:

- Simple track pitch corrections to fix the overall tuning of a track
- Automated and manual micro-tuning to correct bad or out-of-tune notes in a vocal or instrument performance
- Using parameters that fix tuning steps that can create an absolutely "perfect" performance
- Creating an extreme tuning effect that exaggerates the various tuning steps in pitch to give a non-human, robotic-like effect to a vocal or instrumental performance

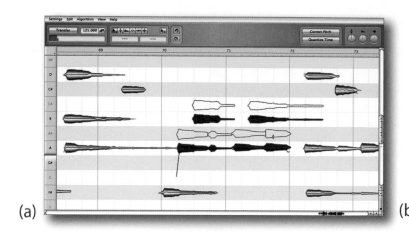

(a)   (b)

FIGURE 15.40

Automatic pitch correction software. (a) Melodyne Editor auto pitch correction system (courtesy of Celemony Software GmbH, www .celemony.com). (b) Auto-Tune pitch correction system (courtesy of Antares Audio Technologies, www. antarestech.com).

Often, these pitch editors allow us to change the pitch in a straightforward, graphical environment according to musical and/or user-defined parameters. Single notes or entire chords can be built up, altered or changed, making it possible to correct a wide range of material. It's even possible to alter individual notes within a polyphonic recording. For example, you could fix a single wrong note in a piano recording or change the notes within a guitar backing track – all after the recording has been made. In short, automatic pitch correction has become practically an art in itself, and several have made a career out of it. Of course, there is too much to fully cover the nuances and intricacies of pitch correction here; fortunately, there are hundreds (maybe thousands) of video tutorials on the subject. Dive into them, then dive into the software – experiment and have fun!

Of course, the use of automatic tuning to fix a track isn't without its critics. In 1998, Cher released one of the first popular songs with Auto-Tune, "Believe", and since then, pitch correction has become a mainstream tool for pop, country, hip hop, you name it. As a tool, pitch correction can be carefully used to make a vocal or track sound more in tune; however, to many, its use (some would say, overuse) takes away from the artistry and humanity of a performance. As with most tools of the effects trade, it's wise to take the time to become familiar with the details, depth and consequences of pitch correction and then talk with the musicians and producer about their intentions.

Multiple-Effects Devices

Since most digital signal processors are by nature multifunctional chameleons, it follows that most hardware and certain plug-in processors can be easily programmed to perform various functions. For this reason, many digital systems have been designed to perform as multiple-effects devices (Figure 15.41). Multiple effects, in this case, can have several basic meanings:

- A single device might offer a wide range of processing functions but allow only one effect to be called up at a time.

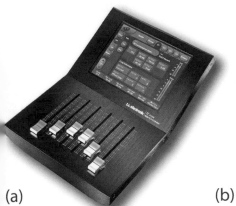
(a)

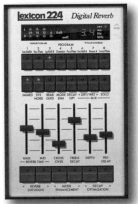
(b)

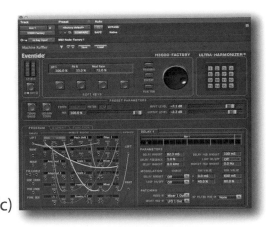
(c)

FIGURE 15.41
Multi-effects processing devices. (a) Controller for tc electronic System 6000 digital effects processor (courtesy of tc electronic, www.tcelectronic.com). (b) Lexicon 224 reverb plug-in for the Apollo and the UAD effects processing card (courtesy of Universal Audio, www.uaudio.com © 2022 Universal Audio, Inc. All rights reserved. Used with permission). (c) Eventide H3000 Factory Ultra-Harmonizer Plug-in (courtesy of Eventide Inc., www.eventideaudio .com).

- A single device might offer a range of processing functions that can be "stacked" to perform a number of simultaneous effects.
- An effects device might have multiple ins and outs, each of which can perform several processing functions (effectively giving you multiple processors that can be used in a multichannel mixdown environment).

Dynamic Effects Automation and Editing

One of the joys of working with effects is the ability to manipulate and vary effects parameters in real time over the duration of a song or audio program. By altering parameters, changing settings and mixing effects levels, the subtle variations in expression can add a great deal of interest to a project.

TRY THIS: DAW EFFECTS AUTOMATION

The vast majority of plug-in effects can be directly and dynamically automated within the computer's DAW program. It's my greatest hope that you'll:

- Take a look at your favorite DAW manual and start reading!
- Open up a tutorial session, or better yet, make your own

- Call up some effects on various tracks
- Call up some automation control parameters and start grabbing controls
- If you have a controller section on your keyboard or other MIDI device, use it!
- Learn how to edit these automation functions so as to be able to finesse your effects in new and interesting ways

The ability to dynamically automate effects settings can be accomplished in any number of ways, including:

- Via MIDI control and parameter change messages
- Via an external hardware controller
- Via DAW automation or other form of automation control

In closing, an almost unlimited degree of effects control is available to us through most high-level (and many entry-level) digital audio workstations. Through the use of any of the readily available hardware controllers or on-screen automation controls, it's possible to manipulate and automate effects within a DAW session with an amazing degree of sophistication and ease. Go ahead, get hold of these fun and effective tools, and experiment your heart out!

CHAPTER 16

The Art and Technology of Monitoring

Throughout the entire recording process, our ability to judge and adjust sound is primarily based on what's heard through the monitor speakers in a project studio or professional control room environment (Figure 16.1). In fact, within the audio and video industries, the word "monitor" refers to a device that acts as a subjective professional standard or reference by which program material can be critically evaluated.

Despite steady advances in design, speakers are still one of the weakest links in the audio chain. This weakness is generally due to potential nonlinearities that can exist in a speaker system's frequency response. In addition, interactions with a room's acoustic nature and frequency response can often lead to peaks and dips that affect a speaker's sonic character in ways that are difficult to predict. Add to this the factor of personal "tastes" in the sound, size and design types of the countless speaker systems that are available, and you'll quickly find that they're also one of the most subjective tools in a production environment.

So, at the outset, a dilemma that refers to the naming of this chapter should be pointed out. This debate refers to whether it should simply be named "Monitoring" or should be elevated to "The Art and Technology of Monitoring". Well, you can see which title won out … and for one simple reason. Monitoring is so important, so subjective and so varied between each and every one of us in the field of music and media production that it deserves to be given special consideration in a way that simply can't be underestimated.

ACTIVE VERSUS PASSIVE LISTENING

So, one would think that in a chapter about monitoring, the most important tool would be our choice of acoustics, speakers, headphone set or software to help us hear the most accurate sound possible … and we would argue that you would probably be wrong. Although these are important, as we'll see later, the most important tool rests on the side of your head and more important than that, what's between your ears.

DOI: 10.4324/9781003260530-16

FIGURE 16.1

Example of a professional monitoring system: Hit Factory Criteria, Miami, FL (courtesy of Solid State Logic LTD, www.solidstatelogic .com).

In short, the actual process of creative listening is the most important factor in the creation of a music or successful audio production. Between the two of us, both DMH and EC have numerous Grammy nominations to our names. In large part, this came about due to the decades of time that we have collectively put into the process of active listening … the process of concentrating on:

- The level of musicality of the material
- The quality of the signal pickup
- The overall feel of how the tracks combine and blend together to create an overall soundscape
- Plus any other number of conscious and unconscious cues that go into the process of creating quality audio

Before we talk about the process of listening actively, let's take a quick look at the passive process. First off, as audio professionals, it's often difficult to talk about passive listening, as we've spent much of our lives concentrating on sound. Passive listening is just that; it's probably what the general consumer public do when they listen to produced audio. It's basically a sound that's in the background. It might be a song that you really like, dislike or simply don't care about. One of the greatest problems with audio production today is the fact that it's everywhere, practically all the time. You walk into an elevator, a grocery store, a shop, a friend's car, and THERE IT IS … ALWAYS THERE!

Not only does this throw us and most of the population into the space of being a passive listener; it might push us into the realm of blocking sound, music and audio as far into the background as possible. The question becomes: why have music everywhere, at all times? Is it to make us feel like we're not alone? Is it there to engage us, to make us more active, more calm, more productive?

The fact is, this constant companion rests in the background, possibly out of the way as just another sensory input. Active listening during the creative process of producing audio, on the other hand, takes center stage … it actually requires that we concentrate on and think about what we are hearing and that these sounds combine in a creative sense.

How does one hone this craft of becoming a better and more attentive active listener? The same way that you get to Carnegie Hall … by practicing … by sitting down in front of a musician, an instrument and/or a set of speakers and giving it your full creative attention over time. It's a process that can develop over months or years, but with time, these skills will improve. It can take the form of being purely creative and emotional, or it can take on a more technical sense of understanding "I just need to turn it up by 1 dB at 1.25 kHz" … It's our opinion that over time, both the subconscious and the conscious side of your listening skills will almost certainly improve.

In the end, it's this sense of active listening that will improve your system, far more than simply throwing more money at your system and hoping that this will solve all of your troubles. It all starts with a conscious desire to listen and even "feel" deeper into your productions and mixes … not always an easy task, but always worth it.

Tools for Listening

One of the best ways to begin the process of actively listening to audio is to find the most comfortable and most trusted listening environment that you have and take a trip to your favorite desert island. By this, we mean making a desert island collection of songs and/or albums that you'd best like to be stranded with … the bottle of rum could be optional.

The next job would be to put them on and then begin listening to them … really listening (ok, maybe the rum isn't such a good idea, after all).

- How do the various songs/projects differ?
- What's the emotional impact?
- What is it that the producer and artists did to bring about this response?
- How do the vocals sound?
- Is the sound thick and dense, or open and full of space?

After some desert island listening time, you might put on some of your own work. Don't be too hard on yourself, but simply notice how your projects sound after listening to some of your idols. Here are a few of our desert island faves:

EC: Pink Floyd "Dark Side Of The Moon", Dire Straits "On Every Street" (recorded by our good buddy Chuck Anley), Eric Clapton "Slowhand"

DMH: Roxy Music "Avalon" (recorded by Bob Clearmountain), Luis Miguel "Romances" (recorded by our good buddy Rafa Sardina), Steely Dan "Aja" (recorded and mixed by a who's who lineup – Roger Nichols, Elliot Scheiner, Al Schmitt, Bill Schnee and assisted by our good buddy Lenise Bent), Rufus Du Sol "Bloom"

SUBJECTIVITY IN THE AUDIO WORLD

So, here's a tough one … You spend all of this time, blood, sweat and tears on a project, and then you listen to it on a substandard system, or on your studio system on a different day when you're not in your best of moods, or almost any variable that you could possibly think of … and your recently released baby just doesn't sound right. This has happened to anyone who's a music or audio creator.

Also, you might listen to something that you absolutely love, and the person next to you doesn't enjoy it at all … or worse. It's called subjectivity, baby! Not everybody has the same sense of taste in music, and that's a good thing. Now, add to that the fact that almost all speakers, production rooms and toys will change the sound in a way that can alter how the music is perceived. Again, for creators, this is the bane of our existence. Truly, the best that we can ever hope for is to produce a product or work of art that makes us happy, hopefully makes others happy and is produced in such a way that it comes across in a pleasing way, over a wide range of reproduction systems. If you've accomplished that goal, then you've succeeded in your personal mission … Then comes the hard part … getting other people to also listen to it.

> Trust me, you'll never come across a sound device that's as subjective and variable as your own sense of hearing perception.
>
> **DMH**

ROOM, SPEAKERS AND OTHER IMPORTANT CONSIDERATIONS

Now that we've lightly touched base on our ears and that perceptive tool that rests between them, let's talk about some of the tips and tricks that can help you to maximize the way in which you and your ears integrate with your production system.

A "Trusted" Space

One of the most important aspects of a production space is to create an environment that you can trust … to know with some degree of certainty that what you are listening to will actually be what is being recorded, mixed, mastered, etc. Without this degree of trust, it's all too easy to be flying blind during the process … not a very nice prospect.

We've already touched upon the most primary aspect of the monitoring process; this relates to the fact that (as with all things artistic) it is very subjective in nature. What might be the right type of speakers for one person might not be right for the next. One person might place a subwoofer into the playback system to boost or make the bass sound more present, while another might cringe at this thought. Some might go for a set of super, high-end accurate audio file–type speakers, while others might go for bookshelves that don't sound that "pretty" but translate well over a wide range of playback speakers. It's the old "you say tom-eh-to and I say tom-ah-to" type of thing. There is no right and no wrong. There is only what is working for you at this point in your journey (you'll find that your tastes and monitoring styles might well change over time … I know that they have for DMH).

Now that we can move past the personal taste argument, the next part of the equation relates to the actual acoustics of the playback environment. Although variations between production rooms often play a huge role in giving a facility its own particular sound, extreme variations in a room's frequency response can lead to difficulties that can definitely be heard in the final product. For this reason, certain basic principles (which are covered in the Control Room Design section in Chapter 3) have become common knowledge to many who attempt the art of project studio and control room design. A few examples include:

- Reducing standing waves to help lessen erratic frequency response characteristics within a room
- Reducing excessive bass buildup in room corners through the use of bass traps
- Keeping the room's acoustic layout symmetrical so that the left/right and front/rear imaging is as consistent as possible
- Using a careful balance of absorptive, reflective and diffusion surfaces to help "shape" a room's sonic character

Fortunately for us, the basic understanding of how project, production and mixdown rooms can best be designed and/or acoustically adjusted has greatly improved over the last several decades. This is largely due to the increased availability of quality acoustical products and a better understanding of general acoustics and how its effects can help shape a room's sound.

Beyond careful acoustic design and construction, a professional or project space "might" choose to further reduce variations in frequency response by tuning (equalizing) its speakers to the room's acoustics so that the adjusted frequency response curve will be as flat as reasonably possible and therefore, reasonably compatible with most other control rooms. Tuning a speaker system to a room can be carried out in one of several ways:

- Altering the settings on the speaker itself
- Using external equalizers (or equalization [EQ] software) to smooth out the monitor output lines
- Using a speaker system that can "self-tune" to match its response to the room

One of the simplest ways to alter the acoustic and frequency response of a speaker system is through the careful control of the basic EQ and system setting controls that are found on most actively powered speaker systems (Figure 16.2).

(a)

(b)

FIGURE 16.2

Rear speaker controls. (a) Mackie HR824mk2 active monitor speaker (courtesy of Loud Technologies, Inc., www.mackie.com). (b) JBL Series 3 Mk2 (courtesy of JBL Professional, a Division of Harmon, www.jblpro.com).

These simple controls let the user roughly match level and EQ settings to best fit their application or placement layout. Often, these settings can be used to:

- Finely match audio balance levels within a stereo and surround system
- Allow basic high- and low-end tuning
- Partially compensate for bass buildup (whenever speakers are placed in or near a corner or other large boundary)
- Offer various speaker "emulation" modes

Larger, passive monitors (often a far-field pair) can be tuned by placing a 1/3-octave bandwidth graphic equalizer between each of the console's control room monitor outputs and the power amplifier. Of course, there are various ways to fine-tune a speaker system and room response to improve a studio's overall monitoring conditions. The simplest approach is to place a high-quality omnidirectional mic at the center listening position and insert it into a channel strip on your digital audio workstation (DAW). By recording a loop of pink noise (search Wikipedia for the "colors of noise") and playing it back equally to each speaker in the system, basic level matching measurements can be carefully taken. Frequency measurements can be taken through equipment that measures sweeps from 20 Hz to 20 kHz that are picked up by a calibration microphone.

Stand-alone software is available to help with measuring such variables as level, EQ and time delay reflections in the control room monitor chain. If you have the correct measurement and hardware tools – and know what you're doing or are carefully researching your steps – you can proceed to analize your room with caution. If not, you might seek out experienced, professional help. Keep in mind that these measurements are often best interpreted by those who are well versed in acoustics, studio design and the fine art of common sense – in other words; careful fine-tuning might be best left to a competent professional or someone who has a professional understanding of acoustics and design. In addition, certain pro and consumer audio speaker systems ship with a measurement mic and built-in soft-/hardware that can automatically analyze and tune the system's response to the room. Such an approach might or might not be an answer to your room's problems.

In the end, it's always wise to start by correcting any problems that exist with a space directly by tackling the acoustical and layout problems that you are facing first. If the room is improperly laid out (for symmetry, improper reflections, absorption, bass buildups at wall boundaries), no amount of automated correction will truly help. However, once you've tackled the room's basic acoustical nightmares, then you can go about the task of making smaller (but carefully made) setting adjustments. Also, keep in mind that these corrections and "Ah-HA! insights" don't often happen overnight … It is usually an ongoing process that improves as your experience and knowledge of your room grows – as always, patience is a virtue.

Monitor Speaker Types

Since the buying public will be listening to your mixes over a very wide range of speaker types under an infinite number of listening conditions, it's often wise to listen to a mix over several standardized speaker types during a recording and/or mixdown session. Quite often, a console or monitor control system will let you select between speaker/monitor types, with each set commonly having its own associated controls for level matching. These types generally include far-field, near-field, small-speaker and headphone monitoring systems.

FAR-FIELD MONITORING

Far-field monitors (Figure 16.3) often involve large, multi-driver loudspeakers that are capable of delivering relatively accurate sound at moderate to high volume levels. Because of their large size and basic design, the enclosures are generally soffit mounted (built into the control room wall to reduce reflections around and behind the enclosure and to increase overall speaker efficiency. An introduction to soffit design and construction can be found in Chapter 3). These large-driver systems are sometimes used during the recording phase because of their ability to safely handle high volume levels (which can come in handy should a microphone drop or a vocalist decide to be cute and scream into an open mic … ouch).

For obvious reasons, this monitor type has never caught on in project studios and has lost popularity to the easier-to-use and less expensive near-field bookshelf system, which is common in most professional studios as well. That's not to say that they can't be found in many studios as a rock-'n'-roll, hip-hop, "how does it sound loud?" reference. Of course, it's important that we all stay aware of the dangers that can come with long-term exposure to such sound levels –not to

FIGURE 16.3

Far-field monitor speakers. (a) Genelec 1236 main reference monitor (courtesy of Genelec Inc., www.genelec.com). (b) PMC QB1-A active monitor speakers (courtesy of PMC Ltd., www.pmc-speakers.com). (c) ATC SCM45A Pro active monitor speakers (courtesy of ATC Loudspeaker Technology Ltd., www.atc.audio).

(a)

(b)

(c)

mention the problems that can come with listening to mixes that were mixed at high levels but played back at moderate-to-low levels (more on this later).

NEAR-FIELD MONITORING

Although far-field monitors are useful for listening at high levels, few home systems are equipped with speakers that can deliver "clean" sound at such high sound pressure levels (SPLs). For this reason, most professional and project studios use near-field monitors that more realistically represent the type of listening environment that John H. and Jill Q. Public will most likely have.

The term *near-field* refers to the placement of small to medium-sized bookshelf speakers on each side of a desktop working environment or on (or slightly behind) the metering bridge of a production console. These speakers (Figure 16.4) are generally placed at closer working distances, allowing us to hear more of the direct sound and less of the room's overall acoustics.

In recent times, near-fields have become an accepted standard for monitoring in almost all areas that relate to audio production for the following reasons:

- Quality near-field monitors more accurately represent the sound that would be reproduced by the average home speaker system.
- The placement of these speakers at a position closer to the listening position reduces unwanted room reflections and resonances. In the case of an untuned room, this helps to create a more accurate monitoring environment.
- These moderate-sized speaker systems cost significantly less than their larger studio reference counterparts (not to mention the reduced amplifier cost because less wattage is needed).

One other consideration that should be taken into account when using near-field speakers on a table, desk or console top is the natural resonances that can be transmitted from the speaker onto the desktop. These resonances can add a distinct coloration and alter a speaker's overall sound. For this reason, several

FIGURE 16.4
Near-field monitor speakers. (a) Yamaha HS8 studio monitor speakers (courtesy of Yamaha Corporation of America, www.yamaha.com). (b) JBL Series 3 Mk2 Active Studio Monitor (courtesy of JBL Professional, a Division of Harmon, www.jblpro.com). (c) Ocean Way Audio HR5 active near-field monitors (courtesy of Ocean Way Audio, www.oceanwayaudio .com).

(a)

(b)

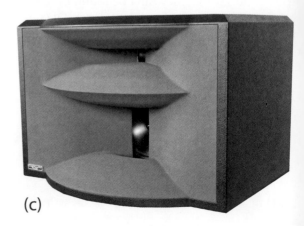

(c)

companies have begun to develop foam and other material isolation pads (Figure 16.5) that act to decouple the speakers from the surface, thereby reducing resonances. I would like to add that there are other options for decoupling speakers from working surfaces. For example, I actually hang my speakers (using standard plumbing pipe fixtures and carefully designed brackets) from the ceiling. This has the advantage of eliminating surface resonances, frees up tons of workspace and can look really amazing if you're careful about it.

As with any type of speaker system, near-fields vary widely in both construction and fundamental design philosophy. It almost goes without saying that extreme care should be taken when choosing the speaker system that best fits your production needs and personal tastes – speakers are possibly the most important tools in your production room (other than you).

SMALL SPEAKERS

Because radio, television, computer and cell phone playback are huge market forces in audio production and are key in the distribution and sales of recordings, it's often good to monitor your final mix through a small, inexpensive speaker set (Figure 16.6). These mimic the nonlinearities, distortion and poor bass response of these media (although the general specs of such speakers continue to greatly improve over the years). Making active decisions at low levels over small speakers will often reveal shortcomings within a mix that can actually help improve the mix's low-end response and overall presence, and you know, having your mix stand out over grocery store speakers or on your laptop can help the bottom line.

Before listening to a mix over such small speakers or over laptop speakers, it's often a smart idea to take a break in order to allow your ears and your brain to recover from the prolonged exposure of listening to higher sound levels over larger speakers. Taking the time to regain your perspective is definitely worth it.

FIGURE 16.5
Speaker isolation pads can help to reduce speaker/stand resonances. (a) Auralex MoPAD Monitor Speaker Isolation Pads (courtesy of Auralex Acoustics, www.auralex.com). (b) Primacoustic Recoil Stabilizer pad (courtesy of Primacoustic, www.prima-coustic.com). (c) IsoAcoustics ISO-L8R155 stands (courtesy of IsoAcoustics Inc., www.isoacoustics.com).

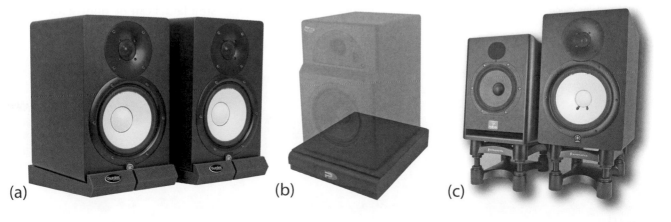

(a)   (b)   (c)

(a)

(b)

(c)

FIGURE 16.6

Small-speaker systems.
(a) Bose SoundLink®
speaker III (courtesy of Bose
Corporation, www.bose.com).
(b) Avantone Active MixCubes
Full-Range Mini Reference
Monitors (courtesy of
Avantone Pro, www.avanto-
nepro.com). (c) Even a laptop
can serve as a "reference"
(courtesy of Lenovo, www.
lenovo.com).

HEADPHONES

Headphones (Figure 16.7) are also an important monitoring tool, as they remove you from the room's acoustic environment. Headphones offer excellent spatial positioning in that they let the artist, engineer or producer place a sound source at critical positions within the stereo field without reflections or other environmental interference from the room. Because they're portable, you can take your favorite headphones with you to quickly and easily check out a mix in an unfamiliar environment.

It should be noted that while headphones eliminate the acoustics of a room from the monitoring situation, they don't always give a true representation of how sounds will behave through loudspeakers. Monitoring through headphones might emphasize low-level sounds like reverb and other effects more than loudspeakers in a room. As a result, listening to a mix over both monitor types is usually a good idea.

Headphone and Monitor Environment Simulation Plug-ins

FIGURE 16.7

Listening to the mix over
headphones can give you
a different, and sometimes
more accurate, perspec-
tive (especially if you're not
familiar with the room).

There are a number of plug-ins currently on the market that are able to help with the equalization, spatialization and psychoacoustic aspects of improving how a pair of headphones might sound. By applying inverse EQ and/or binaural spatialization, a wide range of headphones can be altered to sound flatter (more acoustically accurate) and even simulate the overall stereo and immersive environment of a professional mix room or concert hall (Figure 16.8). Quite surprisingly, these frequency-response and room simulators can be quite good

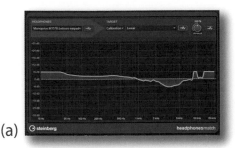
(a)

(b)

(c)

in providing an alternative to one's own monitoring environment or helping to have the most accurate environment when on the road. Further info on these acoustic room simulation plug-ins can be found in Chapter 21 on Immersive Audio.

FIGURE 16.8
Headphone and monitor environment simulation plug-ins. (a) Steinberg Headphonematch (courtesy of Steinberg Media Technologies GmbH, www .steinberg.net). (b) Dear VR Monitor acoustic simulator and (c) Dear VR Monitor headphone compensation (courtesy of Dear Reality, www.dear-reality.com).

EARBUDS

Almost certainly, the most listened-to media speaker device in the world is the earbud. They're portable, they're inexpensive (well, at least most of them are), and they are a relative standard that will usually guarantee a certain level of audio quality. Of course, it almost goes without saying that earbuds (like wireless buds or standard Apple buds) can come in handy in helping to determine how a project will sound to the masses.

IN-EAR MONITORING

One step beyond the concept of either headphones or earbuds is the concept of the professional in-ear monitor. These carefully tuned devices can range from being a generic design that can fit anyone all the way to those that are highly customized and personalized (Figure 16.9). The latter are molded to precisely fit the customer's ear and come in a variety of driver configurations that offer a frequency-response curve that matches the user's preferences and needs. Such a device can also greatly reduce ambient and stage noise, allowing them to be used on stage for professional monitoring purposes. These high-quality, on-stage or on-the-go devices can range into the thousands of dollars, contain 12 or more drivers for each ear, and be custom fitted and acoustically tailored to work only

FIGURE 16.9
Professional in-ear monitors in an on-stage setting (courtesy of 64 Audio, www. 64audio.com).

with your ears. I know of one world-class mastering engineer who uses in-ear monitors to master platinum artists on a regular basis.

YOUR CAR

Last, but not least, your (or your friend's) car could be a big help in determining how a mix will sound in another of modern society's most popular listening environments. You might take your mix out for a spin on a basic car system as well as a supped-up, window-shakin' bass bomb (Figure 16.10).

ALTERNATE ENVIRONMENTS

For all these monitoring types, it's almost always a good idea to take your recent mixes for a spin in another environment. Taking it over to a friend's studio can be, if not fun, hopefully helpful. Listening to the mix over your cell phone/ handy/mobile (both over buds and through its internal tinny speakers) can give helpful cues. However, we're still going to refer you back to your own mix room, as it is the environment that you're most used to.

In addition to these environments, there is one other trick that can really help when you're trying to decide if the vocals or any instruments are too high or low. The idea here is to keep your control room or project door open and simply step outside the room while listening to the playback. By getting some literal distance from your speakers, you'll be surprised how much easier it is to make a simple mix decision that otherwise might drive you nuts. Go ahead, try it out.

Spectral Reference

In addition to our best set of tools – our experience, judgment and ears – a visual tool known as a spectral analyzer can be somewhat useful as a visual cue to an audio program's overall frequency balance at any point in time. These hardware devices or (more likely) software or plug-in applications give a graphic readout of a signal's level at various frequencies throughout the audible band (Figure 16.11). Obviously, such a tool can help an engineer or producer to zero in on an offending or deficient frequency and/or bandwidth simply by looking

FIGURE 16.10
Kickin' it in the ride – preferably on Sunset Boulevard with the sunroof open!

(a)
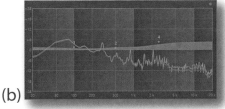
(b)
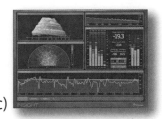
(c)

at the display over time. During both the record and mix phases, these tools can help point out and avoid potential spectral problems.

FIGURE 16.11
Spectral analyzer displays. (a) Within the Cubase/Nuendo EQ window. (b) Supervision (courtesy of Steinberg Media Technologies GmbH, www. steinberg.net). (c) iZotope's Insight Essential Metering Suite (courtesy of iZotope Inc., www.izotope.com).

Speaker Design

Just as the sound of a speaker system will vary when heard in different acoustic environments, speakers of different designs and operating types will usually sound very different from one another, even when heard in the same room. Enclosure size, number of components and driver size, crossover frequencies and design philosophy contribute greatly to these differences in sound quality.

Monitor speakers are available with high-frequency drivers that are made of various types of hard and soft plastic domes, metal domes, and even ones made with corrugated metal or plastic ribbon materials. Bass drivers can also be made from various materials, having various sizes that best match the enclosure size. Enclosures can incorporate an air suspension (an airtight system that seals the air in its interior from the outside environment) or a bass reflex design (which uses a tuned bass porthole that's designed into the front or rear of the speaker enclosure). Air suspension is often used in smaller "bookshelf" designs, often producing a strong, "tight" bass response that rolls off at the extreme low end. Bass reflex vented designs, on the other hand, allow the air mass inside the enclosure to mix freely with the outside air in such a way as to act as a tuned resonator (which serves to acoustically boost the speaker's output at the extreme lower octaves).

Because there are so many variables in speaker and amplifier design, it quickly becomes clear that there's no such thing as the "perfect" monitor system. The final choice is often more a matter of personal taste and current marketing trends than one of subjective measurements. Monitors that are widely favored over a long period of time tend to become regarded as the industry standard; however, this can easily change as preferences vary. Again, the best judge of what works best for you should be your own ears and personal sense of style. The overall goal is to:

- Choose the best speaker type, make and sound that best matches your way of working
- Match it as best you can to the acoustics of the room
- Become accustomed to the sound, and use that reference to make the highest-quality productions that you can

ACTIVE POWERED VERSUS PASSIVE SPEAKER DESIGN

As you might expect, most of the more popular monitor types that are in use today incorporate an actively powered amplifier into their design. These cost-effective systems have become widely accepted by the professionals and project communities due to their:

- Compact design
- High-quality sound (often, these systems use bi- or tri-amplified circuits)
- Expandability (additional speakers can be cost-effectively added for immersive monitoring)
- Cost-effectiveness and simplicity (as there's no need for an external power amplifier)

For these reasons, these systems are often ideal for project- and DAW-based facilities and are steadily increasing in popularity.

Electronic crossover networks (Figure 16.12), called active crossovers, use complex analog and digital circuitry to split the incoming line-level audio signal into various frequency bands. Each equalized signal is then fed to its own power amp, which in turn, is used to drive the respective bass, mid- and/or high-driver elements. Such a system is generally referred to as being bi-amplified or tri-amplified (depending on the number of crossovers and power amps that are required by the design). These systems have several advantages:

- The crossover signals are low in level, meaning that inductors (which can introduce audible ringing and intermodulation distortion) can be eliminated from the design.
- Power losses (due to the inductive resistance within the passive crossover network) can also be eliminated.
- Each amp is band-limited (meaning that each speaker will only need to output frequencies for which it was designed – highs to the tweeter, lows to the woofer, etc.), thereby increasing the speaker's overall design and power efficiency.

FIGURE 16.12

Example of a passive two-way crossover system, showing a crossover frequency of 1500 Hz.

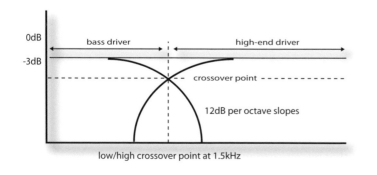

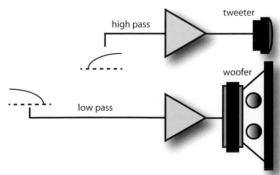

SPEAKER POLARITY

A common oversight that can drastically affect the sound of a multi-speaker system is for the cables to be wired out of phase with respect to each other. Speaker polarity is said to be electrically in phase whenever a signal that's equally applied to both speakers causes their cones to properly move in the same direction (either positively or negatively). When the speakers are wired out of phase, one speaker cone will move in one direction, while the other will move in the opposite direction, causing an incoherent signal that has poor center imaging at the listener's position – not a good thing.

Speaker polarities can be easily tested by applying a mono signal to both or all of the speakers at the same level. If the signal's image appears to originate from a single, localized point directly between the speakers, they have been properly wired (in phase). If the image is hard to locate and appears to originate beyond the outer boundaries of a stereo speaker pair, or shifts as the listener moves his or her head, it's a good bet that the speakers have been improperly wired (out of phase).

An out-of-phase speaker condition can be easily corrected by checking the speaker wire polarities. Using powered speakers, you can check the polarity of the cable using a polarity tester box, or you could get out a volt-ohm meter (every production room should have one) and test that the positive and neutral leads are wired properly. On a passive speaker design, the "hot" lead (+ or red post) leading from each amp channel should be secured to the same lead on its respective speaker/amp connection. Likewise, the negative lead (– or black post) should be connected to its respective lead for each speaker in the system.

MONITORING

When mixing, it's important that the engineer be seated as closely as possible to the center of the sound field (making allowances for the producer, musicians and others who are also doing their best to be in the "sweet spot") and that all the speaker volumes are adjusted equally. For example, if the engineer is closer to one speaker than another, that speaker will sound louder, and the engineer may be tempted either to pan the instruments toward the far speaker or to boost that entire side of the mix to equalize the volumes. The resulting mix would sound properly centered when played in that room, but in another environment, the mix might be off-center.

Here are a few additional pointers that can help you get the best sound from your control room monitors:

- Keep all room boundaries and reflections in the room as symmetrical as possible along the L/R and front/back axis of the mixing sound field.
- Keep large, direct reflections to a minimum within the room. This can be done by introducing absorptive (25–30%) and diffusive (25–30%) materials into the room. These will help reduce reflections to a level that's at least 20 dB down from the direct signal.

- It's often wise to place absorptive materials behind the speakers (in the front section of the room). This reduces reflections that emanate from the rear of the speakers and keeps them from reflecting back to the listener's position.
- Diffusion is often helpful at the sides and especially in the rear of the room to help break up reflections that will return to the listener's position.
- Angle the monitors symmetrically toward the listening position in both the horizontal and vertical planes.
- Whenever near-field monitors are used, you might consider placing them on medium-density foam blocks to reduce console- and desk-borne vibrations.

A sizable number of the world's top mixing engineers recommend the following:

- Balance your speaker levels, whether the system is stereo or immersive.
- Monitor at low levels (commonly at 75–80 dB). This can have a positive impact on how your mixes will sound. This is especially true for monitoring and room systems that are acoustically compromised, as monitoring at lower levels will reduce the effect that room reflections will have on the overall sound.
- Listen to a mix from another room (i.e., just outside the control room door). This can actually give you an alternative viewpoint as to how the mix will sound. Literally stepping out and away from the mix can be a huge help toward giving you a new understanding of how it sounds.

Balancing Speaker Levels

Another important factor that is often overlooked (but should always be taken into consideration) is the need for balancing your speaker levels so that monitoring and playback occur at equal and matched levels at the listening position. If the speaker level settings (at the passive or active power amp) are improperly set, your final mix levels could end up being off-balance – meaning that your center and overall imaging might lean to the left or right. This will result in a final mix that, if not caught in the mastering phase, will be off-center. You'd be surprised how many times mastering engineers run into this problem, even with high-profile releases.

The best way to calibrate your speaker system is to:

- Record a minute or so of pink noise, and copy this same file into your workstation in a loop fashion onto as many channel tracks as you have speakers in your system (just in case you'll need to calibrate an immersive speakers setup).
- Place a sound level meter on a camera stand directly at the listener's position and height (you could use an SPL meter app on your phone or pad, or an inexpensive sound level meter that is bought online will also do the

trick, as it's always nice when the meter has a thread for mounting the unit directly onto a mic stand or camera tripod). Point the meter at the left speaker (Figure 16.13).

- Play the pink noise tracks one at a time (with each track being assigned to its associated speaker) at a level of about 80–85 dB^{SPL} (C weighted), and adjust the sensitivity and gain on both the meter and master monitor gain so that the meter reads at "0" on an analog meter or at a specified level on a digital readout.

- Swivel the meter so that it points to the right speaker. Then, set its level to match until all of the speakers are at equal level at the listening position.

- You're now ready to start mixing or listening, knowing that your system is properly level matched.

- I almost forgot to mention. Remember to "switch the SPL meter off" – I can't tell you how many times DMH has drained the battery this way.

Once done, you can finally be sure that the speaker levels are properly matched. Whenever changes are made at a later time, or if you're in doubt, just call up the noise "session" and repeat the setup test. It's always better to be safe than sorry.

MONITOR VOLUME

Before continuing, we'd like to revisit another important factor – volume. During the record and mixdown stage, it's important to keep in mind that the Fletcher–Munson curve (see Chapter 2) will always have a direct effect on the frequency balance of a mix. Because our ears perceive sound differently at various monitoring levels, our ears will easily perceive the extreme high and low frequencies in the mix when monitoring at loud levels (sounds good, doesn't it?). However, when the mix is played back at lower levels (such as over the radio, TV or computer), our ears will be much less sensitive to these frequencies, and the bass and extreme highs will probably be deficient, which can leave the mix sounding dull, distant and lifeless.

FIGURE 16.13
General setup for balancing your speaker levels.

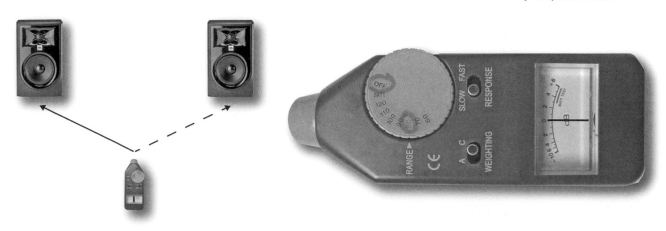

Unlike in earlier years, when excruciatingly high SPLs tended to be the rule in most studios, recent decades have seen the reduction of monitor levels to a more moderate 75 to 90 dB^{SPL} (a good rule of thumb is that if you have to shout (some would say even raise your voice) to communicate in a room, you're probably monitoring too loud). In fact, a number of the top mix engineers and producers (and an ever-growing number of individuals) are mixing top flight productions at levels that are quite low. The general idea behind this method is that if the project sounds really good at low levels, it will sound good at low-to-moderate levels (over laptops, TV, the Internet, etc.,) but when they are turned up, the Fletcher–Munson curve will boost levels on the highs and lows, making it sound even better.

One of the side effects that relate to prolonged exposure to sound (generally when monitoring at higher SPL levels) is ear fatigue. Listening for hours on end to a sound can literally tire both you and your ears out. When this happens, it's easy to start doubting what you're hearing, and basically, you can start to make subtle to not-so-subtle mistakes that you might not otherwise make. If it's at all possible, simply take the time to take a break, preferably outside in a lower-noise environment … get some sun, go for a short walk, ride your skateboard … anything that might be fun and take your mind off your work. Once you're back in the hot seat, you might be surprised at how much easier the decision-making process can be.

Along these lines, some of you might actually take a longer period of time off (if you have it) to literally get away from a mix. Quite often, DMH will actually mix a project, set it aside for a week and then come back to it for any final decisions that might need to be made. It's a way that one might use to gain a fresh perspective on a mix before printing the final version.

No matter what your own preferences for mixing are, it's important to be aware of potential problems with ear fatigue and hearing damage due to prolonged exposure to high SPLs. For more information on safe monitor levels and hearing conservation, contact the House Institute Foundation at www.hifla.org or H.E.A.R.® at www.hearnet.com.

MONITOR LEVEL CONTROL

As monitor systems grow to accommodate various production and playback formats, controlling the monitor level, trim adjustments and switching between speaker sets can become problematic. Many multiple-output consoles and high-end DAW systems are capable of handling the monitoring requirements of the various monitor formats (including multiple speaker selection and surround sound); however, even these can fall short when multiple sources, level trims and straightforward level control are taken into consideration (although newer interface designs are finally offering multichannel level control). For this reason, many have turned to using a high-quality interface or dedicated studio monitor management system in order to handle the monitoring needs of a professional or project studio.

In the case of a studio monitor controller (Figure 16.14a), a single input level is then distributed out to multiple monitor output chains that offer switching, level matching and muting to multiple speaker sets within the control room.

In many cases, a DAW will offer multiple outs for a stereo or immersive bus to multiple speaker sets (Figure 16.14b), enabling the same degree of control over output monitoring through the various outputs of a multi-output audio interface. Certain DAWs will also offer a monitoring chain that is completely independent from the main stereo or immersive output bus (as is the case with Cubase/Nuendo).

In addition, it might be important to consider whether or not the main volume control would be capable of acting as a ganged level control over all of the interface's outputs. Even though certain DAWs offer a main control room level that's independent of your main output bus, this often overlooked feature could definitely come back to bite you if your interface doesn't offer this feature (and truthfully, most interfaces don't).

Monitoring Configurations in the Studio

It's important to remember that a large percentage of your potential customers may first hear your mix over a TV, computer or AM/FM radio in mono. Therefore, if a recording sounds good in stereo but poor in mono, it might not sell as well, because it failed to take these media into account. The same might go for a surround-sound mix of a music video or feature release film in which proper attention wasn't paid to phase cancellation problems in mono and/or stereo (or vice versa). The moral of this story is simply this:

To prevent potential problems, a mix should be carefully checked in all its release formats in order to ensure that it sounds good and that no out-of-phase components are included that would cancel out instruments and potentially degrade the balance.

The most commonly accepted speaker configurations are mono and stereo (further reading on other immersive audio standards can be found in Chapter 19).

1.0 MONO

Even in this day and age, a number of people will first experience a mix in monaural (mono) sound (Figure 16.15a). That's to say, they'll hear your song on their phone, in an elevator, over a small radio or an iPad, etc. For this reason, it's often a good idea to listen to a mix with the stereo channels combined into mono for overall sound, phase and compatibility. The days of doing a special mono mix are pretty much behind us (given the fact that most people experience music over headphones or computer-related devices), but it's always wise to check for mono compatibility. Phase cancellations can cause instruments or

(a) (b)

FIGURE 16.14
Studio monitor management system types: (a) iD22 audio interface and monitoring system (courtesy of Audient Limited, www.audient.com). (b) Cubase/Nuendo "Control Room" monitor/control output (side and cut view).

frequencies in the spectrum to simply disappear whenever a mix is summed to mono. The best tools for reducing phase errors are good mic technique, a phase plug-in or display, and of course, your ears.

2.0 STEREO

Ever since the practical development of the 45°/45° record cutting process, stereophonic (stereo) sound has ruled the turntable. Of course, over the years, stereo has also grown to rule the computer, Internet, FM radio, the CD player, TV, auto and the cell phone. For these reasons, the creation of a quality stereo mix is extremely important with relation to L/R balance (Figure 16.15b), overall frequency balance, dynamics, depth and effects.

Some of the basic rules for monitoring in stereo are:

- Try to make your mixing environment as acoustically and physically symmetrical (within reason) in order to ensure that the L/R balance, effects balance and overall imagery are accurate within the stereo sound field.
- Balance your speaker levels carefully so that they match in level.
- Use your ears, be inventive and have fun!

2 + 1 (STEREO + SUB)

FIGURE 16.15
Basic monitoring configurations: (a) Mono. (b) Stereo.

With the advent of hip hop and other similar genres that rely heavily upon the BOOM and prominent bass in a project, many producers have begun to make use of a +1 subwoofer arrangement in the mix room.

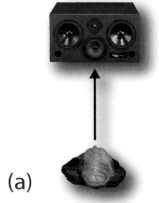

(a)

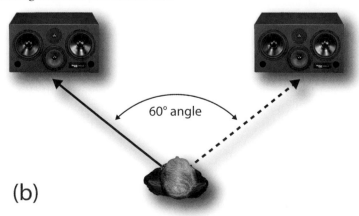

60° angle

(b)

In actuality, 2+1 isn't really a standard but makes use of a set of bandpass filters (often within the sub itself) in order to split the stereo bands into the lows (often extreme lows with a cutoff frequency between 80 and 160 Hz or even higher). The low-pass signal will be sent to the sub, while the high-pass band will be routed to the standard set of stereo speakers . This has the added advantage of taking the pressure of playing back the extreme bass signals through the standard speakers … often allowing the overall acoustic SPL level to be increased. Those of you who remember the Fletcher–Munsen curve in Chapter 2 might recall that this can be a good thing and it can be a bad thing. This is because our insensitivity to bass and highs at low levels means that a project mixed at extreme levels could sound thin and lifeless when played at low volume levels.

It's extremely important to keep in mind that the "+1" represents the addition of a powered sub to a set of stereo speakers and is **NOT an LFE** (.1) channel. Instead of providing an additional sonic boom "effect" that a .1 LFE is meant to add to a visual track, the subwoofer is actually used to extend and to help define the music's low end. A definition and better understanding of the LFE in an immersive system can be found in Chapter 20.

Extreme care should always be taken when setting up such a system, as a loud and improperly set-up sub and crossover combination can easily create a low end that can sound great but be extremely inaccurate. In such a setting, it would be easy to create a system that sounds totally badass in your room but might be inaccurate and actually result in mixes that are bass shy and/or boxy (not quite what you had in mind for your special mix).

The proper use and setup of a subwoofer can help extend the low end while adding a tight definition to the bass that can actually bring a monitor playback system to life. As you might expect, the response and overall sound of a sub are often heavily influenced by the acoustics of the playback environment. In addition to using a software frequency and acoustics analyzer (which is capable of displaying phase relationships as well as frequency response), one of the best ways to set up and tune a sub is through careful experimentation. When placing the subwoofer, the following concepts should be kept in mind:

- Because low-frequency audio is basically non-directional, some feel that a sub can be placed in any spot in the front of the room. However, a centrally placed position is most likely preferable, as a difference can be heard with upper-bass definition and image localization.
- Although a sub's output will be greater when placed near a wall boundary, a "muddy" or "boomy" sound might result due to unfavorable room and corner reflections.
- Most active subs will include level and crossover frequency adjustments. These settings should be carefully set up to closely match your sub to the chosen speaker set or sets.
- Active subs will often include a phase switch that can match the phase (driver motion) to your main speaker set. Using an analyzer or your ears, set the phase to a position that sounds best in your room.

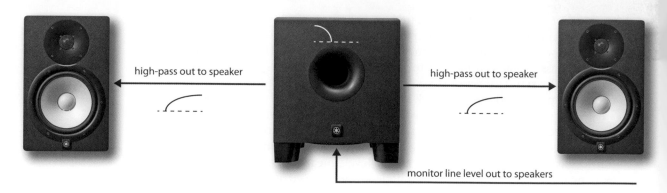

high-pass out to speaker

high-pass out to speaker

monitor line level out to speakers

FIGURE 16.16

Example of a 2+1 system that uses crossovers within the sub to split the band-limited signals going to the sub and monitor speakers.

Although it's not commonly known, there are two ways to set up a 2.1 system. Through careful adjustment and care, either method can work:

- Connect the stereo signal to the sub's inputs and use the sub's internal crossover outputs to feed a bass-limited signal to the main stereo speakers (Figure 16.16). This traditional system will band-limit the stereo speaker's low end and should be carefully tuned using the sub's crossover and level settings.
- The active sub (which usually has two inputs) can be combined with the stereo signal by using a simple "y" connection or summing network. This method will NOT band-limit the stereo speaker's low end and should be tuned with VERY special care, because interference and interactions between the bass drivers can definitely cause problems.

All of this simply leads to the fact that sub placement and setup are super-critical for getting the best sound possible – and it's worth mentioning again that imbalanced crossover and balance settings can cause really big problems for you, your room and your reputation.

CHAPTER 17

The Art and Technology of Recording

Before we delve into the tools, tricks, basics and foundation of one of the most critical aspects of audio production … the recording process, we'd like to rewind just a bit and take a moment to look at two important concepts that are central to music: sound, electronics and the art of sound recording. If any conceptual tools can help you to understand the technological and human underpinnings of sound, art and the process of recording, these are probably it!

THE FUNDAMENTAL BASICS

Before diving into the nuts, bolts and general aspects of the recording process, we thought that we'd like to touch on three aspects that are rather critical towards the process of becoming the best that you can at understanding and carrying out "the craft". These often unspoken aspects are:

- The Good Rule
- The concept of the transducer
- A good attitude
- Active versus passive listening

The "Good Rule"

As was stressed back in Chapter 4 on microphones, it can't be stressed enough that whenever possible, it's best to use the "Good Rule". This rule refers to the fact that a captured performance will only be as good as the performer, instrument, mic, mic placement, studio acoustics and the entire signal chain that follows it. If any of these elements falls short of its potential, the track will suffer accordingly. However, if all of these links are the best that they can be, the recording will almost always be something that you can be proud of!

DOI: 10.4324/9781003260530-17

The "Good Rule"

Good musician + good instrument + good performance + good acoustics + good mic + good placement = good sound.

The miking of vocals and instruments (both in the studio and on stage) is definitely an art form. It's often a balancing act to get the most out of the Good Rule. Sometimes, you'll have the best of all elements; at others, you'll have to work hard to make lemonade out of a situational lemon. The best rule of all is to use common sense, prepare for what's to come your way, trust your own instincts … and strive to always do your best.

The Transducer

One of the next steps toward getting the best sound in the studio is to know how the various stages in the process work. Of these, one of the most important insights that can help is to have a good insight into how electro-acoustic devices in the studio work and interact with their studio environments. In other words … It's a good idea to have at least a good understanding of the various transducers that exist within the pro and project studio, and how they work.

Quite simply, a transducer is any mechanical or electrical device that changes one form of energy into another corresponding form of energy. For example, a guitar is a transducer in that it takes the vibrations of picked or strummed strings (the medium), amplifies them through a body of wood, and converts these vibrations into corresponding sound-pressure waves, which are then perceived as sound (Figure 17.1).

A microphone is another example of a transducer. Here, sound-pressure waves (the medium) act on the mic's diaphragm and are converted into corresponding electrical voltages. The electrical signal from the microphone can then be amplified (not a process of transduction, because the medium stays in its electrical form) and fed to a recording device. A recorder is a device that changes electrical voltages into analogous magnetic flux signals on magnetic tape or into representative digital data that can be encoded onto tape, hard disk or other types of

FIGURE 17.1
The guitar and microphone as transducers.

Table 17.1	Media Used by Transducers in the Studio to Transfer Energy	
Transducer	**From**	**To**
Ear	Sound waves in air	Nerve impulses in the brain
Microphone	Sound waves in air	Electrical signals in wires
Analog record head	Electrical signals in wires	Magnetic flux on tape
Analog playback head	Magnetic flux on tape	Electrical signals in wire
Phonograph cartridge	Grooves cut in disk surface	Electrical signals in wires
Speaker	Electrical signals in wires	Sound waves in air

digital media. On playback, the stored magnetic signals or digital data are converted back to their original electrical form, amplified, mixed and then fed to a speaker system. The speakers convert the electrical signal back into a mechanical motion (by way of magnetic induction), which, in turn, recreates the original air-pressure variations that were picked up by the microphone, and … ta-da … we have a group of transducers that work together to produce sound!

As can be gathered from Table 17.1, transducers can be found practically everywhere in the audio environment, and they tend to have two overriding characteristics:

- It's interesting to note that transducers (and the media they use) tend to be the weakest link in any audio system chain. In general, this process of changing the energy in one medium into a corresponding form of energy in another medium can't be accomplished perfectly (although hi-res digital coding gets close). Noise, distortion and (often) coloration of the sound are introduced to some degree, and unfortunately, these effects can only be minimized, not eliminated.
- Differences in design are another major factor that can affect sound quality. Even a slight design variation between two microphones, speaker systems, digital audio converters, guitar pickups or any other transducer systems can cause them to sound quite different. This is due to the fact that they've been dreamt up, designed and built by humans, each having their own ideas of what materials should be used, what method of operation should be used and how it will sound. In the end, it's all about subjectivity.

These factors, combined with the complexity of music and acoustics, help make the field of recording the subjective and personal art form that it is … both artistically and technically.

It's also interesting to note that fewer transducers are used in an all or largely digital recording system. In this situation, the acoustic waveforms that are picked up by a microphone are converted into electrical signals and then quickly converted into digital form by an analog-to-digital (A/D) converter. The A/D converter changes these continuous electrical waveforms into corresponding discrete numeric values that represent the waveform's instantaneous, analogous voltage levels. Arguably, digital information has a distinct advantage over analog in that data can be transferred between electrical, magnetic and optical media with little or no degradation in quality. This is because the information is stored in its original, discrete binary form, and no transduction process is involved (i.e., only the medium changes, while the data representing the actual information stays in its original digital form). Does this mean that digital's better? Not necessarily. It's just another way of expressing sound through a medium, which in the end, is simply one of the many possible artistic and technological choices in the making and recording of sound and music. Beauty is, indeed, in the ear of the beholder.

A Good Attitude

Both EC and DMH have had the really good fortune of being able to call some of the greatest producers and engineers in the business, friends. What almost always stands out about these guys and gals is that they're really, really fun to be around. They're usually very open and happy to share their opinions and thoughts on almost anything that pops up, industry-wise or not. In fact, in our experience, the more accomplished the person is, the more likely it is that we wouldn't be talking about the process of recording at all. We'd be too busy talking about this and that, and simply having fun!

This is often by design. If you ask any of these "greats", they will tell you that a good attitude is one of the primary ingredients towards making a successful session happen. Not only does this mean a good professional attitude; it also means being actually interested in your fellow co-working musicians, producers, engineers, managers, bottle washers, etc. Many top producers and engineers, in addition to being really good at their jobs, will be very quick to tell you that keeping things light and fun goes a long way towards being called back for the next project.

Active Versus Passive Listening

In the twenty-first century, one of the things that we've gotten really good at is having media readily at hand. Quite often, these media are omnipresent and totally disposable. You walk into a store – there's music! A car passes you by – there can be loud music! You go practically anywhere … well, you know the deal all too well. My point is that we're all too used to being passive in the listening process. We've become really good at tuning sound out as background sounds or even noise.

The recording process, on the other hand, absolutely requires that we become an "active" listener. In short, we must heighten and tune our senses of hearing, tuning, artistry and psychology so as to be able to help translate the music or recorded sounds in the best way possible, using the technological tools of our trade.

THE RECORDING PROCESS

One of the most important insights to be gained towards understanding the recording process is the fact that there are no rules for the process of recording. This rule holds true insofar as inventiveness and freshness tend to play a major role in keeping the creative process of making music alive and exciting. However, there are definitely guidelines and procedures that can help you have a smooth-flowing, professional-sounding recording session, or at the very least, help you solve potential problems when used in conjunction with the following tools for guiding you toward a successful project.

Preparation

We've all heard the old adage "The devil's in the details". This points to the fact that it's important to remember that the hallmark of both a good production and a good production facility rests with the nitty-gritty small stuff that will help you rise above the crowd. The glory goes not so much to those who simply do the job, but to those who take the extra time to get the details of creating a quality project right (both technically and musically). OK, let's take some time to look at some of the details that can help your projects shine. Probably, the most important step toward ensuring that a recording project will become successful and marketable is careful preparation and planning (Figure 17.2).

By far, the biggest mistake that a musician or group can make is to go into the studio without practice and preparation, spend a lot of money and time, release their project on iTunes and YouTube, make a template website – then sit back, fully expecting an adoring audience to spring out of thin air! It ain't gonna happen! Beyond a good dose of business reality and added experience, these artists will have the dubious distinction of joining the throngs who have simply been passed over in the ever-present noise of the Web. Truth is, getting your "product" heard takes hard work, persistence, marketing and luck.

FIGURE 17.2
Workin' out the kinks beforehand in the practice space (courtesy of Yamaha Corporation of America, www.yamaha.com).

One of the most important aspects of the recording process occurs even before the artist and production team step into any type of studio. Questions like the following should be addressed long before you start in order to lay the foundation of a successful project:

- What is the overall goal?
- What is the budget?
- What are the estimated studio costs?
- Will there be enough time to work on tracking vocals, mixing and other important issues before running out of time or money?
- Is the group or artist practiced enough?
- Will there be a producer?
- Who will be the producer?
- If the project doesn't have a producer, who will speak for the group when the going gets rough (essentially making him/her the step-in producer)?
- Are the instruments and voices ready for the task ahead?
- Will additional musicians be needed?
- Will the project be released as physical media or only as a download?
- What is the preferred sample and bit rate for the project? If the project is being recorded for a client, this is a question that should be asked beforehand.
- How much will it cost to manufacture physical product(s), if any?
- Who will do the artwork, and what is that budget?
- What will the advertising costs be?
- Are there any legal issues to consider (an important question for any project)?
- How and when will the website be made, and up and running?
- How will the music be distributed and sold? And to whom?

These questions and a whole lot more will need to be addressed before it comes time to press the big red record button.

THE POWER OF PREPARATION

Whenever a producer is asked about the importance of preparation, more often than not, they'll most likely place preparation at or near the top of the list for capturing a project's sound, feel and performance. Having a plan (both musically and technically) to get the job done in a timely fashion, with the highest degree of artistic motivation and technical preparedness, will almost certainly get the project off to a good start. Even easily overlooked details like making sure that the instruments are tuned and in top form are also extremely important to take care of before the red light goes on. Making sure that the musicians are well practiced, relaxed and rested doesn't hurt either, and don't forget to have water, fruit and food on hand to keep everyone at their best. You'd be surprised how the little things can add to the success of a project – preparation, dude or dudette! From an artist's standpoint, here are a few pointers that might be addressed in the preparation phase:

- Create a "mission statement" for yourself/your group and the project. This can help clue your audience into what you are trying to communicate through your art and music (in general and on this project). This can greatly benefit your marketing goals. For example, you might want to answer such questions as these: Who am I/who are we? What am I/are we trying to communicate through our music? How should the finished project sound? What emotions should it evoke?

- Practice, practice and more practice – and while you're at it, you might want to record your practices to get used to the process (some of these tracks could be used on your website as bonus tracks, for the "making of" music videos … or they might even make it into the final master takes).

- Start working on the project's artwork, packaging and website ASAP. Do you want to tackle this yourself or hire a professional or a qualified friend who wants to help or could use some extra $$$?

- Copyright your songs. Government forms are readily available for the copyrighting of your music (identification and protection of intellectual property). The Library of Congress Form PA is used for the registration of "published or unpublished works of the performing arts". This class includes works prepared for the purpose of being "performed" directly before an audience or indirectly "by means of any device or process". Works of the performing arts include: (1) musical works and any accompanying words; (2) dramatic works, including any accompanying music; (3) pantomimes and choreographic works; and (4) motion pictures and other audiovisual works. In short, it is used to copyright a work that is intended for public performance or display. Form SR is used for the registration of published or unpublished sound recordings. This should be used when the copyright claim is limited to the sound recording itself, and it may also be used where the same copyright claimant is seeking simultaneous registration of the underlying musical, dramatic or literary work embodied in the "phonorecord". In other words, it's used to copyright the recording itself while also protecting the underlying performance that is recorded onto the media. These and other forms can be found at www.copyright.gov /forms or by searching the Library of Congress at www.loc.gov. Again, this might be a good time to discuss these matters with a music lawyer.

- Should you wish to use the services of a professional studio during the recording and/or mixdown phase, it's ALWAYS wise to take the time to check out several studios and available engineers in your area. Take the time to ask about others' experiences in that studio, and listen to tracks that have come out of it as well as those that have been recorded by the engineer. Finding out which studio best fits your style, budget and level of professionalism is both musically and financially an important decision – the time taken to find the best match for you and your music could be the difference between a happy and a potentially not-so-fun experience.

- Taking the time for the band, artist and producer to meet up with the engineering staff and to visit the studio beforehand can also help make the project successful. In this way, everyone can be as prepared as possible while having the added benefit of being familiar with each other and a bit more at ease when things get underway.

WHAT'S A PRODUCER, AND DO YOU REALLY NEED ONE?

One of the first steps that can help ensure the success of a project is to seek the advice and expertise of those who have extensive experience in their chosen fields. This might include seeking legal counsel for help and advice with legal matters or business and record label contacts. Another important "advisor" can come in the form of that all-important title, *producer*.

The producer of a project can fill one of two roles:

- The first type can be likened to a film director, in that his or her role is to be an artistic, psychological and technical guide who can help the band or artist reach their intended goals of obtaining the best possible song, album, remix, film score, etc. It's his or her job to stand back and objectively look at the big picture, and to help organize the various production and recording stages. Their role can also be to offer up suggestions as to how to shape and guide the performance, as well as to direct the artist or group in directions that will result in the best possible final product.
- The second type also encompasses the directorial role but might have the added responsibilities of being an executive producer. He or she would be charged with many of the business responsibilities of overall session budgeting, making arrangements for all studio and session activities, contracting (should additional musicians and arrangers be needed on the project), etc. This type of producer might join with a music lawyer to help find the best sales avenue for the artist and negotiate contact relations with potential record companies or distributors. If you find such a person who fits with your artistic vision, count yourself as being fortunate!

From this, you can see that a producer's role can be either limited or broad in scope. This role should be carefully discussed and agreed on long before any record button is pressed. The importance of finding a producer who can work best with your particular personalities, musical style and business/marketing needs can't be stressed enough, and finding the right match can be a rewarding experience. Here are a few tips to prepare you for the hunt:

- Will the producer be a member of the group? (If there's to be no producer on the project, it's often wise to pick a spokesman for the group who has the best production "chops".)
- Alternatively, will the production be done as a group effort?
- In what way will the producer be paid (as an upfront payment and/or as a percentage of sales)?

Here are just a few of the questions to ask when searching for a producer:

- Does he or she openly discuss ideas and alternate paths that contribute to growth and better artistic expression?
- Is he or she a team player, or are the rules laid out in a dictator-like fashion?
- Does the producer know the difference between a creative endeavor and one that wastes time in the studio?
- Does he or she say "Why?" a lot more often than "Why not give it a try?"
- Check out the liner notes of groups or musicians that you love and admire. You never know – their producer just might be interested in taking you on!
- Find a local up-and-coming producer who might be right for your music. This could help fast-track your reputation.
- Talk with other groups, musicians or even label execs (if you can get in touch with one). They might be able to recommend someone.

Although many engineers have spent most of their lives with their ears wide open and have gained a great deal of musical, production and in-studio experience, it's generally not a good idea to always assume that the engineer will or can automatically fill the role of a producer. For starters, he or she will probably be unfamiliar with the group and their sound, or might not even appreciate or like their music (although the studio and engineer were hopefully chosen for their interest in the band/artist's particular style)! For these and other reasons, it's always best to seek out a producer who is familiar with you, your goals and your style (or is contacted early enough in the game that he or she has time to become familiar).

Going Into the Studio

Before beginning a recording session (possibly a week or more before), it's always good to mentally prepare yourself for what lies ahead by creating a basic checklist that can help answer:

- What type of instruments and equipment will be needed?
- What number and type of musicians/instruments are needed for the session?
- Will any particular miking technique (if any) be used, and where will they be placed?

The best way to do this is for you, your group and the producer (if there is one) to sit down with the engineer and discuss instrumentation, studio layout, musical styles and production techniques. This meeting lets everyone know what to expect during the session and lets everyone become familiar with the engineer, studio and staff. This is always time well spent, as it will invariably come in handy during the studio setup and will help get the session off to a good start. The following tips can also be immensely valuable:

- Record your songs during live gigs or rehearsals. It doesn't matter if you record them professionally or not; however, keep the "always press the record button" adage in mind. If the setup meets basic professional standards, it's always possible to import all or part of a "magical" take into the final project.

- You might want to audition the session's song list before a live audience.
- If possible, work out all of the musical and vocal parts *before* going into the studio. Unrehearsed music can leave the session standing on shaky ground; however, leave yourself open to exploring new avenues and surprises that can be the lifeblood of a magical session.
- Try to leave plenty of time for laying down the final vocal tracks. Many a project has been compromised by spending too much time on "tracking" the basic instruments and then running short on time and money when it comes time to lay down the vocals. This almost always leads to increased tensions and a rushed vocal track. Don't let this happen to you, as vocals are often the central focus of a song (read this point again … yes, it's that important).
- Rehearse more songs than you plan to record. You never know which songs will end up sounding best or will have the strongest impact.
- Again, meet with the engineer beforehand. Take time for the producer and/ or group to get to know him or her, so you'll both know what to expect on the day of the session.
- Prepare and edit any sequenced, sampled or pre-recorded material beforehand. In short, be as prepared as possible with your musical instrument digital interface (MIDI) tools and toys.
- If it fits the musical style, try working to a metronome (click track) if timing is an issue. Not using a click track is also totally OK; just be aware of any timing shortfalls.
- Make sure that the instruments are in good working condition (i.e., bring new strings); also, make sure that the drums are in good condition and are in tune with the song. Forgetting this step can definitely lead to a great deal of lost time, or worse, a substandard sound.
- Create a checklist of all of the small but important details that can make or break a session: for example, extension cords, tuners, extra instrument cords, drum oil, drum tuning lugs, your favorite good luck charm, comfortable shorts or jammies – you name it!
- Take care of your body. Try to relax and get enough sleep before and during the session. Eat the foods that are best for you (you might bring some healthy food, fruit and plenty of liquids to keep your energy up). Be aware of your energy levels, so that low- or high-blood sugar problems don't become a factor (grrrrrrrr …).
- Don't fatigue your ears before a session; keep them rested and clear.

In short, it's always a good idea to plan out the session's technical and musical arrangement so as to most efficiently budget your time and well-being.

Again, it's always wise to confer with the producer and/or engineer about the way in which the musicians are to be tracked. Will the tracks be recorded in a traditional multitrack fashion (with the basic rhythm tracks being laid down first, followed by overdub and sweetening tracks and finishing with vocals), or will a different production style work best for your particular group's taste and organizational way of doing things? Communicating these details to all those involved in the session will help smooth the studio setup process and maintain a basic game plan that can help keep the session reasonably on track.

SETTING UP

Once the musicians have shown up at the studio, it's extremely important that all of the technical, musical and emotional preparation be put into practice in order to get the session off to a good start. Here are a few tips that can help:

- Show up at the studio on time or reasonably early. At some studios, the billing clock starts on time (whether you're there or not). Ask about their setup policies: is there another session before yours? Is there adequate setup time to get prepared? Are there any charges for setup? What is the studio's cancellation policy in case of illness or unforeseen things that could go wrong?
- Use new strings, chords, drumsticks and head – and bring spares. It's also a good idea to know the location, phone number and hours of a local music store, just in case.
- Tune up before the session starts, and stay in tune regularly thereafter.
- Don't use new or unfamiliar equipment (musical, hardware-wise and especially not software-wise). Taking the time to troubleshoot or become familiar with new equipment and software can cost you time and money. The frustration could even result in a lost vibe, or worse! If you must use a new toy or tool, it's best to become *very* familiar with it beforehand.
- Take the time to make the studio a comfortable place in which to work. You might want to adjust the lighting to match your mood ring, lay down a favorite rug and/or bean bag, turn on your lava love light or put your favorite stuffed toys on the furniture. Within reason, the place is actually yours to have fun with!

Tips 'n' Tricks

Acoustics play an essential role in music production and recording from both a technical and a creative point of view. For years, acousticians and music enthusiasts in general have been looking for ways to create exciting new environments in which to capture sounds, which has led to a variety of different styles of studio building and philosophy. Some of the most sought-after recording studios are very appealing for that very same reason: their acoustical properties for sound recording. Think of Capitol Studio B, East West Studio 1 or Ocean Way. All of them are unique and have a particular aesthetic.

With the advent of the project (home) studio, there have been several compromises due to the necessity of being confined to a fixed construction, and so on … but I like to think about them as creative challenges: in other words, the process of creating the sounds that you envision simply by using what you have around you. Creativity and experimentation are the name of the game. One evening at the studio of my dear friend Rafa Sardina, he and I were chatting over a glass of wine about what were his essentials when traveling for a recording project abroad. His answer really surprised me! I was expecting something like: I need this or that microphone, or that preamp … but one of the first things he shared with me is that he usually requests different boards of wood, of different types, dimensions and thicknesses, to create acoustical reflections around the recording environment.

Imagine that you're recording an acoustic guitar in a dull and dead environment. How can you best bring both the instrument and the room to life? First off, you listen. How is the sound interacting with the room? A good experiment would be to get a friend to start moving a wooden flat or large reflective surface around the musician while listening to the recording. You can then listen to the instrument and judge how you'd like your friend to move it to find the sweet spot for what you have in mind in terms of acoustical reflection and sound. You might experiment with this quick and easy technique to get the best sound. Maybe the floor has carpet all over? Try placing wood planks on top of it with different finishes, and as always: listen and follow your instinct

A WORD ABOUT ISOLATION

During the setup (as well as the planning stage), it's always wise to consider how leakage might play a factor in the upcoming recording process. For example, you might consider whether the entire group is to be recorded as a cohesive, acoustic group (as might be the case in a folk or classical session) or whether it is better to further isolate the instruments and vocals from each other. The latter might be the best approach should overdubs be needed, as leakage from the original instrument track into other live tracks in the studio would be a logistical nightmare to fix at a later stage. Questions like the following might be best to consider:

- Should the drums be isolated from the other instruments in order to best control any leakage?
- Should the piano, vocalist or any other instrument be isolated in order to control leakage?
- Are there any instruments or vocal tracks that might need to be overdubbed and/or fixed at a later time? Would leakage cause any problems in this case?

These and other questions like them should definitely be brought up during the planning stage (or at worst, during setup).

There are, of course, so many ways in which leakage and isolation can be controlled:

- Through microphone placement and directionality
- Through the use of acoustical control (live/dead surroundings)
- Through physical separation
- Through the use of iso-booths and/or iso-rooms that will physically isolate the artist from the other instruments

The last could be a good thing from a technical point of view but not so great from a connected, artistic stance. Care should always be taken that the use of isolation doesn't get in the way of the performance, by making sure that the artist can see others and can hear themselves and the others in the headphone mix. In short, it's always wise to check in with the artists to make sure that they are comfortable and able to do their best for the project.

Recording

Obviously, every recording project will be different. However, as always, there are general guidelines that can smooth the process of getting your or our band's music out as a final, marketable product.

As with most things, the road to a successful product is packed with potholes that can only be filled in with dedication, networking, hard work, seeking out the advice and/or help of those who are professionals or have successfully walked the path – and let's not forget the most important ingredients – talent and blind, dumb luck! Beyond these things, the use of the general guidelines outlined in this chapter can help keep you on the right path towards a successful project.

To begin, it's the engineer's job to help capture a project's sound to digital audio workstation (DAW) or tape during the recording phase. Recording the best possible sound (both musically and technically) with a clear vision as to the musical, technical and time needs will definitely help start the project off on the right track.

In the recording phase, one or more sound sources are picked up by a microphone or directly captured as an electrical signal, which is then recorded to a track or series of tracks. Because the recorded tracks are isolated from each other, any number of instruments can be recorded and re-recorded without affecting the other instruments (with a disk-based DAW offering an almost unlimited track count). Of course, the biggest advantage to having individual tracks is that they can be altered, added, combined and edited at any time and in any way to best suit the production.

By recording a single instrument to a dedicated track (or group of tracks), it's possible to vary the level, spatial positioning (such as left/right or immersive panning), equalization (EQ), and signal processing and routing later during mixdown without affecting the level or tonal qualities of other instruments that are being recorded (Figure 17.3). This isolation allows leakage from nearby instruments or mic pickups to be reduced to such an insignificant level that individual tracks can be manipulated, re-recorded and/or processed at a later time without affecting the overall mix.

The recording phase involves the physical process of capturing live or sequenced instruments onto a recorded medium (disk, tape or whatever). Logistically, this process can be carried out in a number of ways:

- All the instruments to be used in a song or concert can be recorded in a single live pass, either on the stage or in the studio.

FIGURE 17.3

Basic representation of how isolated sound sources can be recorded using a console.

- Musicians can be used to lay down the basic tracks (usually rhythm) of a song, thereby forming the basic foundation tracks, whereby additional instruments can be overdubbed and added to at a later time.
- Electronic or groove instruments can be performed or programmed into the sound file and/or MIDI sequenced tracks of a DAW in such a way as to build up the foundation of a song.

The first process in this list could involve the recording of a group of musicians as a single ensemble. This is often the case when recording a classical concert in a large hall, where the ensemble and the acoustics of the room are treated as a single entity, which is to be picked up by a microphone array. The recording of a jazz ensemble would often be recorded in a single, live pass; however, in this case, the mics will often be placed closer to the instruments so as to allow greater control in the mixdown or further production phase. Other styles, such as a rock ensemble, might also record in a single live pass, although this would most often be recorded using close-mic techniques so as to take advantage of a maximum amount of isolation during mixdown and further production.

When a recording is made "in the box" using a DAW, the mixdown process is often streamlined, since the tracks, mixer and effects are all integrated into the software system. In fact, much of the preparation has probably been long underway, as basic mix moves were made and saved into the session during the recording and overdub phases. When working "outside the box", the process can be quite different (depending upon the type of system that you're working with). Newer hardware console systems include facilities to control and communicate directly with the DAW recording system. This integration can be fluid and straightforward, with many or all of the console level, routing and effects settings being recalled directly onto the console's surface.

When approaching the production process from an electronic music standpoint, all bets are off, as there are literally so many possible ways that an artist can build a song, using loops, MIDI instruments and external hardware/software systems, that the process often becomes quite personal. This popular form of production can best be understood by having a better understanding of the DAW (Chapter 7), groove tools and techniques (Chapter 8) and obviously, MIDI (Chapter 9).

A DEEPER UNDERSTANDING OF A CONSOLE'S OR DAW'S RECORDING PATH

At this point, let's start our quest into the hardware side of mixing by taking a conceptual look at various functional stages within an analog console or mixing system. This approach is best, as it gives a set of block-by-block insights into both the hardware and virtual digital recording and mixing environment.

In a traditional hardware mixer (which also goes by the name of board, desk or console) design, the signal flow for each input travels vertically down a plug-in strip known as an I/O module (Figure 17.4) in a manner that generally flows in a logical progression from an input to an individual or combined output.:

Although the layout of a traditional analog hardware mixer generally doesn't match the graphical user interface (GUI) layout of a virtual DAW mixer, the signal flow will follow along the same or similar paths. Therefore, grasping the concept of an analog console's signal chain will also be extremely useful for grasping the general signal flow concept for smaller mixers and virtual DAW mixers. Each system type is built from numerous building-block components, having an input (source) that flows through the signal chain to an output (destination). The output of each source device must be literally or virtually connected to the input of the device that follows it, and so on until the end of the audio path is reached. Keeping this simple concept in mind is important when paths, plug-ins and virtual paths seem to meld together into a ball of confusion. When the going gets rough, slow down, take a deep breath, read the manual (if

FIGURE 17.4

General anatomies of input strips on an analog mixing console. (a) Mackie Onyx 4-bus (courtesy of Loud Technologies, Inc., www. mackie.com). (b) SSL AWS δelta large-format recording console/DAW controller (courtesy of Solid State Logic, www.solidstatelogic.com).

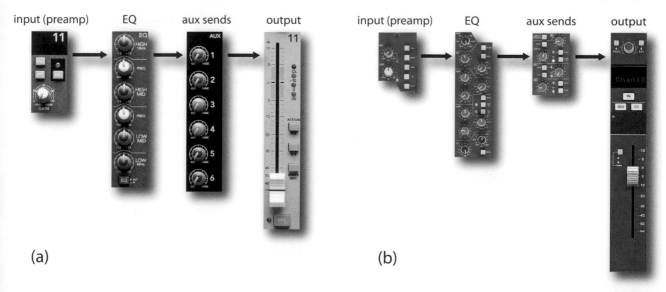

input (preamp)   EQ   aux sends   output   input (preamp)   EQ   aux sends   output

(a)   (b)

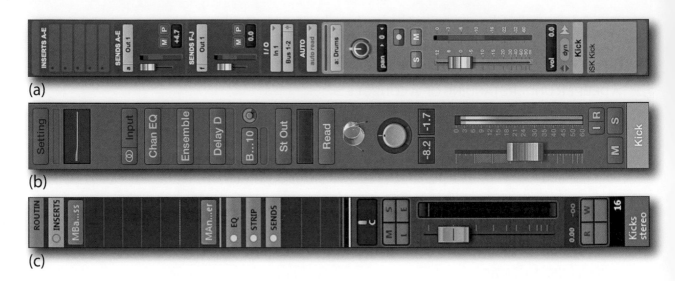

(a)

(b)

(c)

FIGURE 17.5
Virtual mixer strip layouts. (a)
Pro Tools (courtesy of Avid
Technology, Inc., www.avid
.com). (b) Logic (courtesy
of Apple Inc., www.apple
.com). (c) Steinberg's Cubase
and Nuendo virtual mixer
(courtesy of Steinberg Media
Technologies GmbH, a division of Yamaha Corporation,
www.steinberg.net).

you have the time and inclination) – and above all, be patient and keep your wits about you.

Figure 17.5 shows the general I/O stages of three virtual mixing systems. It's important that you take the time to familiarize yourself with the inner workings of your own DAW (or those that you might come in contact with) by reading the manual, pushing buttons, and by diving in and having fun with your own projects.

Gain Level Optimization

As we enter our discussion on console and mixer layouts, it's extremely important that we touch base on the concept of signal flow or gain level optimization.
In fact, the idea of optimizing levels as they pass from one device to another or
from one functional block in an input strip to the next is one of the more important concepts to be grasped in order to create a professional-quality recording.
Although it's possible to go into a great deal of math in this section, I feel that it's
far more important that you understand the underlying principles of level optimization, internalize them in everyday practice and let common sense be your
guide. For example, it's easy to see that if a mic that's plugged into an input strip
is overdriven to the point of distortion, the signal following down the entire
path will be distorted. By the same notion, driving the mic preamp at too low
a signal will require that it be excessively boosted at a later point in the chain,
resulting in increased noise. From this, it follows that the best course of action is
to optimize the signal levels at each point along the chain (regardless of whether
the signal path is within an input strip or pertains to input/output [I/O] levels as
they pass from one device to another throughout the studio).

So, now that we have a fundamental idea of how a hardware and software mixing system is laid out, let's discuss the various stages in greater detail as they

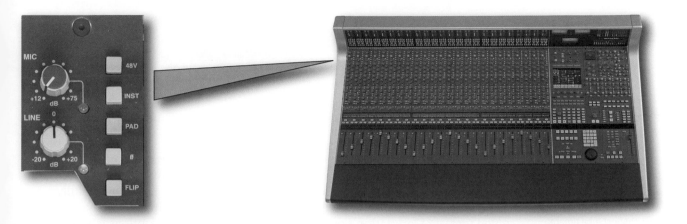

FIGURE 17.6
Channel input section of the Solid State Logic AWS δelta Console (courtesy of Solid State Logic, www.solidstate-logic.com).

flow through the process, starting with a channel's input, through the various processing and send stages, and then out to the final mix bus destination.

Channel Input (Preamp)

The first link in the input chain is the channel input (Figure 17.6). This serves as a preamp section to optimize the signal gain levels at the input of an I/O module before the signal is processed and routed. On a hardware console, mixer or audio interface that has built-in mic preamps, either a mic or a line input can be selected to be the signal source (Figure 17.7). Although these values vary between designs, mic trims are typically capable of boosting a signal over a range of +20 to +70 dB, while a line trim can be varied in gain over a range of –15 (15-dB pad) to +45 dB or more. Gain trims are a necessary component in the signal path, because the output level of a microphone is typically very low (–45 to –55 dB) and requires that a high-quality, low-noise amp be used to raise and/ or match the various mic levels in order for the signal to be passed throughout

FIGURE 17.7
Analog and DAW interface input sections (courtesy of Loud Technologies, Inc., www.mackie.com and Steinberg Media Technologies GmbH, www.steinberg.net).

analog input

DAW and interface inputs

(a)   (b)

FIGURE 17.8
Microphone preamps. (a) Presonus ADL-600 high-voltage tube preamp (courtesy of Presonus Audio Electronics, Inc., www.presonus.com). (b) Universal Audio 2–610S dual channel tube preamp (courtesy of Universal Audio, www.uaudio.com © 2022 Universal Audio, Inc. All rights reserved. Used with permission).

the console or DAW at an optimum level (as determined by the system's design and standard operating levels).

Of course, a mic preamp can take many forms. It might be integrated into a console or mixers input strip (as referred to earlier); it might exist as external hardware pres (pronounced "preeze", as shown in Figure 17.8) that are carefully chosen for their pristine or special sound character; or it might be directly integrated into our workstation's own audio interface. Any of these options is a valid way of boosting the mic's signal to a level that can be manipulated, monitored and/or recorded.

When asked about her recommendations for getting those vintage, quirky analog-type sounds, Sylvia Massy (Prince, Johnny Cash, System of a Down) said: "Try some different things, especially with the mic pres, the front end. The recorder is sorted out, but it's the front-end that's the challenge".

Whenever a mic or line signal is boosted to levels that cause the preamp's output to be overdriven, severe clipping distortion will almost certainly occur. To avoid the dreaded LED overload light, the input gain must be reduced (by simply turning down the gain trim or by inserting an attenuation pad into the circuit). Conversely, signals that are too low in level will unnecessarily add noise into the signal path. Finding the right levels is often a matter of knowing your equipment, watching the meter/overload displays and using your experience.

Input attenuation pads that are used to reduce a signal by a specific amount (e.g., –10 or –20 dB) may be inserted ahead of the preamp in order to prevent input overload. On many consoles, the preamp outputs may be phase-reversed via the "Ø" button. This is used to change the signal's phase by 180° in order to compensate for polarity problems in mic placement or in cable wiring. High- and low-pass filters may also follow the preamp, allowing extraneous signals such as amp/tape hiss or subsonic floor rumble to be filtered out.

From a practical standpoint, level adjustments usually begin at the mic preamp. If the pre has LED or other types of metering, setting your gain at a reasonable level (while being careful not to overload this important first major gain stage in the process) will generally get you off to a good start. Alternatively, you could set the gain on the main strip fader to 0 dB (unity gain). While monitoring levels for that channel or channel grouping, turn the mic preamp up until an acceptable gain is reached. Should the input overload LED light up, back off on the input level, and adjust the output gain structure accordingly. Care should be taken when inserting devices into the signal chain at a direct insert point, making sure that the in and out signals are also working at or near their optimum level. In addition, it's important to keep in mind that the EQ section can also cause level overload problems whenever a signal is overly boosted within a frequency range.

Insert Point for a Hardware Console/Mixer

Many mixer and certain audio interface designs provide a break in the signal chain that occurs after the channel input. A direct send/return or insert point (often referred to simply as direct or insert) can be used to send the strip's line-level audio signal out to an external gain or effects processing device. The external device's output signal can then be inserted back into the signal path, where it can be mixed back into the audio program. Access to an insert point on a hardware console or mixer can be found either at a marked set of jacks on the studio's patch bay or at the rear of some console/mixers themselves (Figure 17.9). It's important to note that plugging a signal processor into an insert point will only affect the audio on that channel.

Virtual DAW Insert Point

Within a workstation environment, inserts are extremely important in that they allow audio or MIDI processing/effects plug-ins to be directly inserted into the virtual path of that channel (Figure 17.10). Often, a workstation allows multiple plug-ins to be inserted into a channel in a stacked fashion, enabling complex and unique effects to be built up. Of course, keep in mind that the extensive use of insert plug-ins can eat up processing power. Should the stacking of multiple plug-ins become a drain on your CPU (something that can be monitored by watching your processor usage meter – aka "busy bar"), many DAWs allow the track to be frozen (committed), meaning that the total sum of the effects can be written to an audio file, allowing the track + effects to be played back without causing any undue strain on the CPU.

Auxiliary Send Section

In a hardware or virtual setting, the auxiliary (aux) sends are used to route and mix signals from multiple input strip channels to a single effect device/plug-in

FIGURE 17.9
Direct send/return signal paths. (a) Two jacks can be used to send signals to and return signals from an external device. (b) A single TRS (stereo) jack can be used to insert an external device into an input strip's path.

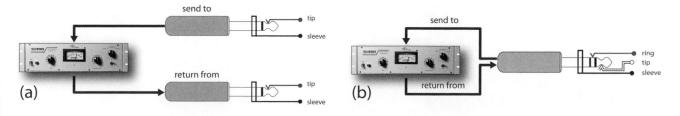

(a) send to — tip, sleeve / return from — tip, sleeve

(b) send to / return from — ring, tip, sleeve

FIGURE 17.10
An effects plug-in can be easily inserted into the virtual path of a DAW's channel strip (courtesy of Avid Technology, Inc., www.avid.com, Steinberg Media Technologies GmbH, www.steinberg.net, and Universal Audio, www.uaudio.com).

FIGURE 17.11
Although a hardware mixer's input path generally flows vertically from top to bottom, an aux send's path flows in a horizontal fashion, in that the various channel signals are mixed together to feed a mono or stereo send bus. The combined mix can then be sent to any device (courtesy of Solid State Logic, www.solidstatelogic.com).

and/or monitor/headphone cue send from within the console or DAW. This section is used to create a mono or stereo sub-mix that's derived from multiple console or DAW input signals and then "send" it to a signal processing, monitoring or recording destination (Figure 17.11).

It's not uncommon for six or more individual aux sends to be found on a hardware or virtual input strip. An auxiliary send can serve many purposes. For example, one send could be used to drive a reverb unit, signal processor, etc., while another could be used to drive a speaker that's placed in that great-sounding bathroom down the hall. A pair of sends (or a stereo send) could also be used to provide a headphone mix for several musicians in the studio, while another send could feed a separate mix to the drummer who's having a hard time hearing the lead guitar. From these and countless other situations, you can see how a send can be used for virtually any signal routing, effects processing and/or monitoring task that needs to be handled. How you make use of a send is up to you, your needs and your creativity.

Virtual DAW Send

With regard to a workstation, using an aux send is a great way to make use of processing effects while keeping the processing load on the CPU to a minimum (Figure 17.12). For example, let's say that we wanted to make wide use of a reverb plug-in that's generally known to be a CPU hog. Instead of separately plugging this reverb into a large number of tracks as inserts, we can greatly save

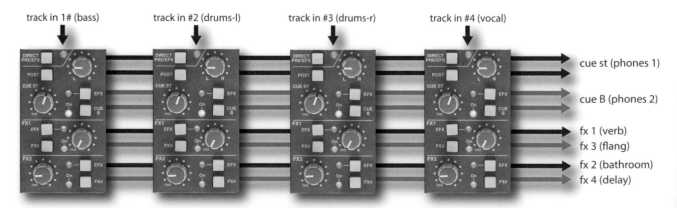

on processing power by plugging the reverb into an aux send bus. This lets us selectively route and mix audio signals from any number of tracks and then send the summed (mixed) signals to a single plug-in that can then be mixed back into the master output bus. In effect, we've cut down on our power requirements by routing any number of audio tracks to a single effects device. Knowing the functional difference between an insert and a send can be a powerful engineering tool.

EQUALIZATION

The most common form of signal processing is EQ. The audio equalizer (Figure 17.13) is a device or processing circuit that lets us control the relative amplitude of various frequencies within the audible bandwidth. Like the auxiliary sends, it derives its feed on a hardware console directly from the channel input section. In short, it exercises tonal control over the harmonic or timbral content of an input signal. EQ may need to be applied to a single recorded channel, to a group of channels or to an entire program (often as a step in the mastering process) for any number of other reasons, including:

- To correct for specific problems in an instrument or in the recorded sound (possibly to restore a sound to its natural tone)

FIGURE 17.12

An effects plug-in can be inserted into an effects send bus, allowing multiple channels to share the same effects processor (courtesy of Avid Technology, Inc., www .avid.com, Steinberg Media Technologies GmbH, www. steinberg.net and Universal Audio, www.uaudio.com).

FIGURE 17.13

Equalizer examples (courtesy of Solid State Logic, www.solidstatelogic .com, Steinberg Media Technologies GmbH, www. steinberg.net and Universal Audio, www.uaudio.com).

hardware EQ

DAW EQ

plug-in EQ

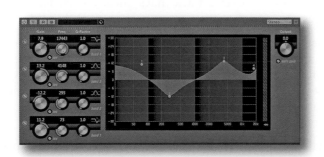

- To overcome deficiencies in the frequency response of a mic or in the sound of an instrument
- To allow contrasting sounds from several instruments or recorded tracks to better blend together in a mix
- To alter a sound purely for musical or creative reasons

When you get right down to it, EQ is all about compensating for deficiencies in a sound pickup, "shaping" the sound of an instrument so that it doesn't interfere with other instruments in a mix, or reducing extraneous sounds that make their way into a track. To start our discussion on how to apply EQ, let's take another look at the "Good Rule".

The "Good Rule"

Good musician + good instrument + good performance + good acoustics + good mic + good placement = good sound.

Let's say that at some point in the "good" chain, something falls short – like, a mic was placed in a bad spot for a particular instrument during a session that's still in progress. Using this example, we now have two options. We can change the mic position and overdub the track or re-record the entire song – or, we can decide to compensate by applying EQ. These choices represent an important philosophy that's held by many producers and engineers (including myself): whenever possible, EQ should NOT be used as a bandage to doctor a track or session after it's been completed. By this, it is meant that it's often a good idea to correct a problem on the spot during the recording (i.e., change the mic or mic position) rather than rely on the hope that you (or someone else) can fix it later in the mix using EQ and other corrective methods.

Although it's usually better to deal with problems as they occur, this simply isn't always possible. When a track needs fixing after it's already been recorded, EQ can be a good option when:

- There's no time, money or possibility to redo the track
- The existing take was simply magical, and too much feeling would be lost if the track were to be redone
- You have no control over a track that's already been recorded during a previous session, and the artists are touring on the other side of the planet

Whenever EQ is applied to a track, bus or signal, the whole idea is to take out the bad and leave the good. If the signal is excessively EQed, the signal will often degrade and lead to volumes that often creep up in level. Thus, it's often a good idea to use EQ to take away a deficiency in the signal and not simply boost the desirable part of the track. Just a few examples of using EQ to cut offensive sounds might include:

- Reducing the high end on a bass guitar instead of boosting its primary bass notes

- Using a peak filter to pull out the ring of a snare drum (a perfect example of a problem that should've been corrected during the session by dampening the drumhead)
- Pulling out a small portion of a vocalist's upper-mid range to reduce any problematic nasal sounds

Using EQ might or might not always be the best course of action. Just like life, use your best judgment – nothing's ever absolute. A complete explanation of equalization can be found in Chapter 15 (Signal Processing).

> **It's always a good idea to be patient with your "EQ style"** – especially when you're just starting out. Learning how to EQ properly (i.e., not over EQ) takes time and practice.

DYNAMICS SECTION

Many top-of-the-line analog consoles offer a dynamics section on each of their I/O modules (Figure 17.14), and of course, dynamic plug-ins are readily available for all DAWs. This allows individual signals to be dynamically processed more easily without the need to scrounge up tons of outboard devices. Often, a full complement of compression, limiting and expansion (including gating) is also provided. A complete explanation of dynamic control can be found in Chapter 15 (Signal Processing).

MONITOR SECTION

During the recording phase, since the audio signals are commonly recorded to DAW or tape at their optimum levels (without regard to the relative musical balance on other tracks), a means for creating a separate monitor mix in the control room is necessary in order to hear a musically balanced version of the production. Therefore, a separate monitor section (or aux bus) can be used to provide varying degrees of control over each input's level and possibly panning, effects, etc. This mix can be routed to the master control room volume control as well as monitor switching between various speaker sets and between mono,

FIGURE 17.14
Dynamics section of a Solid State Logic Duality Console (courtesy of Solid State Logic, www.solidstatelogic.com).

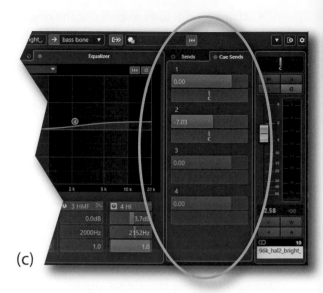

(a)   (b)   (c)

FIGURE 17.15

Monitor mix sections. (a)
Legacy model SSL XL9000K
monitor mix section (courtesy
of Solid State Logic, www
.solid-statelogic.com).
(b) Solid State Logic AWS
δelta main fader/monitor
mix section (courtesy of
Solid State Logic, www
.solidstatelogic.com). (c)
Software monitor "cue"
section (at the right side of
the channel strip) within the
Cubase/Nuendo virtual mixer
(courtesy of Steinberg Media
Technologies GmbH, www.
steinberg.net).

stereo or surround modes in the control room. The approach and techniques
for monitoring tracks during a recording will often vary from mixer to mixer (as
well as among individuals). For example, Figure 17.15a shows a monitor mix
level, pan and control section that is specifically dedicated to this function, while
Figure 17.15b shows a multifunction fader and virtual control that can be used
for main mix levels or monitor mix controls (and even combinations between
the two), depending upon the console's functional status. Figure 17.15c shows
one of the various examples of how monitor mix controls can be displayed on a
DAW. Again, no approach is right or wrong. It simply depends on what type of
equipment you're working with, and on your own personal working style.

Note that during the overdub and general production phases on a large-scale
console or DAW, this idea of using a separate monitor section can be easily
passed over in favor of mixing the tracks directly using the main faders in a
standard mixdown environment. This straightforward process helps us by setting up a rough mix all through the production phase, allowing us to finesse the
mix during production under automated recall. By the time the final mix rolls
around, many of your "mix as you go" levels and automation kinks might easily
have been worked out.

In-Line Monitoring

Many larger console and controller designs incorporate an I/O small fader section that can be used to directly feed its source signal to the monitor mix or
directly to the DAW or analog tape recorder (depending on its selected operating
mode). In the standard monitor mix mode (Figure 17.16a), the small fader is
used to adjust the monitor level for the associated recording track. In the flipped

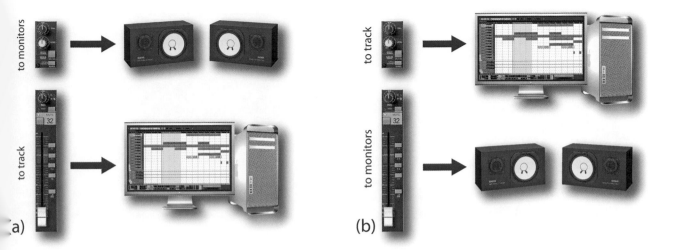

to monitors

to track

(a)

to track

to monitors

(b)

mode (Figure 17.16b), the small fader is used to control the signal level that's being sent to the recording device, while the larger, main fader is used to control the monitor mix levels. This useful function allows multitrack record levels (which aren't often changed during a session) to be located out of the way, while the more frequently used monitor levels are assigned to the larger, more accessible master faders.

FIGURE 17.16
Flipped-fader monitor modes:
(a) Standard monitor mode.
(b) Flipped monitor mode.

Direct Insert Monitoring

Another flavor of in-line monitoring that has gained favor over the years is known as direct insert monitoring. This method (which often makes the most sense on large-format consoles) makes use of the direct send/returns of each input strip to insert the recording device (such as a DAW) directly into the input strip's signal path. Using this approach (which is closely tied to the Insert Point section from earlier in the chapter):

- The insert send output for each associated channel (which can be inserted either before or after the EQ section) is routed to its associated track on a multitrack DAW or ATR.
- The insert return signal is then routed from the recording device's output back into the console's input strip return path, where it's injected back into the channel strip's effects send, pan and main fader path.

With this approach, the input signal directly following the mic/line preamp will be fed to the DAW or ATR input (with record levels being adjusted by the preamp's gain trim). The return path (from the DAW or ATR) is then fed back into the input strip's signal path so it can be mixed (along with volume, pan, effects, sends, etc.) without regard for the levels that are being recorded to tape, disk or other medium. This system greatly simplifies the process, since playing back the track won't affect the overall monitor mix at all, because the recorder's outputs are already being used to drive the console's monitor mix signals.

FIGURE 17.17
Older-style British consoles may have a separate monitor section, which is driven by the console's multitrack output and/or tape return buses (courtesy of Buttermilk Records, www.buttermilkrecords.com).

Separate Monitor Section

Certain British consoles (particularly those of older design) incorporate a separate mixing section that's dedicated specifically to the task of sending a mix to the monitor feed. Generally located on the console's right-hand side (Figure 17.17), the inputs to this section are driven by the console's multitrack output and tape return buses, and offer level, pan, effects and "foldback" (an older British word for headphone monitor control). During mixdown, this type of design has the distinct advantage of offering a large number of extra inputs that can be assigned to the main output buses for use with effects returns, electronic instrument inputs and so on. During a complex recording session, this monitoring approach will often require an extra amount of effort and concentration to avoid confusing the inputs that are being sent to tape or DAW with the corresponding return strips that are being used for monitoring. This is especially true when the channel and track numbers do not agree (which is probably why this design style has fallen out of favor in modern console designs).

Obviously, one of the more important parts of helping to make the musicians feel comfortable and at ease is to create an environment where they can easily hear themselves and each other. This truly can't be stressed enough, and as such, it's extremely important that time be taken to create a monitor mix that can be fully heard and is comfortable for any and all of the musicians involved. Later in this chapter, you'll find a section titled "Monitoring the Mix" that will outline this process in more detail. Paying attention to the potentially complex details of helping the artist(s) to better hear themselves and others will always be truly appreciated and will definitely pay off in the end.

CHANNEL FADER

Each input strip contains an associated channel fader, which determines the strip's bus output level (Figure 17.18) and pan pot setting, which is often designed into or near the fader and determines the signal's left/right placement

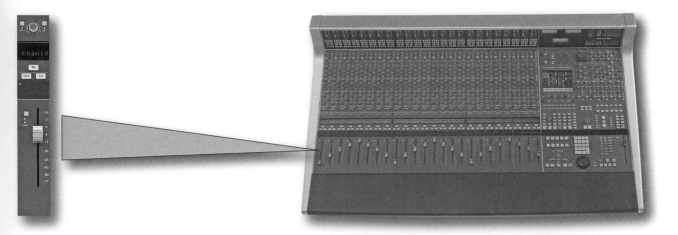

FIGURE 17.18
Output fader section of a Solid State Logic AWS δelta Console (courtesy of Solid State Logic, www.solidstate-logic.com).

in the stereo and/or surround field (Figure 17.19). Generally, this section also includes a solo/mute feature, which performs the following functions:

- Solo: when pressed, the monitor outputs for all other channels will be muted, allowing the listener to monitor only the selected channel (or soloed channels) without affecting the multitrack or main stereo outputs during the recording or mixdown process.
- Mute: this function is basically the opposite of the solo button, as when it is pressed, the selected channel is cut or muted from the main and/or monitor outputs.

Depending on the hardware mixer or controller interface design, the channel fader might be motorized, allowing automation moves to be recorded and played back in the physical motion of moving faders. In the case of some of the high-end console and audio interface/controller designs, a flip fader mode can be called up that literally reassigns the control of the monitor section's fader to that of the main channel fader (as was seen in Figure 17.18). This "flip" allows the monitoring of levels during the recording process to be controlled from the larger, long-throw faders. In addition to swapping monitor/channel fader functions, certain audio interface/controller designs allow a number of functions,

FIGURE 17.19
Example of various hardware and software pan pot configurations.

stereo hardware panner

stereo software panner

L ⟵⟶ R

surround panner

such as panning, EQ, effects sends, etc., to be swapped with the main fader, allowing these controls to be finely tuned under motorized control.

OUTPUT SECTION

In addition to the concept of the signal path as it follows through the chain, there's another important signal concept that should be understood: output bus. From the preceding input strip discussion, we've seen that a channel's audio signal by and large follows a downward path from its top to the bottom; however, when we take the time to follow this path, it's easy to spot where audio is sometimes routed off the strip and onto a horizontal output path. Conceptually (and sometimes literally), we can think of this path (or bus) as a single electrical conduit that runs the horizontal length of a console or mixer (Figure 17.20). Signals can be inserted onto or routed off this bus at multiple points.

Much like a city transit bus, this signal path follows a specific route and allows audio signals to get on or off the line at any point along its path. Aux sends, monitor sends, channel assignments and main outputs are all examples of signals that are taken from their associated input strips and are injected into buses for routing to one or more output destinations. The main stereo or surround buses (which are used to feed the channel faders and pan positioners) are then fed to the mixer's main output bus, where they are combined with the various effects return signals and finally routed to the recording device and/or monitor speakers.

Channel Assignment

After the channel output fader on a console or larger mixer, the signal is often routed to the strip's track assignment matrix (Figure 17.21), which is used to distribute the signal to any or all tracks of a connected multitrack DAW or ATR recorder. Although this section electrically follows either the main or the small fader section (depending on the channel's monitor mode), the track assign buttons will often be either located at the top of the input strip or designed into the main output fader (often placed at the fader's right-hand side). Functionally, pressing any or all assignment buttons will route the input strip's main signal to the corresponding track output buses. For example, if a vocal mic is plugged into channel 14, the engineer might assign the signal to track 14 by pressing

FIGURE 17.20

Example of a master output bus, where multiple inputs are mixed and routed to a master output fader.

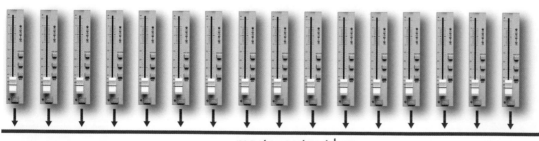

master output fader

master output bus

(a)

(b)

(c)

(you guessed it) the "14" button on the matrix. If a quick overdub on track 15 is also needed, all the engineer has to do is unpress the "14" button and reassign the signal to track 15.

Many newer consoles offer only a single button for even- and odd-paired tracks, which can then be individually assigned by using the strip's main output pan pot. For example, pressing the button marked "5/6" and panning to the left routes the signal only to output bus 5, while panning to the right routes it to bus 6. This simple approach accomplishes two things:

- Fewer buttons need to be designed into the input strip (lowering production costs and reducing the number of moving parts)
- Panning instruments within a stereo sound field and then assigning their outputs to a pair of tracks on the multitrack recorder can easily be used to build up a stereo sub-mix

Of course, the system for assigning a track or channel on a DAW varies with each design, but it's often straightforward. As always, it's best to check with the manual and become familiar with the software before beginning a session.

Grouping

Many DAWs, consoles and professional mixing systems allow any number of input channels to be organized into groups. Such groupings allow the overall relative levels of a series of channels to be interlinked into organized groups according to instrument or scene change type. This important feature makes it possible for multiple instruments to retain their relative level balance while offering control over their overall group level from a single fader or stereo fader pair. Individual group bus faders often have two functions. They can:

- Vary the overall level of a grouped signal that's being sent to a recorded track
- Vary the overall sub-mix level of a grouped signal that's being routed to the master mix bus during mixdown

FIGURE 17.21

Channel assignment sections.
(a) Onyx 4-bus (courtesy of Loud Technologies, Inc., www.mackie.com). (b) API 1608 Console (courtesy of Automated Processes, Inc., www.apiaudio.com).
(c) ProTools (courtesy of Digidesign, a division of Avid Technology, Inc., www. digidesign.com).

The obvious advantage to grouping channels is that it makes it possible to avoid the dreaded and unpredictable need to manually change each channel volume individually. Why try to move 20 faders when you can adjust their overall levels from just one? For example, the numerous tracks of a string ensemble and a drum mix could each be varied in relative level by assigning them to their own stereo or surround sound groupings and then moving a single fader – ahhhh, much easier and far more accurate!

It's important to keep in mind that there are two methods for grouping signals together when using either a hardware console or a DAW:

- Signals can be grouped together onto the same, combined audio bus.
- An external control signal (such as a DC voltage or digital control) can be used to control the relative level of various grouped faders, etc., while keeping the signals separate and discrete.

Using the first method, the grouped signals are physically assigned to the same audio bus, where they are combined into a composite mono, stereo or multi-channel track. Of course, this means that all of your mix decisions have to be made at that time, as all of the signals will be electrically combined together, with little or no recourse for hitting an "'undo' button" at a later time. This physical grouping together of channel signals can be easily done by assigning the involved channels in the desired group to their own output bus (Figure 17.22). During mixdown, each instrument group bus will be routed to the main stereo or surround output through the use of pan pots or L/R assignment buttons.

The second grouping system doesn't combine the signals but makes use of a single voltage or digital control signal to control the "relative" levels of tracks that are assigned to a group bus. This method (Figure 17.23), allows the signals to be discreetly isolated and separated from each other, but their relative levels can be changed in a way that keeps the relative mix balance intact. Quite simply, it's a way in which multiple faders can be either controlled from a single "group fader" or ganged together to move in relative tandem with each other.

FIGURE 17.22
Simplified anatomy of the output grouping section on the Mackie Onyx 4-bus analog console, whereby the signals (in this case, drum tracks, which are grouped to groups 1 and 2) can be combined into a physical group output bus (courtesy of Loud Technologies, Inc., www.mackie.com).

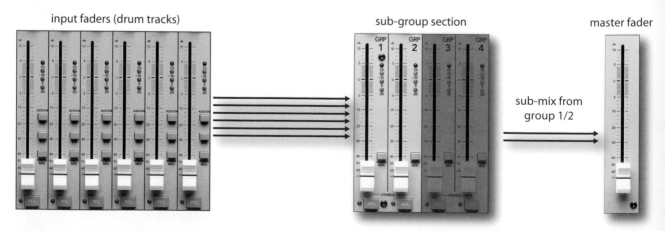

input faders (drum tracks)　　　　　　　　　　　　sub-group section　　　　　master fader

sub-mix from group 1/2

As you might have noticed, the preceding section focused largely upon the idea of a hardware console or mixer. Newer DAWs, on the other hand, now have several options for grouping a signal. Each type has its subtle strengths and should be tried out to see which one works best for you, or in what situation it would work best.

1. Grouping track: by using a separate group track, the digitally summed grouping signal will be routed through a separate group fader. Controlling this group fader will control the relative volumes of the tracks that are assigned to that group.

2. VCA Grouping: using this system, the grouped tracks will not be assigned to an assigned output but will be virtually locked in tandem by a control signal. Moving any fader in that grouping will move the relative volumes of all the faders that are assigned to that group.

3. Using color coding: here, all of the faders in a grouping can be given a specific track color. Of course, this is not a grouping method at all; however, I mention this as an additional way to easily keep track of a grouped set of tracks.

Main Output Mix Bus

The main output bus of an analog console acts to physically or electronically sum (add) the signals from all of the channel faders, pan, effects return and any other signal routing output together into a final mono, stereo or surround audio program output. The output from this audio source can then be processed in a final stage (using compression, limiting or EQ) and sent to a master recording device.

In the digital domain, the main output bus likewise serves to digitally combine the sample levels from all of the channel fader, pan, effects return and any other signal routing outputs together into a final mono, stereo or surround digital audio stream that represents the final audio program data stream.

At this stage, the primary concern that needs to be kept in mind is the careful handling of the program's overall dynamic range. As you might expect, a main output signal that is too low in level could result in increased noise, while excessive output levels will almost certainly result in a distortion. This logic, however, begins to blur when we consider that most of the output signal levels are digital in nature and make use of higher bit depths. For example, a signal could have an overall signal-to-noise dynamic range of up to 144 dB for a 24-bit, 192 dB

FIGURE 17.23
Tracks can be assigned as a group, so that the relative track levels can change, while keeping the track signals separate in a mix.

group master   grouped faders

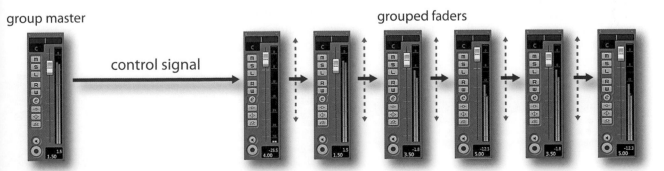

control signal

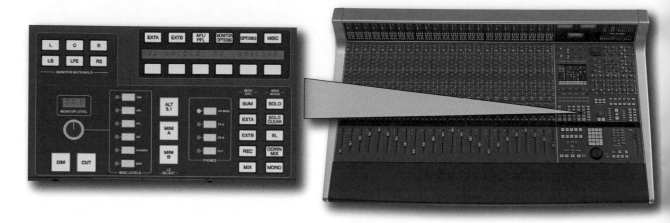

FIGURE 17.24
Monitor level section of a
Solid State Logic AWS δelta
Console (courtesy of Solid
State Logic, www.solidstate-
logic.com).

for a 32-bit and 385 dB for a 64-bit system, meaning that the main output could be at lower levels without adding significant noise to the mix. This would have an added advantage of ensuring that the levels will be low enough that digital clipping distortion will not occur – something to be kept in mind should you know that a mastering engineer will be using their skills to finesse the mix into a final form.

Main Monitor Level Section

Most console, mixing and DAW systems include a central monitor section that controls levels for the various monitoring functions (such as control-room level, studio level, headphone levels and talkback). This section (Figure 17.24) often makes it possible to easily switch between multiple speaker sets and can also provide switching between the various inputs, recording device sources and various output formats (e.g., surround, stereo and mono output buses; tape returns; aux send monitoring; solo monitoring).

FIGURE 17.25
The patch bay. (a) Ultrapatch
PX3000 patch bay (courtesy
of Behringer International
GmbH, www.behringer.de).
(b) Rough example of a
labeled patch bay layout.

PATCH BAY

A patch bay (Figure 17.25) is a panel found in the control room and on larger consoles that contains accessible jacks that correspond to the various inputs and outputs of every access point within a mixer or recording console. Most professional patch bays (also known as patch panels) offer centralized I/O access to most of the recording, effects and monitoring devices or system blocks within

the production facility (as well as access points that can be used to connect between different production rooms).

Patch bay systems come in a number of plug and jack types as well as wiring layouts. For example, prefabricated patch bays are available using tip-ring-sleeve (balanced) or tip-sleeve (unbalanced) 1/4-inch phone configurations. These models will often place interconnected jacks at the panel's front and rear, so that studio users can reconfigure the panel simply by rearranging the plugs at the rear access points. Other professional systems using the professional telephone-type (TT or mini Bantam-TT) plugs might require that you hand-wire the connections in order to configure or reconfigure a bay (usually an amazing feat of patience, concentration and stamina).

Patch jacks can be configured in a number of ways to allow several signal connection options among inputs, outputs and external devices (Figure 17.26):

- Open: when no plugs are inserted, each I/O connection entering or leaving the panel is independent of the other and has no electrical connection.
- Half-normaled: when no plugs are inserted, each I/O connection entering the panel is electrically connected (with the input being routed to the output). When a jack is inserted into the top jack, the in/out connection is still left intact, allowing you to tap into the signal path. When a jack is inserted into the bottom jack, the in/out connection is broken, allowing only the inserted signal to pass to the input.
- Normaled: when no plugs are inserted, each I/O connection entering the panel is electrically connected (with the input routing to the output). When a jack is inserted into the top jack, the in/out connection is broken, allowing the output signal to pass to the cable. When a jack is inserted into the bottom jack, the in/out connection is broken, allowing the input signal to pass through the inserted cable connection.

FIGURE 17.26
Typical patch bay signal routing schemes (courtesy of Behringer International GmbH, www.behringer.de).

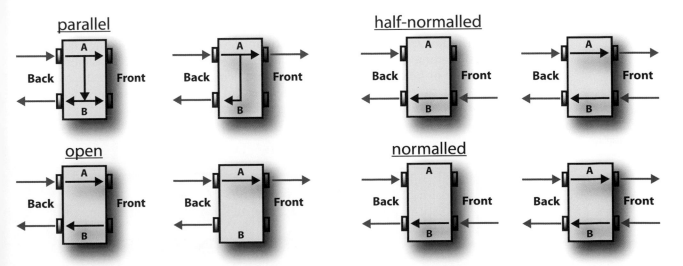

- Parallel: in this mode, each I/O connection entering the panel is electrically connected (with the input routing to the output). When a jack is inserted into either the top or the bottom jack, the in/out connection will still be intact, allowing you to tap into both the signal path's inputs and outputs.

Breaking a normaled connection allows an engineer to patch different or additional pieces of equipment into a circuit path. For example, a compressor might be temporarily patched between a mic preamp output and an equalizer input. The same preamp/EQ patch point could be used to insert an effect or other device type. These useful access points can also be used to bypass a defective component or to change a signal path order. Versatility is definitely the name of the game here!

METERING

The level of a signal's strength at an input, output bus and other console level point is often measured by visual meter display (Figure 17.27). Meter and indicator types will often vary from system to system. For example, banks of readouts that indicate console bus output and tape return levels might use VU metering, peak program meters (PPMs, found in European designs) or digital/software readouts. It's also not uncommon to find LED overload indicators on an input strip's preamp, which give quick and easy peak indications as to whether you've approached or reached the component's headroom limits (a sure sign to back off your levels).

The basic rule regarding levels isn't nearly as rigid as you might think and will often vary depending on whether the device or recording medium is analog or digital. In short, if the signal level is too low, tape, amp and even digital noise could be a problem, because the levels throughout the signal chain might not be optimized. If the level is too high, overloaded preamps, saturated tape or clipped digital converters will often result in a distorted signal. Here are a few rules of thumb:

- In analog recording, the proper recording level is achieved when the highest reading on the meter is near the zero level, although levels slightly above or below this might not be a problem, as shown in Figure 17.28. In fact, overdriving some analog devices and tape machines (slightly) will often result in a sound that's "rough" and "gutsy".

FIGURE 17.27
A set of LED, light-bar and VU meter displays.

too low

too high

just right

- When recording digitally, noise is often less of a practical concern (especially when higher bit depths are used). It's often a good idea to keep levels at a reasonable level (i.e., peaking at about –12 on the meter) while keeping a respectful distance from the dreaded clip or "over" indicator. Unlike analog, digital is generally unforgiving of signals that are clipped and will generate a grunge sound that's guaranteed to make you cringe! Because there is no real standard for digital metering levels beyond these guidelines, many feel that giving a headroom margin that's 12 dB below "0" full scale is usually a wise precaution – especially when recording at a 24-bit or higher depth.

FIGURE 17.28
VU meter readings for analog devices that are too low, too high and just right.

It's important to keep in mind that recorded distortion isn't always easy (or possible) to fix. You might not be able to see the peak levels that could distort your sound files, but your ears might hear them after the damage is done. When in doubt, back off and give your levels some breathing room.

The Finer Points of Metering

Amplifiers, magnetic tape and even digital media are limited in the range of signal levels that they can pass without distortion. As a result, audio engineers need a basic standard to help determine whether the signals they're working with will be stored or transmitted without distortion. The most convenient way to do this is to use a visual level display, such as a meter. Two types of metering ballistics (active response times) are encountered in recording sound to either analog or digital media:

- Average (rms)
- Peak

From Chapter 2, we know that the root-mean-square (rms) value was developed to determine a meaningful average level of a waveform over time. Since humans perceive loudness according to a signal's average value (in a way that doesn't bear much relationship to a signal's instantaneous peak level), the displays of many meters will indicate an average signal-level readout. The total amplitude measurement of the positive and negative peak signal levels is called the peak-to-peak value. A readout that measures the maximum amplitude fluctuations of a waveform is a peak-indicating meter.

One could definitely argue that both average and peak readout displays have their own sets of advantages. For example, the ear's perception of loudness is largely proportional to the rms (average) value of a signal, not its peak value. On the other hand, a peak readout displays the actual amplitude at a particular point in time and not its overall perceived level. For this reason, a peak meter might show readings that are noticeably higher at a particular point in the program than the averaged rms counterpart (Figure 17.29). Such a reading will alert you when short-term peaks are at levels that are above the clipping point while the average signal is below the maximum limits. Under such conditions (where short-duration peaks are above the distortion limit), you might or might not hear distortion, as it often depends on the makeup of the signal that's being recorded; for example, the clipped peaks of a bass guitar will not be nearly as noticeable as the clipped high-end peaks of a cymbal. The recording medium often plays a part in how a meter display will relate to sonic reality; for example, recording a signal with clipped peaks onto a tube analog tape machine might be barely noticeable (because the tubes and the tape medium act to smooth over these distortions), whereas a DAW or digital recorder might churn out a hash that's as ugly as the night (or your current session) is long.

Getting to the heart of the matter, it goes without saying that whenever the signal is too high (hot), it's an indication for you to grab hold of the channel's mic trim, output fader or whatever level control is the culprit in the chain and turn it down. In doing so, you've actually become a dynamic range-changing device. Additionally, the main channel fader (which can be controlling an input level during recording or a tape track's level during mixdown) is by far the most intuitive and most often used dynamic gain-changing device in the studio.

FIGURE 17.29
A peak meter reads higher at point A than at point B, even though the average loudness level is the same.

In practice, the difference between the maximum level that can be handled without incurring distortion and the average operating level of the system is called headroom. Some studio-quality preamplifiers are capable of signal outputs as high as 26 dB above 0 VU and thus, are said to have 26 dB of headroom. With regard to analog tape, the 3% distortion level for analog magnetic tape is

typically only 8 dB above 0 VU. For this reason, the best recording level for most program material is around 0 VU (although higher levels are possible, provided that short-term peak levels aren't excessively high). In some circumstances (i.e., when using higher-bias, low-noise/high-output analog tape), it's actually possible to record at higher levels without distortion, because the analog tape formulation is capable of handling higher magnetic flux levels. With regard to digital media, the guidelines are often far less precise and will often depend on your currently chosen bit depth. Since a higher bit depth (e.g., 24 or 32 bits) directly translates into a wider dynamic range, it's often a good idea to back off from the maximum level, because noise generally isn't a problem.

Now that we've gotten a few of the basic concepts out of the way, let's take a brief look at two of the most common meter readout displays.

The VU Meter

The traditional signal-level indicator for analog equipment is the VU meter (see previous Figure 17.28). The scale chosen for this device is calibrated in "Volume Units" (hence its name) and is designed to display a signal's average rms level over time. The standard operating level for most consoles, mixers and analog tape machines is considered to be 0 VU. Although VU meters do the job of indicating rms volume levels, they ignore the short-term peaks that can overload a track. This means that the professional console systems must often be designed so that unacceptable distortion doesn't occur until at least 14 dB above 0 VU. Typical VU meter specifications are listed in Table 17.2.

Since recording is an art form, I have to rise to the defense of those who prefer to record certain instruments (particularly drums and percussion) at levels that bounce or even "pin" VU needles at higher levels than 0 VU. When recording to a professional analog machine, this might actually give a track a "gutsy" feel that can add impact to a performance. This is rarely a good idea when recording instruments that contain high-frequency/high-level signals (such as a snare or cymbals), because the peak transients will probably distort in a way that's hardly pleasing – and it's almost never a good idea to overload the meters when

Table 17.2 VU Meter Specifications	
Characteristic	**Specification**
Sensitivity	Reads 0 VU when fed a +4-dBm signal (1.228 V into a 600-Ω circuit)
Frequency response	±0.2 dB from 35 Hz to 10 kHz; ±0.5 dB from 25 Hz to 16 kHz
Overload capability	Can withstand 5 times 0-VU level (approximately +18 dBm) continuously and 10 times 0-VU level (+24 dBm) for 0.5 sec

recording to a digital system. Always be aware that you can often add distortion to a track at a later time (using any number of ingenious tricks), but you can't easily remove it from an existing track (software does exist that can help remove some distortion after the fact, but it should never be relied upon). As always, it's wise to talk such moves over with the producer and artist beforehand.

The Average/Peak Meter

While many analog devices display levels using the traditional VU meter, most digital hardware and software devices will often display a combination of VU and peak program-level metering using a liquid crystal display (LCD) or on-screen readout. This best-of-both-worlds system makes sense in the digital world, as it gives us a traditional readout that visually corresponds to what our ears are hearing while providing a quick and easy display of the peak levels at any point in time. Often, the peak readout is frozen in position for a few seconds before resetting to a new level (making it easier to spot the maximum levels), or it can be permanently held at that level until a higher level comes along to bump it up. Of course, should a peak level approach the clipping level, a red "clip" indicator will often light, showing that it's time to back off the levels.

In recent years, a number of high-quality sound level meters have come onto the market that allow the user to visually inspect the peak, loudness and spectral levels within a recording. These tools can be extremely useful for obtaining the best overall level of a mix as well as the careful balancing of relative song levels within a project.

Another point that should be pointed out is that there are a number of prominent engineers and producers who advocate that we spend less time looking at the level meters, waveforms and faders … and more time concentrating on the overall musical and sound quality of the recording and/or mix. Just food for thought.

DIGITAL CONSOLE AND DAW MIXER/CONTROLLER TECHNOLOGY

In the last few decades, console design, signal processing and signal routing technology have undergone a tremendous design revolution with the advent of digital technology. DAWs, digital consoles and mixers, the iPad and controller design have found their ways into professional, project and audio production facilities at an amazing pace. These systems use large-scale integrated circuits and central processors to convert, process, route and interface to external audio and computer-related devices with relative ease. In addition, this technology makes it possible for many of the costly and potentially faulty discrete switches and level controls that are required for such functions as track selection, gain and EQ to be replaced by a touch-controlled system. The big bonus, however, is that since routing and other control functions are digitally encoded, it becomes a simple matter for level, patch and automation settings to be saved into memory for instantaneous recall at any time.

The Virtual Input Strip

From a functional standpoint, the basic signal flow of a digital console's or DAW's virtual mixing environment is conceptually similar to that of an analog console, in that the input must first be boosted in level by a mic/line preamp (where it will then be converted into digital data). From this point, the signals can pass through EQ, dynamics and other signal processing blocks, and various effects and monitor sends, volume, routing and other sections that you might expect, such as main and group output fader controls. From a control standpoint, however, sections of a digital system might be laid out in a manner that looks and feels completely different.

By the very nature of analog strip design, all of the physical controls must be duplicated for each strip along the console. This is the biggest contributing factor to cost and reliability problems in traditional analog design. Since most (if not all) of the functions of a digital console pass through the device's central processor, this duplication is neither necessary, cost-effective nor reliable. As a result, designers have often opted to keep the most commonly used controls (such as the pan, solo, mute and channel volume fader) in their traditional input strip position. However, controls such as EQ, input signal processing, effects sends, monitor levels and (possibly) track assignment have been designed into a central, virtual control panel (Figure 17.30) that can be used to focus on and vary a particular channel's setting parameters. These controls can be quickly and easily assigned to a particular input strip by pressing the "select" button on the relevant input strip channel and then making the necessary changes to that channel.

In certain digital console, mixer and hardware controller designs, each "virtual input strip" may be fitted with a virtual pot (V-pot) that can be assigned to a particular parameter for instant access. In others, the main parameter panel can be multipurpose in its operation, allowing a number of controls and readouts to be reconfigured in a chameleon-like fashion to fit the task at hand. In other cases, touch-screen displays can be used to provide an infinite degree of user control, while some might use software "soft" buttons that can easily reconfigure the form and function of the various buttons and dial controls in the control panel.

FIGURE 17.30
Presonus StudioLive 16.0.2 Digital mixing console with channel parameter section (courtesy of Presonus Audio Electronics, Inc., www.presonus.com).

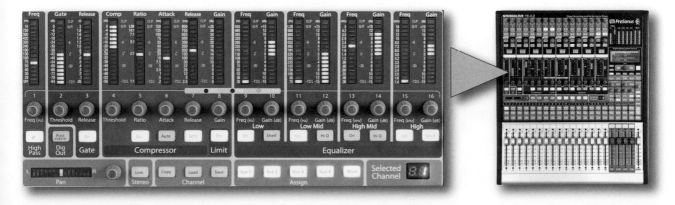

From this, it's easy to see that digital consoles and hardware controllers often vary widely in their layout, ease of operation and cost-effectiveness. As with any major system device, it's often a good idea to familiarize yourself with the layout and basic functions before buying or taking on a project that involves such a beastie. As you might expect, operating manuals and basic tutorials are often available on the various manufacturers' websites.

The DAW Software Mixer Surface

Of course, in this new millennium, the vast majority of mixers that exist in the world are software mixers that are integrated into our DAW of choice. With the increased power of the computer-based digital audio workstation, mixing "in the box" has become the norm rather than the exception.

Through the use of a traditional (and sometimes not so traditional) user graphical interface, these mixers offer functional on-screen control over levels, panning, EQ, effects, DSP, mix automation and a host of functions that are simply too long to list here. Often, these software mixers (Figure 17.31) emulate their hardware counterparts by offering basic controls (such as fader, solo, mute, select and routing) in the virtual input strip. Likewise, selecting a channel track will assign many of the channel's functions to a wide range of virtual parameters that can be controlled from a virtual strip or control panel.

When using a software mixer in conjunction with a DAW's waveform/edit window on a single monitor screen, it's easy to feel squeezed by the lack of visual "real estate". For this reason, many opt for a dual-monitor display arrangement. Whether you are working in a PC or Mac environment, this working arrangement is easier and more cost-effective than you might think, and the benefits will immediately become evident, no matter what audio, graphics or program environment you're working on.

FIGURE 17.31
Example of virtual, on-screen DAW mixers. (a) ProTools (courtesy of Avid Technology, Inc., www.avid.com). (b) Logic (courtesy of Apple Inc., www.apple.com).

When using a DAW controller, the flexibility of the DAW's software environment can also be combined with the hands-on control and automation of a hardware or iPad-based controller surface (Figure 17.32). More information on the DAW software mixing environment and hardware controllers can be found in Chapter 7.

(a)

(b)

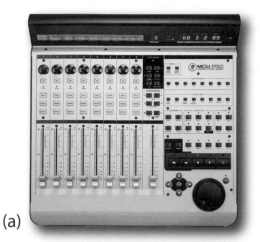

(a)

(b)

OK, BACK TO THE TASK OF RECORDING

Now that we've looked more closely at the internal stages of a console and DAW, let's turn our attention back to the recording process itself.

As was said, the beauty behind the modern recording process is that the various instruments can be recorded separately onto tracks of a DAW, thereby allowing as much isolation between the separately recorded tracks as possible. This is extremely important, as the name of the game is to capture the best performance with the highest possible quality, while achieving optimum isolation between these tracks, and at optimum signal levels (often without regard to level balances on the other tracks).

This idea of isolation can also be a point of discussion. When recording a classical ensemble, distant miking techniques rely upon the fact that there is little or no isolation between the instruments and their acoustic environment. Recording a classical ensemble for film, for example, will often make use of mic placements that are at a closer, more compromised distance, allowing the mics to pick up leakage (bleed) from other instruments, but to a lesser extent, so as to allow greater control during mixdown. One of the top engineers in the world will often use omnidirectional mics during jazz sessions at a semi-distance to pick up the instruments, nearby leakage and the room, stating that leakage can often be a good thing, adding to the "liveness" of the recording. The moral to these examples is that mic distance and isolation are often situational, relying upon the style and intended effect that the room and pickup technique might have upon the music itself – it's all part of the art of recording.

The last two of these procedures are most commonly encountered within the recording of modern, popular music. In the second approach, the resulting foundation tracks (to which other tracks can be added at a later time) are called basic, rhythm or bed tracks. These consist of instruments that provide the rhythmic foundations of a song and often include drums, bass, rhythm guitar and keyboards (or any combination thereof). An optional vocal guide (scratch track)

FIGURE 17.32
DAW controllers. (a) Mackie Universal Control (courtesy of Loud Technologies, Inc., www.mackie.com). (b) V-Control Pro controller for the iPad (courtesy of Neyrinck Audio, www.neyrinck.com).

can also be recorded at this time to help the musicians and vocalists capture the proper tempo and that all-important feel of a song.

From a technical standpoint, the microphones for each instrument are selected either by experience or by experimentation and are then connected to their respective console or audio interface inputs. Once done, the mic type and track selection should be noted within the DAW track channels, notepads and/or a track sheet (paper or not) for easy input and track assignment in the studio. Mic choice and placement can also be noted as a reference during subsequent over-dub sessions, while taking photos and screenshots of any complicated setups (which can be saved within the session's directory) can also come in handy and might save your butt later on during production.

A Word on Miking Techniques

As we've all seen from Chapter 4, the way in which mics are used to pick up instruments (as well as the way the mics and instruments react with their acoustic environments) is of paramount importance as to how the overall sound will be translated in the recording process.

We would simply ask that you keep in mind how the use of pickup distance, room acoustics and the use of additionally recorded room mics can be used as an artistic tool towards injecting life into an overall captured sound. Of course, it all depends upon the type of music, as well as the final intended result.

Some engineers find it convenient to standardize on an organized system that uses the same console mic input and DAW/tape track number for an instrument type at every session. For example, an engineer might consistently plug their favorite kick drum mic into input number 1 and record it onto track 1, the snare mic onto number 2, and so on. This way, the engineer instinctively knows which track belongs to a particular instrument without having to think too much about it. When recording to a DAW, track names, groupings and favorite identifying track colors can also be super helpful towards easily identifying the instrument or group type.

Once the instruments, baffles (an optional sound-isolating panel) and mics have been roughly placed, and headphones that are equipped with enough extra cord to allow free movement have been distributed to each player, the engineer can now get down to the business of labeling each input strip on the console (if there is one) with the name of the corresponding instrument.

At this time, the mic/line channels can be assigned to their respective tracks. Make sure you fully document the assignments and other session info on the song or project's track sheet (Figure 17.33). If a DAW is used, make sure each track in the session is properly named for easy reference and track identification.

If a large-scale controller or digital console is used, labeling tracks might be as simple as naming the track on the DAW, which will then be shown on the

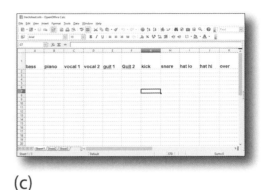

(a)   (b)   (c)

FIGURE 17.33
Studio track sheets. (a)
Capitol Records (courtesy
of Capitol Records, www
.capitolrecords.com). (b)
John Vanderslice (courtesy
of John Vanderslice and Tiny
Telephone, www.johnvander-
slice.com). (c) Home-made
sheet using a spreadsheet
program.

console's channel readout display. If you're working in the analog world, label strips (which are often provided just below each channel input fader) can be marked with an erasable felt marker, or the age-old tactic of rolling out and marking a piece of paper masking tape could be used. (Ideally, the tape should be the type that doesn't leave a tacky residue on the console surface.)

After all of the assignments and labeling have been completed, the engineer then can begin the process of setting levels for each instrument and mic input by asking each musician to play solo or by asking for a complete run-through of the song and listening to one input at a time (using the solo function). By placing each of the channel and master output faders at their unity (0 dB) setting and starting with the EQ settings at the flat position, the engineer can then check each of the track meter readings and adjust the mic preamp gains to their optimum level while listening for potential preamp overload. If necessary, a gain pad can be inserted into the path in order to help eliminate distortion at the pickup.

After these levels have been set, a rough headphone mix can be made so that the musicians can properly hear themselves (never underestimate the importance of getting the headphone mix right). Mic choice and/or placements can be changed, or EQ settings can be adjusted, if necessary, to obtain the best sound for each instrument. Also, dynamic limiting or compression could be carefully inserted and adjusted for any channel that requires dynamic attention. It's important to keep in mind that it's easier to change the EQ settings and/or dynamics of a track later during mixdown (particularly if the session is being recorded digitally) than to undo any changes that have been made during the recording phase.

Once this is done, the engineer and producer can again listen for tuning and extraneous sounds (such as buzzes or hum from guitar amplifiers or squeaks from drum pedals) and adjust or eliminate them. Soloing the individual tracks can ease the process of selectively listening for such unwanted sounds and getting the best sound from an instrument without any listening to the other tracks. Soloing a track will not distract the musicians, as the monitor feed won't be affected. If several mics are to be grouped into one or more tracks, the balance between them should be set with extreme care at this time. When recording to a

DAW, organizing your individually recorded tracks into groups, folders, adding color track identifiers, etc. while making session notes is relatively easy and can be changed at any later time.

After this procedure has been followed for all the instruments, the musicians might do a couple of practice rundown songs so that the engineer and producer can listen to how the instruments sound together before being recorded. (As disk space usually isn't a concern, you might strongly consider recording these practice tracks, as they might turn out to be the best takes – you just never know!) During the rundown, you might also consider soloing the various instruments and instrument combinations as a final check and then monitor all of the instruments together to hear how the combined tracks sound. Careful changes in EQ might (if necessary) be made at this time, making sure to note these changes in the DAW notepad or track sheet for future reference. These changes should be made sparingly, however, because final balance compensations are probably better made during the final mixdown phase.

During the practice rundown, it's also a good idea to ask the musician(s) to play through the entire song so you'll know where the breaks, bridges and any other point of particular importance might be. Making notes and even writing down or entering the timing numbers (into a DAW using session markers or into an analog recorder's auto-locator) can help speed up the process of finding a section during a take or overdub. You can also pinpoint the loud sections at this time so as to avoid any overloads. If compression or limiting is used, you might keep an ear open to ensure that the instruments don't trigger an undue amount of gain reduction (again, if the tracks are recorded digitally at a high bit rate, you might consider applying gain reduction later during mixdown). Even though an engineer might ask each musician to play as loudly as possible, they'll often play even louder when performing together. This well-known fact might require further changes in the mic preamp gain, recording level and compression/limiting thresholds. Soloing each mic and listening for leakage can also help to check for isolation problems between the instruments. If necessary, the relative positions of mics, instruments and baffles can be changed one last time.

ADDITIONAL THOUGHTS

It's obviously a foregone conclusion that no two recording sessions will ever be exactly alike. In fact, in keeping with the "no rules" rule, they're often radically different from each other. During the recording session, the engineer watches the level indicators and (if necessary) controls the faders to keep from getting overload distortion. It's also an engineer's job to act as another set of production ears by listening for both performance and quality factors. If the producer doesn't notice a mistake in the performance, the engineer just might catch it and point it out. The engineer should always try to be helpful and remember that the producer and/or the band will have the final say, and that their final judgment of the quality of a performance or recording must be accepted.

From both the engineer's and the musician's standpoint, here are a few additional pointers that can help the session go more smoothly:

- It's always best to get the right sound and vibe onto disk or tape during the session. If you need to do another take, do it! If you need to change a mic, change it. Getting the right sound and vibe onto the track will almost always result in less stress and a better final product rather than trying to "fix in the mix" at a later time.
- You might want to record "run-throughs", just in case they're magical.
- Know when to quit! If you're pushing too hard or are tired, it'll often show.
- Technology doesn't always make a good track; feeling, emotion and musicality always do.
- Beware of adding extra parts or tracks onto a piece that doesn't need it. Remember, too much can sometimes be simply too much! Musicians and techno-geeks alike often don't know when to say "it's done ... let's move on".
- Leave plenty of time for the vocal track(s). It's not uncommon for a group to spend most of their time and budget on getting the perfect drum or guitar sound. It takes time and a clear focus to get the vocals right – be prepared for Murphy's law.
- If you mess up on a part, keep going; you might be able to fix the bad part by punching in. If it's really that bad, the engineer or producer will hopefully stop you.

In his *EQ Magazine* article "The Performance Curve: How Do You Know Which Take Is the One?", my buddy Craig Anderton laid out his experiences of how different musicians will deal with the process of delivering a performance over time. Being in front of a mic isn't always easy, and we all deal with it differently. Here's a basic outline of his findings:

- Curves up ahead: with this type of performer, the first couple of takes are pretty good, then start to go downhill before ramping back up again, until they hit their peak before going downhill really fast.
- The quick starter: this type starts out strong and doesn't improve over time in later performances. Live performers often fall into this category because they're conditioned to get it right the first time.
- The long ramp-up: these musicians often take a while to warm up to a performance. After they hit their stride, you might get a killer take or a few great ones that can be composited together into the perfect take.
- Anything goes: this category can vary widely within a performance. Often, snippets can be taken from several takes into a single composite. You want to record everything with this type of performer, because you just never know what gem (or bad take) you'll end up with.
- Rock steady: this one represents the consummate pro who is fully practiced and delivers a performance that doesn't waver from one take to the

next; however, you might record several takes to see which one has the most feeling.

From these examples, we can quickly draw the obvious conclusion that there are all types of performers and that it takes a qualified and experienced producer and/or engineer to intuit just which type is in front of the mic and to draw the best possible performance from him or her.

FIXING IT IN THE MIX

The term "fix it in the mix" stems from the 1980s, when multitrack recording had just started to hit its full stride. It refers to the idea that if there's a mistake or something that's not quite right in the recording, "don't worry about it; we don't have to make that decision right now; we'll just fix it later in the mix". Although to some degree, this might (or might not) be true, the fact is that this mentality can come back to haunt you (and your career) if care isn't taken.

Preparing for the mix stage definitely begins in the recording process. For example, if the "fix" isn't dealt with beforehand, it might indeed be possible to deal with it later. The real problem, however, happens when multiple "fixes" that are meant to be dealt with at a later time begin to creep into the project. If this happens, the mix can take on a complicated life as something that needs to be wrestled to the ground in order to sound right, instead of being pre-sculpted into a form that will simply need a final polishing in order to shine.

Although each project has a life of its own, just a few of the ways that a mix can be prepared is to:

- Strive to capture the artist and performance to disk or tape in a way that best reflects everyone's artistic intentions. Indeed, decisions such as EQ, EFX, etc. can be made during the mix – but if the life, spirit and essence of the musical expression isn't captured, no amount of processing during the mix will help.
- Whenever possible, deal with the musical or technical problem as it occurs, during the recording or production phase. This cannot be stressed enough.
- During a punch-in or comp session (combining multiple takes into a single best take), take care to match the tracks as carefully as possible. This might involve documenting the mic and instrument that was used, its placement and distance in the room from the artist, as well as any other detail that will ensure that the tracks will properly blend.
- Begin to create a rough mix during the recording phase that will help you get started toward the final mix. This is especially easy to do in this age of the DAW (indeed, it's quite common), as the mix can begin to take its rough form in a way that can be saved and recalled within the session file.

A WORD ON LEVELS

Generally speaking, capturing the best possible sound onto any recording medium means recording at a level that is free of distortion. In truth, the way that one deals with level when recording a sound source will vary depending upon the medium and also the general style in which one likes to work ... but no matter what the medium, the goal is still the same: to capture a source in its best and most truthful way.

When recording signals to analog tape, or when using an analog device, the goal is to capture the sound in a fashion that best suits the track being recorded and then to record the signal to a track at as high a level as possible (Figure 17.34a), without introducing an unpleasant degree of distortion (see Chapter 5 for more info).

- For a DAW or digital device, the use of a mic and preamp that are low in noise is probably the best practice. The goal here is to keep the source levels and overall signal chain optimized so as to keep noise levels to a minimum. Recorded levels can be approached in either of two ways:

- Whenever low-noise recording systems are used in conjunction with low-noise mics and preamps, it's usually a good idea to record the signal with a high bit depth (24 bits or higher). This will result in a recording that has the widest degree of dynamic range. In this way, the noise floor is so low that track levels can be maxed out at about −12 dB full scale (Figure 17.34b), while no noise will be added by the recording system. Some classical engineers might even record at a lower level because the overall dynamic range of the recording system is so wide that introduced noise ceases to be an overriding issue.

- Another way to look at levels within a DAW relates to digital systems that make use of 32- and 64-bit internal processing architecture. It's becoming increasingly common for a DAW or digital recording system to process audio internally at these high bit depths. Using this approach, the mix bus dynamics will be so wide that both noise (at low levels) and distortion (at high digital levels) are kept to a minimum. In this way, it's actually possible that signals that show at full scale on a recorded or group track might not yet sound distorted. This latter type should not be viewed as a way to record and mix audio within a DAW but rather, as insurance headroom against distortion ... it's a far better practice to keep signals at lower levels in order to ensure that you're working at the best (and not necessarily the highest) levels.

FIGURE 17.34
Signal levels. (a) Traditionally, analog levels were recorded at higher levels in order to avoid excessive noise. (b) Due to higher bit depths, digital signals can often be recorded at lower levels without fear of noise.

(a)

-12dB

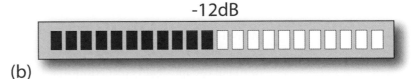

(b)

SESSION DOCUMENTATION

There are few things more frustrating than going back to an archived session and finding that no information exists as to what instrument patch, mic type or outboard effect was used on a session (even one that's DAW-based). The importance of documenting a session in a separate written document or within your DAW's notepad apps can't be overemphasized. The basic documentation that relates to the who, what, where and when of a recording, mixdown, mastering and duplication session should include such information as:

- Artists, engineers, musical and support staff who were involved with the project (addresses, contracts, check photos or scans, studio invoices – anything of importance).
- Session calendar dates and locations.
- Individual song tempos, signatures, etc.
- Mic choice and placement (this might include photos, floor-plan drawings, etc. for future overdub reference).
- Outboard equipment types and their settings (place these notes within the DAW scratchpads, or take photos and place them in the session directory).
- Plug-in effects and their settings or general descriptions (you never know if they'll be available at a future time, so a description can help you to duplicate them with another app). If it's a hard-to-duplicate or specialized effect, you might consider printing the effect onto its own, separate track.
- Again, remember to record and save all MIDI-related tracks. Documenting the instrument, patch and any other info can be really helpful. Printing any MIDI instruments as an audio track will often make things easier, later within the mix.

To ease this process, the artist or producer might pull out a camera or camera phone and start snapping pictures to document the event for posterity as well as for technical reasons. Also, it's often a wise idea to bring along a high-quality video camera to quietly document the setup or actual session for your fans as extra "behind-the-scenes" online video content.

The more information that can be archived with a session (and its backups), the better the chance that you'll be able to duplicate the session in greater detail at some point in the future. Just remember that it's up to us to save and document the music of today for the fans and future playback/mix technologies of tomorrow. Basic session documentation guidelines can be found in the Guidelines & Recommendation section at www.recordingacademy.com/producers-engineers -wing

MONITORING THE MIX

It almost goes without saying that one of the more important parts of helping to make the musicians feel comfortable and at ease is to create an environment where they can easily hear themselves and each other. Without this, you're pretty

much assured that it's going to be a bumpy ride. Therefore, a great deal of prepa-
rational care should go into the process of getting your studio monitoring ducks
in a row. Everyone will thank you for it.

As you might expect, the main benefit of recording individual instruments or
instrument groupings onto isolated tracks at optimum recording levels lies
in the fact that project decisions over relative volumes, effects and placement
changes can be made at any time during the production and/or final mixdown
stage.

Since the instruments have been recorded at levels that probably won't relate to
the program's final balance, a separate mix must be made in order for the artists,
producer and engineer to hear the instruments in their proper musical perspective; for this, a separate mix is often set up for monitoring. As you'll learn later
in this chapter, a multitrack performance can be monitored in several ways. No
particular method is right or wrong; rather, it's best to choose a method that
matches your own personal production style or one that matches the current
needs of the session and its musicians. This monitor mix can be created and
made use of in the control room and in the studio (as a feed to the musicians)
in several ways, such as:

- CR Monitor Mix: the engineer will create a rough mix version of what is
  being recorded in the studio so that he or she can hear the tracks in a way
  that's musically balanced. If the session is being monitored from within a
  DAW, digital console, or console with automated recall over mix parameters, this ever-evolving control-room mix will often improve over the
  course of the project, sometimes to the point that the mix improvements
  can begin to represent the final mix – often making the final mixdown
  process just that much easier.
- Studio Monitor Mix: quite often, this control-room mix will be sufficient
  for the musicians in the studio to hear themselves over headphones in a
  properly balanced way.
- Special Studio Cue Mix: of course, there are times when the musicians as a
  whole or as individuals will need a special, separate mix in order to properly hear themselves or another instrument from which they will need to
  take their cues. For example, a musician's headphone "cue mix" might call
  for more drums so the artist can better hear the song's tempo, or it might
  call for more piano so that the vocalist can better stay in tune. It's totally
  situational, and at times, multiple cue mixes might be called for. In short,
  the idea is to do whatever is needed to assist the musicians so that they can
  best hear themselves in the studio. If they can't hear themselves in a setting
  that's comfortable and best fits their needs, it follows that it will be that
  much harder to deliver their best performance, and that, in the end, is the
  job of a good engineer and producer.

To begin, while the song is being run down, final adjustments to the recording levels can be made. During and/or after the initial levels have been set, the

process of setting up the headphone monitor mix (or mixes) can be undertaken, which can be done in one of several ways.

Monitoring over headphones in the studio is by far the most common way to monitor sound during a recording session.

When recording, it's often best to use sealed headphones to prevent or minimize the monitor feed from leaking back into the newly recorded track. Sometimes, however, this isn't the best way to go. Vocalists and other performers will often prefer to wear only one side of the headphones so that they can hear the natural sound and room acoustics, their own voice or instrument along with the recorded track. Other musicians might bring along their own headphones or in-ear monitors so that they can best hear the mix in a trusted environment.

The number of headphones will vary with the particular session requirements. An orchestral film overdub, for example, could easily use upward of 50 or more headphone pairs during a single session. The next session booking might be an overdub that will call for a single pair; however, sessions that call for several headphones with a single cue mix or a couple of simple mix versions are far more likely to be the norm when recording. Truly, the life of an engineer or setup assistant is rarely a dull one.

Here are a few headphone monitor setup scenarios:

- A single headphone mix (sometimes referred to as a cue mix) can be set up by plugging a pair of headphones into the monitor bus and listening in. This has the advantage of allowing the engineer to hear exactly what the musicians are hearing.
- Another approach is routing the monitor mix to the control-room speakers.
- For those who wish to allow the musicians to create their own mix, a growing number of headphone mix/distribution systems (Figure 17.35) exist that can take the various direct or sub-group mixes of a DAW or console and send these individual feeds directly to the musician's headphone amp/mix station. Here, the musician can have individual control over volume, pan and instrument mix within his or her headphone mix.
- Another method that can be used by a DAW makes use of individual monitor sends that can be routed from within your DAW directly to an output on a multichannel audio interface for routing to the studio's headphone

FIGURE 17.35

Headphone personal mix/ distribution systems. (a) HRM-16 16 channel personal headphone mixing station (courtesy of Furman Sound, Inc., www.furmansound .com). (b) Hear Technologies Hear Back Personal Monitor Mixer System (courtesy of Hear Technologies, www. heartechnologies.com).

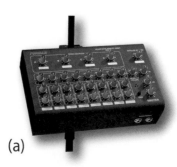

(a)

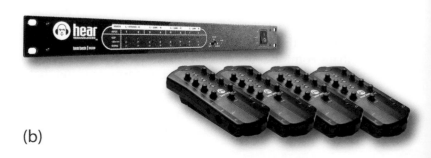

(b)

distribution amps. In this way, the engineer can assign tracks, groups or any combination within the session to the headphones, all with total recall at a later time. With the advent of wireless control over a DAW from a phone or pad, the musician can also simply download the app that integrates with the DAW, call up the headphone mix sends, and dial in his or her own headphone mix wirelessly from their playing position (Figure 17.36).

FIGURE 17.36
Wireless iDevices can be used in the studio to directly control a personal headphone mix.

The importance of proper headphone levels and a good cue balance can't be stressed enough, as they can either help or hinder a musician's overall performance. The same situation exists in the control room with respect to high monitor–speaker levels: some instruments might sound out of tune even when they aren't, and ear fatigue can easily impair your ability to properly judge sounds and their relative balance.

If the musicians can't hear themselves properly, the mix should be changed to satisfy their monitoring needs (fortunately, this can usually be done without regard to the recorded track levels themselves). If several cue systems are available, multiple headphone mixes might need to be built up to satisfy those with different balance needs. During a loud session, the musicians might ask you to turn up their level (or the overall headphone mix) so they can hear themselves above the ambient room leakage. It's important to note that high sound-pressure levels can cause the pitch of instruments to sound flat, so musicians might have trouble tuning or even singing with their headphones on. To avoid these problems, tuning shouldn't be done while listening through phones. The musicians should play their instruments at levels that they're accustomed to and adjust their headphone levels accordingly.

Setup variations also place demands on the distribution of headphone power and the number of required feeds. As you might imagine, the power that would be needed to run 50 headphones (during a classical film score session, for example) could be quite considerable, requiring that a power amplifier be used to drive the headphone distribution boxes throughout the room (Figure 17.37a). On the other hand, the power that would be required to drive one or two headphones in a project studio might be so small that they could be driven from the mixer/interface's internal headphone amp (Figure 17.37b).

(a) (b)

FIGURE 17.37
Headphone distribution amplifiers. (a) Powerplay Pro-8 8-channel headphone distribution amp (courtesy of Behringer Intl GmbH, www .behringer.com). (b) Presonus Audiobox iTwo audio interface (courtesy of Presonus Audio Electronics, www. presonus.com).

Lastly, for those musicians who would like to leave their headphones on the shelf and create a vocal or instrumental take using speakers, it's actually possible to place a microphone directly between two equally spaced speakers (either in the studio or in the control room) that have been intentionally placed out of phase with each other (further info can be found in the section on speaker polarity within Chapter 16). Under ideal conditions, this will result in a 180° phase cancellation at the mic position, meaning that the speaker cue signal can be heard but will cancel out or be greatly reduced in level at the mic's pickup position.

LATENCY

One of the ongoing struggles that are encountered when recording audio in this age of the computer is that audio is continuous and instant in nature, while the computer works in digital chunks that introduce delay in the form of latency. With all of the various digital processing stages that go into even the most simple of recording setups, it's possible for these delays to build up to such a degree that there is an audible delay in the monitoring path. Such a delay could really be disconcerting to a musician who is trying to concentrate on his or her performance while having to deal with a soul-destroying delay in the headphones. Obviously, not a good scenario.

Fortunately, these latency delays can be often reduced or eliminated through the use of "direct monitoring", which takes the direct signal that's being recorded and sends this to the monitor bus without going through any delay-inducing digital paths.

Working in a Combined Recording/Mix Space

Before moving on, it should be noted that the vast majority of project studios (and an ever-growing number of pro studios) are increasingly making use of a single space where recording, production and mixing take place.

Of course, in such rooms, monitoring often takes the form of a fluid balance between speakers in the room and headphones that are used for recording. In short, not only are these setups more cost-effective by necessity but they're also more "homey" and comfortable for both the musicians and those in production.

Overdubbing

Once the basic tracks have been laid down, additional instrument and/or vocal parts can be added later in a process known as overdubbing. During this phase,

kick drum
snare
drums over L
drums over R
guitar
keyboards
vocal (overdub)
vocal (original)

DAW tracks

additional tracks are added by monitoring the previously recorded tracks (usually over headphones) while simultaneously recording new, doubled or augmented instruments and/or vocals onto one or more available tracks of a DAW or recorder (Figure 17.38). During the overdub phase, individual parts are added to an existing project until the song or soundtrack is complete. If the artist makes a mistake, no problem! Simply re-cue the DAW or rewind the tape to where the instrument begins and repeat the process until you've captured the best possible take. If a take goes almost perfectly except for a bad line or a few flubbed notes, it's possible to go back and re-record the offending segment onto the same or a different track in a process known as punching in. If the musician lays down his or her part properly, and the engineer dropped in and out of record at the correct times (either manually or under automation), the listener won't even know that the part was recorded in multiple takes – such is the magic of the recording process!

While recording onto the same track might make sense when an analog recorder is used, a DAW, on the other hand, doesn't have any real track limitations. Therefore, it's usually easy to record the overdub onto another track and then combine the edited tracks into a single track or set of separate grouped tracks. If multiple takes are needed to get the best performance, it's possible to record the different takes to a set of new tracks either manually or under a continuously looped automation function. Once done, the engineer, producer and/or artist can sit down and pick the best parts from the various takes and combine them into a final performance in a process that's known as "comping" (making a composite track, as described in Chapter 7 on the DAW).

In an overdub session where instruments are to be replaced, the same procedure is followed for mic selection, placement, EQ and level, as they occurred during the original recording session (now, you're beginning to see the need for good documentation). Obviously, it's important that any instrument being replaced should have little or no leakage from the original performance tracks. Avoiding this comes under the "good preparation" category when the studio is first being laid out. Care should be taken to ensure that the headphones aren't so loud or improperly placed on the artist's head that excessive noise leaks into the artist's mic.

FIGURE 17.38
Overdubbing allows instruments and/or vocals to be added at a later time to existing tracks on a multitrack recording medium.

Of course, the natural ambience of the session should be taken into account during an overdub. If the original tracks were made from a natural, roomy ensemble, it could be very distracting to hear an added track that was obviously laid down in a different (usually closer and deader) room environment.

OPTIONS IN OVERDUBBING

At first glance, the process of overdubbing seems pretty cut and dried – put a mic out in the studio, put headphones on the performer and start recording. In fact, the process can present as many unique challenges and opportunities as the recording process. For starters, getting the right sound during the process is extremely important. The "Good Principle" applies as much here as within any phase. Not asking the following or other questions might cause headaches later in the game:

- Is the performer in the right headspace to do the overdub?
- Is the instrument properly tuned to the track?
- Are the mic's choice, placement and distance appropriate to the track?
- If an overdub is being made over a previous overdub track, do all of these placements and settings match the previous session (again, stressing the need for pictures and documentation)?

One other possibility that's available to the artist, producer and engineer during an overdub is the wide number of pickup options that can be used in the session to add effects and various mixdown options to the recording. I'm referring to the various ways that an instrument can be recorded to additional, separate tracks to add dramatic options during mixdown. For example:

- A standard close-mic pickup can be used to record the overdub.
- If the instrument is electric or electronic, it can also be recorded "direct" to capture the pure instrument sound.
- Another mic could be placed at a 6' to 10' distance from the instrument (i.e., electric guitar amp) to get a fuller, more distant sound.
- It's also possible to place an additional distant (stereo or even quad) pickup mic setup further back in the room to capture the room's overall acoustic/reverberant sound.
- If the instrument is electronic and has a MIDI out jack, it's generally wise to additionally capture the performance to a MIDI track on the DAW. This option would be invaluable should you need to go back and alter the performance at a later time.

So, why go through all this additional work? Well, let's say that at a later time, the mix engineer calls up your additional distant mics and adds them to the track for extra ambience, and the producer goes wild – she just loves it! – or the track gets picked up as the soundtrack to a feature film and needs to be remixed in immersive. By placing the distant room sounds in the rear or height speakers, the whole guitar sound just got super-huge, in a way that fills out the track amazingly well. On the flip side, let's say that the producer likes the overdub but decides that amp isn't at all right for the track – all you'd have to do is play the direct signal back through that new Vox stack in the big studio (possibly with new a whole new set of room mics as well) and everybody's super happy – or, you captured the MIDI tracks from a synth, which makes it possible to change the patch to get a whole new sound without having to re-perform the track. Are you starting to get the idea that the name of the game is versatility in production and/or mixdown? It's always good to "be prepared".

To Punch or Not to Punch

You've all heard the age-old adage "$%& happens". Well, it happens in the studio – a lot! Whenever a mistake or bad line occurs during a multitrack session, it's often (but not always) possible to punch in on a specific track or set of tracks. Instead of going back and re-recording an entire song or overdub, performing a punch involves going back and re-recording over a specific section in order to fix a bad note, musical line, you name it. This process is done by cueing the tape at a logical point before the bad section and then pressing play. Just before the section to be fixed, pressing the record button (or entering record under automation) will place the track into record mode. At the section's end, pressing the play button again will cause the track to smoothly fall back out of record, thereby preserving the section following the punch. From a monitor standpoint, the recorder begins playback in the sync mode; once placed in record, the track switches to monitor the input source. This lets the performers hear themselves during the punch while listening to playback both before and after the take.

When performing a punch, it's often far better to "fix" the track immediately after the take has been recorded, while the levels, mic positions and performance vibe are still the same. This also makes it easier to go back and re-record the entire song or a larger section should the punch not work. If the punch can't be performed at that time, however, it's always a good idea to take detailed notes about mic selection, placement, preamps and so on to re-create the session's setup without having to guess the details from memory.

As any experienced engineer/producer knows, performing a punch can be tricky. In certain situations, it's a complete no-brainer – for example, when a stretch of silence the size of a Mack Truck exists both before and after the bad section, you'll have plenty of space to punch in and out. At other times, a punch can be very tight or problematic (e.g., if there's very little time to punch in or out, when trying to keep vocal lines fluid and in context, when it's hard to feel the beat of a song or if it has a fast rhythm). In short, punching in shouldn't be taken too lightly or taken so seriously that you're afraid of the process. Talk it over with the producer and/or musicians. Is this an easy punch? Does the section really need fixing? Do we have the time right now? Or, is it better just to redo the song? In short, the process is totally situational and requires attention, skill, experience and sometimes, a great deal of luck.

- Before committing the punch to a track, it's often a wise idea to rehearse the punch without actually committing the fix to tape. This has the advantage of giving both you and the performer a chance to practice beforehand.
- Almost all DAWs will let you enter the punch-in and punch-out times under automation, thereby allowing the punch to be automatically performed with greater precision.
- If you're recording onto the same track, a fudged punch may leave you with few options other than to re-record the entire song or section of a song. An alternative to this dilemma would be to record the fix into a separate track and then switch between tracks in mixdown (a process known as compositing or simply "comping").

Reamping It!

As was seen in Chapter 4 on Miking, another way to alter the sound of a track that has already been recorded (particularly an overdubbed track) is through the use of reamping. This process lets us play back any recorded signal through any type of speaker setup (high-quality speakers, guitar amps, a leslie … you name it) and then re-record the room sound to a new set of tracks that can be used to add a greater sense of space and dimension to the originally recorded track.

Although the concept of recording an instrument directly and playing the track back through a miked amp at a later time is relatively new, the idea of using a room's sound to fill out the sound of a track or mix isn't. The reamp concept takes this idea a bit further by letting you go as wild as you like later in the production process. For example, you could re-record a single, close-miked guitar amp and then go back at a later time and layer a larger, distant stack on top of the original track. Of course, this process works really well for other types of instruments and vocal tracks as well.

Whenever a track is being overdubbed within a larger acoustic space, this use of extra, distant room mics to possibly fill out the sound in a stereo or immersive mix can be used during the overdub session itself.

CHAPTER 18

The Art and Technology of Mixing

In the past, almost all commercial music was mixed by an experienced professional recording engineer under the supervision of a producer and/or artist. Although this is still true at many levels of high-end production, with the emergence of the project studio, the vast majority of facilities have become much more personal and cost-effective in nature. Additionally, with the maturation of the digital revolution, artists, individuals, labels and enthusiasts are taking the time to gain experience in the artistry, techniques and work habits of creative and commercial mixing in their own production workspaces.

Within music, audio-for-visual and audio production, it's a well-known fact that professional mixers have to earn their "ears" by logging countless hours behind the console. Although there's no substitute for this experience, the mixing abilities and ears of producers and musicians outside the pro studio environment are also steadily improving as equipment quality gets better and as practitioners become more knowledgeable about proper mixing environments and techniques – quite often by mixing their own projects and compositions.

THE ART OF MIXING

Remember the old music joke: Q: How do you get to Carnegie Hall? A: Practice, kid, practice! Well, the same goes for the process of learning how to mix. In short, mixing is:

- First and foremost, the art of listening.
- The art of making decisions based upon what you hear and then acting upon these artistic decisions.
- The process of blending art and a knowledge of audio technology and signal flow to turn these decisions into a technological reality so as to create an artistic vision.
- A very subjective and personal artform. There is no right or wrong way, and no two people will mix in exactly the same way. The object of the process is to create a mix that "frames" (shows off) the music in the best possible light .

DOI: 10.4324/9781003260530-18

Ear Training

It's a simple fact that no one learns to play an instrument overnight – at least, not well. The same goes for the art of listening and making complex judgments within the mixing process. Yet for some reason, we sometimes expect ourselves to sit down at a console or workstation and instantly make a masterpiece. I (DMH in this case) personally remember sitting down at my first recording console (a Neve) and having a go at my first attempt at a mix. I was really nervous and completely unsure of what to do next. I remember moving the faders and saying to myself – Ok, now what? Oh yeah, EQ [equalizer], then I'd start blindly fiddling with the EQ. Well, we're here to tell you that this is totally normal! You can't expect to be an expert on your first try, or your second, or your hundredth. Taking time to learn your "instrument" takes time, patience and persistence. After all of this work, you'll begin the process of gaining enough experience to follow your gut instincts towards creating a professional mix.

In addition to practicing with your own mixes, there are many websites that offer free session downloads that'll quickly let you get your hands on many types of projects that have been recorded at various levels of professionalism.

Another way to get insights into the creation of a mix is to sit down with your favorite desert island songs, albums or LPs and play them in your production room. How do they sound when you actively listen to them? Later, you might take the time to listen to them over your favorite headphones. How does this change the experience for you? What can you learn from the music and their mixes?

ACTIVE LISTENING

It was said in the previous chapter that it's critically important that we be an active listener in the recording process. Obviously, this holds equally true in the process of mixing sounds into a final, cohesive song, project or media production.

The act of mixing is, in and of itself, an undertaking of combining sounds in a way that blends well, is artistically pleasing and engaging, and relates the "story" of the project to the listener in the best possible light. This "requires" that we take the time to delve into the subtle combinations of sounds over the entire sonic spectrum as they combine, interact and (in the end) relate their own truth to the listener.

Of course, the art of mixing requires that we take the time to dive in and actively listen. One of the main things that a mix engineer will be doing throughout his or her career is listening to a mix, often over and over. This allows us to become familiar with the nuances of the song and/or project. In this modern age, the instant recall aspect of a digital audio workstation (DAW) gives us the ability to keep going back to a mix and then improving it ad infinitum. This can be a good thing and potentially a bad thing, as many an artist who mixes their own work can attest. To this "it's never done until it's perfect" aspect of mixing, we can only say … go easy on yourself. Music is often a process of self-discovery and ongoing self-expression.

In the end, just as with learning an instrument or doing anything well, the fact remains that as you mix, mix and mix again, you will get better at your listening and mixing skills. It's a matter of experience matched with a desire to do your best.

Preparation

Just as preparation is one of the best ways to ensure that a recording session will go well, the idea of preparing for a mix can also help make the process go more smoothly and be more enjoyable for all.

PREPARING FOR THE MIX

There are no rules for approaching a mix; however, there are definitely guidelines. For example, when listening to a mix of a song, it is often best to listen to its overall blend, texture and "feel". A common mistake amongst those who are just beginning their journey into mixing would be to take each instrument in isolation, EQ it and try to sculpt its sound while listening to that track alone. When this is done, it's quite possible to make each instrument sound absolutely perfect on its own, but when combined into the overall mix, the blend might not work at all. This is because of all the subtle interactions that occur when all of the elements are combined. Thus, it's often a good idea to first listen to the tracks within the context of the full song, and then, you can go about making any mix changes that might best serve it.

> Remember: The mix always has to support the song. It should bring an energy to the performance in a way that allows the musical strengths and statements to shine through.

Preparing for a mix can come in many forms, each of which can save a great deal of setup time and frustration and can help with the overall outcome of the project. Here are a few things that can be thought through beforehand:

- Will you be mixing your own project, or will someone else be mixing? If it's your own project and it's in-house, then you're probably technically prepared. If the mix will take place elsewhere, then further thought might be put into the overall process. For example, will the other studio happen to have the outboard gear that you might or will need? Do they have the plug-ins that you're used to or need for the session? If not, then it's your job to make sure that you bring the installs and authorizations to get the session up and running smoothly, or print the effects to another track.
- If you'll be mixing for another artist and/or producer, it's often helpful to fully discuss the project with them. Is there a particular sonic style that they're after? Should the mix be aggressive or smooth sounding? Is there a particular approach to compression and effects that should be taken?

- The band or artist might also consider providing the mix engineer with a copy of their own rough or demo mix. This might give insights into how the song might best be mixed, effected or approached.
- Have the speaker levels been balanced, such that the center and other panning positions will be accurate? This can be done by sending pink noise equally to the speakers and then adjusting them for equal level.
- Is the mix and/or project to be professionally mastered? If so, are there any instructions from the mastering engineer? For example, he/she might ask that the final mix be delivered in two versions … one with final bus EQ and compression/limiting and another version without any final processing (allowing the mastering engineer to add his or her own final processing).

Given the fact that mixing a song or project is an art, by its very nature, it is a subjective process. This means that our outlook and the very way that we perceive a song will affect our workflow as we approach the mix. Therefore, it's often a good idea to take care of ourselves and our bodies throughout the process.

- Try to be prepared and rested as you start the mix process. Working yourself too hard during the recording phase and then jumping right into the mix just might not be the best approach at that point in time.
- By the same token, over-mixing a song by slaving behind the board for hours and hours on end can definitely affect the way that you perceive a mix. If you've gone all blurry-eyed and your ears are tired (the "I can't hear anything anymore" syndrome) – obviously, the mix could easily suffer.
- You might want to take breaks – sometimes loooooooongg ones. If you're not under any major time constraint, you might even consider coming back to the mix a day or even a week later. This can give a fresh perspective without ear fatigue or any short-term thoughts that might cloud your perception. If this is an option, you might try it out and see if it helps.
- Regarding monitoring, it's often a good idea to use reference monitors that you trust. As such, it's common practice for a mix engineer and/or artist to request their favorite speakers or to bring their own into the studio for the final mix.
- Unlike during the 1970s, when excruciatingly high sound-pressure levels (SPLs) tended to be the rule in most studios, recent decades have seen the reduction of monitor levels to a more moderate 75- to 90-dB^{SPL}. A good rule of thumb is that if you have to shout to communicate in a room, you're probably monitoring too loudly. From time to time, you can always jack it up to 11 to check the mix at higher volumes and then turn it back down to a moderate level. Taking occasional breaks isn't a bad idea either.
- Listen to several speaker types – at home, in your car, on your iPod/phone. Jot down any thoughts and comments that might come in handy should you need to go back and make adjustments. Strangely enough, leaving the room and listening to the mix from further away (with the door open) will often give clues as to how a mix will sound.

- If you have the time (i.e., are working in your own project room), you might want to take a week off and then go back and listen to the mixes with a fresh perspective. You'd be surprised how much you'll miss when you're under pressure.

A dear friend within the Grammy organization once told DMH: "Dave, the one thing that I've found amongst all engineers, is the fact that they are seeking that 'perfect sound' ... it's something that they 'hear' in their heads, but are never quite able to reach that Holy Grail". From a personal standpoint, I can say that this is true. I'm always sonically reaching for that killer sound that's always just beyond reach.

This brings us to: "Wherever you may be, there you are!" By this, I mean: we all have to start somewhere. If you're just starting out, your level of mix sophistication will definitely be different than after you've had several years of intensive mixing experience under your belt. Be patient with yourself and your abilities while always striving to better yourself – always a fine line to walk.

From an equipment point of view, it's obvious that you're not going to start out with your dream system. We all have to start the process by learning what speakers, mics, DAW, etc. will best work for us at our current stage of knowledge, budget and development. Later, your tastes in tools, studio layout and production techniques will surely change – this is all part of growing. Your own personal growth will definitely affect your choice of tools and toys as well as the way that you integrate with them and your acoustic environment. It's all part of the personal journey that is called "your life and career". Be patient with yourself, and enjoy the journey.

THE TECHNOLOGY OF MIXING

Now that all of the tracks of a project have been recorded, assembled and edited, the time has come to make use of an audio production console, mixer or DAW mixer to mix the song or program material into a final media form (Figure 18.1) by combining the overall levels and overall tone character of the various tracks in a production with respect to:

- Relative level
- Spatial positioning (the physical placement of a sound within a stereo or immersive field)
- Equalization (affecting the relative frequency balance and blending abilities of a track)
- Dynamics processing (altering the dynamic range of a track, group or output bus to optimize levels or to alter the dynamics of a track so that it "fits" better within a mix)
- Effects processing (adding reverb, delay or pitch-related effects to a mix in order to augment or alter the piece in a way that's natural, unnatural or just plain interesting)

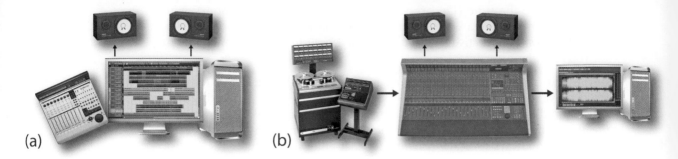

(a)   (b)

FIGURE 18.1
Very basic representation of the mixdown process. (a) Mixing in the box using a DAW. (b) Using an analog console or large-scale controller.

This control over volume, tone, blending and spatial positioning for any or all signals that are applied to its inputs from microphones, electronic instruments, effects devices, recording systems and other audio devices are then combined together into a single, cohesive mix using these various types of physical and virtual mix surfaces.

The mix surfaces basically occur in four forms:

- It can be completely analog in nature (Figure 18.2), whereby all signals passing through the console or mixer are analog.
- It can be completely digital, whereby all signals are converted from analog to digital (or fully stay in the digital domain) and pass through the console or mixer in purely digital form.
- It can be a controller surface (Figure 18.3), whereby no signal will pass through the console surface but instead, will transmit MIDI or other controller messages to remotely control the virtual mixer within a DAW.
- It can be completely virtual in nature (Figure 18.4), allowing all mix and routing functions to be fully carried out in the digital domain over a DAW's mix surface.

FIGURE 18.2
At the analog mixing console. (a) DMH and Martin Skibba at nhow hotel, Berlin (courtesy of nhow Berlin, www .nhowhotels.com/berlin/en, photo courtesy of Sash). (b) DMH and Emiliano Caballero at Galaxy (courtesy of Galaxy Studios, Mol Belgium, www .galaxy.be).

An *audio production console* (which also goes by the name of board, desk or mixer) should also provide a straightforward way to quickly and reliably route these signals to any appropriate device in the studio or control room so they can be recorded, monitored and/or mixed into a final product. A console or mixer can be likened to an artist's palette in that it provides a creative control surface that allows an engineer to experiment and blend all the possible variables onto a sonic canvas. Of course, in this day and age, a DAW is most often used as this palette to create the final mix, which is made "in the box".

(a)   (b)

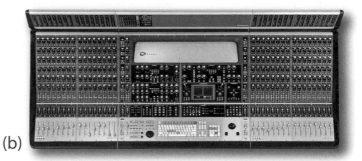

(a)

(b)

Understanding the Underlying Concept of "the Mixing Surface"

In order to understand the process of mixing, it's important that we understand one of the most important concepts in all of audio technology: the *signal chain* (also known as the *signal path*). As is true with literally any audio system, a recording mixer or console (DAW, digital or otherwise) can be broken down into functional components that are chained together into a larger (and hopefully manageable) number of signal paths. By identifying and examining the individual components that work together to form this chain, it becomes easier to understand the basic layout of any mixing system, no matter how large or complex. In order to gain insights into the layout of a mixer, let's start with the concept that it's built of numerous building-block components, each having an input that moves to its output, and then to the input of the next functional block to its output, and so forth down the signal path until the end of the chain is reached. Here are a few important things to keep in mind regarding any audio signal chain path:

- Naturally, whenever a link in this source-to-destination path is broken (or incorrectly routed), no signal will pass. In the heat of production, it's easy to route something into the wrong input or output. Following the "gozintagozouta" approach might seem like a simple concept; however, keeping it always in mind can save your sanity and your butt when paths, devices and cables that look like tangled piles of spaghetti get out of hand

FIGURE 18.3

Consoles as a controller surface. (a) Solid State Logic Duality Console, functioning in its DAW controller mode (courtesy of Solid State Logic, www.solid-state-logic.com). (b) Digidesign ICON D-Command Integrated console (courtesy of Avid Technology, Inc., www.avid.com).

FIGURE 18.4

DAW on-screen mixer. (a) ProTools on-screen mixer (courtesy of Avid Technology, Inc., www.avid.com). (b) Nuendo on-screen mixer (courtesy of Steinberg Media Technologies GmbH, a division of Yamaha Corporation, www.steinberg.net).

(a)

(b)

(fortunately, most or all of these problems have been eliminated in a DAW's signal flow).

- Try to run each block in the signal chain at its optimum gain level. Too little signal will result in increased noise, while too much will result in increased distortion.
- The "Good Rule" definitely applies to the audio mix signal path. That's to say that the overall signal path will be no better than its weakest link. Just something to keep in mind.

The Mixer/Console Signal Path

In the previous chapter on recording, we took a look at the console/mixer's signal path from the standpoint of a hardware mixer. This time, let's start our quest into the virtual side of mixing by taking a conceptual look at the various systems from a DAW's point of view.

In a traditional mixer design, the signal flow for each input travels vertically down a plug-in strip known as an I/O module. Figure 18.5 shows a virtual input strip that flows (much like its hardware counterpart) from top to bottom in the following way.

Although the graphical user interface (GUI) layout of a virtual DAW mixer generally won't be the same from one piece of software to the next, the signal flow will follow along the same or similar paths. Therefore, grasping the concept of the traditional signal chain will be extremely useful for grasping the overall signal flow concept.

The numerous building-block components within the chain will likewise flow from a source signal through the strip to an output (destination) mix bus. The output of each source device must be literally or virtually connected to the input of the device that follows it, and so on until the end of the audio path is reached. Although the signal chain is greatly simplified within a virtual DAW strip, keeping this simple concept in mind is still very important. Now, let's begin to work our way through Figure 18.5's signal path.

CHANNEL INPUT

FIGURE 18.5

ProTools virtual on-screen mixer strip (courtesy of Avid Technology, Inc., www.avid .com).

The first link in the input chain is the channel input, with the source signal being selected at the I/O (input/output) dialog box. In a virtual DAW setting, these levels have generally been optimized at the time of recording, although most DAWs

inserts   sends   routing auto- group   pan   fader   gain/levels
                                        mation

will allow these signal levels to be trimmed up or down to best match the best overall channel mix levels before being sent through to the input strip.

INSERT POINTS

The Insert Points on a strip allow signal processing effects to be "inserted" directly into the path of the strip in such a way that a plug-in can alter or augment the track's audio signal (Figure 18.6). Plug-ins that are inserted into a channel track will most often (but not always) be amplitude-based in nature. That's to say that the effect will often be used to change the amplitude nature of the track (i.e. EQ, compression, limiting).

SEND POINTS

Unlike an insert point (where an effects or processing plug-in is directly inserted into a strip's audio path), the sends are used to route and mix signals from multiple input strip channel sources to a single effect device/plug-in and/or monitor/headphone destination. This combined feed is then used to create a mono or stereo sub-mix that's then sent to a signal processing, monitoring or recording destination (Figure 18.7). Plug-ins that are sent to a device from multiple sources will most often (but not always) be time-based in nature. That's to say that the effect will often be used to change the temporal nature of the track (i.e. delay, reverb).

With regard to a workstation, using an aux send is a great way to make use of processing effects while keeping the processing load on the CPU to a minimum. For example, let's say that we wanted to make wide use of a reverb plug-in that's generally known to be a CPU hog. Instead of separately plugging this reverb into a large number of tracks as inserts, we can greatly save on processing power by plugging the reverb into a send bus. This lets us selectively route and mix audio signals from any number of tracks and then send the summed (mixed) signals to a single plug-in that can then be mixed back into the master output bus. In effect, we've cut down on our power requirements by routing any number of audio tracks to a single processing device. Knowing the functional difference between an insert and a send can be a powerful engineering tool.

AUTOMATION

One of the greatest strengths of the digital age is how easily all of the mix and effects parameters can be automated and recalled within a mix. Although many large-scale consoles are capable of automation, computer-based DAW systems are particularly strong in this area, allowing complete control over all parameters

FIGURE 18.6
An effects plug-in can be "inserted" directly into a channel's signal path.

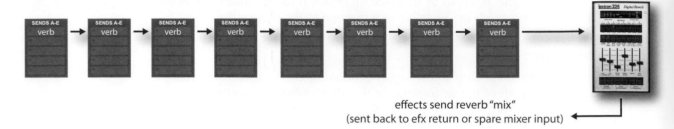

effects send reverb "mix"
(sent back to efx return or spare mixer input)

FIGURE 18.7
An effects plug-in can be "sent" from multiple sources to a single processing device that can then be mixed back into the final mix bus.

throughout all phases of a production. The beauty of being able to set basic levels within a mix or to mix under automation is that a mix can be built up over time, allowing multiple mix versions to be saved, so that we can go back and explore other production avenues or correct any potential problem. In short, the job of mixing becomes much less of a chore, giving us the time to pursue less of the technology of mixing and more of the art of mixing … at least, in theory. What could be bad about that?

Although terminologies and functional offerings will differ from one system to the next, control over the basic automation functions will be carried out in one of two operating modes (Figure 18.8):

- Write mode
- Read mode

FIGURE 18.8
Automation mode selections. (a) Pro Tools showing auto selectors within the edit and mixer screens (courtesy of Avid Technology, Inc., www.avid.com). (b) Cubase/Nuendo showing auto selectors within the edit and mixer screens (courtesy of Steinberg Media Technologies GmbH, www.steinberg.net).

Write Mode

Once the mixdown process has gotten under way, the process of writing the automation data into the system's memory can begin (actually, that's not entirely true, because basic mix moves could easily begin during the recording or overdub phase). When in the write mode, the system will begin the process of encoding mix moves for the selected channel or channels in real time. This mode is used to record all of the settings and moves that are made on a selected strip or strips (allowing track mixes to be built up individually) or on all of the input strips (in essence, storing all of the mix moves live and in one pass). The first approach can help us to focus all of our attention on a difficult or particularly important part or passage. Once that channel has been finessed, another

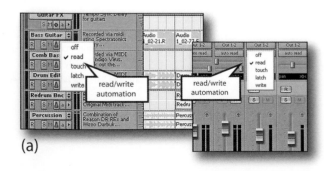

(a)

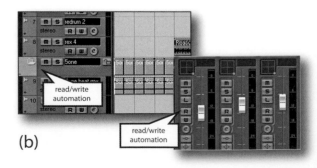

(b)

channel or group of channels can then be written into the system's memory – and then another, until an initial mix is built up.

Often, modern automation will let us update previously written automation data by simply grabbing the fader (either on screen or on the console/controller) and moving it to the newly desired position. Once the updated move has been made, the automation will remain at that level or position until a previously written automation move is initiated, at which point the values will revert to the existing automation settings.

Read Mode

An automated console or DAW that has been placed into the read mode will play the mix information from the system's automation data, allowing the on-screen and moving faders to follow or match the written mix moves in real time. Once the final mix has been achieved, all you need to do is press play, sit back and listen to the mix.

Drawn (Rubber Band) Automation

In addition to physically moving on-screen and controller faders under read/write automation control, one of the most accurate ways to control various automation parameters on a DAW is through the drawing and editing of on-screen rubber bands. These useful tools offer a simple, graphic form of automation that lets us draw fades and complicated mix automation moves over time.

This user interface is so named because the graphic lines (which represent the relative fade, volume, pan and other parameters) can be bent, stretched and contorted like a rubber band (Figure 18.9).

Commonly, all that's needed to define a new mix point is to click on a point on the rubber band (at which point a box handle will appear) and drag it to the desired position. You can change a move simply by clicking on an existing handle (or range of handles) and moving it to a new position.

Before we leave the section on automation, it's important to point out that it's always wise to save your mix at various stages of its development. A mix can be saved under different names (songname01, songname02, 02mybestmix,

FIGURE 18.9

Automation rubber bands. (a) Pro Tools (courtesy of Avid Technology, Inc., www.avid. com). (b) Cubase/Nuendo (courtesy of Steinberg Media Technologies GmbH, www. steinberg.net).

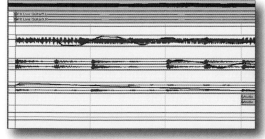
(a)

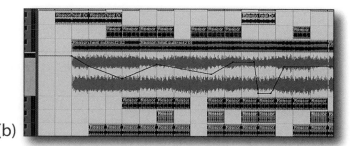
(b)

03jimsnew approach, etc. … personally, DMH saves these versions within the project's directory known as "songname_mixbacks"). Saving previous versions never hurts, as it makes it easier to return to the point should you make a mistake or simply need to take a step back in time.

GROUPING

As we read in the previous chapter, virtually every DAW, console and professional mixing system allows any number of input channels to be organized into groups. This feature can come in very handy during the recording process, but it can often be downright indispensable within the mixing process.

A group is essentially a created sub-mix that consists of instruments, vocal tracks, etc. that are then assigned to a single group track, which gives the mixer control over overall levels, overall effects parameters and plug-in instance capabilities. Giving you immediate control over a collection of tracks from a single, master channel is what this super-useful function is all about. For example, in Figure 18.10, a basic mix screen has been set up that includes four groups … drums, piano, guitar and vocals. It's easy to see that the individual tracks have been assigned to their respective groups (at the top), thereby allowing the group channels (at the left) to act as relative level sub-masters. You can also see that the group channels are then routed out to the final stereo output bus. Now that you've learned a bit about grouping, we urge you to get out your DAW of choice and begin to experiment with this powerful mix tool.

A word about adding effects to a group:

One of the nice side benefits to grouping is the ability to add any type of effects processing only to a specific group. For example, dynamics could be applied only to the drum group so as to tame its levels without affecting any other group or track. This example would help to keep excessive dynamics from triggering any dynamics that might be applied to the main output bus.

PANNING

The pan control within the signal flow path is able to provide:

- The panning of a mono signal between the left and right sides of the stereo field
- The balancing of a l/r stereo signal or immersive set of tracks so as to steer or better position the signal within the soundfield
- Width control over a l/r stereo signal or immersive set of tracks so as to affect the width (size) of the stereo or immersive signal within the soundfield

More information on the pan and positioning controls that are used within immersive audio can be found within Chapter 20 on Immersive Audio.

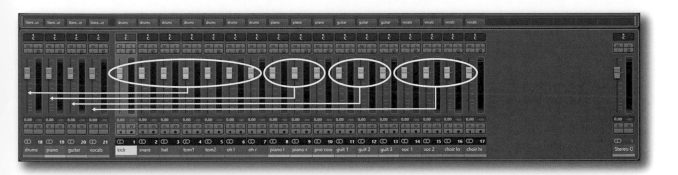

FIGURE 18.10
Groups can be created that allow the relative levels of assigned tracks to be controlled from a single group sub-master.

CHANNEL FADER

The channel fader, of course, is used to vary the final signal levels as they exit the input strip. The "solo" button is used to mute all other channels within the monitoring path, allowing us to hear only that channel's signal … while the "mute" button performs the inverse function of muting only the selected track, allowing us to exclude it from the monitor path.

GAIN LEVELS

The gain and metering levels indicate the exact current fader gain level (generally shown as 0.00 or dB and hundredths of a dB) … while the level indicator displays the overall peak value of the track passage that's being played.

CHANNEL NAME

An important identifier that deserves greater attention when recording or building a project, as it's a lot easier to look for a track that's called "ECs killer rif" than "track0015".

Gain Structure, Baby!

Actually, we lied when we said that there are no rules to the art of mixing. There's one big one – watching your gain structure. Just as with any analog path, any link in the digital chain that exceeds full scale will lead to a distorted signal.

When dealing with the concept of avoiding excessive levels along a signal path, it's obviously a wise idea to be aware of the overall levels as they pass throughout an input strip. This is especially true in this day and age of making everything sound bigger, better, louder!

In fact, this last concept of keeping levels under control isn't always as simple as it might seem, due to the fact that digital plug-ins almost always are hidden out of sight, under the hood. For example, it's a simple matter to insert a compressor directly into a strip's path. Once inserted, the plug-in will usually show its full GUI screen (letting us see all the levels and controls at a glance), but after that, it's quite common for us to simply close this screen and simply forget that it's quietly (or not so quietly) doing its job in the background. Once the mix has gotten underway, any number of factors could come into play that will cause the inserted plug-in to output too high a signal. In fact, gain isn't the only insert culprit that could potentially cause problems with a mix ... changes to a noise gate or a limiter with too short an attack might be among the possible causes that could be giving you trouble. Now, multiply this by any number of project tracks, and you'll begin to see why occasionally taking the time to look under the hood at the levels and settings at which these inserts are working can definitely begin to pay off.

At this stage, the primary concern that needs to be kept in mind is the careful handling of the program's overall dynamic range ... essentially, the idea of keeping your signals at optimum operating levels so as not to introduce noise or undue distortion. This logic, however, begins to blur a bit when we consider that most of the output signal levels are digital in nature and employ internal computational processing that makes use of a 32- or 64-bit architecture. Therefore, although a DAW working with 24-bit track files could have an overall dynamic range of up to 144 dB, the internal dynamic range for a floating-point processor could have a much higher dynamic range. What does this mean in practice? Basically, it means that the gain structure can be much more forgiving (with certain DAW systems), allowing a greater internal processing dynamic range. It should be pointed out, however, that this in no way should be used as an excuse for failing to follow good gain structure practices within a mix. Mixing, like most things of importance, is all about the details as well as the art.

MIXING AND BALANCING BASICS

Once all of the tracks of a project have been recorded, assembled and edited, it's time to put our understanding of this technology to use to mix your tracks into its final media form. The goal of this process is to combine audio, MIDI, instrument, group and effects tracks into a pleasing form using such traditional tools as:

- Relative levels
- Spatial positioning (the physical panned placement of a sound within a stereo or immersive field)
- Equalization (affecting the relative frequency balance of each track)
- Dynamics processing (altering the dynamic range of a track, group or output bus to optimize levels or to alter the dynamics of a track so it fits better within a mix)
- Effects processing (the adding of reverb, delay or pitch-related effects to a mix in order to augment or alter the piece in a way that's natural, unnatural or just plain interesting)

Within a well-produced project, sounds can be built up and placed into a sonic stage through the use of natural, psychoacoustic and processed signal cues to create a pleasing, interesting and balanced soundscape. It's pretty evident that volume can be used to move sound forward and backward within the sound field and that these relative levels can be used to position a sound within that field. It's less obvious, however, that changes in timbre and delay cues can also be introduced through EQ, delay and reverb to move sounds within the stereo or immersive field. Of course, all of this sounds simple enough; however, in practice, the dedication that's required to hone your skills within this evolving art is what mixing careers are made of.

Before we get started, I'd like to first point you in a direction that just might help your overall mixing technique:

- Let's assume that you're familiar with your mixing environment and that it has been properly set up with regard to speaker positioning, balancing and any other adjustments that need to be made … and yes, this might include setting up your favorite lighting and seating scenario, etc.
- Next, once you've imported the tracks into the session, you might go about setting up the edit and mix screens in a way that best fits your way of working. Would you be making use of an external fader/mix controller? Would setting up various group tracks help? Do you want to set up your effects tracks in advance or as you go?

Okay, now let's take a moment to understand the process better by walking through a simple mix. At this point, let's keep it simple! If there's a trick you can use to make your project go more smoothly, use it. At the outset, let's concentrate strictly on the various channel strips in the session (tracks, groups, effects, instrument tracks, etc.) … we'll dive in a more in-depth fashion on the main output bus towards the end of this chapter.

Remember, there's no right or wrong way to mix as long as you watch your levels along the signal path.

With the introduction of the DAW, never before has the process of recording and mixing been more blurred. This is especially true in modern and electronic music production, where the recording process can take place over days, weeks, months or longer. In reality, with the use of sample libraries and electronic instruments, a project is never really recorded … It's sculpted and produced over time using sonic components that combine as an ongoing process.

As such, it's important to realize that the mixdown phase actually began on the very first day that the project was born … as an ongoing, evolving process. Once finished, a final mix can be actually more of a finesse of a creation that has taken place over time. Of course, this process will all depend upon the musical style, the artists involved and the chosen process. That's the nice thing about art … it can be created in an infinite number of ways and styles.

1. If you don't happen to have a multitrack session of your own, you can begin your journey by searching the Web for "free multitrack sessions". There are several sites that have a number of recorded sessions that exist in various styles.

2. Next, let's begin building a mix by setting the output volumes on each of the instruments to a level that's acceptable to your main mixer or console. From a practical standpoint (if a basic mix wasn't built during the production phase), you might want to start with all of the faders down. You can then start to bring the basic tracks up one at a time until a desirable blend begins to take shape. You could also take an entirely different approach by setting them all at unity gain and then adjusting them until a mix begins to develop.

    A good rule of thumb is to begin your mixes at a consistent monitoring level (85 dB^{SPL} C weighted is often recommended). In this way, your headroom, dynamics and overall balance will be at their optimum balance levels.

3. The next step is to repeatedly listen to the project as it begins to take shape, making any fader and pan changes that are necessary until it starts to come together. This allows us to get accustomed to the song, soundtrack or audio program so that we can make the appropriate artistic decisions. If there's a vocal track, it could be brought into the mix at this time to see how the overall blend works, or you could work on the instrument tracks and bring the vocal track in later. In any case, the idea is to begin the process of shaping the mix in a way that sounds pleasing and reinforces the song's musical style and intentions.

4. EQ can be applied to any problematic track that doesn't quite sound right or has problems fitting into the mix properly. A general consensus on EQ is that it's often best to equalize a track within the context of the mix (i.e., not by listening to it as a lone, soloed track). If a problem track does arise, try listening to the offending instrument on its own, making any corrections, and then go back to listening to the entire track with the correction. Did the correction help the mix?

    Another approach to sound shaping using EQ is called "carving". This approach uses EQ to restrict frequencies that are present on a track so that they don't conflict with other important instrument tracks that exist in the same range. For example, a synth track that has a great deal of additional low-frequency energy might be rolled off at the low end so that it doesn't "muddy" up the low end and/or conflict with the kick or other bass instruments.

    As with all things sound, letting your ears be your EQ guide takes time and experience – be patient with yourself. Just remember, when starting out, the temptation is really strong to reach for the EQ and other controls without thinking it through. Many of the industry greats strive to do the adjustments during the recording phase so that few adjustments will be needed during mixdown. You might also want to watch out for getting rid of too much low end across the session; that is one of the more common mistakes when learning how to mix.

5. Effects can be inserted into any track that might benefit from dynamics control (compression, limiting, gating, etc.), delay processing or any other type

of plug-in that might benefit a particular track. You might want to pay special attention to gain staging throughout the path.

6. As was said earlier, you might want to group various instrumental sections together within the mix. The time savings of this process really can't be overstated. For example, by grouping instrument types together:

   • You can reduce the volume on several tracks by simply grabbing a single fader instead of moving each track individually.

   • When a session contains lots of tracks, soloing the group tracks can help you find a track much more quickly than searching through the tracks – it really does!

   • Effects (and even reverbs) can be quickly and easily added to an instrument section at the group level (either as an insert or as a send).

7. Should the mix levels need to be changed at any point from their initial settings, you might turn the automation on for any needed track and begin building up the mix with its help. Remember, this is an option that's open to you when it's needed, not something that you absolutely must use.

It should also be pointed out that track fader and rubber band automation isn't the only form of mix automation that's available to us. Most DAWs allow defined segments within the edit window to be altered in gain simply by grabbing the segment at its top boundary and moving it up or down. For example, if a single note were too loud in a passage, you could simply locate the section, define its in and out boundaries and then drag it up/down to its desired new volume.

8. At this (or any other time), you might want to introduce a few sends into the project. Using sends to add a favorite reverb and other processors at this point can help add life to the mix without unnecessarily using too much processing power.

Just as reverb, delay and other time-based effects can be added to the mix, should you wish to add a bit of controlled "grit" or "dirt" to a mix, an effects send could be set up to add a degree of compression, tape emulation saturation or whatever you might want to add to the mix. Once created, this dirt, glue or whatever you want to call it can be tastefully added to a single channel or grouped set of tracks by mixing it in at the send level. The same goes for room ambience, as a room simulation plug-in (or actual mics in the studio) can be entered into the mix at the send level to add a degree of openness and acoustic realism to the mix.

Main Output Mix Bus

The main output bus of a DAW (or any type of mixing system) acts to combine (sum) the output signals from all of the channel, group, effects and instrument strips together into a final mono, stereo or immersive output signal. This combined signal then constitutes your final mix.

The first aspect of the main output bus that can be looked at is output levels. As with all things audio, the concept of how loud to make the final mix is a very personal one, and there are a number of things that might be considered towards making your choice:

- Will this be the final master? If so, the choice of output levels is up to you or the client.
- Will it be mastered by another? If so, the choice of final levels (and whether further processing will be added) might be requested by the mastering engineer.
- Personal preferences … whether to make the mix levels high (most likely with extra dynamic processing) or more moderate in level (allowing the natural dynamics of the material to come through).

As always, this decision is best made with the intent of the song or program material in mind … it should match the style and general feel of the project.

Of course, the next big topic on everyone's mind with regard to the mix bus is the idea that it can be further processed using plug-in inserts that can further finesse the sound into its best final form … the final mix bus (or some might call it mastering) chain (Figure 18.11).

This final signal processing chain might easily include dynamic processors (compressors and/or limiters), sound shaping processors (tape emulators, distortion plug-in, etc.), self-mastering plug-ins (which often contain all of the above within an integrated plug-in) as well as EQ and/or spectral sculpting tools (Figure 18.12).

Once you're happy with your basic mix, you'll first want to save the mix and then export to a file within your DAW. This export (bounce) can often be done in real time or non-real time, giving you a final version of the mix that can be used in the latest project or production.

FIGURE 18.11

Main output bus (highlighted) showing various plug-ins that can be used to "master" the track (courtesy of Steinberg Media Technologies GmbH, a division of Yamaha Corporation, www.steinberg .net).

SOME FINAL WORDS

Of course, the process of making a truly professional mix is one that comes with time and experience. Once you've gotten to know your system and your own personal mixing style, you'll eventually arrive at a "sound" that's all yours. So, now you've put in the time on your latest project, and you've done your absolute

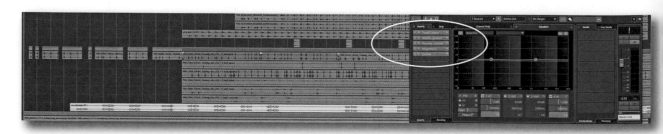

Tube Compressor   Multiband Compressor   Sound Shaper   Tape Emulator   Limiter

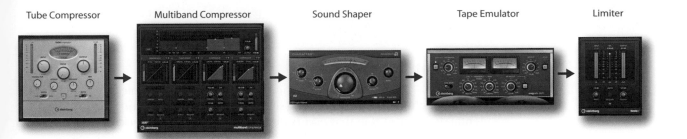

best; hopefully, you'll be able to sit back and say, "Yeah, that's it!" Once you've finished, you might want to:

- Listen to your mixes on various other systems (hopefully, ones you can trust), just to hear how they sound
- Listen to a track next to some of your favorite "desert island" mixes
- Also get the opinions of people you know and trust

FIGURE 18.12
DMH's favorite output bus tools. This is his favorite, but you'll almost certainly have your own favorite sound shaping tools in order to get the best results for you.

In addition to the concept of capturing the pure music, feel and artistry of a production, one of the primary goals during the course of recording a project is the overriding concept that the final product should have a certain "sound". This sound might be "clean", "punchy", "gutsy" or any other sonic adjective that you, the artist, the producer or the label might be striving for. If the final master(s) don't have a sound that makes everyone happy, you, the band and/or the producer will be faced with the following:

1. You might say, "Well, that's the best we can do". Of course, this is always an option; however, before you sign off on the project, ask yourself if it will be something that you and the team will be proud of once it's out to the public.
2. If you need to keep working (and time permits), you might take a bit of a break, get some ideas, opinions and a fresh perspective, and then put your nose to the grindstone until you and your team feel that it's the best that it can be – or at least know you've given it your best. All of this brings up an important point.

Getting Too Close to the Mix

With all these tools and techniques, combined with the time that it takes to see a good recording and mix through to the end, there is one danger that lurks in the minds of all musicians (more than anybody, if it's your own project), producers and engineers – the fact that we might simply be too close to the mix. I (DMH) speak from experience when I say that by the time I've neared the finish of a project and have made so many careful adjustments to get that right production and sound, there are definitely times when I have to throw my hands up and say, "Is it better? I'm so close to the mix and music that I simply can't tell!" When this happens to you, don't fret (too much) – it's all part of being in the production club. You might simply need a break, a walk, some time on a sailboat or just to get away from it, or you might have to just plow through to get it done on time. Just realize that this is totally normal.

Quality Control (QC)

With this aspect of being too close to the mix, it's definitely possible to overlook things that might go wrong in a mix (such as distortion in a short section, an accidental drop-out, a pop that occurs during a quiet section). Heck, even when you're fresh and feeling on top of the world in a mix, something can go wrong and bite you in the butt.

This is why it's ALWAYS important to QC a mix once it's done. Believe us, it happens to the best of them (although less often, because they're usually smarter than that). After the final export, take the time to listen to the entire piece before you send it out. The reputation that you might save just might be your own, and your client's. Don't say that we didn't warn you!

Wherever You May Be, There You Are

For those of you who are just beginning, our best advice is to be patient with yourself. Learning how to make a truly excellent mix takes time, patience and practice. For those of you who are seasoned pros, we'll offer up the same advice … it's an art form … and art will often take on a life of its own. Good or bad, the process of mixing will almost always be a learning (and possibly humbling) experience.

Actually, the topic of the art of mixing could easily fill a book (and we're sure it has); however, we'd simply like to point out the fact that it is indeed an art form and as such, is a very personal process. On a personal DMH footnote, I remember the first time I sat down at a console (an older Neve 1604). I was truly petrified and at a loss as to how to approach the making of my first mix. Am I over-equalizing? Does it sound right? Will I ever get used to this sea of knobs? Well, folks, as with all things – the answers come to you when you simply sit down and mix, mix, mix! It's always a good idea to watch others in the process of practicing their art and take the time to listen to the work of others (both the known and the not-so-well known). With practice, it's a foregone conclusion that you'll begin to develop your own sense of the art and style of mixing, which, after all, is what it's all about.

CHAPTER 19

The Art and Technology of Mastering

The *mastering* process is an art form that uses specialized, high-quality audio gear in conjunction with one or more sets of critical ears to help the artist, producer and/or record label attain a particular sound and feel before the recording is made into a finished, manufactured product. Working with such tools, a mastering engineer or experienced user can go about the task of shaping and arranging the various cuts of a project into a final form that can be replicated into a salable product (Figures 19.1 and 19.2).

In past decades, when vinyl records first ruled the airwaves and spun on everyone's sound system, the art of transferring high-quality sound from a master tape onto a vinyl master disc was as much (if not more) a carefully guarded art form as it was technology. Because this field of expertise was (and still is) well beyond the abilities and affordability of most engineers and producers, the field of vinyl mastering was left to a very select, magical few (and still is).

The art of preparing sound for transfer to disc, optical media and downloadable media is also still very much an art form that's often best left to those who are familiar with the tools and trade of getting the best sound out of a project. However, recent advances in computer and effects processing technology have also made it much easier for producers, engineers and musicians to own high-quality hardware and software tools that are specifically crafted to help create a professional-sounding final product in the studio or on a desk/laptop computer.

In short, the process of mastering a finished mix refers to the judgments and artistic decisions that are made to previously recorded material with respect to:

- Relative level balancing between songs within the project
- Dynamic level (altering the dynamics of a song so as to maximize its level for the intended media or to tighten up the dynamic balance overall or within certain frequency bands)
- Equalization
- Overall level
- Song sequencing

DOI: 10.4324/9781003260530-19

FIGURE 19.1
Darcy Proper in her mastering room at Valhalla Studios, New York (courtesy of Valhalla Studios, www.valhallastudiosny.com).

In the previous chapter, we have seen that once done, the next step towards completing an edited project mixdown might have one of three fates:

- The song or project is not mastered, meaning that the song or project is ready for action.
- The song or project might be self-mastered by the engineer and/or producer.
- The song or project might be sent to a professional mastering engineer for finishing touches.

Each of these possibilities, when left in the right hands, could be a valid choice. However, the reasoning for any of these choices should be thoroughly discussed with the producer and artist in the pre-planning phase, allowing for any on-the-spot change of plans should you require the services of a professional.

WHAT IS MASTERING?

In essence, the job of a qualified mastering engineer is to present the final, recorded product in the best possible sonic "light" for its intended media form and to smooth over any level and spectral imbalances within a project.

FIGURE 19.2
Gavin Lurssen at the mastering desk (courtesy of Lurssen Mastering, LA, www.lurssenmastering.com).

In order to answer the question as to whether you need a mastering engineer, it's always best to understand exactly what a mastering engineer does (or potentially

can do) for a project. In essence, the process of successfully mastering a project can:

Help a Project to Sound "Right"

At the outset, it's important to mention that a final product should have a certain "sound". That sound could be completely transparent (as might be sought after in a clean jazz or faithfully captured classical recording), or it could be as rough and raunchy as possible so as to grab the latest elektro-garage-steampunk band's audience by their sonic throats. It's all about the character, attitude and intent of the sound. In mastering, this is often accomplished through the use of careful level matching, equalization (EQ) matching and dynamics processing. As was also previously mentioned, this process not only takes the right set of processing gear but also requires experienced ears, which are working in a familiar environment and intuitively know how the project will most likely sound under a wide range of playback conditions.

Just how much a mastering engineer might be involved in shaping the actual sound can vary depending upon their level of involvement. The mastering engineer may simply affirm that the musicians/producer/engineer did a great job (i.e., "I hardly touched a thing – the original master was practically perfect!"), or moderate to heavy processing may be required in order to tame the sound into being what everyone wants. In short, if the final mix or mixes aren't up to snuff, having an experienced mastering engineer help you to "shape" the project's sonic character through the careful use of level balancing, dynamics and EQ could help save your sonic Technicolor day.

Help Match the Levels and Overall Character Within a Project

The basic truth is that it takes years for a mix engineer to learn the fine art of blending the songs within a mix so that they seamlessly combine together into a project that has a cohesive feel (if that's what's called for in the mix). A less experienced engineer, mixer and/or producer might come up with an album's worth of songs that abruptly changes in character and "feel" over the course of the project.

Another factor that can affect a project's sound is the reality that recordings may have been recorded and/or mixed in several studios, living rooms, bedrooms and/or basements over the course of several months or years. This could mean that the cuts could actually sound quite different from each other.

In situations like these, where a unified, smooth sound might be hard to attain, it's even more important that someone who's experienced at the art of mastering be sought out. This might involve the careful balancing of levels and dynamics from song to song, but it will almost certainly involve the careful matching of EQ to smooth out the sonic imbalances between the cuts so they don't stick out like a sore thumb.

Help With Overall Levels

The subject of how high to set the levels (or more accurately, relative levels, dynamics, compression/limiting, etc.) is a subject that can be (and has been) the subject of much heated debate. The major question here is "Can it ever be too loud?" In actuality, the question that we're really asking here is "How much dynamics can we take out of the music before it becomes too much to bear?"

Traditionally, the industry as a whole tends to set the average level of a project at the highest possible value. This is often due to the fact that record companies will always want their music to "stand out" above the rest when played on the TV, radio or mobile phone or on the Web. This is usually accomplished by applying compression to the track or overall project. Again, this is an artistic technique that often requires experience to handle appropriately.

I'm going to stay clear of opinions here, as each project will have its own special needs and "desires", but several years ago, the industry became very aware of a top-selling album that was released with virtually no dynamic range at all. All of the peaks and dynamics were lost, and it was pretty much a "flat wall" of sound. This brought to light that maybe, just maybe, there might be limits as to how loud a project should be. Of course, everyone wants to be heard in the playlist in an elevator, on the phone, etc. In reality, some degree of dynamics control is often necessary to keep your mix sounding present and punchy; however, a good mastering engineer can also help you be aware that over-compression can lead to audible artifacts that can "squash" the life out of your hard-earned sound. In fact, light to moderate compression or limiting might be of help for certain types of music, while classical music lovers often spend big bucks to hear a project's full dynamic range. In the end, it all depends on the style, the content and the message.

Help With the Concept of Song Ordering

Have you ever noticed that some of your favorite desert island recordings just seem to have a natural flow to them? An album that takes you on a specific journey is called a "Concept Album", and one of the signature callings of such an album is that one song will follow another in a way that just feels right.

This idea of choosing a project's song order is an art form that's best performed by the artist and/or producer to convey the overall "feel" of a project; however, if the production team doesn't have a good feel for how the album should "flow" (a rather sad idea), they might seek the mastering engineer's opinion.

Without much doubt, the running order in which the songs of a project are played will often play a role in the overall flow and tone of a project. The considerations for song order are infinitely varied and can only be garnered from experience and having an artistic "feel" for how their order and interactions will affect the final listening experience. Of course, sequence decisions are probably best made by the artist and/or producer, as they have the best feel for the project.

A number of variables that can directly affect the sequenced order of a project include:

- *Total length:* how many songs will be included on the disc or album? If you've recorded extra songs, could it include the bonus tracks on the disc? Is it worth adding a weaker song just to fill up the CD? Will you need to drop songs for the vinyl release?

- *Running order:* which song should start? Which should close? What order feels best and supports the overall mood and intention of the project?

- *Transitions:* altering the transition times between songs can actually make the difference between an awkward silence that jostles the mood and a transition that upholds the pace and feel of the project. The Red Book CD standard calls for 2 seconds of silence as a default setting between tracks. Although this is necessary before the beginning of the first track, it isn't at all the law for spacing that falls after that. Most editing programs will let you alter the index space timings between tracks from 00 seconds (butt splice) to longer gaps that help to maintain the appropriate mood.

- *Cross-fades:* in certain situations, the transition from one song to the next is best served by cross-fading from one track directly to the next. Such a fade could seamlessly intertwine the two pieces, providing the glue that can help convey any number of emotional ties.

Help With the Concept of Song Timing

When listening to your favorite desert island album, have you ever noticed how one song just seems to flow into another, or that an abrupt cut followed by a fast pickup of the next song just seems to fit perfectly with the music? This was almost certainly no accident. The concept of song timing is an intuitive process of setting the gap or cross-fade times between songs. It can make the difference between a project that has an awkward pause and one that "flows" smoothly from one cut to the next. Just remember that automatically setting the "gaps" to their default 2-second settings (when burning a CD) might not be in the project's best interest.

It should be noted that although the various tracks within a project can be assembled into a digital audio workstation (DAW), it's also possible to place the gap in and out times of a song directly within the song's session itself. In this way, the song can be exported and then imported into any standard CD burning program or download playlist without adding any manually programmed breaks.

Again, if the band/producer/engineer doesn't have a good sense of how this timing should go (I'm going to cut them a break here, as it's actually an art that's not that easy), they might consult with the mastering engineer (or at least make him/her aware that you would like to carefully address the timing issue).

TO MASTER OR NOT TO MASTER – WAS THAT THE QUESTION?

As was mentioned, the process of mastering often requires specialized technical skills, audiophile equipment, a carefully tuned listening environment, and talented, experienced ears in order to pull a sonic rabbit out of a problematic hat, or even one that could simply use some simple dressing up.

A few years back, DMH had the good fortune of sitting around a big table at a top LA restaurant with some of the best mastering engineers in the US and the UK. As you might expect, the general consensus was that it's never a good idea for artists to master their own project – that artists and even producers are just too close to the project to be objective. Although these arguments make a strong case for extreme caution, I'm not sure I agree that this is *always* the case. However, when approaching the question of whether to master a project yourself or to have a project professionally mastered, it's important that you objectively consider the following questions:

- Are you objectively removed enough from the project to have a critical ear for the sound and general requirements of the project, or are you just too close to it emotionally, musically and technically (realizing that the "sound" of a particular project might follow you throughout your career)?
- Is your equipment and/or listening environment adequate for the task?
- Is the final mix adequate (or more than adequate) for its intended purpose, or would the project benefit most from an outside set of professional ears?
- Does the budget allow for the project to be professionally mastered?

After these questions have been carefully weighted and answered, if the services of a professional mastering engineer are sought, the next question to ask is: "Who'll do the mastering?" In this instance, it's important that you take a long, hard look at the experience level and musical/technical tastes of the person who will be doing the job and make it your job to familiarize yourself with that person's work and personal style. In fact, it's probably best to borrow from the traditional business practice of finding three of the most appropriate mastering house/engineer facilities and following due diligence in making your decision by considering the following:

- Are you familiar with examples of the mastering engineer's work? If not, you should definitely ask for a client list and recent examples of their work and then, have a critical listening session with the producer and/or band members. In fact, one of the better ways to search out the best mastering engineer for the job is to start by seeing who mastered some of your favorite projects.
- Are they familiar with your music genre as well as the type of "sound" that you're hoping to achieve?

- What are their hourly or project rates? Do they fit your budget?
- Are they willing to do a complimentary test mastering session on one of your cuts, or at least discuss how they might approach the process, based upon a track that you've sent them?

MASTERING THE DETAILS OF A PROJECT

As with mixing, mastering is a process of balancing all of the variables that go into making a project sound as good as it can be. Both DMH and EC have the honor of calling some of the best mastering engineers in the world close friends. One thing that we've learned from them about their craft is that it is "all about the details". They'll be the first to tell you that it's keeping on top of the large number of variables that will get you to the finish line. Now, let's take a closer look into some of the details that can help shape your project into a final master that is ready for distribution, no matter who will be doing the final sonic sculpting. On to the details:

"Pre"paration

Just as one of the best ways to make the recording and mixdown process go more smoothly is to fix most of your technical problems and issues *before* you sit down at the console or mixer. Likewise, one of the best ways to ensure that a mastering session has as few problems as possible is to ask the right questions and deal with all the technical issues *before* the mastering engineer even receives your sound files. By far the best way to avoid problems during this phase is to ask questions ahead of time.

The mastering engineer should be willing to sit down with you or your team to discuss your needs and pre-planning issues (or at least, direct you to a document checklist that can help you through the process). During this introductory "getting to know you and your technical requirements" session, here are just a few sample questions that you might ask:

- What should the final master sample rate and bit rate be (usually, the highest sample/bit rate that the session was actually produced at)?
- What should the master's maximum level be (often −7 to −12 dB of full scale)?
- Should all master compression be turned off on the mix bus (usually yes)? Should you supply an extra mix copy with the bus compression turned on so that the group's/mixer's intended ideas can be heard?
- Would they like separate instrument/vocal stem tracks so they can be treated separately?
- Are there any special requirements that the mix engineer should be aware of, or other issues that he/she didn't think of?

Providing Stems

A more recent development in how final masters are delivered to the mastering engineer involves the idea of *stems*. In short, stems are major musical sub-groups of the final master mix. This might be as simple as providing the mastering engineer with the music tracks of the song and with a separated vocal track, or it might be more complex, involving a full drum stem, instrument stem, vocal stem, etc. The idea behind this method is that it allows the mix engineer to provide the best individual level, EQ and dynamics processing for the instruments and vocals separately and then combine them at the end.

There are potential merits to this process, but it's easy to understand how the mix engineer and even the producer might have problems with passing control over to the mastering engineer – it begs the question "When does the mix phase actually end, and who has final musical/production control over the mix?"

Providing a Reference Track

As we have read, it's usually best to provide the final pre-mastered mix at a reduced level and with all of the processing effects (EQ, compression, etc.) turned off at the master out bus. However, you might provide documentation and/or screen shots of these processing settings to the mastering engineer so as to give them an idea of what your intentions are. In addition to this, it's often a good idea to provide an additional copy of your final mix with everything turned on. This way, he or she can listen to the mastered and processed studio mix to compare the two side by side. This can often provide invaluable insights into the artist's, producer's and/or engineer's intentions.

To Be There, or Not to Be There

One of the next questions that might be asked of the artist or producer is whether or not they want to sit in on the mastering phase. In the mastering world, this is not a huge question but simply one that should be addressed. The artist/producer might feel strongly about sitting in and becoming part of the process, or they might feel that the mastering engineer was chosen for his or her artistry, experience and sense of detachment and simply choose to let them do their job on their own. Another, middle approach that is often taken is for the mastering engineer to master a song and then quickly upload it to the band or producer before moving on with the project.

No matter which approach is taken, it's important that the artist and client make sure that there are several ears around to listen to the final mastered project and listen over several types of systems. Above all, be patient with yourselves; however, be critical of the project's overall sound and listen to the opinions of others. Sometimes, you get lucky, and the mastering process can be quick and painless; at other times, it takes the right gear, keen ears and lots of careful attention to detail to finish the job right. In the end, you (artist, producer, label) are the ones who will have to live with this song or project – strive to keep your values up during all of the production phases so it'll be a project that you can be proud of.

COMMON TOOLS OF THE TRADE

At the outset of this chapter, we spoke about the general aspects of mastering that go into the production of a song or project. Now, let's take a more detailed look into these tools.

Sound File Volume

While dynamics is a hotly debated topic among mastering and recording engineers, most agree that it's never a good idea to deliver a sound file to a mastering house, or to begin the process of mastering, with a sound file that has already been compressed and raised to levels approaching digital full scale. Doing so obviously reduces the dynamic choices that can be made in the mastering stage. For this reason, most mastering engineers will ask that the project be delivered without any compression or dynamics of any kind (although you might take a screenshot or write down any settings that you had as a possible suggestion) and that the overall session gain be reduced so that it peaks at a lower level (–12 dB, for example). Again, it's always a good idea to discuss these and other details with the mastering engineer beforehand.

Sound File Resolution

In the end, the sample-rate and bit-rate resolution of a sound file is a matter of personal choice, beliefs and working habits. Some believe that sample rates of 44.1k and 48k with a 24 (or even 16)-bit depth will be sufficient to provide quality audio (given that the interface and converters are very high in quality). Others believe that rates of 96k and 192k with a 24-bit depth are required to capture the music with sufficient resolution. We're going to stay out of this argument and let you make the best choices for your own situation. Whatever your chosen session sample rate will be, it's almost always best to deliver the final master recording to the mastering engineer at the original native rate. That's to say, if the session was recorded and mixed at 24/96, the final mixdown resolution should be delivered to the mastering engineer at that rate and bit resolution. If it was recorded at 24/48, then it should be delivered at that rate.

Dither

As was stated in Chapter 6, the addition of small amounts of randomly generated noise to an existing bit stream can actually increase the overall bit resolution (and therefore, the noise level and signal clarity) of a recorded signal when reducing the file's bit depth to a lower depth (i.e., reducing a 24-bit file to one with a 16-bit depth). Through the careful addition of dither, it's actually possible for signals to be encoded at levels that are less than the data's least significant bit level. You heard that right: by adding a small amount of carefully shaped, random noise into the A/D path, the resolution of the conversion process can actually be improved below the least significant bit level and thereby reduce a sound file's harmonic distortion.

Within mastering, dither is often manually applied to sound files that have been recorded at 24-bit depths so as to reduce the effects of lost resolution due to the truncation of the least significant bits. For example, mastering engineers might carefully experiment with applying dither (by inserting a plug-in or by applying dither within the mastering DAW) to a high-resolution file before saving or exporting it as a 16-bit final master (Figure 19.3). In this way, distortion is reduced, and the sound file's overall clarity will be increased.

Relative Volumes

In addition to addressing the volume levels of each song/section within a project, one of the tasks in the mastering process is to smooth out the relative volume differences between songs over the course of the entire project. These differences could occur from a number of sources, including general variations in mixdown and program content levels, as well as levels between projects that have been mixed at different studios (possibly with different engineers).

The best cues that can be used to smooth out the relative rms and peak differences between songs, etc. can be obtained by:

- Using your ears to fine-tune the relative volume levels from song to song
- Carefully watching the master output meters on a recorder or DAW
- Watching the graphic levels of the songs as they line up on a DAW's screen

This process is also part of the art of mastering, as it generally requires judgment calls to set the relative levels properly in a way that makes sense. Contrary to popular belief, the use of a standard DAW normalization tool can't smooth out these level differences, because this process only detects the peak level within a sound file and raises the overall level to a determined value. Since the average (rms) and peak levels will often vary widely between the songs of a project, this tool will often lead you astray (although certain editors and mastering plug-ins provide normalization tools that have more variables, which are more useful and in depth).

FIGURE 19.3
Dither Plug-ins. (a) Sonnox Oxford Limiter plug-in for the Apollo and the UAD effects processing card (courtesy of Universal Audio, www.uaudio.com © 2022 Universal Audio, Inc. All rights reserved. Used with permission). (b) Apogee UV22 dither plug-in within Cubase/Nuendo (courtesy of Steinberg Media Technologies GmbH, a division of Yamaha Corporation, www.steinberg.net).

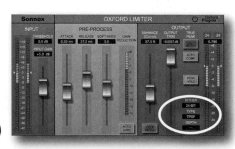

(a)

(b)

EQ

As is the case in the mixdown process, *equalization* is often an extremely important tool for boosting, cutting or tightening up the low end, adding presence to the midrange and tailoring the high end of a song or overall project. EQ can also be used as a tool to smooth out differences between cuts or for making changes that affect the overall character of the entire project. Of course, there is a wide range of hardware and software plug-in EQ systems that are available for applying the final touches within both a professional and a project-based setting (Figure 19.4). It should be stressed that it isn't always necessary to insert a third-party EQ device into the chain in order to get the job done. In the case of a DAW, the EQ on the master output bus can do much of the work (provided that it's musical and works for you).

Dynamics

One of the most commonly used (and overused) tools within the mastering process relates to *dynamics* processing.

Although the traditional name of the game is to achieve the highest overall average level within a song or complete project, care must be taken not to apply so much compression that the life gets dynamically sucked out of the sound (Figure 19.5). As with the first rule in recording – "There are no rules" – the amount of dynamics processing is entirely up to those who are creatively involved in the final mastering process. However, it's important to keep in mind the following guidelines:

- Depending on the program content and genre, the general dynamic trend is toward raising the overall levels to as high a point as possible. It should be noted, however, that this can definitely be pushed too far (so as to be audible and/or harsh). Additionally, there are those in the production community who are calling for a return to allowing the natural dynamics within the music to shine through by pulling back on dynamics processing.

FIGURE 19.4
EQ in mastering. (a) EQ processing in Ozone 7 (courtesy of iZotope Inc., www.izotope.com). (b) Main out channel EQ within Cubase/Nuendo (courtesy of Steinberg Media Technologies GmbH, a division of Yamaha Corporation, www.steinberg.net).

(a)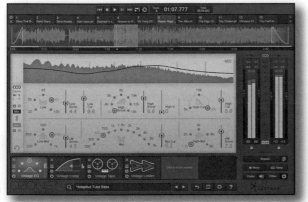

(b)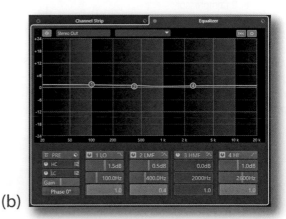

- When pushed to an extreme, compression will often have an intended or unintended side effect of creating a sound that has been "squashed", giving a "wall of sound" character that's thick (potentially a good thing) and/or one that's sonically lifeless (definitely a bad thing).
- When compression is not applied (or little is used), the sound levels will often be lower, thinner (that might be bad) or fuller in dynamic life (that's often extremely good).

From all of this, you'd be correct if you said that the process is entirely subjective and relative! The use of dynamic reduction can help add a strong presence to a recording, while overuse can actually kill its dynamic life; therefore, it's important to use it wisely!

Compression in Mastering

The purpose of *compression* in mastering is, of course, to control the overall dynamics of a song or sound production. A compressor (Figure 19.6) can be used to help control the peaks within the sound file, or it can be used to help tame or control the overall dynamic levels. Before compression (or any other form of dynamic control, for that matter) is applied, it's often a good idea to have a clear (or some form of) vision as to how you want this gain-changing stage to affect the sound. Without a clear idea of how this tool can help, it can be easily overused in a way that can take the dynamic life out of your sound (Figure 19.7).

A compressor doesn't only make things louder; through judicious control, it can be used to make the sound fuller and richer and can make the mix sound "punchier". This is done by setting the compressor's threshold at a point that's just above the music average level. Then by setting the compression ratio at between 2:1 and 8:1, the device will be able to reduce the peak levels without affecting the overall mix's sound. By setting the attack and release times to a fast-acting setting (with a higher ratio) and then setting the threshold so that the

(a)

(b)

FIGURE 19.7
Figure showing the same passage with varying degrees of compression.

compression effect is obvious, you can then go about the task of backing off on the threshold and ratio settings until it starts to sound natural while still registering a degree of compression on the signal reduction meter. Again, the desired sound can range from giving the mix an extra degree of presence or punch, all the way to giving a natural sound, while reducing the level of the program peaks.

Limiting in Mastering

One of the more misunderstood (or possibly misused) dynamics processing functions within the mastering process is limiting. Since, by its very nature, a limiter has a very high input to output gain reduction ratio (up to 100:1), it is used to restrict the program output's signal from exceeding levels above its threshold level. This means that if it is improperly set at threshold levels that are too low, the overall dynamic range can be severely reduced in an unnatural way that can cause signal clipping and distortion.

On the other hand, when a high-quality, fast-acting hardware or software limiter (Figure 19.8) is driven by an input signal that has good dynamics and is set at such a level that any excessive peaks only occasionally occur, then the limiter's threshold can be set at a level that will detect just the peak overloads and reduce them to acceptable levels. Said another way, when a quality limiter is properly set to reduce occasional program peaks, the results will often be transparent and inaudible.

Multiband Dynamic Processing in Mastering

The modern-day mastering process often makes use of multiband dynamic processing (Figure 19.9) in order to break the frequencies of the audio spectrum into bands that can be individually processed. Depending on the system, up to five distinct bands might be available for processing the final signal. Such a hardware or software system could be used to strongly compress the low frequencies of a song using a specific set of parameters while applying only a small amount

FIGURE 19.8
Limiters in the mastering process. (a) Manley Slam! (courtesy of Manley Laboratories, Inc., www.manleylabs. com). (b) Precision Limiter plug-in for the Apollo and the UAD effects processing card (courtesy of Universal Audio, www.uaudio.com © 2022 Universal Audio, Inc. All rights reserved. Used with permission).

(a)    (b)

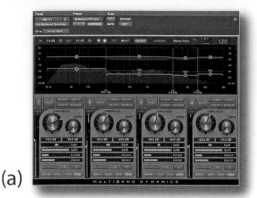

(a)

(b)

FIGURE 19.9

Multiband compression.
(a) ProTools Multiband
Dynamics (courtesy of Avid
Technology, Inc., www.avid
.com). (b) Precision Multiband
Plug-in for the Apollo and
the UAD effects processing
card (courtesy of Universal
Audio, www.uaudio.com©
2022 Universal Audio, Inc. All
rights reserved. Used with
permission).

of compression to the sibilance at its upper end. Alternatively, a multiband processor can be used to add spice to the upper end of the mix while reducing the sharp upper-end attacks, which could be harsh.

Loudness Units Full Scale

Loudness Units Full Scale (or LUFS for short) is used to measure the perceived loudness of a section of music's average level over time. Thus, two musical pieces that have the same LUFS loudness reading will be perceived as equally loud, regardless of the levels that are showing up on the meters.

These readings will always be depicted as a negative number (i.e., −5, −10, etc.), as they are always referenced to a digital full-scale value. In this way, broadcasters can reference their content to a specific LUFS value so that all of the material being streamed will be perceived as equally loud. Mastering engineers can do the same. Even if content has been overly compressed or limited, its overall LUFS value will register as being equally loud as the rest of the program content.

At times, analyzing the LUFS values between differing content can help fine-tune the overall gain levels between programs, tracks or content so as to make sure that they're as consistent as possible.

Mid/Side Processing

Another mastering tool that has been gaining favor is M/S processing (Figure 19.10). This seemingly new technique is actually quite old, being based on Alan Dower Blumlein's 1934 mic technique patent.

In essence, this process is mathematically able to split a stereo signal's sonic image into two components, the Mid (M) and the Side (S) processing channels. The Mid contains all of the in-phase signal, which combines in the (L + R) channel information (containing the direct, non-reverberant, in-your-face sound), while the Side contains the (L + R) + (L − R) information (containing the ambient, reverberant, distant sounds). The fact that this information can be extracted from a stereo signal means that the ratio between the two can be varied, allowing

(a)   (b)

an overall mix to be brought forward (by increasing the Mid over the Side ratio), making it more present, or made to be a wider and more ambient mix (by increasing the Side over the Mid ratio). In addition, since the two information channels are split, they can be individually processed over a range of individual frequency bands in almost any way, allowing a greater degree of flexible control over these aspects of the mix. It should be pointed out, however, that the use of compression in M/S can cause problems with shifting widths in the soundscape – just a note of caution. Finally, M/S processing can be an involved process to understand. For more information, try searching the Web for "What is Mid/Side Processing?" You'll find tons of information on the subject there.

FIGURE 19.10

Plug-ins showing M/S processing functions. (a) Brainworx Modus EQ bx1. (b) K-Stereo Ambience Recovery (courtesy of Universal Audio, www.uaudio.com © 2022 Universal Audio, Inc. All rights reserved. Used with permission).

MID/SIDE EXERCISE:

Place your hand in front of you, and snap your fingers. Now, move your hand towards your right, and now, snap them again! Do you hear the difference in sound location? That is the difference between the center (what's in front of you) and the side (what is either to your L or R).

With this in mind, read the previous section about Mid/Side processing, and then, let's work on a mixing assignment. Let's say we have a stereo mix, where we have a mono Rhodes keyboard that is 100% left, and mono vocals in the center. The goal is to make that keyboard louder without affecting the vocals or panorama of either instrument (remember, you only have a stereo mix file to work from, not a multitrack).

If we place a Mid-Side matrix in the stereo mix, and we solo the "Mid", we'll notice that we still hear the Rhodes keyboard in our Mid channel, together with the vocals. Now, if we boost the "Mid" channel, we will notice that the keyboard, once panned 100% to the left, now feels closer to the center. Why?

Because with Mid-Side techniques, we are controlling the Mid, which in this case is whatever signals are in-phase or common, and we're not controlling signals regarding their localization within the stereo image. The more we boost the Mid channel, the more the keyboard appears to lean towards the center of the stereo field.

Think about it this way. In one fader, you have your keyboard with the panning to the left channel of the mix, with that fader being set to unity gain. On another, separate fader, you have the same keyboard (which is a mono signal) routed to the LR (side) of the matrix, and this fader is all the way down (at infinity). If you now start bringing this LR fader up, you will hear that the keyboard will move towards the center of the mix. That is basically what happens in this case when you use a Mid-Side matrix and alter the gain of the Mid channel. With Center-Side, we can select what signal we want to process depending on its localization in the stereo field. Technology is a beautiful thing!

TO MASTER OR NOT TO MASTER YOURSELF – THAT'S THE NEXT QUESTION!

The question of whether to master the project yourself or to turn your baby over to another person (a highly regarded professional or not) is one that's spawning a whole new industry – self-mastering software and plug-ins.

Obviously, there are different production levels for deciding whether or not to master. Is it a national major release? Is it your first self-produced project? Is there virtually no budget, or did the group blow it all on the release party, or worse? These and countless other questions come into play when choosing how a project is to be mastered.

Having said this, if you decide to master a project (whether it's your own or not), there are a few basic concepts that should be understood first:

- The basic concept of mastering is to present a song or project in its best possible sonic "light". This requires a basic skill set, comfort level, and knowledge of your room's equipment, acoustics and (most of all) any overall downfalls.
- A basic level of detachment is often required during the mastering process. This is a lot harder than it sounds, as sound is a very subjective thing. It's often easy to fool yourself into thinking that your room, setup and sound are absolutely top-notch – and then, when you go to play it on another professional system (or any home system), your hopes can be dashed in an instant. One of the best ways to guard against this is to listen to your mix on as many systems as possible and then ask others to critique your work. How does it sound to them on their system?
- Again, the process of mastering isn't just about adding EQ and compression to make your projects sound "AWESOME, DUDE!" Of course, that would be great, but it's also about a wide range of artistic decisions that need to be carefully finessed in order to master a recording into its final, approved form. Just a few of these decision-making steps include:
  1. The use of level changes to balance the relative track levels and improve the overall "feel" of the project
  2. The application of EQ to improve the sound and overall "tone" of the project
  3. The judicious use of dynamics to balance out the project's sound (over the course of the project) and to increase the project's overall level
  4. Choosing the proper song order

It's worth saying again that the biggest obstacle to self-mastering a project ourselves is usually … ourselves. The problems might lie in our inexperience and understanding of the subtleties of EQ, level and dynamics, our subtle "feel" for timing and sequence order, or any other possible number of things.

We're calling attention to our possible faults and the potential downfalls of making a good master because I feel that the process of mastering should be kept as a sacred process, held by a select few? Absolutely not – we do, however, feel that if you're aware of the potential pitfalls, you just might be able to follow the yellow brick road and learn the basic aspects of self-mastering. We know, because we pretty much did that ourselves!

The Zen of Self Self-Mastering?

So, after having DMH's projects mastered by several engineers, I actually ended up feeling that I could do a better (or at least as good) job, so I made the conscious decision to give it a try. After decades, I've finessed my mixing room into one that I know and trust (as much as anyone can) and have chosen my gear with TLC (I'm particularly happy with my monitor speakers, always a good place to start) in order to do the best job I can.

The "working with my own music" is always the rough part. It's tough enough trying to get the right sound for another band or producer, but when it's your own music, the going really gets tough. This is because you might be soooooooooo close to the music and production that the sounds, the nuances, the "everything" becomes progressively more difficult to second-guess. This is why I allowed myself the luxury of learning self-mastering over the course of several years. I knew that it would not come easily or even overnight – boy, was I right on that one.

One of the first steps is to simply be patient with yourself. The process of understanding the sound that you're after and then going about getting that sound might happen sooner than you think, or it might take a lot longer than you thought. The key is to be patient, to fail, to keep experimenting and eventually, to understand your own system, your sense of art and the sound that you're after.

Desert Island Mixes

The idea of having a personal collection of desert island mixes on hand as a reference keeps popping up in this book. Of course, this points to how important it is to have music that you admire and love on hand as a listening reference, and this also goes for the mastering phase. As you progress, you might compare your own mixes with a favorite desert island project. It used to be that everyone would tell you to listen in your car (always a good idea); however, these days, people spend a lot more time listening to music on the system they know best – their personal phone and headphone/earbud system.

You might listen to one of your fave mixes and then click over to your latest song mix. How does it compare? Does it make you happy, indifferent or disappointed? If it's one of the latter two, then you know you have more work to do. After putting in several years of hard, frustrating labor, I can honestly say that my mixes stand up to many of my favorite mixes – that and four Grammy nominations are things that I can be proud of. If I can do it, I know others can. It just takes true dedication, understanding and learning of the craft (just as it does with anything).

TWO-STEP OR ONE-STEP (INTEGRATED) MASTERING OPTION

One of the concepts of traditional mastering has always felt foreign to me. This lies in the fact that traditional mastering is (by its very nature) a two-step process: mixing and then mastering. The final mix is completed by the artist/engineer/ producer, and then, the masters are sent off to another person to be finalized in a separate stage. I know that this is the original beauty of having a professional who is detached, who can take your "baby" and make it even better. To me, however, the main reason why I started my journey into self-mastering was that I wanted to integrate the two phases into a final DAW project session itself. To me, this makes so much sense:

- It gives us the ability to have complete and total recall over your session, meaning that you can go back and make accurate changes to the final mix/ master version at any time. This can also be a never-ending pitfall if you're not careful.
- It gives us the ability to include song timings within the session itself (this is done by defining your export markings within each song). Whenever each song within the project is exported, these timings will automatically include the proper timings. Just drop it into the project – that's it!
- Most importantly, the final mastering chain will be included within each song, allowing us to add any appropriate amount of final master EQ and compression to each song.

In short, each of these three options allows us to have complete control and repeatability (recall) of not only the recording but also the final mastering phases. However, I'm also a believer that this is not for everybody. If you feel you're not qualified, don't have the time/interest/equipment to make the journey, then hiring a pro is your best bet.

UNDERSTANDING THE SIGNAL CHAIN

Whether you're a professional mastering engineer or self-master your own projects, one of the most important concepts to understand in the process of getting a certain "sound" in the mastering process is the idea of the "signal chain" (Figure 19.11).

In short, a "chain" in the mastering process is the chaining together of a series of hardware and/or software processing stages that combine to shape the final sound of the song or project. It is a very individualized process that will vary from one mastering artist to another. Indeed, it is often their personal signature for shaping how a project might sound.

- In a hardware-based system, this would be the chaining together of hardware EQ devices and dynamics processors in a serial fashion.
- When a separate DAW or mastering software is used as a mastering device, these processing devices are likewise used in a serial plug-in chain to shape the final sound.

| full list | tube compressor | multiband compressor | character (intelligent eq) | Magneto 2 (tape saturation) | limiter |

FIGURE 19.11
Example of a possible DAW mastering signal chain (DMH's favorite) traveling left to right through the main output bus.

- When used within the actual DAW session, a combination of sound and dynamics processors can be inserted into the main output bus channel to shape the final mix/production sound.

If you get the idea that each of these ways of working can achieve similar production results, you would be right. Of course, there will be opinions about which way works best and about how the final results will sound; however, the concept is the same: the signal chain inserts specialized signal processing devices into the final audio chain to achieve a sonic result.

This explanation of the mastering signal chain is where all similarities between one person's signal chain and another's stop. In addition, you'll remember that it was said that mastering takes time. Well, understanding the subtleties (or not so subtle aspects) of how inserting a device and/or processor into a chain will affect the overall sound often takes time. Additionally, the understanding of what's needed to best bring out the best in a piece will often change from project to project. The concept of what it might take for an artist to bring his or her songs out into the sonic light will also take time, and just when you feel that you've found your personal chain for your music, you might very well find that the latest and greatest song of yours might call for something a bit different to make it sound the best it can. You might have found an awesome signal chain that works for your music, but it's wise to realize that it won't always be a hard and fast rule that your signature chain will always be the best one for a particular song. As always, your ears, experience and intuition are the best and most accurate equipment that you have at your disposal.

Mastering Plug-Ins

Of course, with the interest in DIY mastering comes a selection of plug-ins that are specifically designed (or bundled) to handle the processing functions and basic details that go into mastering. These could include separate plug-ins that are specifically tailored to provide the EQ, compression, limiting and multiband needs of finessing a master into its final mastered form. In essence, these tools combine together to create a personalized mastering chain.

Overall mastering stand-alone programs or plug-ins have been designed, which provide an all-in-one environment that often includes EQ, dynamics, M/S processing, restoration and more. These programs/plug-ins (Figure 19.12) provide visual sound file and setup cues, template suggestions and various other tools to help you through the process.

  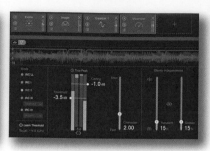

FIGURE 19.12
Izotope Ozone 10 complete set of mastering plug-ins (courtesy of iZotope Inc., www.izotope.com).

Finally, the approach that would most likely be taken by a professional mastering engineer would be to build his or her own final mastering chain of hand-chosen software and hardware options that work together to support the audio program or music. In the end, all of these options are totally valid, depending upon your skill level and/or desire to learn the finer points of the craft.

With regard to inserting a final self-mastering chain into your final mix bus, one last concept should be addressed. This relates to the idea of whether or not to have the final mastering chain turned on while you're mixing. The general convention is that you should absolutely NOT have this chain turned on, and it should only be judiciously turned on once the mix phase has been completed.

In the continuously evolving story of "there are no rules", it should be said that over the years, DMH has become so familiar with the general working of his self-mastering chain that he does keep it turned on with a great deal of success. It's all up to you and your personal working style. However, if you're new to the process, it'll probably be best to keep it turned off during the mix phase.

MASTERING FOR THE INTERNET

From a technical standpoint, mastering for the Internet can either be complicated, requiring professional knowledge and experience, or it can be a simple and straightforward process that can be carried out from any computer. It's a matter of meeting the level of professionalism and development that's required by you and your audience's needs.

Generally, it's a good idea to keep the quality as high as possible. Most online distribution companies will offer guidelines for accepted sound file, sample and bit-rate types and/or will re-encode the sound file to match their own internal codec rates and format (making it a simple, easy-to-upload process).

ON A FINAL NOTE

In closing, it's always a wise idea to budget in some time to QC (quality control) the mastered recording before committing it to being a final product. I've learned that the final process of exporting (bouncing) a mix into its final form is often best done in "real time". This lets us listen to the entire track as it's being played out, allowing us one final chance to catch something that might be technically or sonically wrong in the final mix.

If at all possible, take a week and listen to it in your car, on your boom box, your home theater system, in another studio – virtually everywhere! Then, revisit the mastered mix one last time. As a musician, producer or record label, it will be your calling card for quite some time to come. Once you're satisfied with the sound of the finished product, then you can move from the mastering phase to making the project into a final, salable product.

CHAPTER 20

Immersive Audio

At a special media event that was attended by about 500 people, our good friend George Massenburg asked the crowd, "How many of you truly know the joys of immersive audio playback or production?" In that crowd, where DMH was, about 20 of us raised our hands. He then went about the business of playing back several very well-known mixes that were remixed in 5.1 and talked about the benefits of immersive sound music and audio-for-visual media production (Figure 20.1). Right-on, George!

Truth is, whether you're an advocate or an adversary of the concept of immersive audio, one thing is for sure – it exists in the here and now, and it is certain to play an ever-growing role in the media technologies of tomorrow.

I think that at this point, I (DMH) have to break from my role as a neutral author and state flat out that I'm a HUGE immersive audio fan and a four-times Grammy nominee in this category. For me, the ability to compose and mix in immersive music has been an uplifting and hugely beneficial experience. I clearly remember as a kid, placing two album covers behind my ears and listening as the music came to life around my head (go ahead, give it a try)! The ability to augment music and visual media by placing sounds within a 360° circle has literally opened up new dimensions in mixing and effects-placement technologies in a way that keeps me creatively young.

Most of the people we know who are ideologically closed to the idea of immersive audio haven't worked with or listened to the medium. We urge that you keep an open mind to the process, watch movies and listen to music in surround (in any available format), and if at all possible, take the time and effort to familiarize yourself with the process of producing sound in surround.

Before DMH puts his neutral writer's cap back on, I'd like to present the strongest argument for becoming familiar with the production techniques of immersive: enhanced job opportunities. I have friends living in the technological heart of numerous cities who are completely unfamiliar with any and all forms of immersive. It simply never occurred to them that they could increase their opportunities, client base and perceived prestige in the fields of mixing music,

DOI: 10.4324/9781003260530-20

FIGURE 20.1
Immersive setup at Control A, Synchron Stage Vienna (courtesy of Synchron Stage Vienna, www.synchronstage.com).

soundtracks for movies and gaming by investing in an immersive monitoring system and learning the basic tools and techniques of recording, mixing and mastering media for multichannel sound. If for absolutely no other reason, the ability to understand and work in such new and upcoming technologies can help give your career a marketing edge.

IMMERSIVE AUDIO: PAST TO THE PRESENT

Of course, it all started with the movies. In the pre-"talky" days, movie theaters were anything but silent – organs, chimes and all sorts of percussion clanged from the front, sides and rear parts of the room behind ornate wall features. With the introduction of movie sound and the use of the musical score in the late 1920s, all of this came to a halt when soundtracks were played back by the only known delivery format of the time: mono.

On November 13, 1940, Walt Disney's *Fantasia* opened up the sound field to stereo when it premiered at New York's Broadway Theater. Although it wasn't the first film that was produced using a "multiple channel recording" process, Fantasia was the first to introduce multichannel sound to the public.

The final mix of *Fantasia* was printed onto four master optical tracks for playback using a special RCA system called "Fantasound". (Unlike the two-channel format that was adopted for home playback, film sound started out with, and continues to use, a minimum of four discrete channels.) This multi-speaker setup placed 3 horns behind the screen and 65 smaller rear-side speakers around the walls of the theater. Due to the outlandish setup costs (estimated at about $85,000 for each theater at the time), RCA quickly stopped making this advanced system.

In the early 1950s, the first commercially successful multichannel sound formats came onto the scene with the development of CinemaScope (4-track 35 mm) and Todd-AO (6-track 70 mm). Both of these formats made use of magnetic tracks that existed alongside the release print picture and required that the projector be fitted with special playback heads, amps and speakers.

In the early 1970s, the home consumer stereo market was gaining in popularity and audio quality. With the development of higher-quality amps, speakers and record turntables came new experimentations in systems design that eventually led to the development of Quadraphonic Sound (Quad). This playback system made use of four speakers (Figure 20.2) that were placed in the four corners of a room, which enveloped the home listener in a L/R/Ls/Rs listening experience.

Although analog reel-to-reel and cassette tape machines were used in homes, they were still relatively expensive. By elimination, this meant that playback would have to be carried out by the most popular medium of the day – the LP record.

Reproducing four channels from a record wasn't easy. Often, the task of encoding four channels onto the two walls of a vinyl record was done with relative phase or by using a complex, high-frequency carrier tone that was used to modulate the sum and difference channels. However, the real difficulties lay in the wide assortment of incompatible encode/decode formats that were offered by various manufacturers. Given the fact that your system might not play back the latest and greatest release from another company, and that discs were both expensive and prone to deterioration over a short period of time (the high-frequency signals on modulated records would literally wear away), the Quad revolution quickly died away.

Stereo Comes to Television

Since its inception, surround sound has been used in motion picture soundtrack production with great success. With the introduction of Dolby noise reduction and multichannel audio in the theater, good sound was not only appreciated, it was expected! On the other side of the media tracks, television sound was strictly a lo-fi, mono experience up until the early 1980s and was strictly an afterthought to the visual image. However, with the adoption of the video cassette recorder (VCR) and later, hi-fi stereo sound from a VCR, discriminating audiences began to appreciate the higher-quality audio that accompanied the almighty image. With the dawning of the music video (I want my MTV!), stereo

FIGURE 20.2
Example of a Quadraphonic (quad) speaker setup.

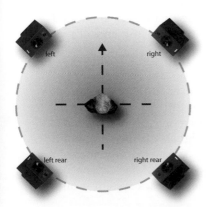

broadcast television and the stereo VCR, TV was finally forced into offering a higher-quality, multichannel, visual experience.

Theaters Hit Home

In 1982, Dolby Labs introduced "Dolby Surround", an extension of their professional Dolby Film Sound Project. By 1987, millions of homes were beginning to be fitted with consumer receivers and high-end audio systems that were integrated with video. With the introduction of Dolby Pro Logic, a simple system was put into place that allowed phase information to be extracted from the two tracks of a stereo program to reproduce the L/R/C/Surround field. In 1992, with the introduction of Dolby's AC3 surround encode/decode system (Dolby Digital), it became possible for discrete 5.1 surround sound to be encoded directly with the new visual entertainment medium of the early twenty-first century – the DVD – and now to the newer Blu-ray disc.

Today's Immersive Audio Experience

Recent advances in immersive audio are still largely being driven by the film industry, but that's not to say that immersive for television and music, and especially the immersive audio experience that's being produced today for the gaming industry, doesn't play a big part in driving the economics of media and music production.

Obviously, one of the biggest markets in immersive technology centers around the home theater and the idea of enjoying high-definition movies in the comfort of your home or even private home screening theater (complete with plush seats, impressive screen and popcorn popping machine). On the audio file front, the sky is also the limit, with the high-resolution playback systems that often accompany high-end home theater systems; 24-bit/96k (and higher) sound files of the latest classical or popular music releases can be downloaded. Last, but hardly least, the latest high-performance/high-resolution gaming consoles have advanced the art of the immersive experience to new levels. Here, high-performance computing teams up with multichannel audio to place sound effects, dialog and music onto a larger-than-life soundscape. Often, newer gaming releases involve huge music budgets that make use of full-size orchestral scores with live instrumentation in top-flight music halls. In the end, it's all about the experience.

After all of this, it would be unthinkable not to mention the latest darling of the film, home theater and music markets … Dolby Atmos. The general concept behind Atmos is the idea of scalability. Let's not get into the details at this point (although we'll definitely touch on this later), but at its core, Atmos does not make use of audio channels per se; rather, the audio signals can be placed and steered throughout an immersive setup according to its speaker complement. If a system was mixed in a full immersive soundfield and is played back over a system with the same speaker setup … no problem. If the playback system is only

5.1, for example, the system will detect this and move the upper speakers down so that it sounds as good as it can under the circumstances. A fully immersive system played back through a Dolby Atmos system in stereo would, in theory, fold all of the audio objects down into stereo with no ill effects.

IMMERSIVE LAYOUTS

In this day and age of changing technologies within the field of immersive audio, we feel that it's best that we begin by understanding the general layout concepts. This is best done by first introducing the descriptor that defines what type of speaker layout is in use. We'll first show this through a basic (if you can call it basic) 5.1.4 system, as shown in Figure 20.3.

In this case, the "5" refers to the numbers of full-range speakers that are placed at or below ear height around the soundfield, and of course, refers to left/right/center/left surround/right surround (or L/R/C/Ls/Rs). The "1" refers to the number of low frequency effect (or LFE) speakers that are used in the system (more on this later). Lastly, the "4" in this case refers to the number of full-range (or near full-range) speakers that are placed above the listener, giving a sense of height. Following this format, here are a few of the more popular current layout schemes:

- 2.0 – The stereo that we all know and love
- 2.1 – Stereo with the addition of an LFE channel (which we'll find later is different from one with a subwoofer)
- 4.0 – A quad playback layout (L/R/Ls/Rs) with no center speaker

FIGURE 20.3
A 5.1.4 speaker setup.

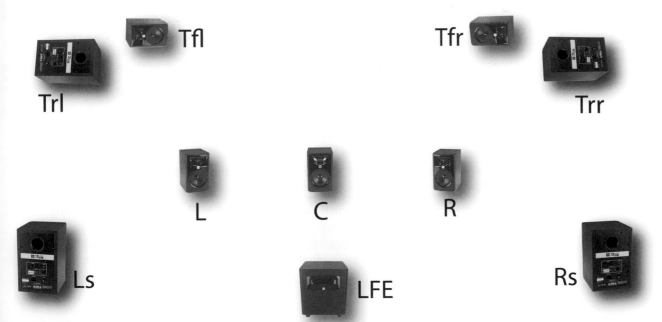

- 5.0 – A surround layout (L/R/C/Ls/Rs) with no LFE channel
- 5.1 – A surround layout (L/R/C/LFE/Ls/Rs) with an LFE channel
- 7.1 – A surround layout (L/R/C/Lw/Rw/LFE/Ls/Rs) with an LFE channel and two side (wide) channels
- 5.1.2 – A surround layout (L/R/C/LFE/Ls/Rs/Tfl/Tfr) with an LFE and two height channels (generally placed above and to the front of the listener)
- 5.1.4 – A surround layout (L/R/C/LFE/Ls/Rs/Tfl/Tfr/Trl/Trr) with an LFE and four height channels
- 7.1.4 – A surround layout (L/R/C/Lw/Rw/LFE/Ls/Rs/Tfl/Tfr/Trl/Trr) with an LFE, two side (wide) and four height channels

Of course, in a movie or special installation, these numbers can go higher, but we think you get the idea that it can get pretty crazy.

The LFE

Within an immersive system, the next aspect of system setup that's best to understand is the concept of the LFE. Contrary to popular belief by the general listening public, LFE is NOT a subwoofer. A stereo system that contains a sub is actually a 2.0 (stereo) system, whereby a low-pass crossover is placed in the circuit (usually an actively powered one). The low bass (possibly 120 Hz and below) is then sent to a mono sub, while frequencies above this will be sent to the stereo speakers in the system. The LFE channel is just that, a separate channel that contains an extra ".1" degree of low-end information (which is usually band-limited to frequencies of 80 Hz and below).

It was never really intended as a bass channel; rather, it was developed by the film industry to give an extra "ummmffff" to the low-bass end ... to add an extra degree of impact, especially during explosions and low rumbles that occur in films (it was actually first used in the film *Earthquake* to add extra realism to the "earth-shaking", or should I say "butt-shaking", score.

The LFE's placement and setup within the room should also be taken seriously, as placing it in a corner or in a position that's affected by a harmonic node in the room could greatly affect its response. If possible, the sub should be placed on the floor, near the front plane of the room, at a reasonable distance from the front wall (to prevent excessive bass buildup). Finally, most active subs will offer full control over gain and crossover frequency, allowing the user to best match the driver's bass response to the room and main speaker set.

With regard to the LFE channel, almost all of these low-end powerhouses are actively self-powered. There are usually controls for setting the crossover frequency, overall LFE speaker gain and a phase switch that can help with the aspects of low-end phase as the signal reacts to its placement within the room.

There's also an ongoing debate among top music mixers as to whether to use the LFE channel at all within music mixing. Almost everyone agrees that all of the speakers within a music mixing scenario should be full range; however, there are those who don't believe in using the LFE channel at all, while most will put some degree of low content in this channel (stating that customers feel cheated out of their well-earned .1 when they don't get some bottom in that channel). Personally, DMH falls within the camp of putting a moderate degree of low-end content in the LFE channel (while being very careful that all setup calibrations have been observed). It's no fun to work hard on a surround project only to have the bass levels play back at too high a level, muddying up your carefully crafted mix.

When in doubt, it's always a good idea to calibrate your surround system according to the Grammy P&E surround setup guidelines at www.recordingacademy.com/producers-engineers-wing/technical-guidelines

It should be noted that critical musical material should never be sent to the LFE, and this channel alone should be band-limited, usually being cut off above 80 Hz (be aware that not all digital audio workstations (DAWs) will automatically do this; caution should definitely be advised). In addition, many feel that it's often wise to add a subsonic filter to eliminate any excessive low end or rumble that might make its way into the LFE channel (Figure 20.4).

On a final note: Be careful of your LFE channel levels! Seriously, if there's one place that you can easily screw up a mix, it's with the LFE. This is especially true when playing back on a high-end system.

Bass Management in a Surround System

At first glance, bass management might look like an LFE channel or a simple subwoofer system (which is a good reason why bottom-end issues can quickly get so confusing for both stereo and immersive audio). In reality, this is not the case at all. A bass management system (which is used in most lower-cost home theater systems and certain high-end auto systems) uses filters to extract low-frequency information from the main channels and then route this bass to the sub speaker, while the highs are sent to the system's tweeters. In short, this method has the advantage of allowing for one bass speaker and multiple easier-to-build (stereo or surround) tweeters that are small, often inexpensive and can be placed in confined spaces (Figure 20.5).

Although these systems are in wide use in home playback and theater systems, it's widely held that these speaker systems aren't suitable for studio monitoring due to their irregularities in midrange response, poor image localization and limited sound quality.

FIGURE 20.4
An LFE roll-off filter that has been routed only to the LFE channel. (a) A sub-frequency roll-off, starting at 24 Hz. (b) A low-pass filter can be used to roll off frequencies above 80 Hz. (c) The combined curve of the subsonics cutoff and the low-pass filter curves.

(a)

(b)

(c)

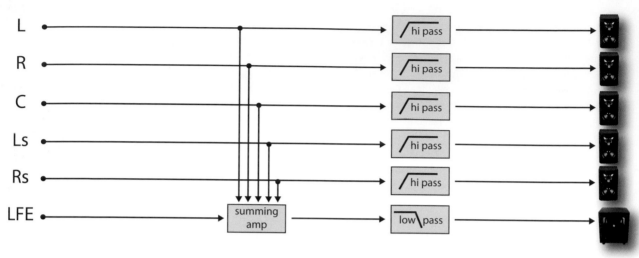

FIGURE 20.5

Basic structure of a bass management system, whereby signals from the main band-limited channels are fed to and combined with the LFE channel (which acts more like a sub than an LFE channel).

FIGURE 20.6

The ITU 5.1 speaker setup diagram (left) and PMC TB2+ 5.1 monitor setup (right) (courtesy of PMC Limited, www.pmcspeakers.com).

Speaker Placement and Setup

When trying to first understand and work with immersive systems, it's always a good idea to begin at a basic starting point. In this case, let's start with the traditional 5.1 surround-sound setup.

As defined by the International Telecommunications Union (ITU), the "official" 5.1 speaker setup is made up of five full-range monitors that are positioned in a circular arc, with the speakers being placed at equal distances to the listener (at the center position). Three of the speakers are placed to the front, with the center speaker being placed dead center (0°) and the left/right speakers being placed at 30° arcs to the center point. The surround speakers are then placed behind the listener at 110° arcs to the center point (Figure 20.6).

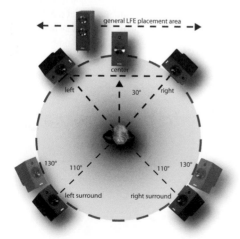

It should be noted that the 110° arc in the rear surround field is a debatable point. During a special playback event, a rather noteworthy group of surround producers and engineers (including myself) compared a 110° arc with a wider 130° arc. We decided on the latter, with popular opinion being that the head and ears provided less of a barrier to the wide-angle sound. Having the tweeters at the outer edge sides will also help. You might want to perform a DIY and check it out for yourself.

From a practical standpoint, DMH has often found 5.1 surround speaker placement to be a bit more forgiving than the "spec" suggests. For example, it's sometimes not practical to place the three front speakers in an equidistant arc on most consoles or DAW desks because there simply isn't room. Usually, this means placing the speakers in a straight line or smaller arc (while angling the speakers for the best overall soundfield coverage).

Placing three matched speakers on the front bridge of a console, on floor/ceiling mounts or flush in a soffit generally isn't difficult. However, placing the center speaker on a DAW desk can sometimes be a challenge, simply because the computer monitor (or monitors) is commonly placed at this center position. If matched speakers are used, where can the center speaker be placed? You might be able to place the center speaker on its side, or you can sometimes get a dedicated, low-profile center speaker that allows the computer monitor to be placed on a shelf that can hold the monitor(s). In DMH's room, his 5.1.4 speakers are hanging from the ceiling, allowing the center speaker to "float" above the monitors (Figure 20.7).

In certain cases where space, budget and time are a consideration, it might not be possible to exactly match the rear speakers to those of the front three (which should be matched). In such cases, use your best judgment to match the general characteristics of the speaker sets. For example, if a company makes a speaker series that comes in various sizes, you might try putting a pair of their smaller

FIGURE 20.7
DMH's 5.1.4 speaker setup.

monitors in the rear. If this isn't an option, intuition and ingenuity are your next best friends.

5.0 SURROUND MINUS AN LFE

Simply put, 5.0 is 5.1 without an LFE big-bang-boom channel. Since LFE is band-limited to the extreme bass frequencies and is intended to provide only an added EFFECT!, many surround professionals argue that it's unnecessary to add the LFE channel to a surround music project at all. Others, on the other hand, argue that consumers expect and deserve to get their .1's worth. As with much of audio production, immersive audio is a new frontier that's totally open to personal interpretation.

7.1 SPEAKER PLACEMENT

Quite simply, 7.1 is carried out by adding two side channels (180° left and right) to the official 5.1 ITU speaker setup. The idea of this format is to add more definition, movement and localization to the surround experience by giving you localized sound that comes directly from the left and right.

ADDING HEIGHT SPEAKERS

In recent years, several systems have cropped up that add height channels (5.1.2, 5.1.4, 7.1.4, etc.) to an already existing immersive system. Having applications for film, game and music, this type of playback makes it possible for jets to fly literally over your head, or reverb and added ambience can float above your head in a mix, adding a dimension to the sound that truly has to be heard to be believed.

Although these immersive audio media systems are primarily in use in a growing number of major-release theater houses around the world, these systems are also strongly making their way into the home and home theater environment as well. Surprisingly, there are a number of audio formats that make use of height channels; however, by far the most widely adopted of these is Dolby Atmos.

IMMERSIVE SOUNDBARS

It almost goes without saying that the biggest obstacles to having a full immersive installation in our homes are:

- Expense (how many speakers will I need?)
- Technical challenges (How hard will this be to get up and running, and will I need to read manuals all day?)
- Where do I put the speakers, and will they be in the way?

These are not small considerations … particularly the last one. Speakers hanging from the ceiling and speakers placed all around the room is for many, a complete show stopper. The modern answer to most of these is the soundbar. A soundbar is generally just that, a sleek, thin bar that fits below and just in front of the TV … out of the way and an all-in-one package that's often fairly easy to set up.

These systems vary in price and configuration, often offering either a wired or a wireless LFE and a wired or wireless set of rear speakers. The height part of the equation is often dealt with by providing upward-firing height speakers that bounce the sound off the ceiling towards the listener.

In practice, of course, the effectiveness of these devices can range from so-so to an experience that can rival a full immersive speaker set. DMH can recall a time in Berlin when he listened to a top-flight bar that was playing back a Dolby Atmos demo disc. Even though it was an audio salesroom, the acoustics in the room were truly awful. Even so, the demo was amazing … with sounds flying behind us (there were no rear speakers … it was all done with psychoacoustics) and above us (from the upward-firing speakers). Truly impressive!

Surround Channel Layouts and Assignments

When creating an immersive multichannel mix for film, music, gaming, etc., it's important to know what channel output configuration should be used (either as a default that's set within your DAW or as a track configuration that's required by outside clients). For example, a surround configuration that's used by Steinberg and other companies has a default channel layout of L/R/C/LFE/Ls/Rs (Left/ Right/Center/LFE/Left surround/Right surround), while Pro Tools defaults to a L/C/R/Ls/Rs/LFE layout.

Of course, it's important that we be aware of these track configurations and that we lay our production rooms out according to that chosen layout. What's more important, however, is that your final exported mixdown tracks be properly named, with the track configuration being clearly identified within the filename (i.e., Project/Songtitle_Left.wav, Project/Songtitle_Right.wav, Project/Songtitle _Center.wav, etc.). Most immersive DAWs are capable of automatically adding these track suffixes to the track names during the final export process.

THE IMMERSIVE AUDIO INTERFACE

One might think that the subject of which audio interface will best serve (or be able to serve) your needs would be a simple one. This is sadly often not the case. The reasoning for this falls beyond the simple need for the number of discrete outputs that you'll need for your system … but deals with how (or if) your interface system will be able to simultaneously control the monitor output mix levels.

Many interfaces (some which actually go out of their way to call themselves a surround- or immersive-ready interface) will not be able to control the overall monitor gain levels from their master gain control. For those who are looking to get into immersive mixing, this is a critical factor that should be researched before buying an interface system. For those who are wishing to mix in 2.0 (stereo), 5.1 or even 7.1, there are a number of interface choices that will provide control over monitoring; however, when dealing with systems with height channels (5.1.4 and 7.1.4 and above) the number of interfaces that can provide overall monitor gain is far more limited.

Let's dive into this a bit further. Quite often, an interface that has 8, 10 or even more discrete output channels will have a main out volume. However, once you've gotten it home, more often than not, you'll find that this main out will only control the monitor gain of your left and right channels … that's it! Others might allow you to assign and "gang" up to eight channels to this volume control. This works great up to 7.1, but leaves you stranded for 5.1.4 and above. For monitor control over these systems, you'll have to research which interface systems will let you assign outputs (including optical outs) to the main gain control (Figure 20.8).

When dealing with monitor levels for an immersive system, there are three general scenarios that might be found:

1. The main monitor gain will have control over the left and right channels and not its additional outputs. This type of system will require that you buy an additional hardware monitor control. These systems can easily cost $5,000 or more … a lot of money for an expensive immersive main volume control.

2. Certain immersive-ready DAWs are able to provide a software main volume control to control the overall system gain. Many people are quite wary of this type of control, thinking that a glitch can cause the overall gain to get out of control gainwise. Although DMH has an interface with a main gain control for his 5.1.4 system, he has never had a situation (using Nuendo) where the software main gain has gotten out of control.

3. As stated before, the last scenario makes use of a well-designed audio interface that offers full control over all its analog outs, providing overall speaker gain control (at no extra cost, as it's simply part of the interface's design).

FIGURE 20.8
Focusrite Scarlett 18i20 interface allows 10 of the unit's outputs to be assigned to and controlled from the main volume control (courtesy of Focusrite Plc, www. focusrite.com).

Speaker Level Calibration

As is the case with any speaker setup in a professional monitoring environment, it's very important that the speakers within a stereo, surround or immersive speaker setup be adjusted so that the acoustic output of each speaker is at the same level at the listening position. This level matching can be done in several ways (Figure 20.8). Although certain surround amp/receiver systems are able to output pink noise (random noise that has equal energy over the audible range), it's far more common to:

- Record a minute or so of pink noise and then copy this same file into your workstation onto as many channel tracks as you have speakers in your system.
- Place a sound-pressure level (SPL) meter on a camera stand directly at the listener's position (you could use an SPL meter app on your phone or pad, or an inexpensive sound level meter that's bought online will also do the trick; it's always nice when the meter has a thread for mounting the unit directly onto a tripod). Point the meter at the first speaker to be tested (Figure 20.9). This will usually be the left speaker.
- Play the pink noise through one speaker at a time (with each track being assigned to its associated speaker) at a level of about 80–85 dB^{SPL}, and adjust the sensitivity and gain on both the meter and the master monitor gain so that the meter reads at "0" on the meter or at a specified level on a digital readout.
- Swivel the meter so that it points to the next speaker being tested. Now, solo that speaker/track and set the gain on that speaker to match to that same level. Then, move around the entire setup, matching levels until all of the speakers are at equal level at the listening position.
- You're now ready to start mixing or listening, knowing that your system is properly level matched.
- I almost forgot to mention, don't forget to "switch the SPL meter off" – I can't tell you how many times I've drained the battery this way.

FIGURE 20.9
Speaker noise calibration tests are used to make sure that your speakers in a stereo, 5.1 or immersive system will play back at their proper, balanced levels.

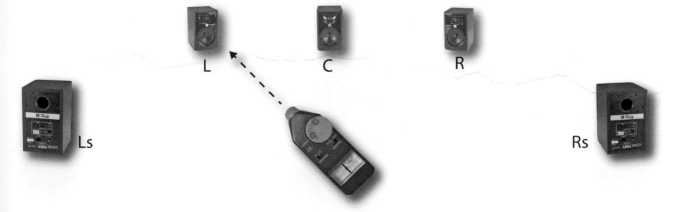

L   C   R

Ls   Rs

It should be noted that setting LFE levels properly is a bit more complicated (most set the LFE at +4 dB^{SPL} relative to the other speaker levels). However, if you're serious about setting up a system properly, I'd strongly recommend that you consult the immersive technical setup guidelines from the Producers and Engineers Wing of NARAS (the Grammy folks). It's well worth downloading this and other freely distributed guidelines from www.recordingacademy.com/producers-engineers-wing/technical-guidelines.

Speaker Time Calibration

It's common knowledge that these speaker level matches must be made in order to set up your immersive system; however, what is less well known (but equally important) is that you should time-align all of the speakers in your system so that the sound from each of the speakers in the system will arrive at your ears at the same, relative time (Figure 20.10).

Such time adjustments can be easily done by measuring the distance from each of the speakers to your head or SPL meter (usually in cm) and writing them down. The speaker that is the furthest from this point will actually be your reference distance (having no introduced delay). The distances from each speaker will then be subtracted from your reference distance (in cm) and will be entered into the speaker time calibration app in your DAW.

Once having made the adjustments, sit down and listen to a mix with the time delays turned on and then off … prepare to be amazed (especially when using height channels)!

FIGURE 20.10
Speaker time calibration is done by measuring the distance from the listener's position to each speaker and creating a delay offset for each distance such that the sound arrives at the listener at the precise same time.

Monitoring in 5.1, 5.1.4, 7.1.4 and Beyond

With the recent proliferation of surround-sound speaker options for both the home and the studio, options exist at all levels of quality, functionality and cost-effectiveness for installing a surround and immersive system into a studio monitoring environment. As with most new technologies, it's important that your existing facility be taken into account so as to maximize control over monitor levels and monitor format choices (discrete surround, stereo and mono) as

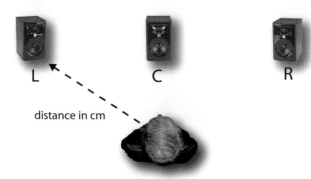

L C R

distance in cm

Ls Rs

well as its integration with your current console and/or DAW system. Before choosing a 5.1 or higher speaker system/setup, it would be extremely wise to consider the following:

- What are the commercial advantages to producing audio in surround? Are there any new clients or business ventures that allow you to make use of this technology?
- What is the budget for such a system?
- Can your existing speakers be integrated into the surround system? As you might recall from Chapter 16, active monitors include a powered amplifier(s) within their design, whereas passive monitors require that an external power amplifier be used to power their drivers. One of the benefits of designing your system around a powered speaker setup is lower system costs, as well as its ability to upgrade a stereo system to an immersive environment by simply adding extra matched monitors to an existing stereo production system.
- Can your console produce audio in surround sound? If your console has six or more output buses (8 bus +), your system can output surround in some manner; however, true surround panning and surround monitor control are often difficult to pull off when using such hardware systems.
- Can your DAW produce audio in surround sound? Certain basic DAWs are capable of routing audio to multiple output buses (5.1, 7.1 or higher); however, most of the more recent immersive-capable workstations are able to integrate tracks, effects and track exporting into a final immersive format with an amazing degree of versatility.
- How do you plan to monitor in immersive? If the console, interface or DAW offers true surround monitor capabilities, you're in luck. If not, a hardware surround monitor control system or immersive-savvy audio interface might be necessary (although you might be able to route the DAW outputs through a console and control all of the monitor levels together in a ganged fashion).
- What types of surround mastering tools should you invest in? Creating a surround-sound mix is only part of the battle. Make no mistake, often the real challenge of mastering an immersive audio project into its final form requires that we have a great deal of knowledge, attention to detail and extreme patience.

Once these and other considerations have been taken into account, and you're ready to make the jump, the task of choosing and installing a surround system into the production control room can get underway. This can be a daunting task, requiring technical expertise and acoustical knowledge, or it can be a straightforward undertaking that requires only basic placement, system setup and a big dose of techno-artistic patience.

THE BASIC FORMATS FOR UNDERSTANDING IMMERSIVE AUDIO

Although there are many formats that are available on this up-and-coming immersive market, there are three overriding formats that will need to be looked at in order to best understand the field of immersive audio:

- Binaural
- Ambisonics
- Dolby Atmos

Binaural Immersive Audio

From Chapter 2 (Acoustics and Hearing), we learned that two ears can discern the direction of a sound's origin. This capability of two ears to localize a sound source within an acoustic space is called spatial or binaural localization. This effect is the result of three acoustic cues that are received by the ears:

- Interaural intensity differences (for localizing cues at mid- to high frequencies)
- Interaural arrival-time differences (for localizing cues at low frequencies)
- The effects of the pinnae (outer ears, which are helpful in localizing height)

From these cues, it's possible for us to hear sounds coming from all around us in a 360° fashion.

A recording format known as binaural audio makes use of two mics that are spaced at a distance of about the size of an average head and can include a simple physical divider that mimics the head, or can be more sophisticated in design to include a dummy head design that more or less precisely mimics a head. By routing the left channel to the left and the right to the right, a simple stereo recording can be made such that by putting on a pair of headphones (open air tends to work best) … you'll hear the recording with all of the binaural cues intact (using a psychoacoustic process known as HRTF or Head Related Transfer Function). All of this means that this relatively straightforward recording process will be – to varying degrees – immersive.

One of DMH's favorite encounters with binaural took place many years ago at a famous radio production house in Berlin. We were listening to a recording of a motorcycle racetrack. You could hear the bikes as they rode around you … however, at one time, a lady started to ask a question behind me and to the right. I turned around to see who was asking the question, and no one was there! Such can be the power of binaural.

Ambisonic Immersive Audio

Ambisonics is an immersive recording and playback format that is able to encode audio information in a 360° horizontal and vertical plane.

Ambisonic information is not carried in a channel-based fashion; instead, it is encoded according to mathematical and phase-based information (known as B-format encoding) that can then be recorded to a set of analog or digital audio tracks. In its basic form, B-format can be thought of as the M/S miking techniques, which include the three-dimensional aspect of height (including an X, Y and Z set of informational axes).

Ambisonic recordings can be made in one of two ways:

- Directly, using an Ambisonic mic fitted with multiple capsules that are designed into an Ambisonic array
- The information can be mixed using specialized (or should we say "spatialized") panners that can output the channel information into a mathematical representation of the B-format

Mathematically, the B-format is available in first-, second- or third-order degree resolutions:

- First-order is encoded over four audio channels to represent the 360° soundfield.
- Second-order is encoded over eight audio channels to represent the 360° soundfield.
- Third-order is encoded over 16 audio channels to represent the 360° soundfield.

As you might expect, as the order (number of mics and/or encoding channels) increases, so does the overall sonic and placement resolution.

With regard to recording with an Ambisonic mic, one of the relative advantages is the ability to alter the mic's level, direction and polar patterns in mixdown. Indeed, it's actually possible to tilt, rotate and move the mic during mixdown with relative ease.

Although it's a bit of a different animal, mixing in Ambisonics can be relatively straightforward. However, with all levels, approaches and styles of Ambisonics, it's important to realize that the final result (be it acoustic or mathematical in nature) will usually be a rendered binaural stereo file that can be listened to over standard headphones.

Immersive Headphone Monitoring

Now that we have a basic understanding of both binaural audio and Ambisonics, let's take a look at one of the more recent advances in monitoring technology – room simulation (Figure 20.11).

Just as it's possible to capture a convolution (impulse response) snapshot of the sonic response, delay characteristics, reverb response, etc. of a room and then import this impulse file into a reverb program so as to give us a reverb unit that precisely mimics this room … it's possible to capture the overall acoustic

(a)

(b)

FIGURE 20.11

Dear Reality Monitor Virtual Mix Room Simulator. (a) Stereo mix room simulation. (b) 5.1.4 mix room simulation (courtesy of Dear Reality, www.dear-reality.com).

convolution response of an acoustic environment. This means that the overall frequency response, room reflections and speaker interaction within that room can be precisely captured within a room simulation plug-in. Effectively, through the use of Ambisonics and of course, binaural placement, it's possible to place the listener (using headphones) directly into a simulation of that acoustic space.

Not only is it possible to accurately capture the sound of a stereo speaker setup within the room, but because the results are based upon the 360° capture of the room, any immersive speaker setup can also be reproduced. Thus, using headphones, it is now a relatively simple matter for the user to use such a plug-in to monitor an immersive mix in 2.0, 5.1, 5.1.4, 7.1.4, as well as other monitor configurations.

Although these simulations differ from the sound of a true multi-speaker immersive setup, the results can be quite surprising. Such systems, of course, can be used on the road to help create an immersive mix and can eliminate the need for a full immersive speaker setup. In fact, these room simulations have grown in popularity to such an extent that many immersive movie and music mixes have been made using headphones. Once the mix has been completed, it can then be taken to a true immersive room so as to do a final quality control check before the mix is then signed off.

Dolby Atmos

Most of the immersive encoding systems that are out there are channel-based in nature, meaning that an overall mix would be made using a DAW, whereby the spatial information panned to the left would be routed to the left mix channel, the right to the right mix channel, etc. … so that at the end, the result would be a mix that contains as many channels as there are in the final mix format. This is not the case with Dolby Atmos, as it doesn't operate in a track-based world but rather, as an "object-based" mix system (Figure 20.12).

Objects can be thought of as points in the immersive field, which are represented as metadata that is encoded as location or dynamic pan automation data. Up to 128 objects can be at play within a mix, which can then be "steered" or moved as three-dimensional coordinates within the immersive soundscape.

At its most basic level, the true beauty of Atmos is its scalability and its flexibility. Since the final mix is represented as object points within a mix, the system is able to scale to any speaker layout that might be encountered. For example, a Blu-ray film release that has been mixed in 7.1.4 would, of course, play back through a full 7.1.4 home theater in all its multichannel glory. However, should that same Atmos mix be played back through a 5.1.4 system, Atmos will be fully aware that the L/R wide channels are not there. As a result, it will read the object metadata for the wide channels and distribute them appropriately between the L/R front and rear speakers, with no loss in channel information. When played back over a 5.1 layout, the same 7.1.4 mix would place the L/R wide channels appropriately within the L/R front and rear soundscape … while also taking the information from the height channels and properly steering them down to the lower immersive speakers. If only stereo playback is available, the metadata will steer all of the mix information so as to be properly placed within the 2.0 (stereo soundfield). Thus, the idea is that all of the objects would be steered to match the most appropriate playback layout … in theory, it's an awesome concept.

We'll now take an overview dive into the basics of Atmos. For those who wish to dig deeper, there's a never-ending trove of detailed information and videos on the web. We would like to provide a simple warning, though … Atmos has become a favorite new format for delivering mixes in a fresh, new form. As such, it's become a favorite topic for many YouTube sites. One of the strengths of this system is that it's so open-ended, meaning that there are often many ways in which to get the job done. As such, there's often not just one way to accomplish a goal. This can be a very good thing, but it can also provide its challenges, as one video or site might have you try one approach to attaining your goals, while another might give an entirely different approach. Both might work, or they might not. In short, everyone is striving to be the authority on the subject while giving out only parts of the techno story for getting the job done. We're just saying, don't always accept one person's approach as the right and only way (not even the info in this section). As always, it's best to do your own research, set up a mix, and try this and other approaches for yourself.

FIGURE 20.12
Dolby Atmos Renderer within Steinberg's Nuendo, showing the ADM Authoring window (left), Nuendo immersive MultiPanner (center) and Renderer (right).

THE BASICS

As of 2022, an Atmos session on a compatible immersive DAW MUST be set and loaded with samples at a rate of 48k (usually 24/48) and MUST be selected to have a system buffer size of 512.

Up to 128 objects are available to a mix, using either of two elements:

- Audio Bed: a "bed" is the "channel-based" main output bus that is position-locked to a standard speaker layout (i.e., 2.0, 2.1, 5.1, 7.1, 7.1.4, etc.). Every session must have at least one bed, as this is where the LFE will be inserted into the mix. As Atmos is, first and foremost, a film and visual audio mixing medium, a bed can be thought of as a sub-group, whereby one bed could be set up for a film's music track, while a second bed could be used for dialog, etc.
- Object: as stated, an "object" is NOT position-locked to a standard speaker layout, and makes use of metadata to control how an audio track or linked set of tracks is panned within an immersive three-dimensional space. As such, these tracks can be panned with great precision within the listener's environment. It should be noted, however, that an object CAN be statically placed into a traditional speaker location, thereby imitating a locked speaker pan position.

ADM Authoring Tool

The Dolby Atmos ADM (Audio Definition Model) Authoring Tool gives us access to:

- Renderer selection
- Auto object connection: for the automatic creation of an object when the appropriate track is created
- Add/Remove objects or beds
- Object/Bed display and routing
- Export ADM: allows exporting of the 24/48 wavefile (includes two metadata file tracks and full session wavefiles)

Renderer Tool

The Dolby Atmos Renderer Tool gives us access to:

- Bed/Object definitions, assignments and output activity
- Trim and downmix capabilities
- Renderer output level

Once the mix has been rendered, the resulting immersive ADM wavefile can be imported into a video editing program or uploaded to an immersive audio distribution site, as the file includes all of the required metadata and wavefile data. In addition, an ADM file can be read into a compatible DAW for playback of the immersive file. In short, once the ADM file is exported, the file is straight-forwardly ready for production and/or playback.

As a final note, the Atmos Renderer (either the Dolby Renderer itself or the renderer that is included with a number of immersive-capable DAWs) is also capable of rendering a stereo-encoded immersive Ambisonic/binaural version of the mix, which can be played over standard headphones for release and upload to the public.

Atmos Levels

It is absolutely crucial to realise that main bus level management in Atmos isn't like a standard mix. The general concensus is that the overall average mix bus levels should not exceed -18LUFS (Loudness Units Full Scale), while the overall peak levels should not exceed -2LUFS. Of course, this flies in the face of the loudness wars idea of making the mix as loud as possible. If these level maximums are not carefully observed, severe peak limiting of your master mix bus could occcur during playback. Trust us, this could get your mixes into VERY BIG trouble.

As Atmos is a new frontier technology, both DMH and EC would definetely recommend that you not take our word as full gospel, as updates and changes could occur at any time. As this field is quite technical, it's best to read up on Atmos and all of it's immersive components as much as possible. As a final note, beware of youtube atmos gurus … you could easily be led astray by someone who professes to know all the answers or have all the latest tricks. It's always best to double-check the facts yourself.

MIXING IN SURROUND

As you might imagine, mixing in immersive requires that the mixer be relatively competent in his/her understanding of the complexities that can go into routing, mixing and creating an immersive track. It can be a massive undertaking, but it's almost always a hell of a lot of fun. As you might remember, one of the main rules in audio is that there are no rules – only guidelines. Well this is very, very true of immersive audio. It's been said by many an immersive mixer that this specialized field of production is living in its Wild West times.

Of course, the direct control over the dialog, music and special effects of a movie or media production within a DAW lets the production team have direct and recallable access to all level adjustments, panning, effects plug-ins and more – all within a single, contained session. This ability to let go and have fun with a project, save it and then call it up again with full recall gives us access to a sonic playground that is a joy to behold.

It's DMH's strong belief that how you choose to convey your music through the art of immersive sound mixing is up to you and the muses. It's an extremely fun medium that should be experimented with, mastered and experienced to its fullest.

REISSUING BACK CATALOG MATERIAL

On a final note, one of the unintended by-products of immersive audio is the fact that many older, classic video and music projects are now being reissued and restored with immersive mixes by numerous media companies. This resurrection, restoration and rescue are helping us to save older analog (and digital) masters that have aged to the point that restoration becomes a monumental task and often, a labor of love. For example, when transferring analog music tracks to high-definition digital archive sound files – it's not uncommon to hear horror stories of 16- and 24-track masters that have to be reconditioned and carefully baked, then played onto an analog machine, only for the iron oxide to shed off the tape and separate from its backing. Quite literally, the tape will be self-destructing before your eyes as it plays off the reel and onto the floor during the actual transfer to DAW. In such a case, you would literally only get one chance to save the master.

Rescue stories like these are varied and awesome (in the truest sense of the word). Sometimes, the film and audio masters have already deteriorated past the playable point, and a safety backup must be used. Others must be tracked down in order to find that "right take" that wasn't properly documented. Suffice it to say that media reissues are definitely doing their part toward helping to keep the original masters alive and kickin' into the twenty-first century, not to mention the fact that they are breathing new life into the visual and music media in the form of a killer, new immersive version with restored sound and possibly a restored picture as well.

In a nutshell, listening to a well-crafted immersive project over killer speakers and video monitors in the studio, home playback or a home theater system can rank way up there with chocolate, motorcycle ridin' and good sex – absolutely the best!

CHAPTER 21

Media Distribution and Manufacturing

Given the huge number of changes that have occurred within the business of music marketing and media distribution, it stands to reason that an equally large number of changes have occurred when it comes to planning and considering how you are going to get your new-born project out to the masses. In short, many of the rules have completely changed, and it's the wise person who takes the time to put their best foot (and plan) forward to make it all shine. So, having said this – what are our best options for reaching people in this day and age? Well, there's:

- Music streaming: the 700-pound gorilla that a lot of online distribution companies have had to come to grips with and finally embrace. The idea is that paying a monthly subscription to a music service will give us instant access to their entire music library in an on-demand fashion – usually without the option of owning a physical or download copy of our (well, actually, their) music. It worked for video streaming (i.e., Netflix), so, why not for music?

- Online (download) distribution: online music sites exist and are even coveted by those of us who actually like to own our music. This 200-pound gorilla (which lets us collect and consume albums or single tracks when and where we want them) is definitely getting smaller by the year.

- Hi-res download distribution: similar to online music download services, high-resolution sites allow us to download our favorite classical, new releases or newly remastered, hi-res versions of our favorite older classics at higher sample rates or in true immersive audio. Most often released as larger, hi-res 24/96 sound files, these special releases offer up the highest-quality versions of a master recording. Again, this is presented to customers who value the quality (or perceived "specialness") of their personal hi-res, personally owned music collection.

- Vinyl: we're all pretty much aware of vinyl's major comeback as a viable music medium. It's the perfect in-hand "see what I just got?" product – both having a huge retro allure and saying to those around you that you care enough about your music to dare to be different. What's also special

DOI: 10.4324/9781003260530-21

about vinyl is the particular perks that go with 180-gram pressings, large-scale liner notes and (last but not least) the satisfaction of a medium that requires that you actually sit down and listen to at least a whole side from beginning to end.

- Compact disc (CD): some would call this a dying breed, but when you walk into a music store, you'll still see them everywhere (although DMH's fave store is now about 85% vinyl). Fact is, a lot of people still like to have their music media "in hand".
- DVD and Blu-ray: still viable visual media (but just barely). Of course, DVDs are good for movies and for music video/live concerts; however, Blu-ray discs tout high-definition video and hopefully lots of extra material along with high-resolution 5.1, 7.1 and Dolby Atmos immersive soundtracks in a compact physical form.
- Free music streaming: one of the biggest music services on the Web is (at the present time) still actually free … well, that is, unless you'd like to opt out of adverts and other ploys to catch your eye. In fact, YouTube is the 700-ton gorilla that has eclipsed most of the streaming services as an outlet for getting music, DJ playlists and music videos out to the public.

With all of these distribution methods, one of the greatest misconceptions surrounding the production of music, visual and other media is the idea that once you finish your project and have the master files in hand, your work is done; that the production process is now over. Of course, this is FAR from being the truth. There are many decisions that need to be made long before the process has been finished (and most often, long before the project has even begun). Questions like:

- Can we define who the market is and take steps to make "the product" known to this market?
- Will it be self-distributed, or will it be released by a music label that knows the ins and outs of the business and how to get "the product" out to the masses?
- Will it be available only online, or will it also be available in physical form?
- Will there be a marketing strategy and budget?
- What IS the marketing strategy?

All of these will definitely need to be addressed. One of the biggest mistakes that can be made during the creation of a project is to think that your adoring public will be clamoring for your product, website and merchandise without any marketing strategy or general outreach strategy.

Early in this book, I told you about the first rule of recording – that "there are no rules, only guidelines." Well, that actually isn't true. There is one rule:

If you don't pre-plan and follow through with your production and marketing strategies, you can be fairly sure that your project will sit in a lost directory on a server or lost on a shelf. Or, worse, you'll have 1,000 CDs sitting in your basement that'll never be heard or downloads that will be lost in the deep digital ocean – a huge, huge shame given all the hard work that went into making them.

PRODUCT DISTRIBUTION

With the establishment of Internet music streaming and distribution, along with the steady breakdown of the traditional record company distribution system (with the possible exception of vinyl), bands and individual artists have begun to produce, market and sell their own music on an ever-increasing scale. This concept of the "grower" selling directly to the consumer is as old as the town square produce market. By using the global Internet economy, independent distribution, fanzines, live concert sales, etc., savvy independent artists are taking matters into their own hands by learning the inner workings of the music business. In short, artists are being forced to take business matters more seriously in order to reap the fruits of their labor and craft – something that has never been and never will be an easy task.

OK, let's begin the journey. The best place to start is with a series of questions that are sort of "cart before the horse" in nature:

- Make a website. There are a vast number of ways that an artist can create their own site using a powerful, intuitive and easy-to-use template or personalized layout system. The days of learning html are pretty much gone (we'll drink to that)!
- Get your music out. Aggregators (such as distrokid and others) that can get your music out to the various streaming platforms are relatively cheap and easy. Others (such as Bandcamp) are able to fairly pay the artist and offer the artist a great degree of distribution flexibility.
- Promote and market the hell out of yourself (more on this later) – never an easy task.
- There is absolutely no way that this book will be able to give you all the answers. For that, we'll leave you to other books and the hundreds of thousands of YouTube videos that can help give you guidance. To this, we'll only say: be true to yourself, be yourself, realize that each artist's journey is unique, it's ok to fail (we all do) … and most important of all, try to enjoy the journey.

Build Your Own Website

If you build it, they will come! This overly simplistic concept definitely doesn't apply to the Web. With an ever-increasing number of dot-whatevers going online every month, expecting people to come to your personal or music site just because it's there simply isn't realistic. Like anything that's worthwhile, it takes connections, persistence, a good product and good ol'-fashioned dumb luck to be seen as well as heard! If you're selling your music, T-shirts or whatever at gigs, on the streets, and to family and friends, cyberspace can help increase sales by making it possible (and even easy) to get your band, music or clients onto several independent music websites that offer up descriptions, downloadable samples, direct sales and a link that goes directly to your main website. Such a site could definitely help to get the word out to a potentially new public – and to help educate your audience about you and your music.

Of course, artists aren't the only ones who have websites. Most folks in fields involving the crafts have a site; heck, even some studios have sites for their studio pets or mascots. Here are a few general guidelines that help your website work its magic for you and/or your organization.

- *Make your site a central hub:* it's always a good idea to direct your social network friends and fans to your personal or band page. A blog or vlog on this site can be an effective way to help keep people involved in your or your band's latest escapades. This helps to inform them about your upcoming projects, older ones that are for sale, current goings-on in your personal and professional life, etc.

- *Give it a strong, uncluttered main page:* a simple front page that makes a statement about you and your work can go a long way towards making a connection with the fans. From there, they can dig deeper and be drawn into "all things you".

- *Keep your site simple and straightforward:* try to make your site navigation easy to use and understand. Don't use problematic animations or excessive programming that can easily confuse your audience.

- *Make your site personal:* speak directly to your fans: Let your fans know what makes you tick, what you're up to, as well as what's happening in your professional life.

- *Keep it up to date:* this is really important. Nothing's worse than going to a site only to see that there's been no activity on it. Nothing says "I don't care" like a neglected site.

Uploading to Stardom

In this day of surfing and streaming media off the Web, it almost goes without saying that the Web has become the most important and effective marketing tool for musicians and labels alike. It lets us cost-effectively upload and promote our songs, projects, promotional materials, touring info and liner notes and distribute them to mass audiences. However, long before the recording and mix phases of a project have been completed (assuming that you're also doing your business and promotion homework), the next and possibly most important steps to take are:

- Target your audience: who are your fans, and what is your message (music, communication style, personal brand)?

- Create a "presence": develop a Web marketing and music distribution presence and then find the best ways to get that message out to your fans.
- Develop a social network system to interact with the fans (if that's what you want).
- Broadcast your music: using terrestrial radio or TV, podcasts and YouTube videos, you can expand your audience reach.
- Perform live: this can be done on the stage or on the Web (in a live or YouTube live-cast platform).
- Distribute your music: using a carefully chosen online distributor, you can create a sales presence that's uniquely your own, or (more likely) you could use one aggregator that will deliver your music to most or all of the major online distribution networks from one central distribution hub. Using these tools, it's easy to make the world your musical oyster.
- BUT – the trick is to get heard above the millions of others who are also vying for everyone's attention. That's the hard part that requires talent, luck, tenacity, showmanship and connections.

One of the more important considerations that should be made before you start the upload process is to decide upon the proper *metadata* for your project, songs or song. Metadata (or *tagged* information) is the inputting of media information relating to:

- Project name
- Artist name
- Song names
- Music genre
- Publisher info
- Copyright info, etc.

Don't fool yourselves, metadata is king. If you're looking for a needle in a haystack, that needle will have to be identified in a way that makes it easy to find; otherwise, it'll be virtually impossible to find amongst the billions of bits of hay.

It's important that you enter this data correctly and with forethought, as the distribution system will use this information within their search database for finding your music in their system, on the Web, etc. Quite simply, inaccurate metadata will bury your hard work so deep that it won't get played.

Streaming

Streaming refers to music and video content that exists only in the cloud (on Web servers) and is available for listening on a 24/7 basis for free or by a monthly fee subscription. The major difference here is that the listener has no ownership over the music and usually can't download the media files directly; we're simply licensed to access and listen to them.

In this day and age, when music is literally and inescapably everywhere, the most popular streaming service is free – of course, we're talking about YouTube.

Known for its vast resource library of video material, YouTube is by far the most popular way for fans to connect with their favorite artist's or group's music. How much does the artist get paid for these free online efforts? Well, nothing or practically nothing … but again, more on this later.

The idea of streaming audio from a network provider source, of course, does not require that the "product" exist as a physical object. It does, however, require that you put in the same amount of preparation and forethought that you would put into a piece of physical media.

Long before releasing your music, the following questions should be fully addressed:

- Who is my/our target audience?
- Is there a marketing budget?
- What is our marketing strategy?
- What will be the Web presence for the release?
- Who will do the artwork and general layout?
- Are there any special considerations for tagging the metadata?

As you might imagine, there are a large number of internet music streaming providers on the Web. Getting your music out to each and every one can be daunting. As a result, a number of aggregators exist that are able to act as gatekeepers for getting your projects out onto all of the various streaming platforms (and for collecting the various income sources). This can take place either for free, with a commission or for a yearly usage fee.

Companies like Spotify, Pandora, Amazon, Tidal, Apple and a few hundred others have been working to perfect a business model that allows the customer to feel comfortable with paying a monthly fee for access to their music database. One of the greatest and highly publicized problems is the notoriously low fees that are paid out to musicians and music labels for the use of their content … but more on this later.

The one thing to be aware of here is that the general per-stream income that gets distributed to the artist from streaming providers is, to put it mildly, low. The general idea here is that the income from the various streaming services can combine to provide a basic income stream over time. As with all things music business, be aware, read all of the fine print and keep abreast of your online activity.

Download Sites

Music download sites differ from streaming services (although many also offer on-demand streaming) in that the music can be auditioned and then downloaded in a way that allows the listener to own the music files, which can then be loaded onto their own computers or personal media devices. Such services often offer download qualities ranging from basic mp3 all the way to uncompressed flac and wav files, which allow a hi-def listening experience.

The same considerations (noted earlier) that go into streaming also apply to download files and should be considered before putting the files online.

It should be noted that a few download sites (notably Bandcamp and Immersive Audio Album) offer direct payments (85% and 80%, respectively) to the label or artists in a way that directly supports the artist in a fair and immediate fashion.

Internet Radio

Due to the increased bandwidth of many Internet connections and improvements in audio streaming technology, almost all of the world's radio stations have begun to broadcast on the Web. In addition to offering a worldwide platform for traditional radio listening audiences, a large number of corporate and independent Web radio stations have begun to spring up that can help increase the fan and listener base of musicians and record labels. Go ahead, get on the Web and listen to your favorite Mexican station, catch the latest dance craze from Berlin, or chill to reggae rhythms streaming on an island breeze (a great way to jump around the world is via www.radio.garden).

MONEY FOR NOTHIN' AND THE CHICKS ...

In the past, there were smaller, separate stores for meats, cheese, candy, etc., and the milk would be delivered to your doorstep. It was all pretty much kept on a smaller, personable scale. Now, in the age of the supermarket, where everything is wholesale, processed, packaged and distributed to a single clearinghouse, there are more options, but the scale is so large that older folks can only shop there with the aid of a motorized shopping cart.

For more than six decades, the music industry has largely worked on a similar principle: find artists who'll fit into an existing marketing formula (or more rarely, create a new marketing image), produce and package them according to that formula and put tons of bucks behind them to get them heard and distributed, and then put them on all of the Mom 'n' Pop music store shelves everywhere. There were fewer artists who were vying for the listeners' attention, and since you heard them on the radio all the time, it was a foregone conclusion that the kids would go out and spend their hard-earned money on their favorite band or artist.

This was not a bad thing in and of itself; however, for independent artists, the struggle has been, and continues to be, one of getting themselves heard, seen and noticed – without the aid of the well-oiled mega-machine. With the creation of cyberspace, not only are established record industry forces able to work their way onto your desktop screen (and into your multimedia speakers), but independent artists now also have a huge medium for getting heard.

Through the creation of streaming services, dedicated music websites, search engines, links from other sites and independent music dot-coms, as well as through creative gigging and marketing, new avenues have begun to open up for the Web-savvy independent artist (Figure 21.1). The only trick is to connect you to the fans so you can get noticed in that great music supermarket in the cloud.

As you might expect, making a living in today's online marketplace is as difficult as it ever was. Sure, it's easy to get your material out there, but getting heard, seen and surfed above the digital fray requires a great deal of knowledge, dedication, artistry, time and money. When you factor in the reality that income on sales for digital streaming can be relatively meager (this is often true with digital download, as well), it's no wonder that most popular artists turn to touring, merchandise and promotional associations to keep their business afloat. Even with a huge fan base and media buying/streaming public, it can be a difficult business that requires a great deal of savvy and determination.

The Moral Question

This is beyond the scope of this book; however, it's a topic that we feel should be addressed in some form. Suppose that you go into a restaurant, sit down and order a full meal, and at the end, you pay the waiter a penny and leave. Just how would that go down? Now, suppose that there was free food on every corner … you'd walk into an elevator, and there would be a buffet at the side, just waiting for you. Food would be everywhere … it would be inescapable! Well, that's pretty much what has happened with music. You walk into a store, you step into that elevator, you turn on the radio (more on this in a sec), you hear a car go by and you simply can't get away from … music. That's what happens when it's everywhere, all the time. DMH remembers being in school in London. Our flat was very near the Victoria and Albert Museum. Inside were rooms that were completely full of objects that were solid gold. After two or three rooms like this … I remember thinking "Ho hum, another room full of gold!" Crazy, but that's what happens to us humans when our senses get overloaded over time.

As a result of this, we've come to fully expect that our music will be free or very close to it. We know that you've worked on your project for the most part of a year and have put all of your blood, sweat, tears and soul into it …. So, give it to me for free, damn it!

FIGURE 21.1
Getting the artist's work out to the masses is hard work. Here's our chance to put our best foot forward (our music sites www.davidmileshuber.com and www.emilianocaballero.com). Moral of the story? Never pass up an opportunity!

On a personal level, DMH has been struggling with this issue for quite some time now. He's pulled back all of his music from streaming and has released the projects solely on Bandcamp and Immersive Audio Album (where the artist gets 85% and 80%, respectively, instead of Spotify's $0.0033 per stream). Now, after several years and some strong convincing from a number of friends, the music is now being re-released onto all of the streaming platforms. Why? So that the music will get out there and be heard. Isn't that why we make music in the first place … first, for ourselves as artists and then, for the greater public to enjoy.

Regarding vinyl, perhaps that's why the medium has made such a comeback. It's such a personal format … you put it on, sit back, read the liner notes and enjoy the ride … all in one go. Plus, the majority of new vinyl releases are made by independent labels and artists, so the artist (hopefully) gets better support in the end. It all seems less corporate.

Royalties and Other Business Issues

In this "business" of music, one of the goals is to get paid for your hard-earned work. With the added complexities of having so many options for getting an artist's music out to the world come the added complexities of calculating and collecting the cash. Again, an exhaustive overview of the topic of the music business is far beyond the scope of this book. Fortunately, many books and online resources have been written about "the biz". However, care must be exercised, and counsel (which can come in many forms) can often be sought to help guide you through these enticing and perilous waters. As always, caveat emptor!

Let's start here:

- If you are the writer of an original composition and have not signed with a publishing company, then you own the publishing (and all rights) to that composition.
- If you have signed with a label or publishing company (one that you or the band hire to *hopefully* take care of the everyday inner working of collecting funds and running your business, while you're busy making and performing music), then the publishing rights to a composition (and thus, the paid-out funds) will traditionally be split between the writer and the publishing company 50/50.
- The "writer" can be a single artist, a singer/songwriter, members of a band (equal split), a single member of a band (possibly the one who originally wrote the song), or it can be shared between the lyricist and the composer – in short, it is open to definition in any way that the production team sees fit.
- A contract is a written legal agreement that can also be written in almost any way – it is always negotiable before it is executed (signed). If there are parts that do not serve your needs and cannot be agreed upon, you probably should seek counsel and think twice (or more times) before signing.

Do You Need a Music Lawyer?

It's important to realize that music in the modern world is a BUSINESS. Once you get to the part of getting your band or your client's band out to the buying public, you'll quickly realize just how true this is. Building and maintaining an audience with an appetite for your product can easily be a full-time business – one where you'll encounter well-intentioned people as well as those who would think absolutely nothing of taking advantage of you and/or your client.

Of course, the subject of legal rights is way beyond the scope of this book, but in this day and age of self-publishing and distribution, it's definitely worth a mention.

Obviously, the first place to begin diving into the complex and (more often than not) misunderstood subject of music law is by researching the topic on the Web and by reading any number of books that are out on the subject. Keep in mind that any info in this book and within any other printed materials on the subject will simply be an opinion and guide. Each artist has his or her own specific legal needs and requirements, which can range from being simple (with answers that are stated in easy-to-understand terms) to extremely complex (having twists and turns that can send the most seasoned music executive to the local pub for lubrication).

This last sentence hints at the question about whether or not you need a music lawyer. Again, it depends upon the situation and your specific needs. If you're uploading your own self-published music into a distribution service, and the terms are fairly straightforward, then you probably won't need help. Of course, you'll want to take the time to carefully read the site's fine print so as to see if there are any pitfalls or snags that might jeopardize your ownership of the publishing rights or other unforeseen events. In this case, you could be just fine researching this legal fine print on your own, with the help of the Web and possibly with trusted advice. If things are more complicated, for example, if a record label will be acting as the publisher and the music's "writer" percentage is split between the band, the lyricist and the producer, then you'd definitely be wise to sit down with a trusted legal source to hammer out details that are equitable and free of confusion. In any case, care and time should *always* be taken before signing your name on the dotted line.

Whether you're selling your products on the street, at gigs, on Apple Music or in the stores through a traditional music label, it's often wise to *at least* entertain the idea of seeking out the counsel of a trusted music lawyer whenever an agreement comes into play. The music industry is fraught with its own special legal and financial language, and having someone on your side who has insight into the language, quirks and inner workings of this unique business can be an extremely valuable asset. Before we move on, it should be pointed out that many metropolitan areas have "Lawyers for the Arts" organizations that regularly offer seminars and events as well as one-on-one consultations with artists who are on a tight budget and have need for legal counsel.

Do You Need a Label?

Both DMH and EC are big believers in the adage "Wherever you may be … there you are". If you're starting from scratch, that's totally OK. Being your own chief bottle washer has its advantages, but it also has all the disadvantages that come with having to do everything (or trying to). It's not easy doing the pre-production, production, mastering, etc. … and then having to put on your marketing cap. We can't be all good at everything; almost none of us are. If you truly believe in the power of your message, then it might be wise to at least entertain the idea of searching out someone or a label that does marketing and has a particular brand of music that has a following.

Ownership of the Masters

If you are the artist/composer/chief bottle washer of a project and are not signed to a label, then you are the owner of your recordings. Once you pass through the gates into the realm of dealing with music publishers and record labels, this can change. Contracts can be worded in such ways that the actual ownership of the "phonorecord", that's to say the actual recorded production, can be transferred to that publishing entity for a specified period of time or forever. In these situations, you might want to read the print *very* carefully and/or seek counsel.

Registering Your Work

Regarding copyright ownership, the truth is that once you've written or recorded an original music work, you then own the copyright; however, unless this work is registered with the Library of Congress, it might be difficult to actually prosecute and seek damages from someone who you know has copied your music.

Copyrighting your music with the Library of Congress can now be done online using their eco online registration system (www.copyright.gov/eco) and can involve one of two submission forms:

FORM SR

Form SR (copyright.gov/forms/formsr.pdf) is used for registration of published or unpublished sound recordings, meaning that you are using this form to register the "phonorecord" or the actual recorded production as it exists in physical or downloadable media form.

FORM PA

Form PA (copyright.gov/forms/formpa.pdf) is used to register published or unpublished works of the performing arts. In the Library's words: "This class includes works prepared for the purpose of being performed directly before an audience or indirectly by means of any device or process." Works of the performing arts include: (1) musical works, including any accompanying words;

(2) dramatic works, including any accompanying music; (3) pantomimes and choreographic works; and (4) motion pictures and other audiovisual works.

Collecting the $$$

Ahhh, the real business side of the industry. Being and staying on top of collecting funds for you, your band or your represented artist is rarely easy or straightforward, to say the least.

Let's start here:

- Direct sales to your fans over a music website or at a performance can put $$$ directly into your or the band's account – very nice.
- Upon uploading your or your band's music to a music download service or aggregator (distributor to multiple download services), this company will then collect information as to how you will get paid (i.e., PayPal, direct deposit, etc.). This is (or should be) fairly straightforward. Care should be taken to read the percentages and how often these proceeds will be paid out. A number of these will also let you split the payouts between band members or collaborators.
- Once the recording is made and is released to the public, then registration with a Performance Rights Organization (PRO) might be in order. PROs such as ASCAP (American Society of Composers, Authors and Publishers), BMI (Broadcast Music Inc.), Soundexchange, Gema, Buma/Stemra and SESAC (Society of European Stage Authors and Composers) are used to collect royalties from music that's played on terrestrial radio stations, restaurants, retail stores and the like.
- It should be noted that companies (such as CD Baby PRO and TuneCore) can actually help with signup and collections from several of the above-mentioned organizations, making the process easier for individuals who want to follow these collection avenues.

From this, it's easy to see that getting paid for your music is not always a simple task. Many artists and labels subscribe to the "multiple trickles from many sources" concept, which holds to the idea that by subscribing to these (and other sources), the payment returns will be small, but they will come in from numerous sources in many ways to add up to a sum that could be worth it.

Cryptocurrency

By now, many of you have heard of blockchain technology (better known as Cryptocurrency, Bitcoin or any number of other fashionable names). But, did you know that this new way of dealing with money was originally created with musicians and other types of artists in mind?

The simplified idea behind Crypto is that financial transactions can take place in an encrypted form over a decentralized web of servers with the data being communicated around the world in real time. There is no bank, there is no off switch … this data exists on the Web in a way that is spread out and beyond the control of any financial oversight entity.

This can best be explained by example. Let's say that the XYZ band has just signed with the ZYX label in Dusseldorf. The contract between the label and artist has now been signed, and a vinyl-only release has been pressed and is ready for worldwide distribution. In the normal world, this media would be distributed to record shops by the label or distribution house, and once sold, the record stores would be responsible for paying the funds out to the label, who would in turn take their cut and then pay the agreed-upon percentages to the band. As you might imagine, this process could easily take six months or longer. It's also a process that's full of holes and potential loopholes that allow the labels, distributors and/or artists to drop the ball and not get paid by the record stores or by the labels in due time (if ever). As you can imagine, it can end up being a financial nightmare that at best can be slow and at worst can easily be abused, allowing money and bad accounting to fall through the cracks.

Crypto, on the other hand, works by allowing the agreed-upon financial transaction and percentages to be digitally agreed and set up in a fashion that works in real time. In this situation, at the outset, the record labels would receive the albums … once sold, the process of paying out any funds would begin at the cash register, whereby the Crypto contract percentages would then be electronically paid out to the distributor, the label, each of the individual band members and any other party that is part of the agreement. In this way, the funds are instantly transmitted into the accounts of the various parties, and the sale is finalized … no wait time, no room for fudging the books, no delay in paying the label and thus, the artists. Time will tell if this new form of payment will become the currency system of the future.

How Crypto Works:

1. A transaction is requested
2. A block is created that represents the $$$ transaction
3. A block is sent to all nodes (servers) in the connected network
4. The transaction data is validated
5. The transaction data is added to the existing blockchain
6. The transaction is complete

PRODUCT MANUFACTURE

Now that we've gotten across the really important idea of pre-planning your strategies, let's have a look at the various media that are involved in the distribution process, how they are made and how they can best be utilized to get the word out to the public.

Although downloadable and streaming media are now dominating the music and media industries, the ability to buy, own and gift a physical media object still warms the hearts of many a media product purchaser. Often containing uncompressed, high-quality music and media, these physical discs and records are held in regard as something that can be held onto, looked at and packaged up with a big, bright gift bow. Their major market-share days may be numbered, but don't count them out any time soon.

The CD

Beyond the process of distributing audio over the Internet (using an online service or from your own site), as of this writing, the CD is still a viable medium for distributing music in a physical form. These 120-mm silvery discs (Figure 21.2a) contain digitally encoded information (in the form of microscopic pits) that's capable of yielding playing times of up to 74 or 80 minutes at a standard sampling rate of 44.1 kHz with a 16-bit depth.

The pit of a CD is approximately half a micrometer wide, and a standard manufactured disc can hold about two billion pits. These pits are encoded onto the disc's surface in a spiraling fashion, similar to that of a record, except that 60 CD spirals can fit into the single groove of a long-playing record. These spirals also differ from a record in that they travel outward from the center of the disc, are embedded into the plastic substrate, and are then covered with a thin coating of aluminum (or occasionally gold) so that the laser light can be reflected back to a receiver. When the disc is placed in a CD player, a low-level infrared laser is alternately reflected or not reflected back to a photosensitive pickup. In this way, the reflected data is modulated so that each pit edge represents a binary 1, and the absence of a pit edge represents a binary 0 (Figure 21.2b). Upon playback, the data is then demodulated and converted back into an analog form.

FIGURE 21.2

The compact disc. (a) Physical disc and package (courtesy of www.davidmileshuber.com). (b) Transitions between a pit edge (binary 1) and the absence of a pit edge (binary 0).

Songs or other types of audio material can be grouped on a CD into tracks known as "indexes". This is done via a subcode channel lookup table, which makes it possible for the player to identify and quickly locate tracks with frame accuracy. Subcodes are event pointers that tell the player how many selections are on the disc and where their beginning address points are located. At present,

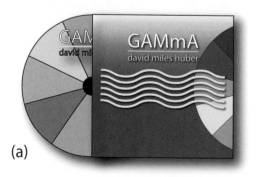

(a)

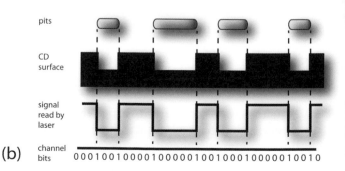

(b)

eight subcode channels are available on the CD format, although only two (the P and Q subcodes) are used.

Functionally, the CD encoding system splits the 16 bits of information into two 8-bit words with error correction (which is applied in order to correct for lost or erroneous signals). In fact, without error correction, the CD playback process would be so fragile and prone to dropouts that it's doubtful if it would've ever become a viable medium. This information is then translated into a data frame using a process known as eight-to-fourteen modulation or EFM. Each data frame contains a frame-synchronization pattern (27 bits) that tells the laser pickup beam where it is on the disc. This is then followed by a 17-bit subcode word, 12 words of audio data (17 bits each), 8 parity words (17 bits each), 12 more words of audio, and a final 8 words of parity (error correction) data.

In order to translate the raw PCM of a music or audio project into a format that can be understood by a CD player, a CD burning system must be used. These come in two flavors:

- Specialized hardware/software that's used by professional mastering and duplication facilities to mass replicate optical media (Figure 21.3).
- Disc-burning hardware/software systems that allow a personal computer to easily and cost-effectively burn individual or small-run CDs.

There are numerous ways in which a CD project can be prepared and burned. For starters, it's a fairly simple matter to prepare and master individual songs within a project and then load them into a program for burning (Figure 21.4). Such a program can be used to burn the audio files in a straightforward manner from beginning to end. Keep in mind that the Red Book CD standard specifies a beginning header silence (pause length) that's 2 seconds long; however, after this initial lead-in, any pause length can be user specified. The default setting for

FIGURE 21.3
Various phases of the CD manufacturing process. (a) The lab, where the CD mastering process begins. (b) Once the graphics are approved, the project's packaging can move onto the printing phase. (c) While the packaging is being printed, the approved master can be burned onto a glass master disc. (d) Next, the master stamper (or stampers) is placed onto the production line for CD pressing. (e) The freshly stamped discs are cooled and checked for data integrity. (f) Labels are then silk-screen printed onto the CDs, whereafter the printed CDs are checked before being inserted into their finished packaging (courtesy of Disc Makers, Inc., www.discmakers.com).

(a)

(b)

(c)

(d)

(e)

(f)

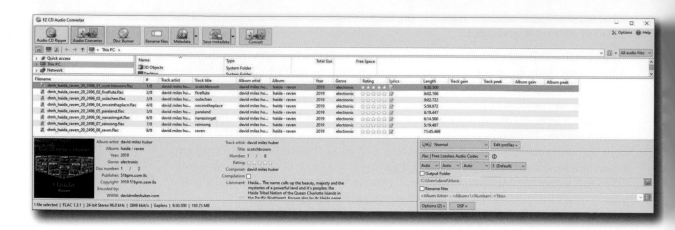

FIGURE 21.4
EZ CD Audio Converter all-in-one multi-format audio converter, CD ripper, meta-data editor and disc burner (courtesy of Poikosoft, www. poikosoft.com).

placing silence between cuts is 2 seconds; however, these lengths might want to vary, as one song might want to flow directly into another, while the next might want a longer pause to help set the proper artistic mood.

Most CD-burning programs will also allow you to enter "CD Text" information (such as title, artist name/copyright and track name field code info), which will then be written directly into the CD's subcode data field. This can be a helpful tool, as important artist, copyright and track identifiers can be directly embedded within the CD itself and will be automatically displayed on most CD hard- and software players. Additionally, database services such as Gracenote can be used to display track and project information to players that are connected to the Web. These databases (which should not be confused with CD Text) allow CD titles, artist info, song titles, project graphics and other info to appear on most media-based music players – an important feature for any artist and label.

As always, careful attention should always be paid to the details when creating a master optical disc, making sure that:

- High-quality media is used
- The media is burned using a stable, high-quality drive
- The media is carefully labeled using a recommended marking pen
- Two copies are delivered to the manufacturer, just in case there are data problems on one of the discs

As with any part of the production process, it's always wise to do a full background check on a production facility and even compare prices and services from at least three manufacturing houses. Give the company a call, request a promo pack (which includes product and art samples, service options and a price sheet), ask questions and try to get a "feel" for their customer service abilities, their willingness to help with layout questions, etc. You'd be surprised just how much you can learn in a short time.

Once you've settled on a manufacturer, it's always a good idea to research what their product and graphic arts needs and specs are before delving into this production phase. When in doubt about anything, give them a call and ask; they are there to help you get the best possible product and to avoid costly or time-consuming mistakes. The absolute last thing that you or the artist wants is to have several thousand discs arrive on your doorstep that are WRONG! Receiving a test pressing and graphic "proof" is almost always well worth the time and money. It's never wise to assume that a manufacturing or duplication process is perfect and doesn't make mistakes. Remember, Murphy's law can pop up anywhere and at any time!

> Whenever possible, it's always wise and extremely important that you be given art proofs and test pressings before the final products are mass duplicated.

With the ever-growing demands of marketing, gigging and general business practices, many independent musicians are also taking on the task of burning, printing, packaging and distributing their own physical products from the home or business workplace. This homespun strategy allows individual or small runs to be made on an "on-demand" basis without tying up financial resources and storage space with unsold CD inventories.

DVD and Blu-ray Burning

Of course, on a basic level, DVD and Blu-ray burning technologies have matured enough to be available and affordable for the Mac or PC. From a technical standpoint, these optical discs differ from the standard CD format in several ways. The most basic of these are:

- An increased data density due to a reduction in pit size (Figure 21.5)
- Double-layer capabilities (due to the laser's ability to focus on two layers of a single side)
- Double-side capabilities (which again doubles the available data size)

In addition to the obvious benefits of increased data density, a DVD (max of 17 Gb) or Blu-ray (max of 50 Gb – dual layer) will also have higher data transfer rates, making them the ideal media for:

- Hi-res video
- Multichannel surround sound
- Data- and access-intensive video games
- High-density data storage

As DVD and Blu-ray writable drives have become commonplace, affordable data backup and mastering software has come onto the market that has brought the art of video and hi-def production to the masses. Even high-level optical media mastering is now possible in a desktop environment. More information on the finer points of codec data compression and related media technologies can be found in Chapter 11.

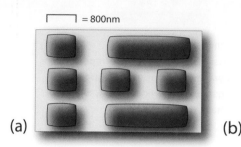

$\Box$ = 800nm

(a)

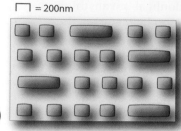

$\Box$ = 200nm

(b)

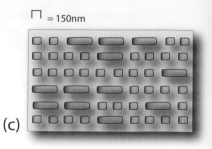

$\Box$ = 150nm

(c)

FIGURE 21.5
Detailed relief showing (a)
Standard CD. (b) DVD and (c)
Blu-ray pit densities.

Optical Disc Handling and Care

Here are a few basic handling tips for optical media (including the recordable versions) from the National Institute of Standards and Technology:

Do:

- Handle the disc by the outer edge or center hole (your fingerprints may be acidic enough to damage the disc over time).
- Use a felt-tip permanent marker to mark the label side of the disc. The marker should be water or alcohol based. In general, these will be labeled as being a non-toxic CD/DVD pen. Stronger solvents may eat through the thin protective layer to the data.
- Keep discs clean. Wipe with a cotton fabric in a straight line from the center of the disc toward the outer edge. If you wipe in a circle, any scratches may follow the disc tracks, rendering them unreadable. Use a disc-cleaning or light detergent to remove stubborn dirt.
- Return discs to their cases immediately after use.
- Store discs upright (book-style) in their cases.
- Open a recordable disc package only when you are ready to record.
- Check the disc surface for scratches, etc. before recording.

Don't:

- Touch the surface of a disc.
- Bend the disc (as this may cause the layers to separate).
- Use adhesive labels (as they may unbalance or warp the disc).
- Expose discs to extreme heat or high humidity; for example, don't leave them in direct sunlight or on a car's dash.
- Expose recordable discs to prolonged sunlight or other sources of ultra-violet light.
- Expose discs to extreme rapid temperature or humidity changes.

Especially Don't:

- Scratch the label side of the disc (it's often more sensitive than the transparent side).
- Use a pen, pencil or fine-tipped marker to write on the disc's label surface.
- Try to peel off or reposition a paper or plastic label (it could destroy the reflective layer and/or unbalance the disc).

Vinyl

Obviously, reports of vinyl's death were very premature. In fact, for consumers ranging from Dance DJ trip-hopsters to die-hard classical buffs, the record is making a popular comeback in record stores around the world. However, the truth remains that only a few record pressing plants are still in existence (they're currently working their overtime butts off), and there are far fewer disc-mastering labs that are capable of cutting "master lacquers". As a result, it may take a bit longer to find a facility that fits your needs, budget and quality standards, but it's definitely not a futile venture.

DISC CUTTING

The first stage of production in the disc manufacturing process is disc cutting. As the master is played from a digital source or analog tape machine, its signal output is fed through a mastering console to a disc-cutting lathe. Here, the electrical signals are converted into the mechanical motions of a stylus and are cut into the surface of a lacquer-coated recording disc.

Unlike the CD, a record rotates at a constant angular velocity, such as 33⅓ or 45 rpm (revolutions per minute), and has a continuous spiral that gradually moves from the disc's outer edge to its center. The recorded time relationship can be reconstructed by playing the disc on a turntable that has the same constant angular velocity as the original disc cutter.

The system that's used for recording a stereo disc is the 45/45 system. The recording stylus cuts a groove into the disc surface at a 90° angle, so that each wall of the groove forms a 45° angle with respect to the vertical axis. Left-channel signals are cut into the inner wall of the groove, and right-channel signals are cut into the outer wall, as shown in Figure 21.6a. The stylus motion is phased so that L/R channels that are in phase (a mono signal or a signal that's centered between the two channels) will produce a lateral groove motion (Figure 21.6b), while out-of-phase signals (containing channel difference information) will produce a vertical motion that changes the groove's depth (Figure 21.5c). Because mono information relies only on lateral groove modulation, an older disc that has been recorded in mono can be accurately reproduced with a stereo playback cartridge.

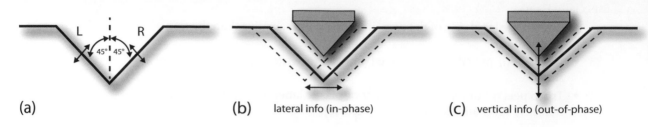

(a)

(b) lateral info (in-phase)

(c) vertical info (out-of-phase)

FIGURE 21.6

The 45/45 cutting system. (a) Stereo waveform signals can be encoded into the grooves of a vinyl record in a 45°/45° vector. (b) In-phase horizontal groove motion. (c) Out-of-phase vertical groove motion.

DISC-CUTTING LATHE

The main components of a vinyl disc-cutting lathe are the turntable, lathe bed and sled, pitch/depth control computer and cutting head. Basically, the lathe (Figure 21.7a) consists of a heavy, shock-mounted steel base (a). A weighted turntable (b) is isolated from the base by an oil-filled coupling (c), which reduces wow and flutter to extremely low levels. The lathe bed (d) allows the cutter suspension (e) and the cutter head (f) to be driven by a screw feed that slowly moves the record mechanism along a sled in a motion that's perpendicular to the turntable.

CUTTING HEAD

FIGURE 21.7

The disc-cutting lathe. (a) Lathe with automatic pitch and depth control (courtesy of Paul Stubblebine Mastering and Michael Romanowski Mastering, San Francisco, CA, www. paulstubblebine.com and www.michaelromanowski. com). (b) Simplified drawing of a stereo cutting head.

The cutting head translates the electrical signals that are applied to it into mechanical motion at the recording stylus. The stylus gradually moves in a straight line toward the disc's center hole as the turntable rotates, creating a spiral groove on the record's surface. This spiral motion is achieved by attaching the cutting head to a sled that runs on a spiral gear (known as the lead screw), which drives the sled in a straight track.

The stereo cutting head (Figure 21.7b) consists of a stylus that's mechanically connected to two drive coils and two feedback coils (which are mounted in a permanent magnetic field) and a stylus heating coil that's wrapped around the tip of the stylus. When a signal is applied to the drive coils, an alternating current flows through them, creating a changing magnetic field that alternately

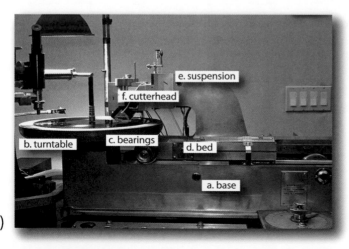

(a)

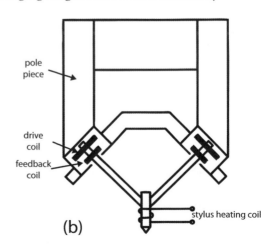

(b)

attracts and repels the permanent magnet. Because the permanent magnet is fixed, the coils move in proportion to a field strength that causes the stylus to move in a plane that's 45° to the left or right of vertical (depending on which coil is being driven).

PITCH CONTROL

The head speed determines the "pitch" of the recording and is measured by the number of grooves, or lines per inch (lpi), that are cut into the disc. As the head speed increases, the number of lpi will decrease, resulting in a corresponding decrease in playing time. Varying the lead screw's rotation can most commonly change groove pitch by changing the motor's speed (a common way to vary the program's pitch in real time).

The space between grooves is called the land. Modulated grooves produce a lateral motion that's proportional to the in-phase signals between the stereo channels. If the cutting pitch is too high (causing too many lines per inch, which closely spaces the grooves), and high-level signals are cut, it's possible for the groove to break through the wall into an adjacent groove (causing overcut) or for the grooves to overlap (twinning). The former is likely to cause the record to skip when played, while the latter causes either distortion or a signal echo from the adjacent groove (due to wall deformations). Groove echo can occur even if the walls don't touch and is directly related to groove width, pitch and level.

These cutting problems can be eliminated either by reducing the cutting level or by reducing the lines per inch. A conflict can arise here, as a louder record will have a reduced playing time but will also sound brighter, punchier and more present (due to the Fletcher–Munson curve effect). Because record companies and producers are always concerned about the competitive levels of their discs relative to those that are cut by others, they're reluctant to reduce the overall cutting level.

The solution to these level problems is to continuously vary the pitch so as to cut more lines per inch during soft passages and fewer lines per inch during loud passages. This is done by splitting the program material into two paths: undelayed and delayed. The undelayed signal is routed to the lathe's pitch/depth control computer (which determines the pitch needed for each program portion and varies the lathe's screw motor speed). The delayed signal (which is usually achieved by using a high-quality digital delay line) is fed to the cutter head, thereby giving the pitch/depth control computer enough time to change the lpi to the appropriate pitch. Although longer playing times can be attained by reducing levels and by varying pitch control, many of the pros recommend keeping the playing times of a "good-sounding" LP to about 18 minutes per side or less.

VINYL DISC PLATING AND PRESSING

When the final master arrives at the plating plant, it is washed to remove any dust particles and then electroplated with nickel. Once the electroplating is complete,

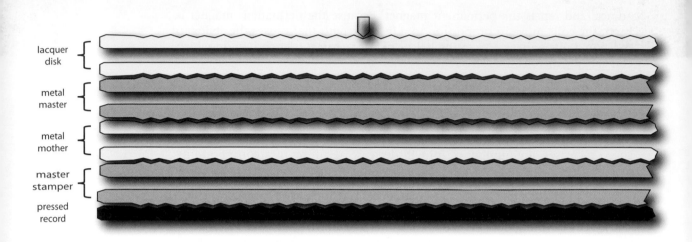

lacquer disk

metal master

metal mother

master stamper

pressed record

FIGURE 21.8
The various stages in the plating and pressing process.

the nickel plate is pulled away from the lacquer. If something goes wrong at this point, the master will be damaged, and the master lacquer must be recut.

The nickel plate that's pulled off the master (called the matrix) is a negative image of the master lacquer (Figure 21.8). This negative image is then electroplated to produce a nickel positive image called a mother. Because the nickel is stronger than the lacquer disc, several mothers can be made from a single matrix. Since the mother is a positive image, it can be played as a test for noise, skips and other defects. If it's accepted, the mother can be electroplated several times, producing stampers that are negative images of the disc (a final plating stage that's used to press the record).

The stampers for the two sides of the record are mounted on the top and bottom plates of a hydraulic press. A lump of vinylite compound (called a biscuit) is placed in the press between the labels for the two sides. The press is then closed and heated by steam to make the vinylite flow around the raised grooves of the stampers. The resulting pressed record is too soft to handle when hot, so cold water is circulated through the press to cool it before the pressure is released. When the press opens, the operator pulls the record off the mold, and the excess (called flash) is trimmed off after the disc is removed from the press. Once done, the disc's edge is buffed smooth, and the product is ready for packaging, distribution and sales.

FURTHER READING

Again, the information in this chapter is simply meant as an introduction. So much information exists on the subject of product manufacture, and even more so on the distribution and business of music, that I urge you to dig deeper so as to gain a better understanding into your own personal "edge" on the ever-changing landscape of the music business.

CHAPTER 22

It's All About the Journey

I'm sure you've heard the phrase "Those were the good old days". I've (DMH) usually found it to be a catchall that refers to a time in one's life that had a sense of great meaning, relevance and all-around fun. Personally, I've never met a group of people who seem to bring that sense of relevance and fun with them into the present more than music and audio professionals, enthusiasts and students. The fact that many of us refer to the tools of our profession as "toys" says a lot about the way we view our work. Fortunately, both Emiliano and I were both born into that clan and have reaped the benefits all our lives. It's our belief that the here and now are always our "good ol' days"!

Music and audio industry professionals, by necessity, tend to keep their noses to the workaday grindstone; however, market forces and personal visions often cause them to keep one eye focused on future technologies. These can be new developments (such as advances in digital, immersive audio and other technologies) or rediscovering retro trends and techniques that are decades old (such as the reemergence of tube technology and the reconditioning of older devices that sound far too good to put out to pasture). Such is the time paradox of a music and audio professional, which leads us to the book's final task: addressing how we navigate our lives and the journey through a career in (or personal love affair with) music, audio production and the task of simply being alive (as if that were ever simple).

WHEREVER YOU MAY BE ... THERE YOU ARE!

Let's start with you and where you are at the moment on your quest towards understanding audio. You might be a student who's focused on beginning your career with a bang, you might be a seasoned pro who's decided to read this book as a refresher (we never stop learning), or you might be a weekend warrior who wants to simply get the most out of their home studio tools and toys. It's all good.

The point is, no matter who you are, what you want or where you intend to be ... you are simply who you are at your current stage of development. You know

DOI: 10.4324/9781003260530-22

what you know in the here and now ... more knowledge and experience comes with practice and time. Our advice to you is to go easy on yourself and not to worry about the gear that you don't have, what you don't know or what you can't do. Your studio, its gear and the level that you're at will grow over time if you put your best effort into learning and practicing your craft.

There's a great video on YouTube called "That Elusive Console Magic", which alludes to the fact that the gear in your bedroom is probably better and more powerful than any of the older, classic consoles that are out there. So, why do those older mixes sound so good? Almost certainly because of "The Good Rule". Quite simply, because the musicians were so good, had a great project ready to bring to life, and had a great studio and talented support staff. All of these ingredients take talent, professionalism, dedication and, of course, time ... and this brings us to:

SHOWING UP IS HUGE!

A good buddy introduced DMH to the phrase "showing up is huge!" many years ago ... and it can't be more true. If you want to get noticed, quite simply, you have to show up! This means being physically present, being a good guy or gal to be around and always striving to put your best foot forward. If you're in an organization and they need someone to help check people in at an event, you might consider volunteering. The last person in the check-in line just might be a music producer who knows a studio that needs help, or some such thing. You just never know! In order to be noticed, you need to stand out, and that generally means getting your butt out there, rolling up your sleeves, and helping to get things done and make things better.

Speaking of showing up ... showing up at the right place at the right time means showing up at the wrong place at the right time or the right place at the wrong time a thousand times. You just never know when all of your personal stars will line up. It all comes down to showing up and putting your best foot forward ... which is huge!

BEING YOURSELF!

Probably the best piece of advice that can be given when entering into this or any other creative field is simply to be yourself. This can never be overstressed! You are the best person you can possibly be, and if you try to be someone else or try to fake it in order to impress folks around you, there's a good chance that it will backfire. The music production field is full of people who are there because they want to be there, generally not because of the glamor (although glamor does occur from time to time, often when you least expect it). It is the general consensus, that if you do your best, are a good guy/gal to work with and try hard ... the rewards will come over time and that you will be appreciated for who you are.

THE VARIOUS "DOORS" OF LIFE

Early in life, I (DMH) tried to check out the scene in Nashville to see if it'd be right for me. I was surprised to quickly get a job with a highly respected on-location recording company and set out to work, going on the road and doing my thing with a field team. Well, it didn't take me long to realize that Nashville wasn't my kind of town (it's my belief that we all have places in the world that best suit our temperament and soul). Not only that, my boss turned out to be a complete and total tyrant. Needless to say, I gave it all a go for a bit, said my goodbyes and quickly left town. Quite simply, it wasn't for me, and I had the smarts to realize it.

OUT OF THIS EXPERIENCE, DMH LEARNED A FEW THINGS ...

- First, leaving a job that's not right for you is OK (if you can afford it at the moment, or maybe even if you can't). Try not to beat yourself up over it.
- Second, realize that the problem just might not be you. Sometimes, a problem might be beyond your control. A job might not be a good fit and/or you might have a boss that simply makes you miserable. Over the years, this has happened to me a couple of times. If something similar happens to you, try not to beat yourself up, roll up your sleeves and move on.
- This brings us to the third and biggest realization. I realized early on that life was full of doors. Let's break this one down by using an analogy. Let's say you're in a room that has 12 doors. By being where you are at life, you find yourself standing in front of door #1. Opening it leads to a corridor that goes for a while and then leads to another door that opens to a dead end. Bummer, we now have to backtrack to our original room of doors. Going through door #5 looks like fun and is full of party lights but eventually winds up in a dead end ... Now, we might be getting frustrated, but this is our life we're talking about, so we backtrack and try another door, and maybe even another. By being at the wrong place at the wrong time a thousand times, we finally open up door#11, and (if we're lucky) we find a room that was made just for us ... and we can begin to build the life that we envisioned for ourselves.
- Fourth ... All of this might easily not last forever, and you'll have to go back to the room and go through door#12.

Moral of the story? There are no rules or guarantees in life. It's all a journey! Navigate the waters in the best way that you can ... which brings us to:

CREATING YOUR OWN REALITY

Unless you have 10 million in the bank, you've come to the realization that no one in this biz is going to hand you a job, a million-selling record or a Grammy once you've graduated from school, university, or mom and dad's place. You're going to have to work for it, have the desire to learn as much as possible and above all … have a talent for your craft. Even with all of this, the "school of hard knocks" might not be done with you just yet. It's important to realize that it will take perseverance and a big dose of luck.

The next part of the puzzle can be a lot more nebulous. We believe that it takes more than hard work, it takes intention … a sense of having a direction that you would like to travel in to achieve your goals. In short, a personal plan can be often set in motion in order to help direct yourself towards your goal.

- Is there a specific company that you'd like to work for? If so, seek them out and make yourself known to them, or better yet, learn how to be of value to their business in a way that best suits your needs, talents and desires.
- If such a place doesn't exist, you might go about ways to create your own business or service that best suits your needs.

For many of us in this techno-artistic field, the journey often began with a huge, personal leap that put us on the road to a new life. None of this is easy, but new directions in life will often begin with one big, personal step.

PERSONAL AND PROFESSIONAL STUMBLING BLOCKS

It happens to all of us. When things go right, it's simply a matter of time until something will go wrong. It's simply human nature and part of the voyage. The really difficult trick is not to let it get to you too badly (we know, easier said than done).

One of the things that have always struck me (DMH) is that I'll be talking with a friend who's in a rut and will then have him or her tell me about what's going on in their lives. Invariably, they'll tell me about this or that in a way that is often quite positive. I'll then go about telling them that things are actually progressing in a way that's way better that the other person thinks. It's not unusual or uncommon to think that things aren't going as fast or as well as they should, when in fact, that person is actually progressing in a way that isn't nearly as bad as all that.

The other side of the coin is that both DMH and EC have long figured out that when the going gets rough, it may be time to be easy on yourself. If things aren't going according to plan, it might be time to reassess your plan, to make a new plan, to gently steer the course … but it's not always the best idea to be hard on yourself or to overly stress. If you hit a stumbling block, give yourself a bit of time to find the best way to pick yourself up without undue stress that might make things worse.

REINVENTING YOURSELF

Continuing a bit on this topic, DMH asked both George Massenburg and the late Phil Ramone (go ahead, Google 'em) whether or not they made a habit of actively reinventing themselves over their careers. On both occasions, they came back with a quick "of course". George offered up that he always tries to step back and see how he can make life the most interesting and most fun as possible and steers the course accordingly.

Personally, we feel that it's always a good idea to actively step back and allow yourself the luxury of actively pursuing those things that will benefit your life and/or bring you the most joy. All of this goes back to the previous section on creating your own reality. Trust us, no one else can truly do this for you. When it comes to inventing or reinventing yourself, thinking outside of the box is often the best way to begin.

PERCEPTION AS IT DEALS WITH THE ART OF HEARING

There's one aspect of sound that often gets overlooked when dealing with the art of production. It's not about the toys, it's not about the sound itself … it's about how we creatives perceive sound. In short, we're talking about how our past experiences, our familiarity (or lack of it) with the room and the recording and/or playback equipment, and the acoustics of the room, combined with our mental state at the time of hearing. Any or all of these factors can affect the outcome of a mix.

I (DMH) am always a bit more than jealous of painters who have created a work of art. I know that lighting and the perception of the viewer will affect how we see the piece. But, playing back a recording will be affected by all of the links in the recording chain (including the perception of the producer/recording and mix engineers) as well as the ever-evolving number of possible changes that are introduced upon playback.

Other complications get added when you consider that the artist (and probably the producer) might have lived with this song or project for so long that their perception becomes clouded by their own perceptions of how it "should" sound. These personal perceptions can become so embedded in our psyche that it can almost become paralyzing. For this reason, it's always a good idea to seek out the advice or service of other professionals who can offer up advice without the added weight of self-doubt and being overly close to the project.

From my (DMH) standpoint, these obstacles don't get easier with time. However, experience can give us the tools to help us to step back and, either personally or collaboratively, find the best way to present the project in its best light. All good things come to those who are willing to put in the time, blood, sweat and tears … within limits … at which time, you probably should call in for help.

I'll close this out with the story of a producer who would tell the band to do this, that and the other … and then he'd simply leave the studio until the tasks were done. Once done, he'd walk back into the room, take a fresh listen and then say … OK, do this, this and this, and then leave again. You could look at this from two angles. You could say that he was a slouch who was never there, or you could say that he was wise enough to walk back into the studio in a way that allowed him to assess the situation from a fresh perspective.

WHEN IS TOO MUCH GEAR SIMPLY TOO MUCH?

Quite simply, no matter how much gear you have or how expensive it is, the true limit to your studio's abilities is only limited by you, your determination and your talent, not by how expensive your mikes or speakers are, nor by how many plug-ins you have available to your DAW .

Don't get us wrong; it's important to do your research and to get the production gear that best fits your budget, your abilities and the level you're currently at. The adage "it's not what you have, it's how you use it" definitely fits here. It's highly unlikely that when you get up on stage to make your Grammy acceptance speech, the audience will be interested in the cost of your gear or how many plug-ins you use … you will be appreciated for your talent, your determination and (most likely) your or your team's marketing skills. In short, the moral of the story is to use what you have to the best of your abilities and realize that the next piece of gear won't necessarily make you better … you will!

THE BANK WORKS FOR YOU, YOU DON'T WORK FOR THE BANK

This topic is off the subject of audio, but both DMH and EC believe that it's one of the core concepts that can help keep you financially independent and happy – in short, freedom, baby!

This relates to the simple idea that:

- If you make more than you spend – you're rich
- If you spend more than you make – you're poor

I'm sure we all have stories of friends who have pulled out a fist-full of credit cards, only to tell us that they're all maxed out and that they have no clue how to get out of debt. I know that this is tough love, but the best way to stay out of debt is not to have any … or to keep it so low that it's completely manageable.

In short, if you follow this rule, keep your debts low and put the rest into savings or into paying off your house/car mortgages early, the fact is that over time, you will begin to build up savings in a way that gives you freedom.

Another way to measure this is the next time you go to a bank and need to sit down with an agent. Do you feel nervous, like they always have the upper hand

and control over you – or are you sitting down with someone who works for you, in that they are helping you to manage your money? If it's the former, you might think about how you can go about controlling your finances without having them control you. Being conscious of this process is one of the most important things that you can possibly do.

IT DOES TAKES TIME

I (EC) remember growing up in Lugo (a small town in the north-west of Spain) listening to my favorite albums and dreaming of making records in big-time cities like Los Angeles, New York or London …

At the time, it seemed almost impossible to fulfill those dreams, but on the other hand, I had no doubts that somehow it would happen. What would it look like? I had no idea, but I knew that I had to trust myself (it got easier with time), follow my gut and do the best I could with what I knew.

Many years and many (interesting) experiences later, I find myself writing these words in this book (which I remember reading as one of the first recording books I bought years ago) in my apartment in Hollywood, CA, during Xmas time.

To everyone discovering their journey, I'd say:

> Be patient with yourself, your skills and your journey. It will be difficult at times, and these words might not resonate, but they will in a not so distant future.

HAPPY TRAILS

Before we wrap up the tenth edition, We'd like to take a moment to honor one of the greatest forces driving humanity today (besides sex) – the dissemination and digestion of information. Through the existence of quality books, trade magazines, university and institute programs, workshops and the Web, a huge base of information on almost any imaginable subject is now available to a greater number of aspiring artists and technicians than ever before. These resources often provide a strong foundation for those who are attending accredited schools as well as those who are attending the street school of hard knocks. No matter what your goals are in life (or in this business of music), I urge you to jump in, read and surf through pages – keep your eyes and ears open for new sounds, ideas, technologies and experiences. The knowledge and skills you gain will always be well worth the expended time and effort.

On a final note, DMH would like to paraphrase Max Ehrmann's "Desiderata", which urges us to "keep interested in our own career, however humble, as it's an important possession in the changing fortunes of time". Through my work as a producer, musician, writer and educator, I've been fortunate enough to know many fascinating, talented and fun people. For some strange reason, I was born with a strong drive to have music and production technology in my life. By "keeping interested in my own career" and working my butt off (while having several brushes with extreme luck), I've been able to turn this passion into a successful career.

To me, all of this comes from following your bliss (as some might call it), listening to reason (both your own and that of others you trust) and doing the best work that you can (whatever it might be). As you know, thousands of able and aspiring bodies are waiting in line to make it as an engineer, a successful musician, a producer, etc. So, how does that one person make it? By following the same directions that it takes to get to Carnegie Hall – practice! Or as the T-shirt says, "Just do it!" Through perseverance, a good attitude and sheer luck, you'll be led through paths and adventures that you never thought were possible. Remember, being in the right place at the right time simply means being in the wrong place a thousand times – "Showing up in life is huge!"

YOU, YOUR ART AND YOUR WELL-BEING

You have come this far, so be proud of yourself and your journey. Making art is a beautiful gift, and as with everything in life, learning how to find the balance between art, business and general life is important.

At times, being the creator, producer and as DMH says, "the chief bottle washer" can be stressful, daunting and make you wonder why you put yourself in that position in the first place.

Remember, any artist will go through ups and downs during their career … some days you will feel inspired to create, other days … maybe not so much, and that is OK.

Your life outside the studio is part of your artistic journey (which will permeate through your creativity in many different ways). Traveling, resting, and having other hobbies and interests will also contribute to your life's work.

Remember: your life is your canvas, and there are many different colors. Go paint in a way that best expresses "you", and always strive to have fun along the way!

Oh … one last thing! If any of you have an interest in the "History of Recorded Sound", DMH has created an online book on the subject that can be found at:

www.davidmileshuber.com/hrs

Both EC and DMH feel that it's important that those of us in production have at least a passing knowledge of the people and technology that came before us. For example, a good understanding of classical and distant miking techniques will give you an edge over those that don't … every time. Learning the foundation of where we came from can provide a better understanding of how to improve our own, personal tool set.

This is where our personal history began … at Galaxy Studios, Mol, Belgium, in January 2012. We truly wish you all the best and lots of fun during your own personal journey.

Dave and Emiliano

Index

Note: Page numbers in *italics* indicate figures, **bold** indicate tables in the text

1/4″ balanced connectors 112, *113*
1.0 Mono monitoring configurations 477–8, *478*
2 + 1 (Stereo + Sub) monitoring configurations 478–80, *480*
2.0 Stereo monitoring configurations 478, *478*
25/25/50 guideline *92*, 92–3
3:1 distance rule 124, *124*
3:2 pulldown rate 384–5
45/45 cutting system *620*
5.0 surround minus an LFE 588
5.1.4 mix room simulation *596*
5.1.4 speaker setup *583*
7-Band Digirack EQIII plug-in for Pro Tools *245*
7.1 speaker placement 588
700-Hz filter *425*

Ableton Live performance audio workstation *265*, 270, 271, *271*, 273, 351, *352*
absorption *86*, 86–91; 25% absorption 92; coefficient 86; coefficients for various materials **87**; flexible surfaces 90–1, *91*; high-frequency 88, *89*; low-frequency 89–90, *89–90*
A/B stereo: miking technique *135*; recording technique 135
AC3 **363**
accelerator processing systems 244, *244*
accent microphone placement 133–4, *134*
accent miking 119, 149
accessing sound files 257
accessories for digital audio workstation (DAW) 259–60
accessorizing: digital audio workstation (DAW) 259–60; iOS in music production 355
AC electrical circuit 408

Acon Digital's Acoustica Digital Audio Editor *448*
acoustics: blockage 59; control room 69–91; defined 65; echo chamber 95–6; guitar 144–5, *145*, 437; partitions 79–80, *79–80*; separation 65, 93, 123; spice 95; trauma 55
acoustics and studio design 65–96; 25/25/50 guideline *92*, 92–3; absorption *86*, 86–91, **87**; audio for visual/gaming 67–8; echo chamber 95–6; isolation 65, 70–80; media production environment 67; partitions 79–80, *79–80*; professional studio 66; project studio 68–9; reflections 83–5, *83–5*, 93–6; reverberation (reverb) 93–5; speaker placement 93; symmetry 91–2; *see also* microphones (mic); microphone placement techniques; sound and hearing
active bass trap 90
active crossovers *see* electronic crossover networks
active listening: mixing 538–9; recording 484–5
active powered speaker design *472*
Active Sensing messages 303–4
AC wiring 408
adaptive filtering 444–6, *445–6*
ADAT optical I/O interconnection 207–8, *208*
Adobe Audition CC *448*
Advanced Audio Coding (AAC) 370, 376
AEA A440 ribbon microphone *100*, 166
AEA R44 ribbon microphone 166
AES67 208
AES/EBU (Audio Engineering Society and the European Broadcast Union) 204–5, *205*
affordability: abilities and 557; editability and 281; iOS in music production 345

.aif 202, 258, 365
AKG 321 *109*
AKG C214 condenser microphone *101*, 167, *167*
AKG C3000 condenser *109*
AKG D112 dynamic microphone 164, *165*
"Alan Sides" approach 116
Alesis SR-18 stereo drum machine *326*
aliasing 194
all-thread 74
alternate environments for monitoring 470; *see also* monitoring
ambience 93; acoustic 6; artificial 6; distant pickup 414; intelligibility and 93; miking and *120*, 124; natural 451, 534; obtrusive background 444; sonic 270; studio or concert hall 136; *see also* room
ambient miking 119, 129
ambient/room surround mics 139, *139*
ambisonic immersive audio 594–5; *see also* immersive audio
ambisonic pickup 140, *140*
American Society of Composers, Authors and Publishers (ASCAP) 612
Ampex ATR-102 1-inch stereo, ATR *174*, 175
Ampex ATR 102 tape emulation plug-in *187*
amplification *399–401*, 399–404; band-limited 472; distribution amp 404, *405*; equalizers 403, *403*; operational amplifier (op-amp) 402, *402*; preamplifiers 402–3; summing amp 403, *404*
amplifier (amp) 399–406; amplification 399–404; guitar 146, *147*; power amplifiers 404–6, *405*; saturation 401
amplitude 37–8, *37–8*, 237; change over frequency *109*; fluctuations 389; frequency ranges of human hearing *37*; in-phase waves *44*; measure of *38*; processing 422; quantization 192; short high *49*; of zero *43*; *see also* equalization (EQ)
analog: multitrack machines *175*; recall 414–15; signal processing *414*, 414–15; sound 182; track configurations *180*
analog-based noises 183
analog tape: emulation plug-in *432*; machine 8–9, *9*; noise 182–3
analog tape recorder (ATR) 171–87; advantage of 172; archive strategies 186–7; backup/archiving 185–7; editing magnetic tape *184*, 184–5;
medium of magnetic recording 173–85; professional analog 174–80; tape, tape speed and head 180–4; tape availability 187; tape emulation plug-ins 187; tape restoration 186
analog-to-digital (A/D) 195–8, 209–11, 224, 432, 484, 566
Anderton, C. 287, 525
anechoic (no reverb) 95
Antares SoundSoap 5 Audio restoration plug-in *447*
anti-alias filtering 194, *194*, 197
Aphex Model 622 Logic-Assisted Expander/Gate *441*
API 1608 Console *509*
API-2500 Compressor Plug-in *246*
Apogee UV22 dither plug-in *196*, 566
Apollo: accelerated plug-ins *416*; audio interfaces 226; delay plug-ins *247*; FireWire/Thunderbolt 226; Lexicon 224 reverb plug-in *457*; Oxide Tape Recorder *432*; Precision Limiter plug-in *569*; Precision Multiband Plug-in *570*; Sonnox Oxford Limiter plug-in *566*; Twin USB 226
Apple 222, 243, 346; Audio Interchange File (AIFF or .aif) 202, 230; Music 610; *see also* iOS in music production
applying equalization 427
archive/archiving: data 257; digital audio workstation (DAW) *254*, 254–7, *256*; strategies 186–7
Argosy Halo desk *261*
Arrangement View 271, *271*
Art Diffusor sound diffusers Model E. *85*
artificial ambience 6; *see also* ambience
asperity noise 182
assistant engineer 24–5
ATC SCM45A Pro active monitor speakers *465*
Atmos Levels 599
ATT 370
attack (A): of compressor 435; in envelope 48, *49*
attenuation devices 89
AU (Audio Units) 243, 347–8
audio: bit rate and file sizes **367**; connectivity 348; cue services 346; driver protocols 225–6; equalizer 423, *423*; file 234, 266, 336, 347, 364–5, 440, 582, 615; for games 19; industry professionals 623; interface 224–5, *225*; production console 542; units 347–8; for video and film 18, *18–19*; for visual/gaming 67–8

AudioBus, iOS in music production 347, *347*

Audio Cyclopedia 65

Audio Definition Model (ADM): Authoring Tool 598; authoring window *597*; wavefile 598–9

audio-for-the-Web 20

Audio Interchange File (AIFF; .aif) 365

audio over Ethernet (AoE) 208, 223–4, *224*

Audio Stream Input/Output (ASIO) architecture 226

AudioSuite 243

Audio-Technica AT5045 condenser instrument microphone 166–7, *167*

audio-to-MIDI 336; groove 267–8, *268*

Audio Video Bridging (AVB) 208

auditory perception 56–7

Auralex MoPAD Monitor Speaker Isolation Pads *467*

Auria Pro *351*

authoring 363

auto-locator devices 9

automatic dialog replacement (ADR) 67

automatic pitch correction 455–6, *456*

automation: digital audio workstation (DAW) 215; dynamic effects 457–8; effects 249–50; mixer/console signal path 545–8, *546–7*; plug-ins 416, *417*; rubber bands *547*

Auto-Tune pitch correction system *456*

auxiliary controllers in toys 319

auxiliary (aux) sends *419*, 499–500, *500*

Avantone Active MixCubes Full-Range Mini Reference Monitors *468*

AVB Control app *224*

average absorption coefficient (A_{ave}) 86, 94

average/peak meter 518

average signal level 432, 440, 515

backup/archiving 185–7; current patch data 305; data 257; digital audio workstation (DAW) 254, 254–7, *256*

backwards compatibility, MIDI 2.0 307

balanced power 411, *411*

balanced/unbalanced lines 112–14

balancing: mixing and 550–4; speaker levels 474–5, *475*

bandpass filter 47, 425, *426*, 439, 479

band website 31

bantam connectors/cables 112, 513

barriers 70–1, 75, 78, 79

basic car system 470

bass guitar 142, 437; *see also* electric bass guitar

bass traps 89–90, *90–1*

Battery 3 virtual drum and loop module *275*

beats 58, 265, 354

beat slicing, groove 266–7, *267*

beats-per-minute (bpm) 332

bedroom studio *281*; *see also* project studio

behind the scenes 27–8

being yourself 624

Bell, A. G. 49

Beyerdynamic M160 ribbon mic *100*

Beyer M-160 ribbon microphone 165–6, *166*

BFD2 acoustic drum library module *275*

bias current 178–9; on recorded linearity *179*

BiCoastal Music 67

bidirectional communication, MIDI 2.0 307, *307*

bidirectional polar pattern 105, *105*

Bill Putnam's Universal Recorders *122*

binary-coded decimal (BCD) 383

binary number system 189

binaural immersive audio 594

binaural localization 59

biscuit 622

Bitcoin 612–13

bit depth 191, 200; 16 200, 565; 20 201; 24 201, 527, 565, 566; 32- and 64 196, 197, 201; internal 192, 196

bleed *see* leakage

Blige, M. J. 137

Blue Book **362**

Blumlein, A. D. 136,

Blumlein technique 95, 131, 136, *136*, 570

Blu-ray 363, **364**, 602

body, protection of 260–1, *261*

bongos 157

bookings, studio management 26

boost/cut curves of shelving equalizer *425*

boost/cut EQ curves *424*

Bose SoundLink® speaker III *468*

boundary effects 42, *131–2*, 131–4

boundary microphone 131, *132*

brain *see* psychoacoustics

Brainworx Modus EQ bx1 *571*

Brand X Model Z synthesizer 304

brass instruments 142–4, 437; French horn 144; trombone 143; trumpet 142–3, *143*; tuba 143–4

Bricasti M7 Stereo Reverb Processor *452*

bricks 73

Bridgeport Music et al. v. Dimension Films 323

Broadcast Music Inc (BMI) 612
Broadcast Wave Format (BWF) 230,
    365, 385
Brooks, E. 386
Brubeck, D. 122
Buma/Stemra 612
Bush, B. *416*
business challenges 609

Caballero, E. 4, *542*
cable, MIDI 285–7, *286–7*
cabling 210, 212, 223, 285
camera 220, 316; adapter *349*, 355; phone
    528; place a sound-pressure level (SPL)
    meter 591; professional 385; sound
    level meter 474; tripod 475
capacitor 101, 111, 403, 423
Capitol Records *2–3*, 125, *523*
Capitol Studios 119
cardioid microphone *108*
cardioid polar patterns 106–*707*
cardioid variation 149
career development 28–32; networking
    29; professionalism 32–3; self-
    motivation 28–9; starting of 29–32
carving 552
cathode 399–400
caulk 72, 73
CD-RFS **362**
CD-UDF **362**
ceilings 76, *76*
cello 159
centralized computer (server) 256, *256*
central processing unit (CPU) 80, 81,
    226; DAW 244, 416, 500; real-time
    processing 242; in toys 318
chamber reverb 452
chamfer 74
change duration only 247
change in both pitch and duration 247
channel assignment 508–9, *509*
channel fader 506–8, *507*, 549
channel input (preamp) *497–8*, *497–9*,
    544–5
channel name 549
Channel Pressure (or Aftertouch)
    messages 298
channels, MIDI 1.0 *293–4*, *293–5*
Channel Voice messages 296–8, *296–8*
charge *(Q)* 102, 104
chorus 246
chorusing 450
chromatic harmonica 163
CinemaScope 580
clarinet 161, *162*

Class A amplifier 406
Class AB amplifier 406
Class B amplifier 406
Class D amplifier 406
classical recording 128
cleanliness 183
click/pop eliminator *448*
click track 332–3, 490
clipping 401
clips 271
close microphone placement 120–5
close miking 119, *120*
cloud computing 360–1
codec services 347
collecting funds 612
Collins, P. 137
combination tones 58
combined recording/mix space 532
combing 449
communication 215, 307, 319
compact disc (CD) 361, **362**, 602, *614–16*,
    614–17; Baby PRO 612; CD-Text 361
competent administration staff 26
complex waves 47, *47*
compositing (comping) 240, 533, 536
compressed codec soundfile formats 365
compression 433, *433–5*, *433–8*; in
    mastering 568–9, *568–9*; mono mix
    437; output ratios of *435*; in sound and
    hearing 36, *37*
computers 80, 81, 316–17, *317*;
    connecting to the peripheral world 316;
    multimedia and 358; network 255–7;
    noise minimizing 253–4; processor
    system 251; timecode routing to and
    from 395, *396*; *see also* hardware;
    software
concave surfaces 41, 42, 83
Concept Album 560
concrete 73
condenser dynamic mic 149
condenser microphone *101*, 101–3, *103*;
    external power supply 102–3; phantom
    power supply 103, *103*; powering of
    102–3
congas 157–8
connector 112–13, *113*, 206, 222
console, recording *495–6*, 495–501;
    auxiliary (aux) sends 499–500, *500*;
    channel input (preamp) *497–8*, *497–9*;
    gain level optimization 496–7; insert
    point for hardware console/mixer 499,
    499; virtual DAW insert point 499, *500*;
    virtual DAW send 500–1, *501*
constant–bit rate (CBR) 368

construction 72–8, 83–4, 91, 165
Continue messages 303
continuous jam sync 389
contracts 12, 611
control and automation plug-ins 416, *417*
Control Change messages 298, *299*
controller ID number 299, *299*
controllers: digital audio workstation (DAW) 228; groove 276–7, *277*; surfaces 9; values in editing 340
control messages, MIDI 2.0 308
control panel 318, *319*, 519, *520*
control room: noise isolation within 80–1; professional studio 7–9, *7–9*; reflections 83–5, *83–5*, 93–6; symmetry in design 81–3; Synchron Stage 8
converter 191–2, 196–7, 347, 484
convex surfaces 41, *41*, 83
copy command 235, *236*
copyright songs 487
CoreAudio 226, 348; basic I/O architecture *346*; iOS in music production *346*, 346–7
core audio clock services 347
corners 42, 73, 581
cost factors 65–6
Creation Station 450 desktop PC *219*
CR monitor mix 529
cross-fade (or X-fade) 237, *238*; song ordering 561
crossover system *472*
Crow, S. 137
Crown XLi800 200W two-channel power amplifier *405*
Cryptocurrency 612–13
Cubase/Nuendo *214*, *238*, *240*, 265, 273; Apogee UV22 dither plug-in *566*; audio production software *330*; automation rubber bands *547*; auto selectors *546*; compressor plug-in *246*; "Control Room" monitor/control output *478*; drum pattern *335*; EQ plug-in for *245*; EQ window *471*; Expander plugin *441*; iC Pro *351*; main out channel EQ within *567*; media production DAW *416*; multiband compressor for *247*; Notepad apps *258*; on-screen mixer *543*; Rex file *267*; sync setup *396*; virtual mixer *504*
Cuniberti, J. 132
curves up ahead 525
cut command 235, *236*
cutoff frequency 425; *see also* frequency
cutting head, vinyl *620*, 620–1

cycle 38
cymbal crash 49, *49*

daisy chain 221, 289–90, *290*
Dante 208
data: archiving and backing 257; current patch 305; management 254, *254*; synchronization 383; transmitting patch data between synths 304–5
DAW *see* Digital Audio Workstation
Dear Reality Monitor Virtual Mix Room Simulator *596*
Dear VR Monitor acoustic simulator *469*
Dear VR Monitor headphone compensation *469*
decay (D): in envelope 48, *49*; time 63, 94–5
Decca tree microphone 138, *138*, *139*, 139–40
decibel (dB) 49–54
DeClick2 *448*
degaussing 183–4
delay 245–6, *247*, 448–51, *449*; 15 to 35 MS *449*, *450*; digital *449*; less than 15 MS *449*; more than 35 MS 451–2; *see also* latency
delivery media 359
denatured (isopropyl) alcohol 183
desert island mixes 573
desk *see* recording
desktop computers 219, *219*
diaphragm capacitance *101*, 102
diatonic harmonica 163
diffraction of sound 42, *43*
diffusers/diffusion 63, 84, *85*, 87, 474; 25% 92; homemade-made 92, *92*
Digidesign DINR plug-in *447*
Digidesign ICON D-Command Integrated console *543*
Digidesign S6 integrated controller/console *230*
digital audio: distribution *209*; levels 203; transmission 203–4; workstations 394–5, *395*
digital audio technology 189–212; ADAT optical I/O interconnection 207–8, *208*; AES67 208; AES/EBU (Audio Engineering Society and the European Broadcast Union) 204–5, *205*; digital audio transmission 203–4; dither 195–6, *196*; fixed-point processing 197; floating-point processing 197; jitter 209–10, *210*; MADI (Multichannel Audio Digital Interface) 206–7, *207*; Nyquist theorem 193–4, *194*;

oversampling 194–5; playback process 198–9; professional sound file formats 202–3; quantization 192, *193*; recording process 197–8, *198*; regarding digital audio levels 203; sampling *191*, 191–2; Serial Copy Management System (SCMS) 206; signal distribution 209, *209*; signal-to-error ratio 195; sound file basics 199–203; sound file bit depths 200–1; sound file sample rates 201–2; S/PDIF (Sony/Philips Digital Interface) 205–6, *206*; TDIF (Tascam Digital InterFace) 208; wordclock 210–12, *211*

digital audio workstation (DAW) 7–9, *9*, 130, 213–61; accessories and accessorizing 259–60; Ambisonic mic 140; audio driver protocols 225–6; audio interface 224–5, *225*; auto-locator devices 9; automation 215; backup, archive and networking strategies *254*, 254–7, *256*; body, protection of 260–1, *261*; communication 215; computer noise minimizing 253–4; computer/processor system 251; controllers 228; controller surfaces of 9; converting audio track to a MIDI track *268*; desktop computers 219, *219*; dither 196; documentation directories 259; documentation within 258, *258*; drivers update 252; dual monitor 252–3; expandability 215; fast memory 251–2; hardware 216–20; hardware, protection of 260; hardware controllers *228*, 228–9; inputs and outputs (I/O) 227; instrument controllers 229, *229*; integration 214–15; investment, protection of 260; keeping production media separate 252; laptop computer 219–20, *220*; large-scale controllers 230, *230*; latency 226–7; looping, groove 269–70, *270*; manuals reading of 252; microphone preamps 114; mixer/controller technology 518–20; on-screen mixer *543*; pitch-shift algorithms *265*; power to the processor 250–61; project studio 69, *69*; session documentation *258*, 258–9; session recall 215; software 232–50; software mixer surface 520, *520–1*; sound file formats 230–2; sound file interchange and compatibility 231–2; sound file sample and bit rates 231; speed and flexibility 215; stock compressor

plug-ins *246*; system interconnectivity 220–4; touch-screen controllers *229*, 229–30; user-friendly operation 215

digital console 518–20

digital DJ software *278*

digital media production, timecode within 385

digital noise reduction 444

digital recording chain 197

digital signal processing (DSP) 213, 264, 415; effects 242–3; plug-ins 243–4

digital-to-analog (D/A) 195, 199

direct insert monitoring 505

directional mic 117

directional response microphones (mic) 104–7

direction injection (DI) 123, *126*

direct monitoring 227

direct signal 452

direct sound 61

direct streaming digital (DSD) audio 365–6

DirectX 243

disc-cutting, vinyl 619, *620*

disc-cutting lathe, vinyl 620, *620*

disc plating and pressing, vinyl 621–2, *622*

distant microphone placement 126–9

distant miking 119, 127, 521

distribution amp 404, *405*

dither 195–6, *196*; mastering 565–6, *566*; plug-ins *566*

DIY sampling, keyboards 323–4

DJs 19–20; iOS in music production and 353; software, groove 277, *278*

DMH 93, *542*, 624–7; output bus tools *555*

DMH's 5.1.4 speaker setup *587*

docking device 348

documentation 258, *258*; directories 259; sequencing 337–8

Dolby Atmos 18, 596–9, *597*; ADM Authoring Tool 598; Atmos Levels 599; basics 598; Renderer Tool 598–9

Dolby Film Sound Project 582

Dolby Labs 370, 582

Dolby Pro Logic 582

Dolby Surround 582

domains 173, *174*

double bass 159

doubling effect 450

drawn (rubber band) automation *547*, 547–8

drivers update 252

drop-frame timecode for color NTSC US video 383–4

drums: button pads 327; electronic instruments, MIDI 325–6, *326*; hand 157–8; loop plug-ins 274–5, *275*; pad controllers 327–8; pattern entry 334, *335*; replacement *328*, 328–9; set 151–3, *152*; *see also* floor-tom; hi-hat
DSP *see* digital signal processing
DTS **363**
dual monitor 252–3
ducking 420
DVD 361–3, **363**, 602, 617
dynamic effects automation and editing 457–8
dynamic microphone 98–9, *99*; *see also* microphones (mic)
dynamic range processors 245, *246–7*, 432–43; compression *433–5*, 433–8; expansion 441–2, *441–2*; limiter plug-ins *439–40*, 439–41; multiband compression 439, *439*; noise gate *442*, 442–3
dynamics: mastering 567–8, *568*; processing in mixing 541; section 503, *503*
DyneOne multiband compressor *439*

ear 54–6; anatomy of *54*; care of 55–6; inner 47, 54, *54*; threshold of feeling 55; threshold of hearing 54–5; threshold of pain 55–6
earbuds 469
early reflections 61–2, 452
Earthquake 584
echo chamber 95–6; *see also* delay
edit decision list (EDL) controller 390
editing: dynamic effects 457–8; magnetic tape *184*, 184–5
editing techniques 338–42; controller values 340; humanization 339–40; mixing sequence 342; playback 341–2; quantization 339; slip time 340; transposition 339
edit tape 184, *184*
effects: automation 249–50; boundary 42, *131–2*, 131–4; digital signal processing (DSP) 242–3; dynamic 457–8; fun with 244; plug-in *500–1*, 545–6; time-based 422, 448–58
effects processors 422–48; hardware and plug-in effects 423–30; mixing 541; noise reduction 443–8; sound-shaping effects devices and plug-ins 431–43
Ehrmann, M. 629
eight-to-fourteen modulation (EFM) 615
electret-condenser microphone 104

electric bass guitar 147–8
electric guitar 146–7, 437; miking the guitar amp 146, *147*; recording direct 147, *147*; *see also* bass guitar
electromagnetic induction (EMI) 411–12; principle of 99; theory of 98, 99
electronic crossover networks *472*
electronic instruments 318–29, 494; drum machine 325–6, *326*; drum replacement *328*, 328–9; MIDI drum controllers *327*, 327–8; MIDI keyboard controller *324*, 324–5, *326*; systems plug-ins and 319–24; in toys 318–19
electronic keyboard instruments 150–1
electrostatic principle 101
embedded metadata *371*
End of Exclusive (EOX) messages 300
engineer 24; assistant 24–5; maintenance 25; mastering 25, 196; producers and 152, 160, 502
ensembles 127–8, 134
envelope 48–9; musical waveform *49*
EQF-100 full range, parametric vacuum tube equalizer *427*
equalization (EQ) 25, 93, 126, 178, *178*, 244, *245*, 403, *403*, 552; applying 427; flat frequency playback curve *178*; hardware and plug-in effects 423–5, *423–6*, 426–7, *426–7*, 428–30; hardware controllers 228; mastering 567, *567*; mixing 541; recording *501*, 501–3
equal-loudness curve *see* Fletcher–Munson equal-loudness contour curves
equivalent noise rating 110
erase head 177–8
error correction 198
Ethernet protocol 227
European Broadcast Union (EBU) 384
even harmonics 46
Eventide H3000 Factory Ultra-Harmonizer Plug-in *457*
excessive dynamic range 160
expandability 215
expanded capabilities, MIDI 2.0 315
expanded resolution, MIDI 2.0 315
expander 441–2, *441–2*, 446
exporting final mixdown to file 250, *250*
extended-range dynamic mic 149
extensive array of DSP 243
external keyed input 419
external power supply 102–3
EZ CD Audio Converter *616*

Fabfilter Pro L-2 limiter plug-in *439*
FabFilter Pro-Q 24-band EQ plug-in *245*

fade-in and fade-out curves *238*
fans 80, 81, 117; noisy 253; portable
    220, 316
Fantasia 580
"Fantasound" 580
Fantom-X7 Workstation Keyboard *321*
far-field monitor speakers 465–6
far-field pair 464
Fast Fourier Transform (FFT)
    446–7, *447*
fast memory 251–2
field-effect transistor (FET) 406
figure-8 pattern 105, *105*
film: acoustic studio design 67; audio for
    18, *18–19*
final mix compression 438
finances controlling 628–9
FireWire® 217, 223, 285
FireWire 400 223
FireWire 800 223
Fitzgerald, E. 137
fixed-point processing 197
"fix it in the mix" 526
FLAC (Free Lossless Audio Codec) 370
flanging 246, 449
flash card 363
flat frequency response curve 40
flats 79, 80, 84, 125, 154
Fletcher–Munson equal-loudness contour
    curves 56–7, *57*
flexible surfaces 90–1, *91*
floating *74*, 74–6, *76*, 197
floors, isolation 74–5, *74–5*
floor tom 157, *157*
flute 162, *162*
flutter echo 84
Focusrite Scarlett 18i20 interface *590*
Foley 67
formants 143, 144, 158, 447
formats: compressed codec soundfile 365;
    immersive audio 594–9; media delivery
    364–70; professional sound file 202–3;
    sound file 230–2; uncompressed sound
    file 364
Form PA 611–12
Form SR 611
frames 381; picture 201; standards
    timecode 383–4
Fraunhofer Institute 370
free music streaming 602
freewheeling 389
French horn 144
frequency *38*, 38–9; of the acoustic
    waveform 99; balance 65, 83–91;
    isolation 75–6; ranges of human

hearing *37*; response curve *39*, 39–40,
    108–9, *109*
front-to-back discrimination 107
full messages 387
fun with effects 244
Furman M-8Lx power conditioner *409*
Furman P-2400 IT 20A Prestige
    Symmetrically Balanced Power
    Conditioner *411*
Furman SS-6B power conditioner/strip *410*

gain level: mixer/console signal path 549;
    optimization 496–7
gain structure 549–50
Galaxy Studios *19*
Galaxy Tape Echo Plug-in *449*
games 67–8; audio for 19
GarageBand™ 263, 354, *354*
Garbage, "Not Your Kind of People" *416*
gate 418, *446*
gate reverb 453
gating 442
Gema 612
Genelec 1236 main reference monitor *465*
General MIDI (GM) 373, **374–5**
gobos *see* flats
Gone with the Wind (Rhett) 234, *235*
good attitude 484, 630
Good Rule 98, 142, 427, 481–2
gozintagozouta approach 543
Gracenote 361
grand piano 148–50, *149*
graphical user interface (GUI) 243, 271,
    342, 343, 495, 544, 549
graphic editing 233, 235
graphic equalizer 426–7, *427*
Green Book **362**
groove tools and techniques 263–78, 494;
    audio-to-MIDI 267–8, *268*; beat slicing
    266–7, *267*; controllers 276–7, *277*;
    DJ software 277, *278*; drum and drum
    loop plug-ins 274–5, *275*; hardware
    268, 268–9; iOS apps 276; loop-based
    plug-ins and 274, *275*; loop files from
    digital wellspring 277–8; pitch-shift
    algorithms *265*, 265–6; pulling loops
    into DAW session 275–6; software
    269–74; sync 264; tempo and length
    264; time- and pitch-change techniques
    264; warping *263*, 266
grounding 407–8
grouping 509–11, *511*, 548, *549*
groups, MIDI 2.0 313–14
guitar 144–5; acoustic guitar 144–5, *145*;
    miking near the sound hole 145; room

and surround guitar miking 145; *see also* bass guitar; electric bass guitar
guitar miking 145
gypsum 72–3

Haddy, A. 138
hall reverb 452
hand drums 157–8
handheld recording, iOS in music production 349, *350*
hard-drive 185, 217–18, 220, 253, 254, *254*, 256
hardware: controllers 228, *228–9*; digital audio workstation (DAW) 216–20; groove 268, *268–9*; mixer *500*; plug-in effects and 423–6, *423–30*; protection of 260; *see also* controllers
harmonica 163, *163*
harmonics 45, *46*; content 45–8, *46–8*; even 46; odd 46
"harp" player 163
HDCD **362**
headphones 468–9, *468–9*; distribution amplifiers *532*; immersive monitoring 595–6, *596*
Head Related Transfer Function (HRTF) 594
hearing 627–8; care of 55–6; conservation 56; *see also* ear; sound and hearing
Hear Technologies Hear Back Personal Monitor Mixer System *530*
Herman Miller Aeron® chair *261*
high-definition audio 199
higher resolution for velocity, MIDI 2.0 308
high-frequency absorption 88, *89*
high-pass filters 425, *425–6*
hi-hat 157
hi-res download distribution 601
Hit Factory Criteria, Miami *460*
Hitpoint markers *266*
home-made sheet *523*
home theater, multimedia and 358–9
HRM-16 16 channel personal headphone mixing station *530*
Huber, D. M. 4
HUM 411–12
human hearing, frequency ranges of *37*
humanization, editing techniques 339–40
hypercardioid polar patterns 106–7, *107*

Ice-T 137
iConnect MIDI4+ 4x4 MIDI interface *349*
iD22 audio interface and monitoring system *478*

iDAWs, iOS in music production 350, *351*
IEEE-1394a 223
IEEE-1394b 223
imaging 81, 82, 136, 218, 463, 473, 474
immersive audio 579–600; ambisonic 594–5; back catalog material 600; bass management in surround system 585, *586*; binaural 594; experience 582–3; formats 594–9; interface 589–93, *590*; layouts *583*, 583–9; LFE 584–5, *585*; mixing in surround 599; past to the present 580–3; speaker placement and setup *586–7*, 586–9; surround channel layouts and assignments 589
immersive Decca tree *139*, 139–40
immersive headphone monitoring 595–6, *596*
immersive miking techniques 134–40
immersive soundbars 588–9
indexes 614
in-ear monitoring *469*, 469–70
in-line monitoring 504–5, *505*
inner ear 47, 54, *54*; *see also* ear
input gain, compressor 434
input (source) mode 179
inputs and outputs (I/O) 227
insert effect's signal path: channel strip *418*; external control over *418*, 418–19
insertion loss 423
insert points: for hardware console/mixer 499, *499*; mixer/console signal path 545, *545*
insert routing 417–19, *418*
instrumental frequency ranges **428**
instrument controllers 229, *229*
integrated hardware sequencers 329
integration 214–15
Intel 222, *222*
interaural arrival-time differences 60, *60*
interaural intensity difference 59, *60*
Inter-Device Audio + MIDI (IDAM) 348
interface 284; audio 224–5, *225*; immersive audio 589–93, *590*; standard audio 348; *see also* graphical user interface; musical instrument digital interface
International Telecommunications Union (ITU) 586
Internet: connection speeds **376**; mastering for the 576; radio 607; streaming audio 376–7
Internet Service Provider (ISP) 360
in-the-box 8, 241, 249, 415, 494, 520, 542, *542*
investment, protection of 260

iOS in music production 345–55; accessorizing 355; affordability 345; AudioBus 347, *347*; audio connectivity 348; audio units for 347–8; Core Audio *346*, 346–7; DJ and 353; groove 276; handheld recording 349, *350*; iDAWs 350, *351*; MIDI connectivity 349; mixing with 350, *350*; mobility 345; multifunctionality 345; as musical instrument *354*, 354–5; music software *354*; recording 349; revolution 13, *14*; touch capabilities 345–6; using on stage 353, *353*

ISO-9660 **362**

IsoAcoustics ISOL8R155 stands *467*

isolated sound sources *494*

isolation 65, 70–80, 95; acoustic partitions 79–80, *79–80*; ceilings 76, *76*; floors 74–5, *74–5*; isolation booths (iso-booths) 79; isolation rooms (iso-rooms) 78, *79*; reducing problems due 161; risers 75–6, *76*; walls 71–4; windows and doors 77–8, *77–8*; *see also* leakage; noise

isolation booths (iso-booths) 79

isolation rooms (iso-rooms) 78, *79*

ITU 5.1 speaker setup *586*

iZotope: Insight Essential Metering Suite *204*, *471*; Ozone 10 *576*; RX9 Audio Repair Software *448*

jacks, MIDI 287–8, *288*

Jackson, M. 125, 137

JavaScript Object Notation (JSON) 312

JBL Series 3 Mk2 *464*; Active Studio Monitor *466*

jitter 209–10, *210*

Jitter Reduction Timestamp, MIDI 2.0 314

Joliet **362**

keeping production media separate 252

keyboard instruments 148–50; controllers *229*; DIY sampling 323–4; grand piano 148–50, *149*; as percussion controller 327; sample libraries 323–4; samplers *321*, 321–3, *322*; synthesizer (or synth) 320–1, *321*; upright piano 150

key input 443

key sidechain input *418*

kick drum *154*, 154–5, 437; signal *421*

Komplete Kontrol S49 keyboard controller *229*

Kontakt Virtual Sampler *322*

K-Stereo Ambience Recovery *137*, *571*

KX49 keyboard controller *229*

label media 611

land 621; *see also* grooves

lane 240

laptop computer 219–20, *220*; *see also* computers

large recording studio 128–9

large-scale controllers 230, *230*

large-scale integrated (LSI) circuit 1

La-Rocc-A-Fella Center *69*

latency 226–7, 532

leakage 71, 73–9, 95, 118, *122*, 122–5; methods for reducing *123*; reducing problems due to 161; *see also* isolation

LEDE (Live End Dead End) 85, *85*, 93

Legacy model SSL XL9000K monitor mix section *504*

Lenovo Legion Y740 laptop *220*

Leslie speaker cabinet 151, *151*

levels: average signal 432, 440, 515; balancing speaker 474–5, *475*; control in monitoring 476–80; digital audio 203; loudness 49–54; signal 527, *527*; timecode 391, **391**

Lexicon 224 reverb plug-in *457*

LFE (subwoofer) 584–5, *585*

Library of Congress 611

lightpipe (ADAT) I/O 227, *227*

lightpipe system 207–8

limiter/limiting plug-ins: dynamic range processors *439–40*, 439–41; in mastering 569, *569*; output ratios of *440*; prevent high-level, high-frequency peaks from distorting analog tape 440; prevent short-term peaks from reducing a program's average signal level 440; prevent signal levels from increasing beyond a specified level 440

Lindberg, M. 8

listening, tools for 461; *see also* speaker

"Live at the Sands" (Sinatra) 122

live performance 152, 278, 282, 327, 352, 452; recordings 17, 17–18; world of touring and 164

live (stage) reverb 452

Livewire 208

local area network (LAN) 256, *256*, 359

Logic *496*, *520*

Logic DAW *215*

London Bridge Studios 15, 67; Neve 8048 console 15

longitudinal timecode (LTC), refresh and jam Sync 388–9, *389*

long ramp-up 525

loop-based plug-ins, software 274

loop files from digital wellspring 277–8

LoopMash HD Groove app *354*
loop plug-ins, drum and drum 274–5, *275*
loudness levels 49–54; logarithm (log) *50*, 50–1; power 52; "simple" heart of the matter *53*, 53–4; sound-pressure level *51*, 51–2; voltage 52; *see also* gain level
Loudness Units Full Scale (LUFS) 570, 599
low-frequency absorption 89–90, *89–90*
low-frequency rumble, microphones (mic) 117
low-pass filters 425, *425–6*, 585
LP *see* vinyl
Lurssen, G. *558*
lyrics 258, 343

M160 ribbon mics 100
M260 ribbon mics 100
MacBook Pro 15" *220*
Macintosh HFS **362**
Mackie DC16 mixing system *353*
Mackie HR824mk2 active monitor speaker *464*
Mackie MCU Pro DAW controller *228*
Mackie Onyx 4-bus *495*, *510*
Mackie Universal Control *521*
Mac Pro™ with Cinema display *219*
MADI (Multichannel Audio Digital Interface) 206–7, *207*
magnetic flux 173
magnetic oxide 173, *173*, 177, 183
magnetic recording 173–85
magnetic tape: heads 176–8, *177*; structural layers of *173*
magnetism 176–8, *177*, 180, 183, 186
Magneto MkII *187*
main output mix bus 511–12, 553–4, *554*
maintenance engineer 25; *see also* engineer
management: of bass in surround system *585*, *586*; bookings, studio 26; data 254, *254*; studio management 25–6
Manley Massive Passive Analog Stereo Equalizer *423*
Manley Slam *569*
Manley Stereo Variable Mu® mastering outboard hardware compressor *568*
manuals reading, digital audio workstation (DAW) 252
"mapping" conventions 311
marimba 158
Maschine groove hardware/software production system *277*
Maschine hardware controller 325
masking 58–9

Massenburg, G. 123, 156, 579, 627
"Mass in C Minor" (Mozart) 143
mastering 557–77; compression in 568–9, *568–9*; defined 557; desert island mixes 573; dither 565–6, *566*; dynamics 567–8, *568*; engineer 25, 196; equalization (EQ) 567, *567*; FabFilter Pro-Q 24-band EQ plug-in *245*; helping match levels and character within project 559; helping project to sound "right" 559; help with overall levels 560; for the Internet 576; limiting in 569, *569*; Loudness Units Full Scale (LUFS) 570; mid/side processing (M/S processing) 570–1, *571*; multiband dynamic processing 569–70, *570*; plug-ins 575–6, *576*; plug-ins, Izotope Ozone 10 *576*; preparation 563; process of 562–3, *572*; reference track 564; self-mastering 572–3; signal chain 574–5, *575*; song ordering 560–1; song timing 561; sound file resolution 565; sound file volume 565; stems 564; two-step or one-step (integrated) 574; volume 566
master/slave relationship 392–3, *393*
materials 71, **72**, 73–7, 80, **87**
mathematics 66, 140
matrix 622
media 601–22; collecting funds 612; Cryptocurrency 612–13; delivery formats 364–70; marketing/finance 607–13; moral question 608–9; music lawyer 610; need of label 611; ownership of masters 611; product distribution 603–7; production environment 67; product manufacture 613–22; royalties and business issues 609; work registration 611
Melodyne Editor auto pitch correction system *456*
memory: in toys 318; USB stick 363; *see also* hard-drive
Message Type, MIDI 2.0 313
metadata 202, 361, 605; *see also* embedded metadata; tagged metadata
metal-oxide-semiconductor FET (MOSFET) 406
meter display of compressor 435
metering *514*, 514–18; finer points of 515–17, *516*; VU meter **517**, 517–18
Meyer, C. 386
microbar 54–5
microphones (mic) 97–140; accent placement 133–4, *134*; balanced/unbalanced lines *112–13*, 112–14;

characteristics of 104–11; choice and placement of 97–8; classification of 104–7; close placement 120–5; condenser *101*, 101–3, *103*; defined 97; design of 98–104; directional response 104–7; distant placement 126–9; dynamic 98–9, *99*; electret-condenser 104; equivalent noise rating 110; frequency-response curve 108–9, *109*; low-frequency rumble 117; modeled condenser mic systems 115, *115*; off-axis pickup 118–19; output characteristics 110; output impedances 111; overload characteristics 110–11; overload distortion 110; pickup issues 117–19; popping 118; preamps *114*, 114–15, *498*; proximity effect 117–18; ribbon 99–100, *100*; room placement of 129–30; selection guidelines for **141**; sensitivity rating 110; techniques 115–19; transient response *109*, 109–10

microphone placement techniques 141–63; brass instruments 142–4; electric bass guitar 147–8; electric guitar 146–7; electronic keyboard instruments 150–1; guitar 144–5; keyboard instruments 148–50; nylon or Spanish Guitar 145–6; percussion instruments 151–7; stringed instruments 158–9; tuned percussion instruments 157–8; voice 159–61; woodwind instruments 161–3

microphone selection 163–9; AEA A440 ribbon microphone 166; AEA R44 ribbon microphone 166; AKG C 214 condenser microphone 167, *167*; AKG D112 dynamic microphone 164, *165*; Audio-Technica AT5045 condenser instrument microphone 166–7, *167*; Beyer M-160 ribbon microphone 165–6, *166*; Neumann TLM 102 condenser microphone 168, *168*; Royer Labs R-121 ribbon microphone 165, *165*; Shure SM57 dynamic microphone 163, *164*; Telefunken C12 condenser microphone 169, *169*; Telefunken ELA M251E condenser microphone 169, *169*; Telefunken M81 dynamic microphone 164, *164*; Telefunken U47 condenser microphone 169, *169*; Warm Audio WA-47 tube microphone 168–9, *169*

Microsoft Wave (.WAV) 385

Microsoft Windows 202, 225, 243, 266

MIDI 1.0 291–2; channels *293–4*, 293–5; Channel Voice messages 296–8, *296–8*, 314, *314*; controller ID number 299, *299*; messages *292*, 292–3; modes 295–6, *296*; System messages 299–305

MIDI 2.0 291–2, 306–16; backwards compatibility 307; bidirectional communication 307, *307*; both protocols 307; channels (256) 309; control messages 308; expanded capabilities 315; expanded resolution 315; future of 315–16; groups 313–14; higher resolution for velocity 308; Jitter Reduction Timestamp 314; Message Type 313; MIDI Capability Inquiry (MIDI-CI) *310*, 310–12; Per-Note messages 309–10; Program Change message 315; Protocol Note Message 314–15, *315*; Three Bs 306–7; Three Ps 311–12; tighter timing 308; Universal MIDI Packet 313, *313*; Virtual Studio Technologies (VST) 3 310

MIDI Echo 288, *288*

MIDI In jack 287–8, *288*

MIDI Note-On message 319

MIDI Out jack 288, *288*

MIDI Thru jack 288, *288*

MIDI timecode (MTC) 299–300, 317, *386*, 386–7, 392–6

mid/side processing (M/S processing) 570–1, *571*

miking: A/B stereo *135*; accent 119, 149; close 119, *120*; drum set *153*, 153–4; for grand piano *149*; guitar amp 146, *147*; immersive techniques 134–40; kickboard area 150; near the sound hole in guitar 145; over the top 150; room 129; upper soundboard area 150; upright piano 150; of violin *159*; of vocals 98; X/Y stereo 136, *136*

minorities in recording industry 26–7

mission statement 487

Mitchell, J. 137

mixdown 249–50

mixer/console signal path *544*, 544–9; automation 545–8, *546–7*; channel fader 549; channel input 544–5; channel name 549; gain level 549; grouping 548, *549*; insert points 545, *545*; panning 548; send points 545, *546*

mixing 1, 79, 82, 537–56; art of 537–41; balancing and 550–4; ear training 538–9; FabFilter Pro-Q 24-band EQ plug-in *245*; forms of *542*; with iOS in music production 350, *350*; preparation 539–41; quality control (QC) 556;

sequence in editing 342; surface 543–4; in surround 599; technology of 541–50, *542*

mobility, iOS in music production 345

Mode 1 (Omni On/Poly) 296

Mode 2 (Omni On/Mono) 296

Mode 3 (Omni Off/Poly) 296

Mode 4 (Omni Off/Mono) 296

modeled condenser mic systems 115, *115*

modes, MIDI 1.0 295–6, *296*

modulation noise 182

monitoring (audio) 459–80; in 5.1, 5.1.4, 7.1.4 and beyond 592–3; active listening 459–61; configurations in studio 477–80; immersive headphone 595–6, *596*; level control 476–80; level section 512, *512*; mix 504, 528–32, *530–2*; modes 179–80, *180*; monitor speaker 465–70; passive listening 459–61; speaker design 471–3; spectral reference 470–1, *471*; subjectivity in the audio world 462; "trusted" space 462–5, *464*; volume *475*, 475–6; *see also* speakers

monitor section 503–6, *504*; direct insert monitoring 505; in-line monitoring 504–5, *505*; separate monitor section 506, *506*

monitor speaker 465–70; alternate environments 470; basic car system 470; earbuds 469; far-field monitor speakers 465–6; headphones 468–9, *468–9*; in-ear monitoring *469*, 469–70; near-field monitor speakers 466–7, *466–7*; simulation plug-ins, monitor environment 468–9, *469*; small speakers 467, *468*

monochrome US video 383

Moog multimode filter *431*

Mopho x4 4-voice analog synth *321*

MOTU Audio System (MAS) 226, 243

MOTU AVB (Audio Video Bridge) Switch *224*

MOTU Ultralite ABV 18.18 USB/AVB audio interface *225*

movies *see* film

Moving Picture Experts Group (MPEG) *see* MP3

Mozart, "Mass in C Minor" 143

MP3 368–9

MP4 369

MPEG-2 **363**

MPEG-4 *see* MP4

M/S (or mid-side) technique 136, *137*

Mullin, J. T. *172*

multiband compression *446*; dynamic range processors 439, *439*; plug-ins *247*

multiband dynamic processing, mastering 569–70, *570*

multi-effects processing devices *457*

multifunctionality, iOS in music production 345

multimedia 357–77; in an era of speed 376–7; environment 358–61; physical media 361–76; web and 20; *see also* media; physical media

multiple-effects devices 456–7, *457*

multiple-phase power 410–11

multiport network 290–1, *291*

multitrack MIDI recording, sequencing 333

multitrack production 1

musical instrument digital interface (MIDI) 1, 18, 130, 213, 279–343, 371–3; audio connections *284*; byte structure of Note-On message *296*; byte structure of Polyphonic Key Pressure 297, *297*; cable 285–7, *286–7*; Capability Inquiry (MIDI-CI) *310*, 310–12; communication 215; communications ports in toys 319; computer and 316–17, *317*; controller mapping 326; converting to audio 335–6; cueing messages 387; described 283–4; drum controllers 327, 327–8, *328*; electronic instruments 318–29; general 373; interface 316–17; iOS in music production 349; jacks 287–8, *288*; keyboard controller *324*, 324–5, *326*; music printing program 342–3, *343*; network connections *283*; phantom power 287; power of 280–3; production environments 280–3, *281–2*; sequencing/scoring *240*, 240–1, 329–42; standard files 372–3; system interconnections 285–91; task not performed by 284–5; velocity *297*; wireless MIDI transmitters 287, *287*

musical octave 46

musical waveform, envelopes *49*

music download sites 606–7

music-for-games setting 129

musicians personal note 32

music industry professionals 623

music law 26

music lawyer 32, 610

music notation program *see* music printing program

music printing program 342–3, *343*

music streaming 601

music studio, changes in 15–17
mute 143–4, 207, 519, 549
Mylar diaphragm 99

Nanologue synth app *354*
National Institute of Standards and
    Technology 618–19
Native Instruments 311, 325, 328;
    Komplete 10 325; S88 mk *324*
near-field monitor speakers 466–7, *466–7*
negative feedback 402
neoprene *74*, 74–5
networking: career development 29;
    computer 255–7; multimedia and 359;
    strategies for DAW *254*, 254–7, *256*
Neumann M50 mics 138
Neumann TLM 102 condenser
    microphone 168, *168*
Neve 33608 Compressor plug-in *568*
New York's Broadway Theater 580
NN-XT sampler module *273*
noise-based distortion 182
noise gate *418*; in duality console *442*;
    dynamic range processors *442*, 442–3;
    plug-in *442*
noise in analog tape 182–3
noise isolation within control room 80–1
noise reduction 172, 443–8; adaptive
    filtering 444–6, *445–6*; digital noise
    reduction 444; Fast Fourier Transform
    (FFT) 446–7, *447*; snap, crackle and
    pop 447–8, *448*; spectral analysis noise
    reduction 447, *447*
nondestructive editing 234, *235*
non-drop code 383
non-drop-frame code 384
non-magnetic gap 177
non-real-time DSP 242
Non-Registered Parameter Number
    (NRPN) 315
normalization 237
normal room (50%) 92–3; *see also* room
notch filters 427
Note-Off messages 297
Note-On messages *296*, 297
Novation 311
NTSC 385
Nuendo Media Production System *214*
Nuendo on-screen mixer *242*
nylon guitar 145–6; *see also specific entries*
Nyquist theorem 193–4, *194*

Ocean Way Audio HR5 active near-field
    monitors *466*
odd harmonics 46

off-axis pickup, microphones (mic)
    118–19
Older Roland MC-303 Groovebox *268*
.omf file extension 231
omnidirectional mic 105, *105*, 117
Omnisphere *321*
one audio-export-only option 231
online (download) distribution 601
Onyx 4-bus *509*
Open Media Framework Interchange
    (OMFI) 231
operational amplifier (op-amp) 402, *402*
optical disc, handling and care 618–19
Orange Book **362**
oriented strand board (OSB) 75
Otari Mx-5050, ATR *174*
out-of-phase 43; speaker 473; *see also*
    phase
output characteristics, microphones
    (mic) 110
output gain, compressor 434
output impedances 111
output section *508*, 508–12; channel
    assignment 508–9, *509*; grouping 509–
    11, *511*; main output mix bus 511–12;
    monitor level section 512, *512*
over-compression 436; *see also*
    compression
overdubbing 79, 125, 532–6, *533*; analog
    tape recording 179
overheads 155–6, *156*
overload characteristics, microphones
    (mic) 110–11
overload distortion 110
oversampling 194–5
overtones 45; acoustic guitar 144; cello
    159; double bass 159; electric guitar
    146; flute 162; partials 46; sound-
    shaping effects devices and plug-ins
    461; structure of a violin 57; trombone
    143; trumpet 142; tuba 144; violin and
    viola 158–9
ownership of masters 611
Oxford Envolution *431*
Oxide Tape Recorder *432*
Ozone 7 *567*

PA (public address) 159, 487, 611–12
panning 60, *61*; mixer/console signal
    path 548
parallel processing 421–2, *422*; *see also*
    sidechain processing
parametric equalizer 426, *427*
partials 45
partitions 79–80, *79–80*

passband 425
passive listening, recording 484–5
passive speaker design *472*
paste command 235, *236*
patch bay *512–13*, 512–14
patch data: in real time 305; between synths 304–5; SysEx dump files 323; from the Web 305
pattern sequencing 334
peak amplitude value 37
peaking equalization curves *424*
peaking filter 424, *424*
peak-to-peak value 37
perception 627–8; of direction 59–60; of space 61–2
perceptual coding 367–70; AAC 370; FLAC (Free Lossless Audio Codec) 370; MP3 368–9; MP4 369; Windows Media Audio (WMA) 369, *369*
percussion controller, keyboard as 327
percussion instruments 151–7; drum set 151–3, *152*; floor tom 157, *157*; hi-hat 157; kick drum *154*, 154–5; miking the drum set *153*, 153–4; overheads 155–6, *156*; snare drum 155, *155*
performance controllers in toys 318
Performance Rights Organization (PRO) 612
period 41
peripherals 216, 233, 316
permanent threshold shift 55
Per-Note messages, MIDI 2.0 309–10
personal stumbling blocks 626
Peter Erskine's studio drum kit *152*
phantom power 100, 103, *103*, 287
phase shift 42–5, *44*
physical media 361–76; Blu-ray 363, **364**; CD 361, **362**; compressed codec soundfile formats 365; direct streaming digital (DSD) audio 365–6; DVD 361–3, *363*; flash card 363; media delivery formats 364–70; memory USB stick 363; musical instrument digital interface (MIDI) 372–3; PCM audio file formats 364–5; perceptual coding 367–70; streaming audio over the Internet 376–7; tagged metadata *371*, 370–1; uncompressed sound file formats 36; *see also* media; multimedia
pickup: Ambisonic 140, *140*; characteristics 119–40; microphones (mic) 117–19; room in the studio 130; *see also* microphones (mic)
Pitch Bend Change messages 298, *298–9*
pitch change and time change 246–8

pitch control, vinyl 621
pitch-shift 246; algorithms, groove *265*, 265–6
plasterboard 73
plate reverb 453
playback: editing techniques 341–2; head 177, *177*; process 198–9
plug-ins 415–16, *416*; analog tape emulation 187, *187*; control and automation 416, *417*; dither *566*; drum and drum 274–5, *275*; Izotope Ozone 10 *576*; loop-based software 274; mastering 575–6, *576*; multiband compression *247*; names 232; signal processor/processing 415–16, *416*; tape emulation 187; UAD2 accelerated *416*
PMC QB1-A active monitor speakers 465
PMC TB2+ 5.1 monitor setup *586*
polarity 104–11
polar pattern 104
Polyphonic Key Pressure 297, *297*
polyvinyl chloride (PVC) 173
popping, microphones (mic) 118
portable drum controllers 327
portable studio *12*, 12–15; iOS revolution 13, *14*; retro revolution 14, *14*
post-production 7, 16, 67, 280–1
PostWork, audio production and post facility 68
power: amplifiers 404–6, *405*; conditioning 409–11, *409–11*; grounding and 407–8; losses 472; musical instrument digital interface (MIDI) 280–3; to the people 20–2; of preparation 486–8; to processor 250–61
Powerplay Pro-8 8-channel headphone distribution amp *532*
preamplifiers (preamps) 103, *103*, 402–3; microphones (mic) *114*, 114–15
Precision Limiter plug-in *569*
Precision Multiband Plug-in *570*
Presonus ADL-600 highvoltage tube preamp *498*
Presonus ATOM Performance Controller *327*
PreSonus DigiMAX D8 8-channel microphone *114*
Presonus Faderport 16 DAW controller *228*
Presonus HP4 4-channel headphone monitor *532*
Presonus Quantum 26x32 Thunderbolt audio interface *225*
Presonus StudioLive 16.0.2 Digital mixing console *519*

Presonus Studio One DAW software *234*
pressure-zone trap 89, *90*
Primacoustic Recoil Stabilizer pad *467*
Primacoustic™TM Polyfuser *91*
PrimacousticTM Razorblade quadratic
  diffuser *85*
printing 116; music notation 342–3
print-through *181*, 181–2
producers 23–4, 31, 488–9
Producers and Engineers Wing (P&E
  Wing) 185, 202
product distribution, media 603–7;
  Internet radio 607; music download
  sites 606–7; streaming 605–6;
  uploading to stardom 604–5; website
  building 603–4
production environments 280–3, *281–2*
production gear 628
product manufacture, media 613–22
professional analog ATR 174–80; bias
  current 178–9, *179*; equalization 178,
  178; magnetic tape heads 176–8, *177*;
  monitoring modes 179–80, *180*; tape
  transport 174–5, *175*
professionalism 32–3
professional sound file formats 202–3
professional studio 4–9, 66; control room
  7–9, *7–9*; described 5–7; design 66, *67*
professional stumbling blocks 626
Profile Configuration, MIDI-CI 311–12
Program Change messages 298,
  298, 315
project studio 9–11, *10–11*; acoustic design
  68–9; self subsidization of 11–12
Propellerhead: Reason (client) 249;
  ReCycle program 266
Proper, D. *558*
Property Exchange, MIDI-CI 312
Protocol Negotiation, MIDI-CI 312
Protocol Note Message 314–15, *315*
ProTools *238*, 273, *330*, *496*, *509*,
  520; 7-Band Digirack EQIII plug-in
  245; automation rubber bands *547*;
  auto selectors *546*; compressor/
  limiter plug-in *246*; hard-disk editing
  workstation *214*; HDX DAW *216*; Mod
  Delay II *449*; multiband compressor
  for *247*; Multiband Dynamics *570*;
  on-screen mixer *242*, 543–4
proximity effect 154; excessive bass boost
  due to 160; microphones (mic) 117–18
psychoacoustics 56–63; auditory
  perception 56–7; beats 58;
  combination tones 58; masking
  58–9; perception of direction 59–60;

perception of space 61–2; reverberation
  62–3, *63*
psychoacoustic enhancement 453–5;
  pitch shifting 454, *454*; time and pitch
  change 454–5
pulling loops into DAW session 275–6
Pulse-Code Modulation (PCM) 198, 363;
  audio file formats **363**, 364–5
Pulse-Density Modulation 365
punching in and out, sequencing *333*,
  333–4
punch-ins 239, *239*
PZM-6D boundary microphone *132*

Q-LAN 208
Quadraphonic Sound (Quad) 581, *581*
quality control (QC) in mixing 556
quality factor (Q) 424
quantization 192, *193*; editing
  techniques 339
quarter-frame messages 387
quarter-wavelength trap 89–90, *90*
quick starter 525

Radial Engineering 132
radio-frequency (RF) 407, 410, 411–12
RAID (redundant array of independent
  disks) 359
Ramone, P. 627
random access memory (RAM) 191, 218,
  251, 318, 321, 448
Rane GE 130 single-channel, 30-band,
  1/3-octave graphic equalizer *427*
rarefaction 36, *37*
Raven MTi Multi-touch Audio Production
  Console *229*
RAVENNA 208
"razor blade" approach 233
RCA 4BX ribbon *109*
RCA coax connection *206*
read mode, automation 547
reality creation 625–6
real-time, on-screen mixing 241–9, *242*;
  accelerator processing systems 244, *244*;
  delay 245–6, *247*; DSP effects 242–3;
  DSP plug-ins 243–4; dynamic range
  processors 245, *246–7*; equalization
  (EQ) 244, *245*; fun with effects 244;
  pitch change and time change 246–8;
  ReWire 248–9; *see also* mixing
Real-Time Audio Suite (RTAS) 243
real-time DSP 242
reamping 536
Reaper DAW software *233*
rear speaker controls *464*

Reason Studios 248, 272; Music Production Software *273*; software instruments *273*
record head 176–7, *177*
recording 481–537; active listening 484–5; channel fader 506–8, *507*; console 7–8, *9*, 147, *404*, *495–6*, 495–501, 512, 538; direct 125–6, 147, *147*; dynamics section 503, *503*; equalization (EQ) *501*, 501–3; "fix it in the mix" 526; Good Attitude 484; Good Rule 481–2; iOS in music production 349; latency 532; metering *514*, 514–18; monitoring the mix 528–32, *530–2*; monitor section 503–6, *504*; output section *508*, 508–12; overdubbing 532–6, *533*; passive listening 484–5; patch bay *512–13*, 512–14; phases of 493–4; preparation *485*, 485–9; process 197–8, *198*; producer 488–9; sequencing, MIDI 331; session documentation 528; single instrument 493; studio 489–93; transducer *482*, 482–4, **483**
Recording Academy 185
Recording Hall 2 63
ReCycle Groove Editing Software *267*
Red Book **362**
reference tones 39, 186
reference track, mastering 564
reflections *41*, 41–2, 83–5, *83–5*, 93–6
regarding digital audio levels 203
regions 234, 237
Registered Parameter Number (RPN) 315
release (R): of compressor 435; in envelope 49, *49*
remixing 1, 7; *see also* mixing
Renderer Tool 598–9
resistors (R) 103
resonances 73–5, 143, 150, 467
Resource Information File Format (RIFF) 202, 230, 365
retro revolution 14, *14*
retro systems 423
reverberation (reverb) 62–3, *63*, 65, 246, 451–3, *452*; reverse 453; studio design 93–5; time 63; types 452–3
ReWire 248–9, *249*; software, groove 273–4
ReWire2 248–9
REX file 266–7, *267*
Rhett's Gone with the Wind 234, *235*
RIAA (Recording Industry Association of America) 206
ribbon microphone 99, *100*; developments in ribbon technology 100

risers 75–6, *76*
"rocking" the tape 175
Rock Ridge **362**
rock steady 525–6
Rockwool 73, *73*, 80, 88–9, *89*, *90*
Rode NT-SF1 Ambisonic mic *140*
Roland 7X7-TR-8 Rhythm Performer *326*
Roland A-88mkII *324*
Roland MC-707 hardware sampler *322*
Roland Octapad spd-30 Digital Percussion Pad *327*
Roland TR-8 Rhythm Performer *268*
Romeo **362**
room: microphone placement 129–30; miking 129; pickup in the studio 130; reverb 452
room modes *see* standing waves
root-mean-square (rms) value 37
royalties 609
Royer Labs R-121 ribbon microphone *100*, *165*, 165
RT60 452
rubber band controls 249
rumble 117
running order, song ordering 561
Rupert Neve Designs Portico 5024 4-channel mic preamp *114*
Rupert Neve Designs RNDI Active Transformer Direct Interface *126*

sample-and-hold (S/H) 196, 198, *199*
sample libraries 257; keyboards 323–4
samplers/sampling *191*, 191–2; keyboards *321*, 321–3, *322*; rate 191, 201–2; systems *322*
saturation 432
saving MIDI files 337
sawtooth waves 47, *47*
saxophones 162–3, *163*
Schmitt, A. 119, 125
scratch vocals 78, 142, 161
S-curved B-flat tenor sax 162
SDDS **363**
SDMI (Secure Digital Music Initiative) 370
selectable frequency equalizer 426, *426*
self-mastering 572–3; *see also* mastering
self-motivation in career development 28–9
self-reinvention 627
sel-sync 179
semi-distant mic *147*
send points, mixer/console signal path 545, *546*
send/return signal paths *499*

send routing 419, *419*
sensitivity rating, microphones (mic) 110
separate monitor section 506, *506*
separation 149–50; acoustic 65, 93, 123; loss 183
sequencing, MIDI 329–42; audio to MIDI 336; changing tempo 332; click track 332–3; documentation 337–8; drum pattern entry 334, *335*; editing 338–42; integrated hardware sequencers 329; MIDI to audio 335–6; multitrack MIDI recording 333; punching in and out *333*, 333–4; recording 331; saving MIDI files 337; session tempo 332; software sequencers 329–30, *330*; step time entry 334
Serato DJ *278*
Serial Copy Management System (SCMS) 206
session documentation *258*, 258–9, 528
session recall 215
session tempo, sequencing 332
Session View 271, *271*
shared web connection 256–7
sharing files 256
shelving filter 425, *425*
Shure 58 dynamic mic *99*
Shure M58 dynamic *109*
Shure MV88 iOS handheld microphone *349*
Shure SM57 dynamic microphone 163, *164*
sibilance 160
sidechain processing 420–1, *421*
Sides, A. 137
signal chain 543; mastering 574–5, *575*
signal distribution 209, *209*
signal levels 527, *527*
signal paths 543
signal paths in effects processing 417–22; insert routing 417–19, *418*; parallel processing 421–2, *422*; send routing 419, *419*; sidechain processing 420–1, *421*; Vive la Difference 420
signal processor/processing 413–58; analog *414*, 414–15; digital 415; effects processors 422–48; plug-ins 415–16, *416*; signal paths in effects processing 417–22; time-based effects 422, 448–58; whatever 413–14
signal-to-noise (S/N) 180, 182, 195
signed artist/superstar approach 12
simple amplifier schematic *400*
simple waves 47, *47*
simulation plug-ins, monitor environment 468–9, *469*

Sinatra, F., "Live at the Sands" 122
single composite track *240*
6th Circuit US Court of Appeals 323
16 bits 200
size, hard-disk requirements 251
Skibba, M. *542*
Skywalker Sound: main control room *18*; scoring stage *282*
slap echo *see* flutter echo
Slate Digital Virtual Tape Machines *187*, *432*
slaves 392–3, *393*
slip time 340
slope 425
slope ratio, compressor 434–5
small speakers 467, *468*
SMPTE/MTC conversion 387, *388*
snap, crackle and pop 447–8, *448*
snare drum 155, *155*, 437
Society of European Stage Authors and Composers (SESAC) 612
Society of Motion Picture and Television Engineers (SMPTE) 381; offset times 390–1; signals, distribution of 391; timecode *388*, 390, *390*
soffits 73–4, 77, 465, 587
software 232–50, 269–74; DAWs looping 269–70, *270*; loop-based audio software 270–4, *271*, *273*; loop-based plug-ins 274; monitor "cue" section *504*; options 231; ReWire 273–4; sequencers 329–30, *330*; synthesizers *321*; *see also* digital signal processing
soldering 113–14
solid state drive (SSD) 218, 252
Solid State Logic AWS δelta *497*, *504*, *507*, *512*
Solid State Logic Duality Console *503*, 543
solo 133–4, 158–9, 335, 507, 549
song ordering 560–1; mastering 560–1
Song Position Pointer (SPP) messages 300
Song Select messages 300
song timing 561
sonic toolbox 97
Sonnox Oxford EQ *423*
Sonnox Oxford Limiter plug-in *566*
Sony 205, 206, 366, 370, 391
sopranino 162
soprano 162
sound and hearing 35–63; basics of 35–6, *36*; compression in 36, *37*; diffraction of 42, *43*; ear 54–6; hole in guitar 145; lock 78, *78*; loudness levels 49–54; psychoacoustics 56–63; rarefaction 36,

37; sound-pressure waves 35; waveform 36–49; wave propagation 36; *see also* ear
Soundexchange 612
sound files: accessing 257; basics 199–203; bit depths 200–1; formats 230–2; interchange and compatibility 231–2; resolution 565; sample and bit rates 201–2, 231; volume 565
sound-pressure level (SPL) 49, 70, 110, 466, 540, 591
sound-pressure variations 99
sound-pressure waves 35
sound recording and editing 233–8, *234–5, 237–8*
sound-shaping effects devices and plug-ins *431–2,* 431–43; dynamic range 431–2; dynamic range processors 432–43, *433*
sound waves 105
SpaceArray sound diffusers *85*
spaced microphones 134–5, *135*
spaced pair 134–5, *135*
Spanish guitar 145–6
Spark Drum Machine *277*
spatial localization 59, 594
spatial positioning, mixing 541
S/PDIF (Sony/Philips Digital Interface) 205–6, *206*
speakers: balancing levels 474–5, *475*; isolation pads *467*; level calibration *590,* 591–2; noise calibration *591*; placement 93; polarity 473; symmetry 81–3; time calibration 592, *592; see also* monitoring (audio); *specific entries*
speaker design 471–3; active powered speaker design *472*; passive speaker design 472, *472*; speaker polarity 473
special studio cue mix 529
spectral analysis noise reduction 447, *447*
Spectral Analysis Software *448*
spectral analyzer, displays *471*
spectral content of sound 422
spectral reference, monitoring (audio) 470–1, *471*
speed: flexibility and 215; hard-disk requirements 251–2
spring reverb 453
square waves *47*
SSL AWS δelta large-format recording console/DAW controller *495*
stable timing reference, video's need 393, *394*
stage, iOS in music production 353, *353*
stand-alone software 464

standard audio interface 348
standard rate for PAL European video 384
standing waves 83–4, *84*
Start messages 300, 303
Steinberg 243, 248; Cubase DAW (host) 249; Cubase virtual mixer *496*; Cubasis *351*; Headphonematch *469*; LoopMash2 Groove plug-in *274*; Multiband Compressor *446*; Nuendo *597*; Nuendo virtual mixer *496*; Reverence reverb plug-in *452*; UR22C 2x2 audio interface *225*; UR22 MK II interface *349*
stems, mastering 564
step input 334
step time entry, sequencing 334
step-up transformer (amp) 99
stereophonic (stereo): microphones 136; miking techniques 134–40; mix room simulation *596*; television and 581–2; waveform signals *620*
Steven Slate Drums Trigger 2 drum replacement plug-in *328*
stopband 425
Stop messages 303
streaming: audio over the Internet 376–7; media 605–6; music 601
stringed instruments 158–9; cello 159; double bass 159; viola 158–9; violin 158–9, *159*
Studer A-800 24-tk, ATR *174*
studio 489–93; isolation 492–3; setting up 491
studio arrangers 22–3
studio furniture *261*
StudioLive 16.4.2 Digital Recording and Performance Mixer with remote iPad app *350*
StudioLive RML32AI wireless mixing system *353*
studio management 25–6
Studio Metronome, live recording audio truck *17*
studio monitor mix 529
studio musicians 22–3
studio production 1, 14, 288
studio track sheets *523*
studs 72, *73*
Stylus RMX real-time groove module *274*
subjectivity in the audio world 462
SubTractor polyphonic synth module *273*
summing amp 403, *404*
Summit Audio TLA-100A Tube Leveling Amplifier *246*
Super Audio CD (SACD) 366

supercardioid polar patterns 106, *107*
surround: bass system 585, *586*; channel layouts and assignments 589; guitar miking 145; miking techniques 138–9; mixing in 599
sustain (S) in envelope 48, *49*
Swedien, B. 125, 137
symmetry 91–2
synchronization (sync) 317, 379–97; groove 264; information data 383; timecode 380–96; trouble prevention 396–7
Synchron Stage Vienna *8*, *580*
sync mode 179, *180*
synthesizer (or synth) 438; keyboards 320–1, *321*; transmitting patch data between 304–5
SysEx: controller 305; dump files 323; dump track 305
System 6000 digital effects processor *457*
System-exclusive (SysEx) messages 304, *304*
System Exclusive messages 312
system interconnections, MIDI 285; cable, MIDI 285–7, *286–7*; configurations 288–91; daisy chain 289–90, *290*; jacks, MIDI 287–8, *288*; MIDI Echo 288, *288*; MIDI In jack 287–8, *288*; MIDI Out jack 288, *288*; MIDI phantom power 287; MIDI Thru jack 288, *288*; multiport network 290–1, *291*; wireless MIDI transmitters 287, *287*
system interconnectivity 220–4
System messages 299–305; End of Exclusive (EOX) messages 300; MIDI timecode (MTC) messages 299–300; Song Position Pointer (SPP) messages 300; Song Select messages 300; Tune Request messages 300
System Real-Time messages: Active Sensing messages 303–4; Continue messages 303; Start messages 300, 303; Stop messages 303; System Reset messages 304; Timing Clock messages 300
System Reset messages 304
system sounds 347
systems plug-ins, electronic instruments 319–24

tagged information 605
tagged metadata *371*, 370–1
tails-out position 181, *181*
"Take Five" (Brubeck) 122
tape 180–4; analog tape noise 182–3; analog track configurations *180*;

availability 187; cleanliness 183; degaussing 183–4; editing block 184, *184*; emulation plug-ins 187, *187*; print-through *181*, 181–2 restoration 186; transport 174–5, *175*
tape-to-head 180
TDIF (Tascam Digital InterFace) 208
technology of mixing 541–50, *542*; gain structure 549–50; mixer/console signal path *544*, 544–9; mixing surface 543–4; *see also* mixing
Telefunken C12 condenser microphone 169, *169*
Telefunken ELA M251E condenser microphone 169, *169*
Telefunken M81 dynamic microphone 99, 164, *164*
Telefunken U47 condenser microphone 169, *169*
television: multimedia and 358–9; stereo and 581–2
tempo: length and 264; sequencing, MIDI 332; *see also* synchronization (sync)
temporal fusion 62
temporary threshold shift 55
"That Elusive Console Magic" 624
theaters 582
Three Bs, MIDI 2.0 306–7
Three Ps, MIDI 2.0 311–12
threshold 433, 434
Thunderbolt® (Mac) 217, *222*, 222–33, *244*, 285
tighter timing, MIDI 2.0 308
timbre 47, 48, 121; perception of 56
time and pitch change: psychoacoustic 454–5; techniques 264
time-based effects 422, 448–58; automatic pitch correction 455–6, *456*; delay 448–51, *449*; dynamic effects automation and editing 457–8; multiple-effects devices 456–7, *457*; psychoacoustic enhancement 453–5; reverb 451–3, *452*; *see also* effects processors
time-base errors 209, *210*
timecode 380–96; 3:2 pulldown rate 384–5; within digital media production 385; frame standards 383–4; levels 391, **391**; MIDI *386*, 386–7; MIDI messages 386–7; production in analog audio and video 388–91; real-world applications 392–6; routing to and from computer 395, *396*; SMPTE 390, *390*; word 382–5
Time Domain Multiplex (TDM) 243
Timing Clock messages 300

Todd-AO 580
tonal balance (timbre) 121
tools for listening 461
Toslink optical connection *206*
total length, song ordering 561
total transport logic (TTL) 175
Touchable Pro *352*
touch capabilities, iOS in music production 345–6
touch-screen controllers *229*, 229–30
Townsend Labs Sphere L22™ microphone system *115*
toys, electronic instruments 318–19
Traktor portable laptop DJ rig *278*
transducer recording 482, 482–4, **483**
transient: cymbals 155; high-frequency 148, 155; instrument 441–2; percussive 266, *266*; proximity effect and 153; quality of the ribbon "sound" 99; reducing the dynamics 568; response in microphones (mic) *109*, 109–10; signals 440; sounds 54, 154, 435, 437; spikes 409; temporal fusion 62
transistor 400, *401*
transitions, song ordering 561
transmission loss (TL) 70–1, *71*
transmitting patch data between synths 304–5
transposition, editing techniques 339
trauma, acoustic 55
triangle waves *47*
triggering 283, 284, 328, 443
triode *400*
Tritschler, J. *14*
trombone 143
trumpet 142–3, *143*
"trusted" space 462–5, *464*
TT (or bantam) connectors/cables 112, 513
tuba 143–4
tumbas 157
TuneCore 612
tuned percussion instruments 157–8; congas and hand drums 157–8; xylophone, vibraphone and marimba 158
Tune Request messages 300
turnover 425
two-step or one-step (integrated), mastering 574
Type 0 MIDI files 337
Type 1 MIDI files 337

UAD2 accelerated plug-ins *416*
UAD-2 DSP PCIe *244*

udio018-05 258
Ultrapatch PX3000 patch bay *512*
uncompressed audio coding *367*
uncompressed sound file formats 364
underlay 74
Uniform Resource Locator (URL) 360
uninterruptible power supply (UPS) 260, 410
Universal Audio (UAD): 2–610S dual channel tube preamp *498*; 4-710d Four Channel Hybrid (Tube/Transistor) Preamplifier *114*; 4-710d Four-Channel Mic Preamplifier *227*; 1176LN limiting amplifier *433*; Lexicon 224 reverb plug-in *457*; Precision Limiter plug-in *439*, *569*; Precision Multiband Compressor/Expander/ Gate plug-in *446*; Precision Multiband plug-in *439*, *570*; Sonnox Oxford Limiter plug-in *566*
Universal MIDI Packet, MIDI 2.0 313, *313*
universal serial bus (USB) 217; characteristics of 221; hubs *222*; USB 2.0 221; USB 3.0 221; USB 3.1 221; USB C 221
unmagnetized tape *174*; *see also* tape
upper partials 45
upright piano 150
user-friendly operation 215

Vanderslice, J. *523*
variable–bit rate (VBR) 368
"vari-speed" mode 176
V-Control Pro DAW controller *229*, *351*, *521*
Velcro™ or tie-straps 261
velocity 40
vibraphone 158
video: acoustic studio design 67; audio for 18, *18–19*; need for stable timing reference 393, *394*; picture sync and 241, *241*; production shoot *380*; recorder *394*; timeline *380*; workstation *394*
video cassette recorder (VCR) 581–2
video tape recorder (VTR) 388
vinyl 601–2, 619–22; cutting head *620*, 620–1; disc-cutting 619, *620*; disc-cutting lathe 620, *620*; disc plating and pressing 621–2, *622*; pitch control 621
vinylite compound 622
viola 158–9
violin 158–9, *159*
virtual DAW insert point *499*, *500*
virtual DAW send 500–1, *501*

virtual input strip *519*, 519–20
virtual instrument 232
virtual mixer strip layouts *496*
virtual pot (V-pot) 519
virtual software drum machines *275*
Virtual Studio Technologies (VST)
    3 226, 310
Vive la Difference 420
vocals 438; booths 79; miking of 98;
    scratch 78, 142, 161
vocoder 419
voice 159–61; call 99; circuitry in toys
    318–19; excessive bass boost due to
    proximity effect 160; excessive dynamic
    range 160; mic tools for 160–1;
    sibilance 160; track 419
volume: mastering 566; monitoring
    (audio) *475*, 475–6
VORTEX WIRELESS 2 Wireless USB/MIDI
    Keytar Controller *287*
VST effect 243
VU meter **517**, 517–18

"wahing" effect 163
Wallace, R. 138
wallboards 72–3, 76
walls 71–4; isolation 71–4
Warm Audio EQP-WA selectable
    frequency tube equalizer *426*
Warm Audio WA-47 tube microphone
    168–9, *169*
Warm Audio WA-412 *227*
warm-sounding 102
warping 248; groove *263*, 266
wattage *see* power
wave (.wav) 202, 230
waveform 36–49; amplitude 37–8, *37–8*;
    defined 36; diffraction of sound 42, *43*;
    envelope 48–9; frequency 38, 38–40,
    39; harmonic content 45–8, *46–8*;
    motion over time *38*; phase 42–5, *44*;
    reflection of sound *41*, 41–2; velocity
    40; wavelength 40–1, *41*
wavelength 40–1, *41*

wave movement in air *36*
wave propagation 36
WDM 225
web, multimedia and 20, 359–60
website: building 603–4; guidelines 604;
    see also Internet
well-being 630
Wells, F. 22
White Book **362**
Williamson, G. *354*
Windows (Microsoft) 202, 225, 243, 266
windows and doors 77–8, *77–8*
Windows Media Audio (WMA)
    369, *369*
wireless iDevices *531*
wireless MIDI transmitters 287, *287*
The Wizard of Oz 283
women in recording industry 26–7
Women's Audio Mission 27
woodwind instruments 161–3; clarinet
    161, *162*; flute 162, *162*; harmonica
    163, *163*; saxophones 162–3, *163*; *see
    also specific types*
wordclock 210–12, *211*
write mode 546–7
WWII vintage German Magnetophones *172*
WYSIWYG ("what you see is what you
    get") 233, 340

XLR connectors 112, *112*, *113*
xylophone 158
X/Y stereo miking 136, *136*
X/Y stereo pair 158

Yamaha HS8 studio monitor speakers *466*
Yamaha Wireless MD-BT01 5-PIN DIN
    MIDI Adapter *287*
Yello Book **362**
YouTube 624
Zappa, F. 137
Z channels 76, *76*

zone 322
zoomed-in edit window 234